# Computational Algorithms for Shallow Water Equations

Eleuterio F. Toro

# Computational Algorithms for Shallow Water Equations

Second Edition

 Springer

Eleuterio F. Toro
DICAM
University of Trento
Trento, Italy

ISBN 978-3-031-61394-4          ISBN 978-3-031-61395-1  (eBook)
https://doi.org/10.1007/978-3-031-61395-1

New edition of the work originally titled Shock-Capturing Methods for Free-Surface Shallow Flows, published by Wiley & Sons, New York, 2001

1st edition: © Wiley 2001
2nd edition: © The Editor(s) (if applicable) and The Author(s), under exclusive license to Springer Nature Switzerland AG 2024

This work is subject to copyright. All rights are solely and exclusively licensed by the Publisher, whether the whole or part of the material is concerned, specifically the rights of reprinting, reuse of illustrations, recitation, broadcasting, reproduction on microfilms or in any other physical way, and transmission or information storage and retrieval, electronic adaptation, computer software, or by similar or dissimilar methodology now known or hereafter developed.
The use of general descriptive names, registered names, trademarks, service marks, etc. in this publication does not imply, even in the absence of a specific statement, that such names are exempt from the relevant protective laws and regulations and therefore free for general use.
The publisher, the authors and the editors are safe to assume that the advice and information in this book are believed to be true and accurate at the date of publication. Neither the publisher nor the authors or the editors give a warranty, expressed or implied, with respect to the material contained herein or for any errors or omissions that may have been made. The publisher remains neutral with regard to jurisdictional claims in published maps and institutional affiliations.

This Springer imprint is published by the registered company Springer Nature Switzerland AG
The registered company address is: Gewerbestrasse 11, 6330 Cham, Switzerland

If disposing of this product, please recycle the paper.

# Preface

## This Book

This book deals with computational methods for solving systems of shallow water equations within the broader field of geophysical fluid dynamics. Such systems, comprised of hyperbolic partial differential equations, serve as mathematical models for simulating the dynamics of water flows in open channels, rivers, lakes, and oceans. Practical applications range from modelling oceanic tides to the breaking of waves on shallow beaches, roll waves in open channels, flood waves in rivers, dam-break waves, tsunami wave propagation, and various environmental fluid dynamics scenarios. Mathematical models of the shallow-water type are also employed for simulating gas dynamics, such as the dispersion of heavy gases over the Earth's surface and atmospheric dynamics. However, the physical complexity of time-dependent, three-dimensional oceanic and atmospheric flows necessitates more comprehensive mathematical models than shallow-water type systems alone. Nevertheless, the basic, one-layer shallow water equations play a crucial role in providing benchmark solutions for assessing methodologies intended for more complex models. Significantly, multi-layer shallow water models are currently undergoing major developments in oceanic and atmospheric dynamics, where the underlying concepts and methodologies from the basic one-layer shallow water equations in this book find applicability. This book offers fundamental concepts and methods essential for constructing advanced high-order computational algorithms within the frameworks of finite volume and discontinuous Galerkin finite element methods. The finite volume methodology is developed in full detail, presenting non-linear, fully discrete high-order upwind and centred methods for solving the non-linear, time-dependent two-dimensional shallow water equations.

# Genesis of This Book

This book departs from the 2001 first edition [1] published by Wiley and Sons Ltd. This second edition, published by Springer, has benefitted from more recent developments in numerical methodology and from the experience gained through its use in teaching contexts over many years. A significant portion of the contents has been utilised in teaching two master courses at the University of Trento: one titled Numerical Methods for Environmental Problems, aimed at environmental engineering students, and another titled Computational Haemodynamics, offered to mathematics students. Moreover, the material has been employed for many years in the annual Winter School for Numerical Methods for Hyperbolic Equations, offered to Ph.D. students and researchers from Europe and beyond. Additionally, the content has served as the basis for numerous lectures and intensive short courses delivered in various locations, including Cranfield, Cambridge, Manchester, Barcelona, Madrid, Santiago de Compostela, Córdoba, Zurich, Oxford Mississippi, Moscow, and Beijing.

# Contents of This Book

The book begins by deriving the shallow water equations for spatial domains with variable bathymetry from the physical principles of conservation of mass and momentum. This derivation is presented in Chap. 1. Chapter 2 follows with fundamental notions on the theory of hyperbolic equations, including characteristics, eigenstructure, hyperbolicity, the Riemann problem, Rankine-Hugoniot conditions, non-uniqueness, and the Lax entropy condition. In Chap. 3, linearised versions of the shallow water equations are examined, with analytical solutions for the general initial-value problem provided alongside a case study. Chapter 4 delves into the salient mathematical properties of the two-dimensional non-linear shallow water equations, including rotational invariance, crucial for designing numerical methods on unstructured meshes. Elementary wave solutions emerging from the Riemann problem for the augmented one-dimensional and split two-dimensional shallow water equations are studied in Chap. 5, introducing rarefaction waves, contact discontinuities, shock waves, or bores. Chapter 6 presents a solver for computing the exact solution through the entire wave structure of the Riemann problem for the non-linear shallow water equations, specifically for cases with a wet bed (no vacuum). Mathematical tools for solving simple dam-break problems, assessing numerical approximations, constructing numerical methods of the Godunov type, and implementing boundary conditions are provided. In Chap. 7, the Riemann problem in the presence of adjacent dry regions (vacuum) is tackled. A suite of test problems with exact solutions is presented in Chap. 8, aiding in understanding the basic physics of shallow water waves and rigorously assessing the performance of numerical methods. Additionally, a computer program for solving the Riemann problem under both wet and dry-bed conditions is provided. Chapter 9 offers a largely self-contained introduction to basic theoretical concepts on numerical analysis for hyperbolic equations, introducing well-known numerical schemes and stating and proving the Godunov theorem, which forms the basis for constructing non-linear methods for hyperbolic equations. Chapter 10 reviews first-order, monotone, upwind and centred numerical methods for non-linear systems of hyperbolic conservation laws, laying the foundation for constructing higher-order schemes for the shallow water equations in subsequent chapters. Chapter 11 presents a range of 10 approximate Riemann solvers for the shallow water equations, applicable to finite volume and discontinuous Galerkin finite element methods, distinguishing between complete and incomplete Riemann solvers. In Chap. 12, non-linear second-order numerical methods for the shallow water equations are constructed following the flux-limiter TVD and ENO approaches, in line with Godunov's theorem. Chapter 13 addresses two important issues: source terms and multiple space dimensions. It presents advection-reaction splitting methods for solving balance laws with source terms, formulates dimensional-split and unsplit methods for the two-dimensional shallow water equations, and discusses the stability of unsplit schemes in two and three space dimensions using the von Neumann method. Chapter 14 introduces advanced, non-linear fully discrete numerical methods of unlimited accuracy within the frameworks of finite volume and discontinuous Galerkin finite element methods, detailing finite volume formulations. The book concludes with two relevant applications. Chapter 15 discusses wave propagation phenomena in dam-break problems, while Chap. 16 studies the physical problem of regular and Mach reflection of bores in tsunami and similar wave-propagation events. A summary and concluding remarks can be found in Chap. 17.

## Who Will Benefit From This Book

This book is primarily intended for environmental scientists, applied mathematicians, and engineers in academia, research laboratories, industry, and consultancy organisations. Senior undergraduate and postgraduate students engaged in mathematical modelling and computational methods for environmental problems will also find value in studying this book. Lecturers can utilise much of the material for courses on numerical methods for wave propagation problems in hydraulics, oceanography, atmospherics, and other geophysical fluid dynamics contexts.

## Acknowledgements

Much of the material in this book has been influenced by long-standing collaboration with several organisations and individuals. Special thanks are due to Kazuyoshi

Takayama for introducing me to the fascinating subject of tsunami wave propagation. The author is indebted to his colleagues at Trento University, including Aronne Armanini, Vincenzo Casulli, Luigi Fraccarollo, Enrico Bertolazzi, Michael Dumbser, Annunziato Siviglia, and Lucas Müller. For feedback on the manuscript, I am grateful to Alistair Borthwick, Ben Rogers, Pilar García-Navarro, Nikolaus Nikiforakis, Stephen Lynch, John Ryan, Guus Stelling, Barbara Zanuttigh, María Elena Vázquez, Annunziato Siviglia, and Arturo Hidalgo. Finally, I would like to express my gratitude to all members of my family for their constant and invaluable support.

March 2024

Eleuterio F. Toro
OBE, Ph.D., Dr. HC(usach),
Dr. HC(uct), Dr. HC(ufro)
Professor Emeritus
University of Trento
Trento, Italy
eleuterio.toro@unitn.it
https://www.eleuteriotoro.com

Academic Visitor
Cavendish Laboratory
Department of Physics
University of Cambridge
Cambridge, UK

## Reference

1. E.F. Toro, *Shock-Capturing Methods for Free-Surface Shallow Flows* (Wiley, 2001)

# Contents

**1  The Shallow Water Equations** .................................... 1
   1.1    Introduction ................................................. 1
   1.2    Conservation Principles ...................................... 3
   1.3    Water Flow with a Free Surface ............................... 4
   1.4    The Shallow Water Equations .................................. 6
   1.5    The Saint Venant Equations for River Flows ................... 10
   1.6    Conclusions ................................................. 12
   References ...................................................... 13

**2  Notions on Hyperbolic Equations** ................................. 15
   2.1    The Linear Advection Equation and Basic Concepts ............ 15
        2.1.1    The Initial-Value Problem ............................ 16
        2.1.2    The Riemann Problem ................................... 19
   2.2    Linear Hyperbolic Systems ................................... 20
        2.2.1    Eigenstructure and Hyperbolicity ...................... 21
        2.2.2    Diagonalization and Characteristic Variables .......... 22
        2.2.3    The Riemann Problem ................................... 25
   2.3    Non-linear Scalar Equations ................................. 27
        2.3.1    Definitions and Examples .............................. 28
        2.3.2    Solution Along Characteristics ........................ 29
        2.3.3    Integral Forms of the Equation ........................ 31
        2.3.4    Generalised Solutions ................................. 33
        2.3.5    Non-uniqueness ........................................ 34
        2.3.6    The Riemann Problem for Burgers's Equation ............ 36
   2.4    First-Order Non-linear Systems .............................. 37
   References ...................................................... 38

**3  Linear Shallow Water Equations** .................................. 39
   3.1    Linearised Models ........................................... 39
   3.2    Eigenstructure and Characteristic Variables ................. 42
   3.3    The General Initial-Value Problem ........................... 45
        3.3.1    Recalling the General IVP for the Scalar Case ......... 45

|  | 3.3.2 | The General IVP for the System Case | 47 |
|---|---|---|---|
| 3.4 | | The Riemann Problem | 49 |
| | 3.4.1 | Recalling the Scalar Case | 49 |
| | 3.4.2 | The System Case | 50 |
| | 3.4.3 | Example | 52 |
| 3.5 | | A Linear Model with Source Terms | 53 |
| 3.6 | | Case Study: Alternative Linearisation | 55 |
| | 3.6.1 | The Linear Equations | 56 |
| | 3.6.2 | The Eigenstructure | 56 |
| | 3.6.3 | Equations in Characteristic Variables | 58 |
| | 3.6.4 | The General Initial-Value Problem | 59 |
| | 3.6.5 | The Riemann Problem | 60 |
| References | | | 63 |

## 4 Properties of the Nonlinear Equations ... 65
  4.1 Recalling the Equations ... 65
  4.2 Eigenstructure in Terms of Conserved Variables ... 66
  4.3 Eigenstructure in Terms of Primitive Variables ... 71
  4.4 Hyperbolic Character of the 2D Equations ... 72
  4.5 Nature of Characteristic Fields ... 74
  4.6 Integral Forms and Rotational Invariance ... 76
  4.7 Steady Supercritical Flow ... 78
  4.8 Concluding Remarks ... 80
  4.9 Suggested Exercises ... 80
  References ... 82

## 5 Elementary Waves in Shallow Water ... 85
  5.1 Elementary Waves in the Riemann Problem ... 85
    5.1.1 Motivation: The Dam-Break Problem ... 86
    5.1.2 The Riemann Problem and Wave Patterns ... 87
    5.1.3 Relations Across the Wave Structure ... 90
  5.2 Single Rarefaction Wave ... 92
    5.2.1 Left Rarefaction Wave ... 92
    5.2.2 Right Rarefaction Wave ... 94
  5.3 Single Shock Wave ... 95
    5.3.1 Right-Facing Shock Wave ... 96
    5.3.2 Left-Facing Shock Wave ... 99
  5.4 Contact Discontinuity and Shear Wave ... 100
  5.5 The Full Wave System ... 101
  5.6 Useful Shock Relations ... 103
    5.6.1 Left Shock ... 104
    5.6.2 Right Shock ... 105
  5.7 Non-conservative Formulation and Shocks ... 107
    5.7.1 Conservative System in Non-conserved Variables ... 107

| | | | |
|---|---|---|---|
| | 5.7.2 | Shock Waves | 108 |
| 5.8 | Concluding Remarks | | 110 |
| 5.9 | Suggested Exercises | | 111 |
| References | | | 112 |

# 6 Exact Riemann Solver: Wet Bed ... 113
- 6.1 Introduction ... 113
- 6.2 The Riemann Problem and Solution Strategy ... 115
- 6.3 Solution in the Star Region ... 116
  - 6.3.1 Non-linear Equation for Water Depth $h_*$ ... 117
  - 6.3.2 Analysis of the Depth Function $f(h)$ ... 118
  - 6.3.3 Iterative Solution for $h_*$ ... 121
- 6.4 Sampling the Complete Solution ... 121
  - 6.4.1 Passive Scalars ... 122
  - 6.4.2 Left of Contact/shear: $S = x/t_* \leq u_*$ ... 123
  - 6.4.3 Right of Contact/shear: $S = x/t_* \geq u_*$ ... 124
- 6.5 Conclusions ... 125
- 6.6 Exercises ... 126
- References ... 127

# 7 Exact Riemann Solver: Dry Bed ... 129
- 7.1 Introduction ... 129
- 7.2 Admissible Wet/Dry Interface Waves ... 130
- 7.3 Dry Bed: Three Possible Cases ... 131
  - 7.3.1 The Dry Bed Is on the Right Side ... 132
  - 7.3.2 The Dry Bed Is on the Left Side ... 134
  - 7.3.3 Generation of Vacuum from Wet-Bed States ... 134
- 7.4 Passive Scalars ... 135
- 7.5 Conclusions ... 137
- References ... 138

# 8 Tests with Exact Solution ... 139
- 8.1 Introduction ... 139
- 8.2 Test 1: Left Critical Rarefaction and Right Shock ... 141
- 8.3 Test 2: Two Rarefactions and Nearly Dry Bed ... 141
- 8.4 Test 3: Right Dry-Bed Riemann Problem ... 142
- 8.5 Test 4: Left Dry-Bed Riemann Problem ... 143
- 8.6 Test 5: Generation of a Dry-Bed Region ... 143
- 8.7 Test Problems with Constant Slope ... 144
- 8.8 Closing Remarks ... 146
- 8.9 Computer Program for the Exact Riemann Solver ... 146
- References ... 162

# 9 Notions on Numerical Methods ... 163
## 9.1 Numerical Approximation of Hyperbolic Equations ... 163
### 9.1.1 Finite Difference Approximation to PDEs ... 164
### 9.1.2 Well-Known Finite Difference Methods ... 166
## 9.2 Basic Properties of Numerical Methods ... 170
### 9.2.1 Forms of Expressing a Numerical Scheme ... 170
### 9.2.2 Monotonicity, Accuracy and Godunov's Theorem ... 172
### 9.2.3 Viscous Form of a Scheme ... 174
### 9.2.4 Conservative Form of a Scheme ... 176
## 9.3 Computational Results ... 179
### 9.3.1 Test Problems, Methods and Parameters ... 179
### 9.3.2 Results and Discussion ... 180
## 9.4 Conclusions and Further Reading ... 186
References ... 187

# 10 First-Order Methods for Systems ... 189
## 10.1 The Finite Volume Framework ... 189
### 10.1.1 Balance Laws in Integral Form ... 189
### 10.1.2 The Finite Volume Formula ... 191
## 10.2 The Godunov Upwind Method ... 193
### 10.2.1 The Numerical Flux from the Riemann Problem ... 193
### 10.2.2 Godunov's Method for the Linear Advection Equation ... 194
### 10.2.3 Godunov's Method and the Source Term ... 196
### 10.2.4 Godunov's Method for Shallow Water ... 198
## 10.3 Initial, Boundary and Stability Conditions ... 200
### 10.3.1 Initial Conditions ... 200
### 10.3.2 Boundary Conditions ... 200
### 10.3.3 Stability Condition ... 202
## 10.4 The Random Choice Method ... 204
## 10.5 Alternative Conservative Schemes ... 206
### 10.5.1 The Flux Vector Splitting Approach ... 206
### 10.5.2 Centred Methods ... 208
## 10.6 Finite Volume Schemes in Multidimensions ... 210
### 10.6.1 Unstructured Meshes ... 210
### 10.6.2 The Numerical Flux ... 212
### 10.6.3 The Cartesian Case ... 214
### 10.6.4 The Telescopic Property ... 215
## 10.7 Numerical Results ... 216
## 10.8 Conclusions ... 220
References ... 221

## Contents

**11 Approximate Riemann Solvers** .................................. 225
    11.1    Recalling the Godunov Upwind Method .................... 225
    11.2    Approximate-State Riemann Solvers ....................... 228
            11.2.1    The Framework .............................. 228
            11.2.2    A Primitive Variable Riemann Solver .............. 229
            11.2.3    Riemann Solver Based on Exact Depth Positivity ......................................... 230
            11.2.4    A Two-Rarefaction Riemann Solver ............... 231
            11.2.5    A Two-Shock Riemann Solver .................... 231
    11.3    HLL Riemann Solvers .................................. 232
    11.4    HLLC Riemann Solvers ................................. 233
    11.5    The Rusanov and Lax-Friedrichs Schemes ................. 235
    11.6    Roe's Approximate Riemann Solver ...................... 236
            11.6.1    The Basic Scheme ............................ 236
            11.6.2    Entropy Fix for the Roe Solver ................... 238
    11.7    The Riemann Solver of Osher and Solomon ............... 239
    11.8    The Dumbser-Osher-Toro Riemann Solver: DOT ........... 243
            11.8.1    Notation ..................................... 243
            11.8.2    The DOT Numerical Flux ...................... 244
    11.9    Path-Conservative Methods ............................. 245
            11.9.1    Non-conservative Methods ...................... 245
            11.9.2    The Framework .............................. 246
            11.9.3    DOT Path-Conservative Scheme ................. 247
            11.9.4    FORCE-$\alpha$ Path-Conservative Scheme ............ 248
            11.9.5    Choosing $\alpha$: Accuracy Versus Size of Time Step .... 249
            11.9.6    FORCE-$\alpha$ in DG Finite Element Methods .......... 250
    11.10  Computation of Wet/Dry Fronts .......................... 251
            11.10.1  Artificial Bed Wetting .......................... 251
            11.10.2  Conservative-Form Induced Errors ................ 254
            11.10.3  Dry-Bed Approximate Riemann Solvers ........... 254
    11.11  Concluding Remarks ................................... 255
    References ................................................... 256

**12 Second-Order Non-linear Methods** ........................... 261
    12.1    Introduction .......................................... 261
    12.2    The Weighted Average Flux (WAF) Method ................ 263
            12.2.1    The Basic WAF Scheme ....................... 263
            12.2.2    TVD Version of the WAF Scheme ................ 265
            12.2.3    Handling Critical Flow ......................... 266
    12.3    The MUSCL-Hancock Scheme .......................... 268
            12.3.1    The Basic Linear Scheme ....................... 268
            12.3.2    TVD Version of the MUSCL-Hancock Scheme ..... 271

|         |        |         |                                                    |     |
|---------|--------|---------|----------------------------------------------------|-----|
|         | 12.3.3 | ENO Version of the MUSCL-Hancock Scheme            | 274 |
| 12.4    | FORCE-Based TVD Schemes: The SLIC Method          || 275 |
| 12.5    | Numerical Results                                 || 276 |
| 12.6    | Conclusions                                       || 280 |
| References |     |                                                   || 280 |

## 13 Sources and Multidimensions ... 283

- 13.1 Introductory Remarks ... 283
- 13.2 Treatment of Source Terms by Splitting ... 285
  - 13.2.1 Preliminary Notions ... 285
  - 13.2.2 Splitting for Systems with Source Terms ... 287
  - 13.2.3 Upwinding and Advection-Reaction Splitting ... 290
- 13.3 Solvers for Ordinary Differential Equations ... 292
  - 13.3.1 First-Order Systems of ODEs ... 293
  - 13.3.2 Conventional Numerical Methods for ODEs ... 294
  - 13.3.3 TVD Runge–Kutta Schemes for ODEs ... 296
  - 13.3.4 A Note on Stability ... 297
- 13.4 Multidimensional Systems of PDEs ... 298
  - 13.4.1 Dimensional-Split Schemes ... 298
  - 13.4.2 Unsplit Finite Volume Schemes ... 301
- 13.5 Stability of Multi-dimensional Schemes ... 302
  - 13.5.1 Von Neumann Linear Stability Analysis ... 302
  - 13.5.2 Examples of Stable Schemes in 2D/3D ... 305
- 13.6 Unsplit 2D/3D Second-Order WAF Schemes ... 308
  - 13.6.1 Construction of Schemes in 2D/3D ... 309
  - 13.6.2 Stability of the Schemes in 2D/3D ... 310
  - 13.6.3 Extensions of WAF Framework ... 311
- 13.7 Concluding Remarks ... 312
- References ... 312

## 14 ADER High-Order Methods ... 317

- 14.1 Introduction ... 317
- 14.2 ADER in the Finite Volume Framework ... 319
  - 14.2.1 Preliminaries ... 319
  - 14.2.2 The ADER Approach to High Order ... 320
- 14.3 The Toro-Titarev Solver for $GRP_m$ ... 322
  - 14.3.1 Flux Leading Term ... 323
  - 14.3.2 Flux Higher-Order Terms ... 323
  - 14.3.3 The Numerical Source ... 324
- 14.4 The HEOC Solver for $GRP_m$ ... 326
  - 14.4.1 Time-Evolution Step for the Flux ... 327
  - 14.4.2 Data Interaction Step for the Flux ... 328
  - 14.4.3 The Numerical Source ... 328
  - 14.4.4 Variations of the HEOC Solver ... 329
- 14.5 The Montecinos-Toro Solver for $GRP_m$ ... 329
- 14.6 Supplementary Topics ... 331

|   |      | 14.6.1 | Other Solvers for $GRP_m$ | 331 |
|---|------|--------|---------------------------|-----|
|   |      | 14.6.2 | A Note on Spatial Reconstruction | 331 |
|   |      | 14.6.3 | ADER in the DG Framework | 335 |
|   | 14.7 | Examples: Second-Order ADER Schemes | | 336 |
|   |      | 14.7.1 | ADER2 with TT Solver for $GRP_1$ | 336 |
|   |      | 14.7.2 | ADER2 with HEOC Solver for $GRP_1$ | 340 |
|   | 14.8 | ADER Applied to Shallow Water Flows | | 343 |
|   | 14.9 | High Accuracy is Efficiency | | 344 |
|   |      | 14.9.1 | Efficiency for the Linear Advection Equation | 344 |
|   |      | 14.9.2 | Efficiency for the Euler Equations | 345 |
|   | 14.10 | Concluding Remarks and Further Reading | | 347 |
|   | References | | | 348 |
| 15 | **DAM-BREAK Modelling** | | | 353 |
|   | 15.1 | Introduction | | 353 |
|   | 15.2 | Circular Dam: Computation of Wave Phenomena | | 355 |
|   |      | 15.2.1 | Geometry, Equations and Methods | 355 |
|   |      | 15.2.2 | Computational Results | 357 |
|   | 15.3 | Physical Models: Experiments and Numerics | | 364 |
|   |      | 15.3.1 | Introduction | 364 |
|   |      | 15.3.2 | The Problem: A Dam with Channel Bend at 45° | 365 |
|   |      | 15.3.3 | Numerical Methods and Computational Geometry | 365 |
|   |      | 15.3.4 | Comparison of Numerical Results with Experiments | 367 |
|   | 15.4 | Conclusions | | 374 |
|   | References | | | 375 |
| 16 | **Mach Reflection in Tsunamis** | | | 377 |
|   | 16.1 | Introduction | | 377 |
|   | 16.2 | The Problem | | 379 |
|   | 16.3 | Analytical Study | | 381 |
|   |      | 16.3.1 | Oblique Bore Relations | 382 |
|   |      | 16.3.2 | Regular Reflection | 383 |
|   |      | 16.3.3 | Transition from Regular to Mach Reflection | 384 |
|   | 16.4 | Numerical Study | | 385 |
|   | 16.5 | Closing Remarks | | 390 |
|   | References | | | 391 |
| 17 | **Concluding Remarks** | | | 393 |
|   | References | | | 398 |
| **Index** | | | | 403 |

# Chapter 1
# The Shallow Water Equations

**Abstract** In this Chapter we derive the time-dependent, non-linear two-dimensional shallow water equations on domains with variable bathymetry. Emphasis is given to the assumptions made in the derivation of the equations, so as to make clear their actual range of applicability. The equations are expressed in conservation-law form, both in differential form and in integral form. The introduced partial differential equations (PDEs) will be the main focus of the rest of this book.

## 1.1 Introduction

A wide variety of physical phenomena are governed by mathematical models of the so-called *shallow-water* type. These systems of hyperbolic partial differential equations serve as mathematical models for the dynamics of water flow in ocean tides, breaking of waves on shallow beaches, roll waves in open channels, flood waves in rivers, surges, dam-break wave modelling, tsunami wave propagation and various specific applications in environmental fluid dynamics disciplines. Shallow-water type models are also used for studying the dispersion of heavy gases over earth surface and the dynamics of the atmosphere. Note however that the physical complexity of three-dimensional oceanographic and atmospheric flows requires more complete mathematical models than shallow-water type systems. Nonetheless, the basic, one-layer shallow water equations still play an important role in providing benchmark solutions to assess methodologies intended for more complex mathematical models.

A key assumption made in the derivation of the approximate shallow water theory concerns the pressure distribution; this is given as in hydrostatics and results from assuming that the vertical acceleration of the water particles has a negligible effect on the pressure. If dissipative effects are neglected, then the resulting shallow water equations are a time-dependent, two-dimensional system of non-linear partial differential equations (PDEs) of hyperbolic type. The same governing equations result as the approximation to lowest order in a perturbation procedure. This involves a formal development of all quantities in powers of a parameter $\epsilon$, which is the ratio of the water depth to some other characteristic length associated with the horizontal direction. Recommended supplementary reading on the shallow water equations is

© The Author(s), under exclusive license to Springer Nature Switzerland AG 2024
E. F. Toro, *Computational Algorithms for Shallow Water Equations*,
https://doi.org/10.1007/978-3-031-61395-1_1

found in the books by Stoker [1], Cunge et al. [2], Lighthill [3], Roberson et al. [4], LéMehauté [5], Henderson [6] and Baines [7]. Numerical aspects of shallow water type equations are discussed in the textbooks of Abbott [8], Weiyan [9], Vreugdenhil [10], and Castro-Orgaz and Hager [11].

In this Chapter, after stating two underlying conservation principles, namely conservation of mass and conservation of momentum, we proceed to derive the shallow water equations for domains of variable bathymetry. The equations are expressed in conservation law form, both in differential form and in integral form. Finally, we consider the case of one-dimensional models for river flows of variable cross-section, the Saint Venant equations.

Before introducing the conservation equations in the next section, we first establish some conventions on the notation used in this Chapter, for the most frequently used symbols. A Cartesian frame of reference $(x, y, z)$ is chosen and the time variable is denoted by $t$. Transformation to other coordinate systems is carried out using the chain rule in the usual way. Any quantity $\phi$ that depends on space and time will be written as $\phi(x, y, z, t)$. In most situations the governing equations will be partial differential equations. Naturally, these will involve partial derivatives for which we use the following notation

$$\phi_t \equiv \frac{\partial \phi}{\partial t} \equiv \partial_t \phi \,, \quad \phi_x \equiv \frac{\partial \phi}{\partial x} \equiv \partial_x \phi \,, \quad \phi_y \equiv \frac{\partial \phi}{\partial y} \equiv \partial_y \phi \,, \quad \phi_z \equiv \frac{\partial \phi}{\partial z} \equiv \partial_z \phi \,. \tag{1.1}$$

We also recall some basic notation involving scalars and vectors. The *dot product* of two vectors $\mathbf{A} = (a_1, a_2, a_3)$ and $\mathbf{B} = (b_1, b_2, b_3)$ is the *scalar* quantity

$$\mathbf{A} \cdot \mathbf{B} \equiv a_1 b_1 + a_2 b_2 + a_3 b_3 \,.$$

Given a scalar quantity $\phi$ that depends on the spatial variables $x, y, z$, the *gradient operator* $\nabla$ as applied to $\phi$ is the vector

$$\text{grad}\, \phi \equiv \nabla \phi = \left( \frac{\partial \phi}{\partial x}, \frac{\partial \phi}{\partial y}, \frac{\partial \phi}{\partial z} \right) \,.$$

The *divergence operator* applies to vectors and the result is a scalar quantity; for a vector $\mathbf{A}$, the divergence of $\mathbf{A}$ is

$$\text{div}\, \mathbf{A} \equiv \nabla \cdot \mathbf{A} = \frac{\partial a_1}{\partial x} + \frac{\partial a_2}{\partial y} + \frac{\partial a_3}{\partial z} \,.$$

In the next section we discuss the physical conservation principles that will be used to derive the basic shallow water equations of interest to us here.

## 1.2 Conservation Principles

The general conservation laws of mass and momentum for a compressible material, written in differential conservation law form, are

$$\rho_t + \nabla \cdot (\rho \mathbf{V}) = 0 \tag{1.2}$$

and

$$\frac{\partial}{\partial t}(\rho \mathbf{V}) + \nabla \cdot \left[\rho \mathbf{V} \otimes \mathbf{V} + p\mathsf{I} - \Pi\right] = \rho \mathbf{g} \, . \tag{1.3}$$

These equations state, respectively, conservation of mass and balance of momentum $\rho \mathbf{V}$. The *independent variables* are $t$ for time and $x$, $y$, $z$ for space. The dependent variables are $\rho$ for density, $\mathbf{V} = (u, v, w)$ for velocity, with $u$, $v$, $w$ the $x$, $y$, $z$ components of velocity, respectively; $p$ is pressure; the vector $\mathbf{g} = (g_1, g_2, g_3)$ is a body force vector; the tensors involved are

$$\left.\begin{aligned}
\mathbf{V} \otimes \mathbf{V} &= \begin{bmatrix} u^2 & uv & uw \\ vu & v^2 & vw \\ wu & wv & w^2 \end{bmatrix}, \\
\mathsf{I} &= \begin{bmatrix} 1 & 0 & 0 \\ 0 & 1 & 0 \\ 0 & 0 & 1 \end{bmatrix}, \\
\Pi &= \begin{bmatrix} \tau^{xx} & \tau^{xy} & \tau^{xz} \\ \tau^{yx} & \tau^{yy} & \tau^{yz} \\ \tau^{zx} & \tau^{zy} & \tau^{zz} \end{bmatrix}.
\end{aligned}\right\} \tag{1.4}$$

Here $\Pi$ is the *viscous stress tensor* and may be defined through *the Newtonian approximation*. The corresponding integral form of the balance laws (1.2), (1.3) is given respectively by

$$\frac{\partial}{\partial t} \int\!\!\int\!\!\int_V \rho \, dV = -\int\!\!\int_\Omega \mathbf{n} \cdot (\rho \mathbf{V}) \, d\Omega \tag{1.5}$$

and

$$\frac{\partial}{\partial t} \int\!\!\int\!\!\int_V (\rho \mathbf{V}) \, dV = -\int\!\!\int_\Omega \left[\mathbf{V}(\mathbf{n} \cdot \rho \mathbf{V}) + p\mathbf{n} - \mathbf{n} \cdot \Pi\right] d\Omega \tag{1.6}$$
$$+ \int\!\!\int\!\!\int_V \rho \mathbf{g} \, dV \, .$$

Here $V$ is an arbitrary *control volume* of an element of fluid in three-dimensional space and $\Omega$ is its boundary. Computationally, $V$ will be a *finite volume* or *computational cell*; see Chap. 9.

Sometimes it is convenient to express the equations in terms of the *primitive* or *physical* variables $\rho$, $u$, $v$, $w$ and $p$ (or some other set). When viscous effects are neglected but body forces are retained via a source term vector, the equations become

$$\left.\begin{aligned}
\rho_t + u\rho_x + v\rho_y + w\rho_z + \rho(u_x + v_y + w_z) &= 0, \\
u_t + uu_x + vu_y + wu_z + \frac{1}{\rho}p_x &= g_1, \\
v_t + uv_x + vv_y + wv_z + \frac{1}{\rho}p_y &= g_2, \\
w_t + uw_x + vw_y + ww_z + \frac{1}{\rho}p_z &= g_3.
\end{aligned}\right\} \quad (1.7)$$

Recall notation for partial derivatives in (1.1). These are the equations we may use to define the free-surface problem of interest here.

## 1.3 Water Flow with a Free Surface

Consider the flow of water with a *free surface* under gravity in a three-dimensional domain, with $x$-$y$ determining a horizontal plane and $z$ defining the vertical direction, which we associate with the free-surface elevation. Figure 1.1 illustrates the physical set up. The region of interest lies above a horizontal datum at $z = 0$. The fluid lies above the channel bottom elevation $b(x) \geq 0$; the slope $b'(x)$ of the bottom elevation gives rise to a source term in the equations; $h(x, t)$ is the flow depth; $u(x, t)$ is the particle velocity; the free surface is under the influence of the acceleration due to gravity $g$. We assume variable bathymetry

$$z = b(x, y) \quad (1.8)$$

and the *free surface* is defined by

$$z = s(x, y, t) = b(x, y) + h(x, y, t), \quad (1.9)$$

where $h(x, y, t)$ is the depth of water, the vertical distance between the bottom and the free-surface position. Figure 1.1 depicts the geometry for a simplified situation for a chosen, constant value of $y$.

Assuming the density of the fluid to be constant, the governing Eq. (1.7) simplify to

$$u_x + v_y + w_z = 0, \quad (1.10)$$

$$u_t + uu_x + vu_y + wu_z = -\frac{1}{\rho}p_x, \quad (1.11)$$

$$v_t + uv_x + vv_y + wv_z = -\frac{1}{\rho}p_y, \quad (1.12)$$

$$w_t + uw_x + vw_y + ww_z = -\frac{1}{\rho}p_z - g. \quad (1.13)$$

## 1.3 Water Flow with a Free Surface

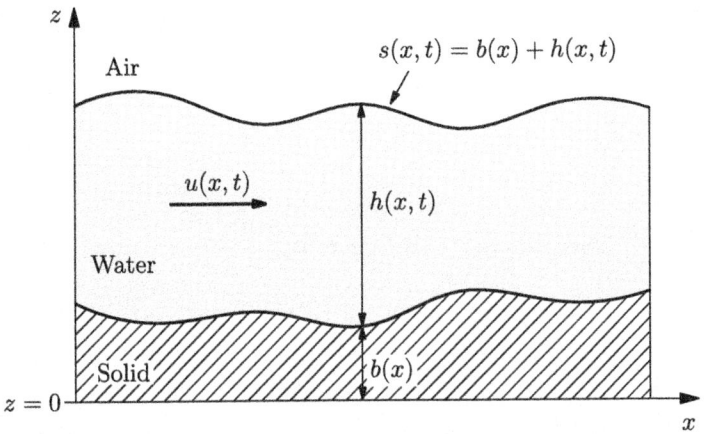

**Fig. 1.1** Side view of the physical set up for the shallow water equations. The water layer lies between the impermeable bottom $b(x)$ above a horizontal datum $z = 0$ and the free surface $s(x, t) = b(x) + h(x, t)$

Here we have assumed that the body force vector is $\mathbf{g} = (0, 0, -g)$, where $g$ is the acceleration due to gravity, taken as $g = 9.81$ m/s$^2$, a constant.

In principle, given initial conditions at time $t = 0$ and boundary conditions on the bottom and the free surface, the solution of the problem is given by the solution of the four Eqs. (1.10)–(1.13) for the four unknowns $p, u, v, w$. As posed, the initial-boundary value problem for (1.10)–(1.13) remains computationally challenging. The main difficulty in solving the full problem is associated with the free surface. This is a boundary and as such, boundary conditions are to be satisfied, but the position of this boundary itself is unknown and therefore the domain on which the equations are to be solved is not known a priori.

Approximate theories leading to simpler problems exist. One such approximate theory assumes that the amplitude of the free-surface disturbance from the rest position is small with respect to a characteristic length, such as wave length. This assumption leads to linear initial-boundary value problems and thus to a *linear theory*. Another approximation, which is the one we adopt in this book, results from the assumption that the depth of water is small with respect to wave length or free-surface curvature, for example. This assumption gives rise to non-linear initial value problems that are analogous to those associated with wave propagation in compressible materials [12–14]. In this book we are mainly concerned with this non-linear shallow water theory. For details see Stoker [1] and Cunge [2], for example. We anticipate that in spite of the strong simplifying assumptions made in deriving the non-linear shallow water theory, the numerical solution of the equations remains a computationally challenging task.

Before deriving the shallow-water theory we discuss boundary conditions for the full problem (1.10)–(1.13). Assume that a boundary is given by the surface

$$\psi(x, y, z, t) = 0 . \tag{1.14}$$

For the free surface we have

$$\text{Free surface:} \quad \psi(x, y, z, t) \equiv z - s(x, y, t) = 0 \tag{1.15}$$

and for the bottom boundary we have

$$\text{Bottom or bed:} \quad \psi(x, y, z, t) \equiv z - b(x, y) = 0 . \tag{1.16}$$

Two boundary conditions are imposed on the free surface $s(x, y, t)$, with $\psi$ given by (1.15), namely the *kinematic condition*

$$\frac{d}{dt}\psi(x, y, z, t) = \psi_t + u\psi_x + v\psi_y + w\psi_z = 0 \tag{1.17}$$

and the *dynamical condition*

$$p(x, y, z, t)|_{z=s(x,y)} = p_{atm} = 0 , \tag{1.18}$$

where $p_{atm}$ is the atmospheric pressure, which for convenience is taken here to be identically zero. For the bottom boundary $b(x, y)$, condition (1.17) also applies, with $\psi$ given by (1.16).

## 1.4 The Shallow Water Equations

The first assumption in the derivation of the shallow water equations is that the *vertical component of acceleration*, given by

$$\frac{dw}{dt} = w_t + uw_x + vw_y + ww_z , \tag{1.19}$$

is negligible. Insertion of this condition $dw/dt = 0$ into Eq. (1.13) gives

$$p_z = -\rho g . \tag{1.20}$$

Given the dynamical condition (1.18) that the atmospheric pressure $p_{atm}$ is zero on the free surface, we obtain

$$p = \rho g(s - z) . \tag{1.21}$$

## 1.4 The Shallow Water Equations

Differentiation of (1.21) with respect to $x$ and $y$ gives

$$p_x = \rho g s_x, \quad p_y = \rho g s_y. \tag{1.22}$$

Note that $p_x$ and $p_y$ are both independent of $z$ and thus the $x$ and $y$ components of the acceleration of water particles $du/dt$ and $dw/dt$ are independent of $z$. Hence the $x$ and $y$ velocity components $u$ and $v$ are also independent of $z$, that is $u_z = v_z = 0$. Therefore, by virtue of the above conditions and by making use of (1.22) in Eqs. (1.11) and (1.12), we obtain

$$u_t + u u_x + v u_y = -g s_x \tag{1.23}$$

and

$$v_t + u v_x + v v_y = -g s_y. \tag{1.24}$$

An important step in deriving the shallow water equations now follows. We integrate the continuity Eq. (1.10) with respect to $z$, the vertical coordinate, between $z = b(x, y)$ (bottom) and $z = s(x, y, t)$ (free surface). That is

$$\int_b^s (u_x + v_y + w_z)\, dz = 0, \tag{1.25}$$

which leads to

$$w|_{z=s} - w|_{z=b} + \int_b^s u_x\, dz + \int_b^s v_y\, dz = 0. \tag{1.26}$$

We now apply boundary conditions in order to determine the first two terms in Eq. (1.26) above. Expanding (1.17) as applied to the free surface (1.15) gives

$$(s_t + u s_x + v s_y - w)\big|_{z=s} = 0. \tag{1.27}$$

Expanding (1.17) as applied to the bottom boundary (1.16) gives

$$(u b_x + v b_y - w)\big|_{z=b} = 0. \tag{1.28}$$

From Eq. (1.27) we obtain

$$w|_{z=s} = (s_t + u s_x + v s_y)\big|_{z=s} \tag{1.29}$$

and from (1.28) we obtain

$$w|_{z=b} = (u b_x + v b_y)\big|_{z=b}. \tag{1.30}$$

Substitution of (1.29) and (1.30) into (1.26) gives

$$(s_t + us_x + vs_y)\big|_{z=s} - (ub_x + vb_y)\big|_{z=b} + \int_b^s u_x \, dz + \int_b^s v_y \, dz = 0 \,. \tag{1.31}$$

To simplify (1.31) further, we apply Leibniz's formula

$$\frac{d}{d\alpha} \int_{\xi_1(\alpha)}^{\xi_2(\alpha)} f(\xi, \alpha) \, d\xi = \int_{\xi_1(\alpha)}^{\xi_2(\alpha)} \frac{\partial f}{\partial \alpha} \, d\xi + f(\xi_2, \alpha) \frac{d\xi_2}{d\alpha} - f(\xi_1, \alpha) \frac{d\xi_1}{d\alpha} \tag{1.32}$$

to the last two integral terms in (1.31), which yields

$$\int_b^s u_x \, dz = \frac{\partial}{\partial x} \int_b^s u \, dz - u\big|_{z=s} \cdot s_x + v\big|_{z=b} \cdot b_x \tag{1.33}$$

and

$$\int_b^s v_y \, dz = \frac{\partial}{\partial y} \int_b^s v \, dz - v\big|_{z=s} \cdot s_y + v\big|_{z=b} \cdot b_y \,. \tag{1.34}$$

Substitution of (1.33) and (1.34) into (1.31) gives

$$s_t + \frac{\partial}{\partial x} \int_b^s u \, dz + \frac{\partial}{\partial y} \int_b^s v \, dz = 0 \,. \tag{1.35}$$

Recall that both $u$ and $v$ are independent of $z$; also $s = b + h$ and $b_t = 0$. Equation (1.35) then simplifies to

$$h_t + (hu)_x + (hv)_y = 0 \,. \tag{1.36}$$

This is the law of *conservation of mass* and is written in differential conservation law form.

The momentum Eqs. (1.11) and (1.12) can also be expressed in differential conservation-law form. To this end we add Eq. (1.36), pre-multiplied by $u$, to equation (1.23), pre-multiplied by $h$; we also make use of a relation that does assume differentiability of the water depth $h$, namely

$$h \frac{\partial h}{\partial x} = \frac{\partial}{\partial x}\left(\frac{1}{2}h^2\right) , \tag{1.37}$$

to obtain

$$(hu)_t + (hu^2 + \tfrac{1}{2}gh^2)_x + (huv)_y = -ghb_x \,. \tag{1.38}$$

Similarly, for the $y$ momentum equation we obtain

$$(hv)_t + (huv)_x + (hv^2 + \tfrac{1}{2}gh^2)_y = -ghb_y \,. \tag{1.39}$$

## 1.4 The Shallow Water Equations

All three partial differential Eqs. (1.36), (1.38) and (1.39) can be written in differential conservation-law form as the single vector equation

$$\partial_t \mathbf{Q} + \partial_x \mathbf{F}(\mathbf{Q}) + \partial_y \mathbf{G}(\mathbf{Q}) = \mathbf{S}(\mathbf{Q}) , \qquad (1.40)$$

with

$$\left.\begin{array}{c} \mathbf{Q} = \begin{bmatrix} h \\ hu \\ hv \end{bmatrix} , \quad \mathbf{F}(\mathbf{Q}) = \begin{bmatrix} hu \\ hu^2 + \frac{1}{2}gh^2 \\ huv \end{bmatrix} , \\[2em] \mathbf{G}(\mathbf{Q}) = \begin{bmatrix} hv \\ hvu \\ hv^2 + \frac{1}{2}gh^2 \end{bmatrix} , \quad \mathbf{S}(\mathbf{Q}) = \begin{bmatrix} s_1 \\ s_2 \\ s_3 \end{bmatrix} = \begin{bmatrix} 0 \\ -gh\partial_x b \\ -gh\partial_y b \end{bmatrix} . \end{array}\right\} \qquad (1.41)$$

In the system of Eq. (1.40) $\mathbf{Q}$ is the vector of conserved variables, $\mathbf{F}(\mathbf{Q})$ and $\mathbf{G}(\mathbf{Q})$ are flux vectors and $\mathbf{S}(\mathbf{Q})$ is the source term vector. For many applications there will be additional terms in the vector $\mathbf{S}(\mathbf{Q})$, such as Coriolis forces, wind forces and bottom friction. Most of this textbook is devoted to the numerical approximation of the differential part of the equations assuming $\mathbf{S}(\mathbf{Q}) = \mathbf{0}$ -the homogeneous version of (1.40). For some problems of practical interest, the numerical treatment of the source terms is a relatively standard procedure and simple numerical methods to deal with source terms are studied in Chap. 13. We note however that there are many situations of practical interest that require sophisticated methods to adequately deal with source terms. Some of these issues are also addressed in Chap. 14.

Equations (1.40) and (1.41) may be recast in integral form as

$$\frac{\partial}{\partial t} \int\!\!\int\!\!\int_V \mathbf{Q} \, dV = - \int\!\!\int_\Omega \mathbf{n} \cdot \mathbf{H}(\mathbf{Q}) \, d\Omega , \qquad (1.42)$$

where $\mathbf{n} \cdot \mathbf{H}(\mathbf{Q})$ is the *normal flux component* through the surface $\Omega$ and $\mathbf{H}(\mathbf{Q}) = (\mathbf{F}, \mathbf{G})$. The integral form (1.42) of the equations admits *discontinuous solutions*, such as *bores*, whereas the differential form (1.40) does not. It is a relatively standard procedure to show that the differential form is a special case of the integral form and is obtained from the integral form by assuming that solutions are sufficiently smooth. See the texbooks of LeVeque [15], Godlewski and Raviart [16], Toro [14] and Dafermos [17] for detailed discussions on smooth and discontinuous (weak) solutions of hyperbolic conservation laws.

From a computational viewpoint, most of the ideas for devising basic schemes can be discussed in terms of the one-dimensional system system of balance laws written in differential form

$$\partial_t \mathbf{Q} + \partial_x \mathbf{F}(\mathbf{Q}) = \mathbf{S}(\mathbf{Q}) , \qquad (1.43)$$

with $\mathbf{Q}$, $\mathbf{F}(\mathbf{Q})$ and $\mathbf{S}(\mathbf{Q})$ given by

$$\mathbf{Q} = \begin{bmatrix} h \\ hu \end{bmatrix}, \quad \mathbf{F} = \begin{bmatrix} hu \\ hu^2 + \frac{1}{2}gh^2 \end{bmatrix}, \quad \mathbf{S} = \begin{bmatrix} s_1 \\ s_2 \end{bmatrix} = \begin{bmatrix} 0 \\ -ghb'(x) \end{bmatrix}. \quad (1.44)$$

Equations of the form (1.43) and (1.44) may also be written in integral form, in order to allow for discontinuous solutions, or generalised solutions or weak solutions. One integral form is obtained by integration of (1.43) in space and time in the rectangular control volume $V$ in the $x$-$t$ plane

$$V = [x_L, x_R] \times [t_1, t_2]. \quad (1.45)$$

The result is the integral form of the conservation laws written as

$$\left. \begin{aligned} \int_{x_L}^{x_R} \mathbf{Q}(x, t_2) dx &= \int_{x_L}^{x_R} \mathbf{Q}(x, t_1) dx \\ &+ \int_{t_1}^{t_2} \mathbf{F}(\mathbf{Q}(x_L, t)) dt - \int_{t_1}^{t_2} \mathbf{F}(\mathbf{Q}(x_R, t)) dt \\ &+ \int_{t_1}^{t_2} \int_{x_L}^{x_R} \mathbf{S}(\mathbf{Q}(x, t)) dx dt. \end{aligned} \right\} \quad (1.46)$$

The reader is encouraged to verify that (1.46) can be obtained from integrating (1.43) exactly in the control volume (1.45). An alternative integral form is

$$\frac{d}{dt} \int_{x_L}^{x_R} \mathbf{Q}(x, t) dx = \mathbf{F}(\mathbf{Q}(x_L, t)) - \mathbf{F}(\mathbf{Q}(x_R, t)) + \int_{x_L}^{x_R} \mathbf{S}(\mathbf{Q}(x, t)) dx. \quad (1.47)$$

Equations (1.46) and (1.47) admit discontinuous solutions while Eqs. (1.43), (1.44) do not. The integral form (1.46) will be used frequently in this book, particularly for constructing finite volume methods.

## 1.5 The Saint Venant Equations for River Flows

Of considerable practical importance are the Saint Venant equations

$$A_t + (Au)_x = 0 \quad (1.48)$$

and

$$(Au)_t + (Au^2)_x + gAh_x = gA(b_x - b_f), \quad (1.49)$$

which are used for modelling river flows. The geometry and quantities involved are illustrated in Fig. 1.2, which depicts a cross-section of the river or channel at a fixed position $x$ along the river. Here $A = A(x, t)$ is the *wetted cross-sectional area* defined as

## 1.5 The Saint Venant Equations for River Flows

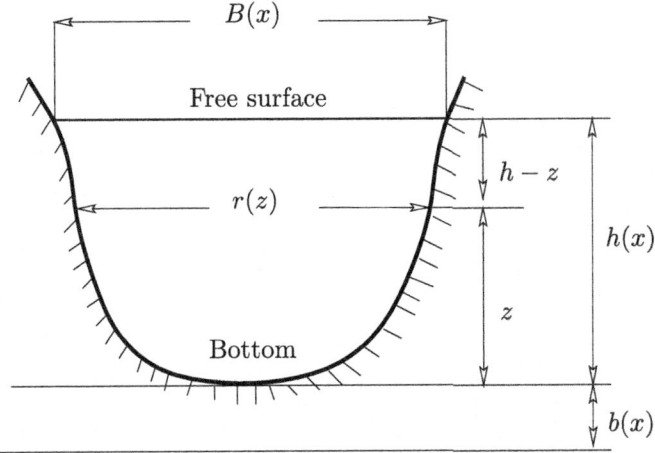

**Fig. 1.2** Cross-section of river and notation for the Saint Venant equations

$$A = \int_0^{h(x)} r(x, z) \, dz \,, \tag{1.50}$$

where $r(x, z)$ is the *cross-sectional width* and $r(x, h) = B(x)$ is the *breadth* of the free surface or free-surface width. The quantity $b_f$ in Eq. (1.49) is called the *friction slope*. Note that $B = \partial A / \partial h$. The volumetric water *discharge* is $Q = Au$. Equations (1.48) and (1.49) may be derived from the two-dimensional shallow water Eqs. (1.40) and (1.41). They can also be derived from first principles; see Cunge et al. [2] for a detailed derivation and discussion of these equations. An important remark on the Saint Venant equations (1.48) and (1.49) is that the momentum equation cannot in general be written in conservation-law form, such as in (1.40) or (1.43). There are then the theoretical and practical difficulties of defining shocks that obey Rankine-Hugoniot conditions, discussed in Chaps. 4, 5 and 6. Another difficulty is that, as they stand, Eqs. (1.48) and (1.49) cannot be solved by a *conservative numerical method*. A possible way around this problem is to assume that if discontinuities are present, these are *weak*, in the sense that jumps across the discontinuities are in some sense small. Then, making use of the chain rule to evaluate $(Ah)_x$, one can rewrite Eq. (1.49) as

$$(Au)_t + (Au^2 + gAh)_x = gA(b_x - b_f) - ghA_x \,. \tag{1.51}$$

This device is not strictly justified but allows the use of conservative methods to the left-hand side of Eq. (1.51). The penalty is that there is a *differential* term on the right-hand side of (1.51) that needs to be accounted for, perhaps in a second stage, as part of the treatment of source terms. We strongly discourage this approach. Alternatively, one may accept the fact that the equations are not written in conservation-law form and use non-conservative methods. In the context of non-conservative methods one may also express the equation in terms of primitive variables, or physical variables.

For instance, we may choose to express the equations in terms of the variables area $A(x, t)$ and velocity $u(x, t)$. By expanding derivatives in (1.48) and (1.49), making use of (1.48) into (1.49), followed by algebraic manipulations, one obtains the system

$$A_t + u A_x + A u_x = 0, \qquad (1.52)$$

$$u_t + u u_x + \frac{g}{B} A_x = g A (b_x - b_f). \qquad (1.53)$$

These equations can then be written in *quasi-linear form* (non-conservative) as

$$\mathbf{W}_t + \mathbf{A} \mathbf{W} x = \mathbf{S}, \qquad (1.54)$$

where the vectors $\mathbf{W}$, $\mathbf{S}$ and the coefficient matrix $\mathbf{A}$ are given by

$$\mathbf{W} = \begin{bmatrix} A \\ u \end{bmatrix}, \quad \mathbf{A} = \begin{bmatrix} u & A \\ g/B & u \end{bmatrix}, \quad \mathbf{S} = \begin{bmatrix} 0 \\ g A (b_x - b_f) \end{bmatrix}. \qquad (1.55)$$

Numerical methods for solving equations in non-conservative form are studied in Chap. 11.

## 1.6 Conclusions

In this Chapter we have derived the basic time-dependent, two-dimensional non-linear shallow water equations with variable bathymetry. These are the equations that are central to the themes of this book. Particular emphasis has been given to the assumptions made in the derivation of the equations, so as to make clear the actual range of applicability of the model and that of similar shallow-water type models. Complex applications will probably require a re-derivation of the equations by the modeller, retaining perhaps more physical effects. Such extra effects may take the form of additional partial differential equations or extra algebraic source or forcing terms. When modelling pollutant transport and biochemistry problems, for instance, one requires additional *species* equations; these are advection-diffusion-reaction type partial differential equations. For the main purpose of this book, which is concerned with the study of computational algorithms for solving the shallow water equations, the governing equations as stated are sufficient.

# References

1. J.J. Stoker, *Water Waves. The Mathematical Theory with Applications* (John Wiley and Sons, 1992)
2. J.A. Cunge, F.M. Holly, A. Verwey, *Practical Aspects of Computational River Hydraulics* (Pitman Publishing Ltd., Reprinted by the University of Iowa, 1994)
3. J. Lighthill, *Waves in Fluids* (Cambridge University Press, 1978)
4. J.A. Roberson, J.J. Cassidy, M.H. Chaudhry, *Hydraulic Engineering* (Houghton Mifflin Company, Boston, 1988)
5. B. LéMehauté, *An Introduction to Hydrodynamics and Water Waves* (Springer-Verlag, New York, Heidelberg, Berlin, 1976)
6. F.M. Henderson, *Open Channel Flow* (MacMillan Publishing Co, 1966)
7. P.G. Baines, *Topographic Effects in Stratified Flows* (Cambridge University Press, 1995)
8. M.B. Abbott, *Computational Hydraulics* (Ashgate Publishing Company, 1992)
9. T. Weiyan, *Shallow Water Hydrodynamics*. Elsevier Oceanography Series, vol. 55 (Elsevier, Amsterdam, 1992)
10. C.B. Vreugdenhil, *Numerical Methods for Shallow Water Flow* (Kluwer Academic Publishers, The Netherlands, 1994)
11. O. Castro-Orgaz, W.H. Hager, *Shallow Water Hydraulics* (Springer, 2019)
12. E.F. Toro, *Riemann Solvers and Numerical Methods for Fluid Dynamics* (Springer–Verlag, 1997)
13. E.F. Toro, *Riemann Solvers and Numerical Methods for Fluid Dynamics*, 2nd edn. (Springer–Verlag, 1999)
14. E.F. Toro, *Riemann Solvers and Numerical Methods for Fluid Dynamics. A Practical Introduction*, 3rd edn. (Springer-Verlag, 2009)
15. R.J. LeVeque, *Finite Volume Methods for Hyperbolic Problems* (Cambridge University Press, 2002)
16. E. Godlewski, P.A. Raviart, *Numerical Approximation of Hyperbolic Systems of Conservation Laws*, 2nd edn. (Springer, 2021)
17. C.M. Dafermos, *Hyperbolic Conservation Laws in Continuum Physics*. Grundlehren der Mathematischen Wissenschaften, vol. 325 (Springer, 2000)

# Chapter 2
# Notions on Hyperbolic Equations

**Abstract** This Chapter is a succinct introduction to basic notions on the theory of hyperbolic equations. The linear advection equation and general linear hyperbolic systems are studied in some detail, including the concepts of characteristics, eigenstructure, hyperbolicity and the Riemann problem. For the non-linear case we focus on scalar equations; the concepts of integral forms, shock formation, Rankine-Hugoniot condition, non-uniqueness of discontinuous solution and the Lax entropy condition are introduced. The complete solution of the Riemann problem for Burgers' equation is given. The material is tailored to the aims of this book; it furnishes the bases for analysing the mathematical and physical character of the non-linear shallow water equations in Chaps. 4 and 3; and for solving the Riemann problem in Chaps. 6 and 7. The contents are also useful for designing and interpreting numerical methods for wave propagation phenomena.

## 2.1 The Linear Advection Equation and Basic Concepts

Many problems in science and engineering, such as wave propagation and transport phenomena, are governed by advection-diffusion-reaction partial differential equations (PDEs). In the scalar case, a single equation, we may write

$$\partial_t q(x,t) + \partial_x f(q(x,t)) = s(x,t,q(x,t)) + \partial_x (\alpha(x,t,q(x,t))\partial_x q(x,t)) , \tag{2.1}$$

where $q(x,t)$ is the *unknown*, called the *dependent variable*; $q(x,t)$ is a function of two *independent variables* $x$ and $t$; $f(q)$ is a prescribed function of $q$ called the *flux*, or *physical flux*; $s(x,t,q)$ is also a prescribed function, called the *source term*. The last term is called the diffusion term, for which $\alpha(x,t,q(x,t))$ is the *diffusion coefficient*. Equation (2.1) is parabolic due to the presence of the viscous term, a second-order term. A particular example of (2.1) is obtained by choosing

$$f(q) = \lambda q , \quad s(q) = 0 , \quad \alpha = 0 , \tag{2.2}$$

with $\lambda$ a constant wave propagation speed. This choice leads to the *linear advection equation* (LAE)

$$\partial_t q + \lambda \partial_x q = 0, \quad -\infty < x < \infty, \quad t > 0. \tag{2.3}$$

## 2.1.1 The Initial-Value Problem

We study the simplest case of (2.1), the linear advection equation, in which the spatial domain is infinite and an initial condition (IC) at the initial time $t = 0$ is prescribed, namely

$$\left. \begin{array}{ll} \text{PDE:} & \partial_t q + \lambda \partial_x q = 0, \quad -\infty < x < \infty, \quad t > 0, \\ \text{IC:} & q(x, 0) = h(x), \quad -\infty < x < \infty, \end{array} \right\} \tag{2.4}$$

where $h(x)$ is a prescribed function of distance $x$. Equation (2.4) defines a pure *initial-value problem (IVP)*, or *Cauchy problem*.

**Characteristic Curves and the Solution**

*Characteristic curves*, or *characteristics*, are functions $x(t)$ in the $x$-$t$ half-plane of independent variables satisfying the IVP for an ordinary differential equation (ODE), namely

$$\left. \begin{array}{ll} \text{ODE:} & \dfrac{dx}{dt} = \lambda, \quad t > 0, \\ \text{IC:} & x(0) = x_0. \end{array} \right\} \tag{2.5}$$

The solution of (2.5) is immediate and reads

$$x = x_0 + \lambda t. \tag{2.6}$$

Figure 2.1 illustrates solution (2.6) in the $t$-$x$ plane. In practice it is more common to represent characteristics in the $x$-$t$ plane. The inclination of the characteristics

**Fig. 2.1** Characteristic $x(t)$ in the $t$-$x$ plane given by (2.6) for positive characteristic speed $\lambda$, where $x_0$ is the foot of the characteristic

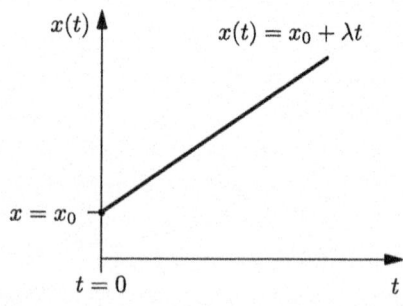

## 2.1 The Linear Advection Equation and Basic Concepts

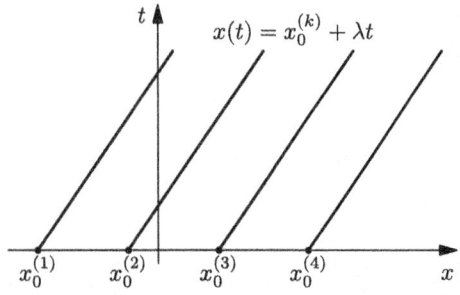

**Fig. 2.2** Family of characteristic curves $x(t)$ in the $x$-$t$ plane, for the case of positive characteristic speed $\lambda$. Compare with Fig. 2.1

depends on the characteristic speed $\lambda$, in fact on $1/\lambda$. In the linear case with constant coefficients, characteristics are all parallel to each other, as seen in Fig. 2.2.

Consider now the time-rate of change (or total derivative) of $q(x(t), t)$ along a characteristic curve $x = x(t)$

$$\frac{dq}{dt} = \frac{\partial q}{\partial t}\frac{dt}{dt} + \frac{\partial q}{\partial x}\frac{dx}{dt} \ . \tag{2.7}$$

But the curve $x(t)$ satisfies the ODE in (2.5), then (2.7) becomes

$$\frac{dq}{dt} = \partial_t q + \lambda \partial_x q = 0 \ . \tag{2.8}$$

That is, $q(x, t)$ is constant along $x = x_0 + \lambda t$. Consequently, the PDE in (2.4) becomes an ODE, namely

$$\frac{dq}{dt} = 0 \text{ along the characteristic } x = x_0 + \lambda t \ .$$

This ODE states that $q(x, t)$ is constant along the characteristic. From the above observations, the value of $q(x, t)$ at a point $(x, t)$ on the characteristic curve passing through $(x, t)$ is equal to the value of $q$ at the point $x_0$ called *the foot of the characteristic*. That is

$$q(x, t) = q(x_0, 0) = h(x_0) \ . \tag{2.9}$$

But from (2.6)

$$x_0 = x - \lambda t$$

and therefore the solution of IVP (2.4) is

$$q(x, t) = h(x - \lambda t) \ , \tag{2.10}$$

**Fig. 2.3** The solution at point $(\hat{x}, \hat{t})$ is found by tracing the characteristic from $(\hat{x}, \hat{t})$ back to its foot $x_0$. There are three possibilities: **a** $\lambda > 0$, **b** $\lambda = 0$, **c** $\lambda < 0$

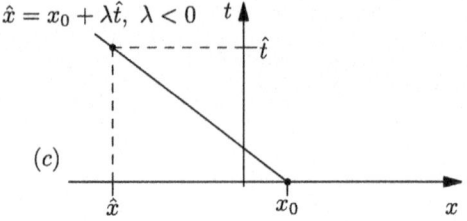

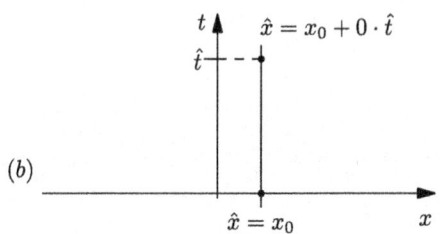

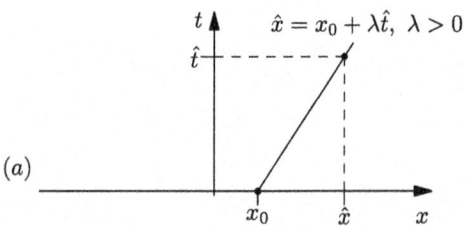

which is the initial condition $h$ in (2.4) evaluated at the position $x - \lambda t$. Figure 2.3 shows the three posible cases that can occur due to the value of the characteristic speed $\lambda$.

**Example 2.1** Here we study in detail the following initial-value problem

$$\begin{aligned} \text{PDE: } & \partial_t q + \lambda \partial_x q = 0, \quad -\infty < x < \infty, \quad t > 0, \\ \text{IC: } & q(x, 0) = h(x) = \begin{cases} 0 & \text{if} \quad x < -1, \\ 1 - x^2 & \text{if} \quad -1 \leq x \leq 1, \\ 0 & \text{if} \quad x > 1. \end{cases} \end{aligned} \quad (2.11)$$

According to formula (2.10) the solution of (2.11) is

$$q(x, t) = h(x - \lambda t) = \begin{cases} 0 & \text{if} \quad x < -1 + \lambda t, \\ 1 - (x - \lambda t)^2 & \text{if} \quad -1 + \lambda t \leq x \leq 1 + \lambda t, \\ 0 & \text{if} \quad x > 1 + \lambda t. \end{cases} \quad (2.12)$$

Note that for a given speed $\lambda$ and a chosen time $t$, the solution is simply a function of $x$, called a profile. See Fig. 2.4. □

## 2.1 The Linear Advection Equation and Basic Concepts

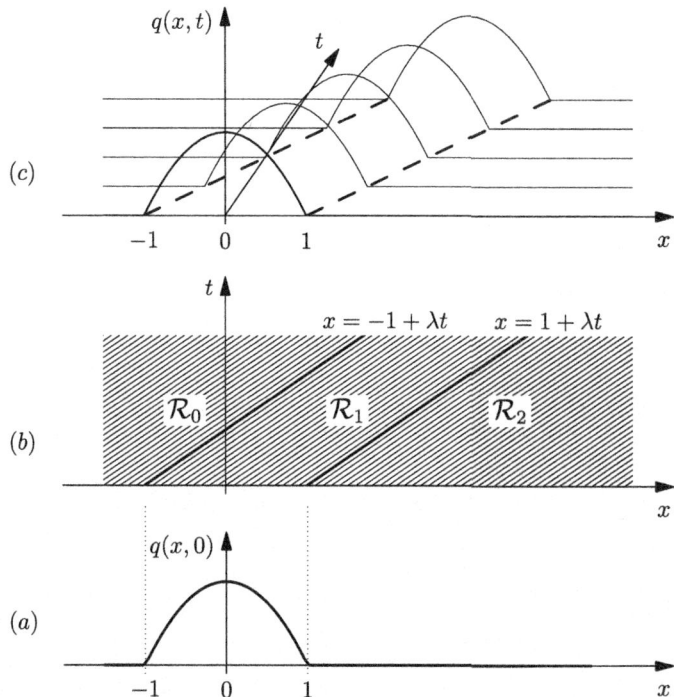

**Fig. 2.4** Solution (2.12) of initial value problem (2.11). Frame **a** displays the initial condition $q(x, 0)$; frame **b** displays picture of characteristics in $x$-$t$ space and frame **c** shows solution profiles $q(x, t_k)$ at different times $t_k$. In frame **b**, two characteristics split the half-plane $x$-$t$ into three distinct regions: $\mathcal{R}_0$, $\mathcal{R}_1$ and $\mathcal{R}_2$, respectively associated with the three lines in Eq. (2.12)

**Remark on notation.** The notation $h(x)$ and $h(x - \lambda t)$ may cause confusion; $h(x)$ means that $h$ is a function of a single spatial variable named $x$, or something else, like $x - \lambda t$, also a spatial quantity.

### 2.1.2 The Riemann Problem

Riemann problem for the linear advection equation is the special IVP

$$\left.\begin{array}{l} \text{PDE:} \ \partial_t q + \lambda \partial_x q = 0 \,, \quad -\infty < x < \infty \,, \ t > 0 \,, \\ \text{IC:} \ \ q(x, 0) = h(x) = \begin{cases} q_L (\text{constant}) & \text{if } x < 0 \,, \\ q_R \ (\text{constant}) & \text{if } x > 0 \,, \end{cases} \end{array}\right\} \quad (2.13)$$

where $q_L$ (left of 0) and $q_R$ (right of 0) are two constants. From (2.10) it is obvious that the solution of the Riemann problem is

**Fig. 2.5** Solution of Riemann problem (2.13). Frame **a** displays piece-wise constant initial condition $q(x, 0)$. Frame **b** displays picture of characteristics in the $x$-$t$ half plane, divided into two regions $\mathcal{R}_0$ and $\mathcal{R}_1$ by the characteristic emerging from the initial discontinuity. Frame **c** shows solution profiles $q(x, t_k)$ at different times $t_k$

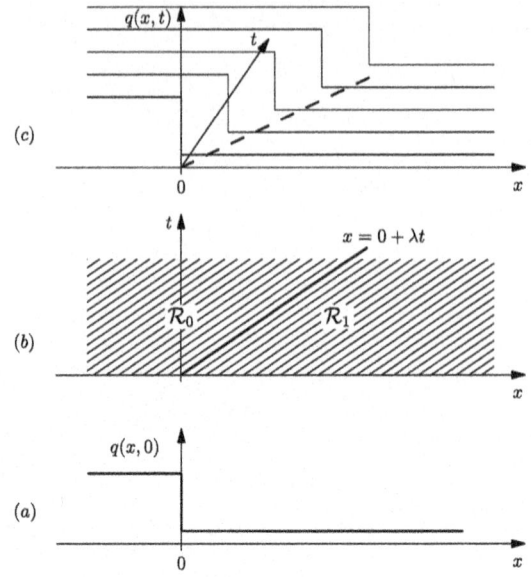

$$q(x, t) = h(x - \lambda t) = \begin{cases} q_L & \text{if } x - \lambda t < 0 \leftrightarrow \frac{x}{t} < \lambda, \\ q_R & \text{if } x - \lambda t > 0 \leftrightarrow \frac{x}{t} > \lambda. \end{cases} \qquad (2.14)$$

Figure 2.5 depicts the solution of the Riemann problem (2.13). Frame (a) illustrates the initial condition. Frame (b) depicts the half-plane $x$-$t$ divided into two regions $\mathcal{R}_0$ and $\mathcal{R}_1$ by the characteristic emerging from the origin; each region $\mathcal{R}_0$, $\mathcal{R}_1$ assumes respectively a constant value of the initial condition, $q_L$ and $q_R$. Frame (c) shows the time evolution of the solution.

## 2.2 Linear Hyperbolic Systems

We now consider a general one-dimensional, time-dependent system of $N$ linear hyperbolic equations with source terms, along with an initial condition, namely

$$\left. \begin{array}{ll} \text{PDEs:} & \partial_t \mathbf{Q} + \mathbf{A} \partial_x \mathbf{Q} = \mathbf{S}(\mathbf{Q}(x, t)) \,, \, -\infty < x < \infty \,, \, t > 0 \,, \\ \text{IC:} & \mathbf{Q}(x, 0) = \mathbf{Q}^{(0)}(x). \end{array} \right\} \qquad (2.15)$$

Here $\mathbf{Q}$ is the vector of unknowns, $\mathbf{A}$ is the matrix of coefficients (constant), $\mathbf{S}(\mathbf{Q})$ is the vector of source terms and $\mathbf{Q}^{(0)}(x)$ is an arbitrary function of distance $x$; these are denoted as follows

## 2.2 Linear Hyperbolic Systems

$$\mathbf{Q} = \begin{bmatrix} q_1 \\ \ldots \\ q_i \\ \ldots \\ q_N \end{bmatrix}, \quad \mathbf{A} = \begin{bmatrix} a_{11} & \ldots & a_{1i} & \ldots & a_{1N} \\ \ldots & \ldots & \ldots & \ldots & \ldots \\ a_{i1} & \ldots & a_{ii} & \ldots & a_{iN} \\ \ldots & \ldots & \ldots & \ldots & \ldots \\ a_{N1} & \ldots & a_{Ni} & \ldots & a_{NN} \end{bmatrix}, \quad \mathbf{S(Q)} = \begin{bmatrix} s_1 \\ \ldots \\ s_i \\ \ldots \\ s_N \end{bmatrix}. \quad (2.16)$$

Note that the linear advection Eq. (2.3) is a special case of (2.15).

### 2.2.1 Eigenstructure and Hyperbolicity

We begin with some conventional definitions regarding the $N \times N$ system of PDEs in (2.15).

**Definition 2.2** (*Eigenvalues*) The *eigenvalues* of the system in (2.15) are the roots of the **characteristic polynomial**

$$P(\lambda) \equiv Det(\mathbf{A} - \lambda \mathbf{I}) = 0, \quad (2.17)$$

where $\mathbf{I}$ is the $N \times N$ unit matrix and $\lambda$ is a parameter.

If the eigenvalues $\lambda_i$ are real numbers, then they are conventionally written in increasing order

$$\lambda_1 \leq \lambda_2 \leq \ldots \leq \lambda_i \leq \ldots \leq \lambda_{N-1} \leq \lambda_N, \quad (2.18)$$

**Definition 2.3** (*Right eigenvector*) A *right eigenvector* $\mathbf{R}_i$ of $\mathbf{A}$ corresponding to the eigenvalue $\lambda_i$ is a column vector

$$\mathbf{R}_i = [r_{1i}, r_{2i}, \ldots, r_{ii}, \ldots, r_{Ni}]^T, \quad (2.19)$$

such that

$$\mathbf{A R}_i = \lambda_i \mathbf{R}_i. \quad (2.20)$$

The full set of $N$ right eigenvectors corresponding to the eigenvalues (2.18) are

$$\mathbf{R}_1, \mathbf{R}_2, \ldots, \mathbf{R}_i, \ldots, \mathbf{R}_{N-1}, \mathbf{R}_N. \quad (2.21)$$

**Definition 2.4** (*Left eigenvector*) A *left eigenvector* $\mathbf{L}_i$ of $\mathbf{A}$ corresponding to $\lambda_i$ is the row vector

$$\mathbf{L}_i = [l_{i1}, l_{i2}, \ldots, l_{ii}, \ldots, l_{iN}], \quad (2.22)$$

such that

$$\mathbf{L}_i \mathbf{A} = \lambda_i \mathbf{L}_i. \quad (2.23)$$

The $N$ eigenvalues (2.18) generate corresponding $N$ left eigenvectors

$$\mathbf{L}_1, \mathbf{L}_2, \ldots, \mathbf{L}_i, \ldots, \mathbf{L}_{N-1}, \mathbf{L}_N. \qquad (2.24)$$

**Definition 2.5** (*Hyperbolic system*) The $N \times N$ system of PDEs in (2.15) is said to be hyperbolic if $\mathbf{A}$ has $N$ real eigenvalues and a corresponding complete set of $N$ linearly independent eigenvectors.

Note that for hyperbolicity, the eigenvalues are not required to be all distinct. What is important is that there is a *complete set of linearly independent eigenvectors, corresponding to the real eigenvalues*.

**Definition 2.6** (*Strictly hyperbolic system*) A hyperbolic system is said to be *strictly hyperbolic* if all eigenvalues of the system are distinct.

A system may have real but not distinct eigenvalues and still be hyperbolic if a *complete set* of linearly independent eigenvectors exists.

**Definition 2.7** (*Weakly hyperbolic system*) If all eigenvalues of the $N \times N$ system of PDEs in (2.15) are real but no complete set of linearly independent eigenvectors exists then the system is called *weakly hyperbolic*.

It is important to remark that a weakly hyperbolic system is not to be mistaken with a non-strictly hyperbolic system.

**Definition 2.8** (*Orthonormality of eigenvectors*) The eigenvectors $\mathbf{L}_i$ and $\mathbf{R}_j$ are orthonormal if

$$\mathbf{L}_i \bullet \mathbf{R}_j = \begin{cases} 1 & \text{if } i = j, \\ 0 & \text{if } i \neq j. \end{cases} \qquad (2.25)$$

### 2.2.2 Diagonalization and Characteristic Variables

Consider the matrix $\mathbf{R} = [\mathbf{R}_1, \ldots, \mathbf{R}_i, \ldots, \mathbf{R}_N]$, whose columns are the right eigenvectors of $\mathbf{A}$ and the diagonal matrix $\Lambda$ formed by the eigenvalues. In full

$$\mathbf{R} = \begin{bmatrix} r_{11} & \ldots & r_{1i} & \ldots & r_{1N} \\ \ldots & \ldots & \ldots & \ldots & \ldots \\ r_{i1} & \ldots & r_{ii} & \ldots & r_{iN} \\ \ldots & \ldots & \ldots & \ldots & \ldots \\ r_{N1} & \ldots & r_{Ni} & \ldots & r_{NN} \end{bmatrix}, \quad \Lambda = \begin{bmatrix} \lambda_1 & \ldots & 0 & \ldots & 0 \\ \ldots & \ldots & \ldots & \ldots & \ldots \\ 0 & \ldots & \lambda_i & \ldots & 0 \\ \ldots & \ldots & \ldots & \ldots & \ldots \\ 0 & \ldots & 0 & \ldots & \lambda_N \end{bmatrix}. \qquad (2.26)$$

## 2.2 Linear Hyperbolic Systems

**Proposition 2.9** *If $\mathbf{A}$ is the coefficient matrix of the hyperbolic system in (2.15) then*

$$\mathbf{A} = \mathbf{R}\Lambda\mathbf{R}^{-1} \ \text{or} \ \Lambda = \mathbf{R}^{-1}\mathbf{A}\mathbf{R} \ . \tag{2.27}$$

*In this case $\mathbf{A}$ is said to be diagonalisable and consequently the system in (2.15) is said to be diagonalisable.*

**Proof** The proof is left as an exercise. □

### Characteristic Variables

The existence of $\mathbf{R}^{-1}$ makes it possible to define the *characteristic variables* $\mathbf{C} = [c_1, c_2, \ldots, c_N]^T$ via

$$\mathbf{C} = \mathbf{R}^{-1}\mathbf{Q} \ \leftrightarrow \ \mathbf{Q} = \mathbf{R}\mathbf{C} \ . \tag{2.28}$$

For calculating the partial derivatives, recall that the coefficient matrix is constant. Therefore

$$\partial_t \mathbf{Q} = \mathbf{R}\partial_t \mathbf{C} \ , \quad \partial_x \mathbf{Q} = \mathbf{R}\partial_x \mathbf{C}$$

and direct substitution of the these expressions into Eq. (2.15) gives

$$\mathbf{R}\partial_t \mathbf{C} + \mathbf{A}\mathbf{R}\partial_x \mathbf{C} = \mathbf{S} \ .$$

Multiplication of this equation from the left by $\mathbf{R}^{-1}$ and use of (2.27) gives

$$\partial_t \mathbf{C} + \Lambda \partial_x \mathbf{C} = \hat{\mathbf{S}} \ , \quad \hat{\mathbf{S}} = \mathbf{R}^{-1}\mathbf{S} \ . \tag{2.29}$$

This is called the *canonical form* or *characteristic form* of system (2.15). Assuming $\hat{\mathbf{S}} = \mathbf{0}$ and writing the equations in full, we have

$$\partial_t \begin{bmatrix} c_1 \\ \ldots \\ c_i \\ \ldots \\ c_N \end{bmatrix} + \begin{bmatrix} \lambda_1 & \ldots & 0 & \ldots & 0 \\ \ldots & \ldots & \ldots & \ldots & \ldots \\ 0 & \ldots & \lambda_i & \ldots & 0 \\ \ldots & \ldots & \ldots & \ldots & \ldots \\ 0 & \ldots & 0 & \ldots & \lambda_N \end{bmatrix} \partial_x \begin{bmatrix} c_1 \\ \ldots \\ c_i \\ \ldots \\ c_N \end{bmatrix} = \begin{bmatrix} 0 \\ \ldots \\ 0 \\ \ldots \\ 0 \end{bmatrix} \ . \tag{2.30}$$

Clearly, each $i$-th equation of this system is of the form

$$\partial_t c_i + \lambda_i \partial_x c_i = 0 \ , \quad i = 1, \ldots, N \tag{2.31}$$

and involves the *single unknown* $c_i(x, t)$, which is decoupled from the remaining variables. Moreover, this equations is identical to the linear advection Eq. (2.3), with characteristic speed $\lambda_i$.

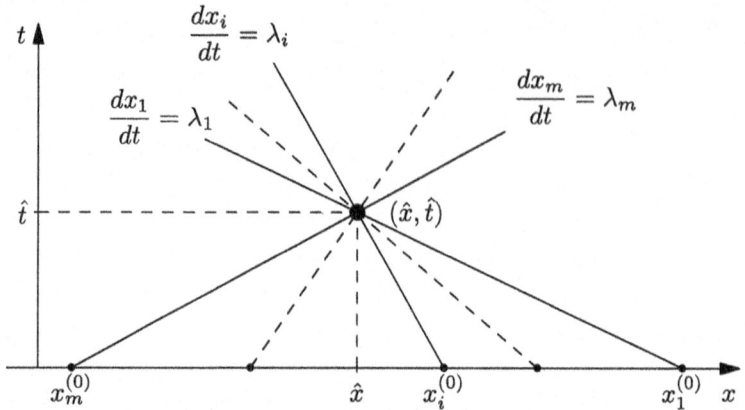

**Fig. 2.6** The solution at a point $(\hat{x}, \hat{t})$ depends on the initial condition at the foot $x_i^{(0)}$ of each characteristic $x_i(t) = x_i^{(0)} + \lambda_i t$

We have $N$ decoupled equations, each one defining a characteristic curve. Thus, at any chosen point $(\hat{x}, \hat{t})$ in the $x$-$t$ half-plane there are $N$ characteristic curves $x_i(t)$ passing through $(\hat{x}, \hat{t})$ and satisfying the $N$ ODEs

$$\frac{dx_i}{dt} = \lambda_i, \quad \text{for } i = 1, \ldots, N, \tag{2.32}$$

as depicted in Fig. 2.6.

**Remarks**:

1. Each characteristic curve $x_i(t) = x_i^{(0)} + \lambda_i t$ intersects the $x$-axis at the point $x_i^{(0)}$, which is the *foot of the characteristic* passing through the point $(\hat{x}, \hat{t})$. The point $x_i^{(0)}$ is given as

$$x_i^{(0)} = \hat{x} - \lambda_i \hat{t}, \quad \text{for } i = 1, 2, \ldots, N. \tag{2.33}$$

See Fig. 2.6.

2. Each Eq. (2.31) is just a linear advection equation whose solution at $(\hat{x}, \hat{t})$ is given by

$$c_i(\hat{x}, \hat{t}) = c_i^{(0)}(x_i^{(0)}) = c_i^{(0)}(\hat{x} - \lambda_i \hat{t}), \quad \text{for } i = 1, 2, \ldots, N, \tag{2.34}$$

where $c_i^{(0)}(x)$ is the initial condition, at the initial time. The initial conditions for the characteristic variables are obtained from the transformation (2.28) applied to the initial condition $\mathbf{Q}(x, 0)$ in (2.5).

3. Given the assumed order (2.18) of the distinct eigenvalues the following inequalities are satisfied.

$$x_N^{(0)} < x_{N-1}^{(0)} < \ldots < x_2^{(0)} < x_1^{(0)}. \tag{2.35}$$

## 2.2 Linear Hyperbolic Systems

**Definition 2.10** (*Domain of dependence*) The interval $[x_N^{(0)}, x_1^{(0)}]$ is called the domain of dependence of the point $(\hat{x}, \hat{t})$. The solution at $(\hat{x}, \hat{t})$ depends exclusively on initial data at points in the interval $[x_N^{(0)}, x_1^{(0)}]$. See Fig. 2.6.

This is a distinguishing feature of hyperbolic equations. The initial data outside the domain of dependence can be changed in any manner we wish but this will not affect the solution at the point $(\hat{x}, \hat{t})$.

**Proposition 2.11** (The general initial-value problem) *The solution of the general IVP for the $N \times N$ linear homogeneous hyperbolic system*

$$\left.\begin{array}{ll} PDEs: & \partial_t \mathbf{Q} + \mathbf{A} \partial_x \mathbf{Q} = \mathbf{0}, \ -\infty < x < \infty, \ t > 0, \\ IC: & \mathbf{Q}(x, 0) = \mathbf{Q}^{(0)}(x) \end{array}\right\} \quad (2.36)$$

*is given by*

$$\mathbf{Q}(x, t) = \sum_{i=1}^{N} c_i(x, t) \mathbf{R}_i . \quad (2.37)$$

*The coefficient $c_i(x, t)$ of the right eigenvector $\mathbf{R}_i$ is a characteristic variable.*

*Proof* The proof is left as an exercise. □

**Remarks**:

1. The function $c_i(x, t)$ is the coefficient of $\mathbf{R}_i$ in an *eigenvector expansion* of the solution vector $\mathbf{Q}(x, t)$.
2. Given a point $(x, t)$ in the $x$-$t$ plane, the solution $\mathbf{Q}(x, t)$ depends only on the initial data at the $N$ points $x_0^{(i)} = x - \lambda_i t$. See Fig. 2.6.
3. These points are the intersections of the characteristics of speed $\lambda_i$ with the $x$-axis.
4. Solution (2.37) represents superposition of $N$ waves of unchanged shape $c_i^{(0)}(x)\mathbf{R}_i$ propagated with speed $\lambda_i$.

### 2.2.3 The Riemann Problem

The Riemann problem is a special case of the IVP (2.36), in which the initial condition is piece-wise constant.

**Proposition 2.12** *The solution of Riemann problem*

$$\left.\begin{array}{ll} PDEs: & \partial_t \mathbf{Q} + \mathbf{A} \partial_x \mathbf{Q} = \mathbf{0}, \ -\infty < x < \infty, \ t > 0, \\ IC: & \mathbf{Q}(x, 0) = \mathbf{Q}^{(0)}(x) = \begin{cases} \mathbf{Q}_L & \text{if } x < 0, \\ \mathbf{Q}_R & \text{if } x > 0, \end{cases} \end{array}\right\} \quad (2.38)$$

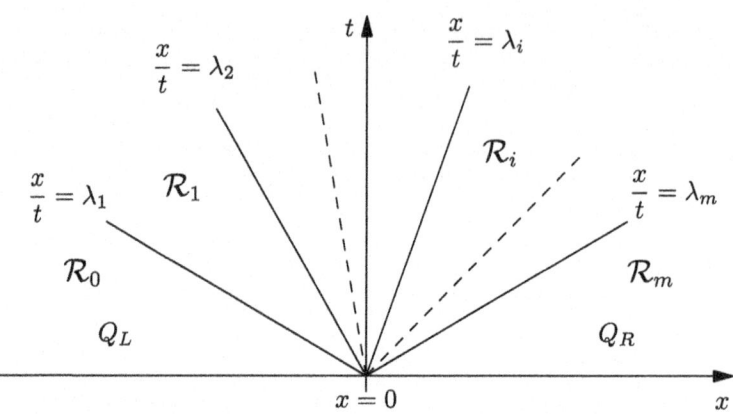

**Fig. 2.7** Structure of the solution of the Riemann problem. There are $N$ waves that divide the half $x$-$t$ plane into $N+1$ regions (wedges) $\mathcal{R}_i$, with $i = 0, 1, \ldots, N$

with $\mathbf{Q}_L$ and $\mathbf{Q}_R$ two constant vectors, is given by

$$\mathbf{Q}(x,t) = \sum_{i=1}^{I} c_{iR}\mathbf{R}_i + \sum_{i=I+1}^{N} c_{iL}\mathbf{R}_i , \tag{2.39}$$

where

$$\sum_{i=1}^{N} c_{iL}\mathbf{R}_i = \mathbf{Q}_L , \quad \sum_{i=1}^{N} c_{iR}\mathbf{R}_i = \mathbf{Q}_R \tag{2.40}$$

and $I = I(x,t)$ is the maximum value of $i$ for which $x - \lambda_i t > 0$. See Fig. 2.8.

**Proof** The proof is left as an exercise. □

**Remarks on the solution of the Riemann problem**:

1. The initial data consists of two constant vectors $\mathbf{Q}_L$ and $\mathbf{Q}_R$, separated by a discontinuity at $x = 0$.
2. This is a special case of IVP (2.36).
3. The structure of the solution of the Riemann problem (2.38) is depicted in Fig. 2.7 in the $x$-$t$ plane.
4. The solution consists of a fan of $N$ waves emanating from the origin, one wave for each eigenvalue $\lambda_i$. The speed of the wave $i$ is the eigenvalue $\lambda_i$.
5. These $N$ waves divide the $x$-$t$ half plane into $N+1$ constant regions

$$\mathcal{R}_i = \left\{ (x,t) / -\infty < x < \infty; \, t \geq 0; \, \lambda_i < \frac{x}{t} < \lambda_{i+1} \right\} , \tag{2.41}$$

for $i = 1, \ldots, N-1$; $\mathcal{R}_0$ corresponds to the initial data $\mathbf{Q}_L$ and $\mathcal{R}_N$ corresponds to the initial data $\mathbf{Q}_R$. See Fig. 2.7.

## 2.3 Non-linear Scalar Equations

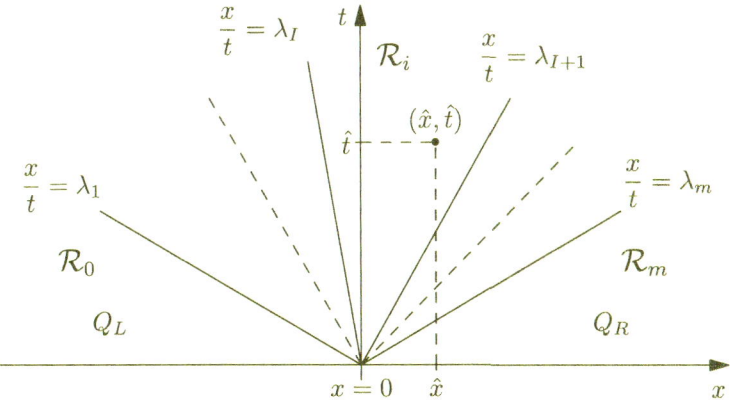

**Fig. 2.8** The solution of the Riemann problem at a point $(\hat{x}, \hat{t})$ depends on the associated index $I = I(\hat{x}, \hat{t})$

Solving the Riemann problem (2.38) means finding constant values for $\mathbf{Q}$ in regions $\mathcal{R}_i$, for $= 1, \ldots, N - 1$.

**Corollary 2.13** *The solution of the Riemann problem may be expressed as*

$$\mathbf{Q}(x, t) = \mathbf{Q}_\mathrm{L} + \sum_{i=1}^{I} \delta_i \mathbf{R}_i = \mathbf{Q}_\mathrm{R} - \sum_{i=I+1}^{N} \delta_i \mathbf{R}_i \;, \tag{2.42}$$

*where the coefficients* $\Delta C = [\delta_1, \ldots, \delta_i, \ldots, \delta_N]^T$ *are the solution to the linear algebraic system*

$$\sum_{i=1}^{N} \delta_i \mathbf{R}_i = \Delta \mathbf{Q} \equiv \mathbf{Q}_R - \mathbf{Q}_L \;. \tag{2.43}$$

*Proof* Proof left as an exercise. □

Form (2.42) is more convenient. We only need to solve one linear system (2.43). Note that there are two choices for calculating the solution, one may begin from the left vector $\mathbf{Q}_\mathrm{L}$ and add $I$ wave jumps or from the right vector $\mathbf{Q}_\mathrm{R}$ and substract $N - I$ wave jumps. Figure 2.8 illustrates the solution at a point $(\hat{x}, \hat{t})$.

## 2.3 Non-linear Scalar Equations

Here we study non-linear first-order PDEs, but restrict ourselves to the scalar case.

### 2.3.1 Definitions and Examples

Consider the first-order PDE for the unknown function $q(x,t)$

$$\partial_t q + \partial_x f(q) = 0 \,. \tag{2.44}$$

This equation is called a *conservation law*, in which $q$ is the *conserved variable*; $f(q)$ is the *flux function* or *physical flux*, a prescribed function of $q$. Equation (2.44) is said to be written in *differential, conservative form*. One may express (2.44) in *quasi-linear form* as

$$\partial_t q + \lambda(q)\partial_x q = 0 \,, \quad \lambda(q) = \frac{d}{dq}f(q) \equiv f'(q) \,. \tag{2.45}$$

Here $\lambda(q)$ is called *characteristic speed*. Equations of the type (2.44) may be characterised by the behaviour of the flux $f(q)$ and its derivative, namely the characteristic speed $\lambda(q) = f'(q)$. There are three cases:

1. **Convex flux**: $\lambda(q)$ is a monotone *increasing* function of $q$, that is

$$\frac{d}{dq}\lambda(q) = \lambda'(q) = f''(q) > 0 \,, \quad \forall q \,. \tag{2.46}$$

2. **Concave flux**: $\lambda(q)$ is a monotone *decreasing* function of $q$, that is

$$\frac{d}{dq}\lambda(q) = \lambda'(q) = f''(q) < 0 \,, \quad \forall q \,. \tag{2.47}$$

3. **Non-convex, non-concave flux**: $\lambda(q)$ vanishes for some $q$, that is

$$\frac{d}{dq}\lambda(q) = \lambda'(q) = f''(q) = 0 \,, \quad \text{for some } q \,. \tag{2.48}$$

**Example 2.14** (*The inviscid Burgers equation*)

$$\left. \begin{aligned} &\partial_t q + \partial_x f(q) = 0 \,, \\ &f(q) = \frac{1}{2}q^2 \,, \\ &\lambda(q) = f'(q) = q \,, \quad \lambda'(q) = f''(q) = 1 > 0 \,, \quad \forall q \,. \end{aligned} \right\} \tag{2.49}$$

The flux is convex; the monotone increasing behaviour of $\lambda(q)$ means that larger values of $q$ propagate faster than smaller values of $q$. This leads to wave distortion and shock formation. We note that the true Burgers equation is viscous, namely

$$\partial_t q + \partial_x f(q) = \alpha \partial_x^{(2)} q \,, \quad f(q) = \frac{1}{2}q^2 \,,$$

## 2.3 Non-linear Scalar Equations

**Fig. 2.9** Characteristic curve $x(t) = x_0 + \lambda(h(x_0))t$ emanating from $x_0$: foot of the characteristic

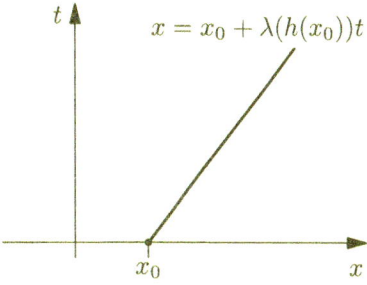

where $\alpha$ is the viscosity (or diffusion) coefficient.

**Example 2.15** (*A traffic flow equation*)

$$\left.\begin{aligned}
&\partial_t q + \partial_x f(q) = 0, \\
&f(q) = u_{max}(1 - q/q_{max})q, \\
&\lambda(q) = f'(q) = u_{max}(1 - 2q/q_{max}), \\
&\lambda'(q) = f''(q) = -2u_{max}/q_{max} < 0, \quad \forall q.
\end{aligned}\right\} \quad (2.50)$$

Here $u_{max} \geq 0$ and $q_{max} > 0$ are two constants, with $0 < q \leq q_{max}$. The flux is concave; larger values of $q$ will propagate more slowly than smaller values of $q$, the opposite behaviour to that of Burgers' equation.

### 2.3.2 Solution Along Characteristics

Consider the initial-value problem (or Cauchy problem)

$$\left.\begin{aligned}
&\text{PDE:} \quad \partial_t q + \partial_x f(q) = 0, \\
&\text{IC:} \quad q(x, 0) = h(x).
\end{aligned}\right\} \quad (2.51)$$

As for LAE, solutions along characteristic curves $x = x(t)$, with

$$x = x_0 + \lambda(h(x_0))t \quad (2.52)$$

can be defined as

$$q(x, t) = h(x_0) = h(x - \lambda(h(x_0))t). \quad (2.53)$$

Figure 2.9 depicts the situation.

For non-linear equations, characteristics are no longer parallel, as in the linear case. Therefore, charateristic curves may cross, as illustrated in Fig. 2.10.

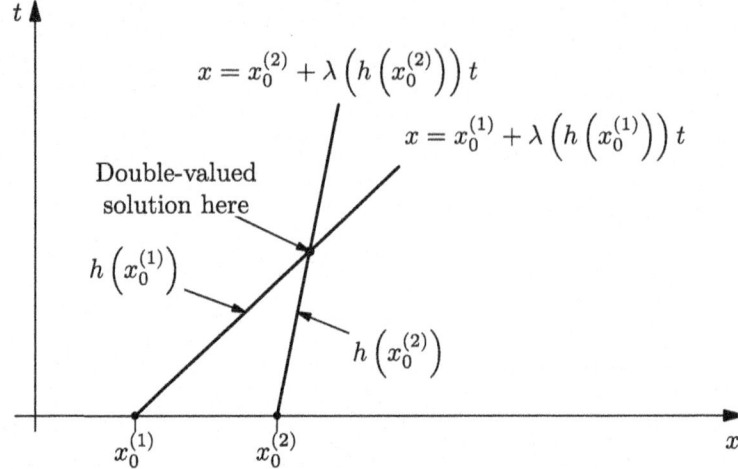

**Fig. 2.10** Crossing characteristics. Characteristics from $x_0^{(1)}$ and $x_0^{(2)}$ carry different initial values $h(x_0^{(1)})$ and $h(x_0^{(2)})$, leading to multi-valued solutions

## Shock Formation: A Numerical Example

For non-linear equations, even if the initial data is continuous, discontinuities may develop in time. This is illustrated in Fig. 2.11, where a sequence of profiles corresponding to an increasing sequence of time values is shown, starting from $t = 0$, the initial condition. The phenomenon of shock formation in non-linear equations calls for the extension of the definition of solution. To this end the equations are

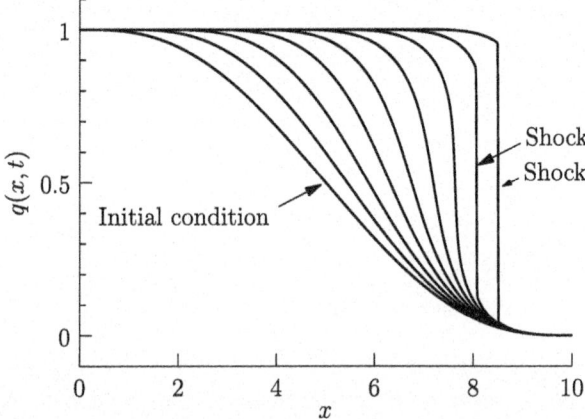

**Fig. 2.11** Shock wave formation from smooth initial condition at time $t = 0$. Burgers' equation solved numerically with the first-order Godunov method on a very fine mesh

## 2.3 Non-linear Scalar Equations

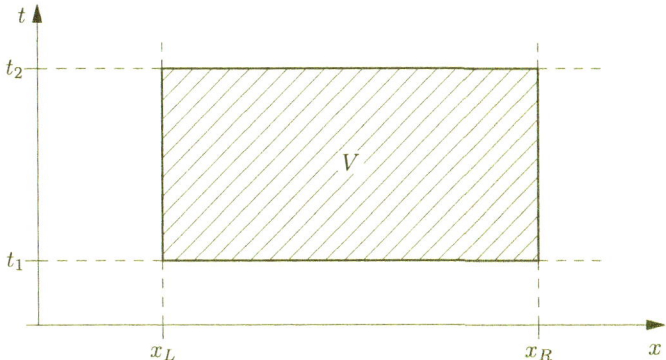

**Fig. 2.12** Control volume $V = [x_L, x_R] \times [t_1, t_2]$ in $x$-$t$ space. Equations will be integrated exactly on this volume to derive integral forms of the conservation law

reformulated in terms of integral relations, which no longer require smoothness of the solution.

### 2.3.3 Integral Forms of the Equation

Consider the general case written in differential conservative form

$$\partial_t q(x,t) + \partial_x f(q(x,t)) = s(q(x,t)) \,. \tag{2.54}$$

This equation includes a source term and is thus called a *balance law*. If $s(q(x,t)) = 0$, then the equation is a conservation law.

Here we study *integral forms*, to accommodate discontinuous solutions. We shall also derive a condition to be satisfied at discontinuities. To this end we consider a *control volume* $V$ in the $x$-$t$ plane, depicted in Fig. 2.12 and defined as

$$V = [x_L, x_R] \times [t_1, t_2] \,. \tag{2.55}$$

We integrate Eq. (2.54) in space and time in the control volume $V$

$$\int_{x_L}^{x_R} \int_{t_1}^{t_2} [\partial_t q(x,t) + \partial_x f(q(x,t))] \, dx dt = \int_{x_L}^{x_R} \int_{t_1}^{t_2} s(q(x,t)) \, dx dt \,. \tag{2.56}$$

On rearranging the space and time integrals we obtain

$$\int_{x_L}^{x_R} \left[ \int_{t_1}^{t_2} \partial_t q(x,t) dt \right] dx = -\int_{t_1}^{t_2} \left[ \int_{x_L}^{x_R} \partial_x f(q(x,t)) dx \right] dt \\ + \int_{x_L}^{x_R} \int_{t_1}^{t_2} s(q(x,t)) dx dt \ . \qquad (2.57)$$

Exact space-time integration gives the *integral form* of the balance law (2.54), namely

$$\int_{x_L}^{x_R} q(x,t_2) dx = \int_{x_L}^{x_R} q(x,t_1) dx - \left[ \int_{t_1}^{t_2} f(q(x_R,t)) dt - \int_{t_1}^{t_2} f(q(x_L,t)) dt \right] \\ + \int_{x_L}^{x_R} \int_{t_1}^{t_2} s(q(x,t)) dx dt \ . \qquad (2.58)$$

In the absence of the source term, the integral form states that *the amount of $q(x,t)$ in the interval $[x_L, x_R]$ at time $t = t_2$ is equal to the amount of $q(x,t)$ in the interval $[x_L, x_R]$ at time $t = t_1$ plus a difference of time integrals of the fluxes at the extreme points*. In the presence of a source term this statement is modified appropriately.

**Finite Volume Formula**

It is also convenient to obtain an averaged version of (2.58). This is accomplished by dividing (2.58) through by the length $\Delta x$, namely

$$\frac{1}{\Delta x} \int_{x_L}^{x_R} q(x,t_2) dx = \frac{1}{\Delta x} \int_{x_L}^{x_R} q(x,t_1) dx \\ - \frac{\Delta t}{\Delta x} \left[ \frac{1}{\Delta t} \int_{t_1}^{t_2} f(q(x_R,t)) dt - \frac{1}{\Delta t} \int_{t_1}^{t_2} f(q(x_L,t)) dt \right] \\ + \frac{\Delta t}{\Delta x \Delta t} \int_{x_L}^{x_R} \int_{t_1}^{t_2} s(q(x,t)) dx dt \ . \qquad (2.59)$$

The averaged integral expression (2.59) can be written as

$$q^{new} = q^{old} - \frac{\Delta t}{\Delta x} \left[ f_{right} - f_{left} \right] + \Delta t \ s_{vol} \ , \qquad (2.60)$$

which is exact, with the following definitions

## 2.3 Non-linear Scalar Equations

$$\left.\begin{array}{c} q^{new} = \dfrac{1}{\Delta x}\int_{x_L}^{x_R} q(x,t_2)dx \, , \quad q^{old} = \dfrac{1}{\Delta x}\int_{x_L}^{x_R} q(x,t_1)dx \, , \\[6pt] f_{right} = \dfrac{1}{\Delta t}\int_{t_1}^{t_2} f(q(x_R,t))dt \, , \quad f_{left} = \dfrac{1}{\Delta t}\int_{t_1}^{t_2} f(q(x_L,t))dt \, , \\[6pt] s_{vol} = \dfrac{1}{\Delta x \Delta t}\int_{x_L}^{x_R}\int_{t_1}^{t_2} s(q(x,t))dxdt \, . \end{array}\right\} \quad (2.61)$$

Numerical methods called finite volume methods, use the *finite volume formula* (2.60) to compute approximate solutions in which $q^{old}$ is a known average of the solution at the previous time level and the remaining terms on the right hand side of (2.60) are found by appropriate approximations of the integrals in (2.61). The computational parameters $\Delta t$ and $\Delta x$ must be prescribed to complete the scheme to compute $q^{new}$. See Chap. 9.

### 2.3.4 Generalised Solutions

A generalised (or weak) solution of the conservation law (2.54) is a function $q(x,t)$ that satisfies the integral form (2.58). Weak solutions admit discontinuities (shocks), which satisfy the *Rankine-Hugoniot jump condition*.

**Proposition 2.16** (Rankine-Hugoniot Condition) *A discontinuity of a weak solution of the conservation law (2.54), no source term, satisfies the Rankine-Hugoniot jump condition across it, namely*

$$f(q(s_R,t)) - f(q(s_L,t)) = [q(s_R,t) - q(s_R,t)]s \, , \quad (2.62)$$

*where $q(s_L,t)$ and $q(s_R,t)$ are limiting values from left and right of the discontinuity; $f(q(s_R,t))$ and $f(q(s_L,t))$ are the corresponding flux values and $s$ is the speed of the discontinuity.*

*Proof* For the proof see [1], for example. □

**Example 2.17** (*Burgers's equation*) Assume a shock wave of speed $s$ with states $q_L$ and $q_R$. The Rankine-Hugoniot condition gives

$$f(q_R) - f(q_L) = \frac{1}{2}q_R^2 - \frac{1}{2}q_L^2 = s(q_R - q_L) \, ,$$

from which the shock speed is given by

$$s = \frac{1}{2}(q_L + q_R) \, . \quad (2.63)$$

This is a very special case. The shock speed is a simple arithmetic average of the characteristic speeds either side of the shock.

**Summarising**: in order to admit discontinuous solutions one needs to formulate the equations in integral form and enforce the Rankine-Hugoniot condition across discontinuities, while in smooth parts of the solution one may formulate the equations in differential form.

## 2.3.5 Non-uniqueness

The enlarged set of solutions of the integral formulation includes smooth (classical) and discontinuous solutions. However, now the set is too large, it contains spurious, non-physical solutions. Hence this requires an admissibility criterion to discard *unphysical shocks*.

**A Non-uniqueness Example**

To illustrate the question of non-uniqueness we consider the following example:

$$\left.\begin{array}{l} \text{PDE: } \partial_t q + \partial_x f(q) = 0, \ f(q) = \tfrac{1}{2}q^2, \\ \text{IC: } \quad q(x,0) = h(x) = \begin{cases} q_L = 0 \text{ if } x < 0, \\ q_R = 1 \text{ if } x > 0. \end{cases} \end{array}\right\} \quad (2.64)$$

**Solution 1: rarefaction wave**. One solution of the problem is the rarefaction wave (smooth)

$$q(x,t) = \begin{cases} q_L = 0 & \text{if } \quad x/t < 0, \\ x/t & \text{if } 0 \le x/t \le 1, \\ q_R = 1 & \text{if } \quad x/t > 1. \end{cases} \quad (2.65)$$

Figure 2.13 illustrates the solution and the corresponding picture of characteristics.

**Solution 2: shock wave**. Another, discontinuous, solution (shock) is given as

$$q(x,t) = \begin{cases} 0 \text{ if } x/t < s = 1/2, \\ 1 \text{ if } x/t > s = 1/2. \end{cases} \quad (2.66)$$

Figure 2.14 shows the shock solution to problem (2.64). Note that characteristics diverge from the shock. So the initial value problem (2.64) has at least two solutions.

**Admissible Shocks: the Lax Entropy Condition**

The proposed solution (2.66), called rarefaction shock, is not accepted as a *physical* solution. *Rarefaction shocks* are excluded. Admissible discontinuities are those aris-

## 2.3 Non-linear Scalar Equations

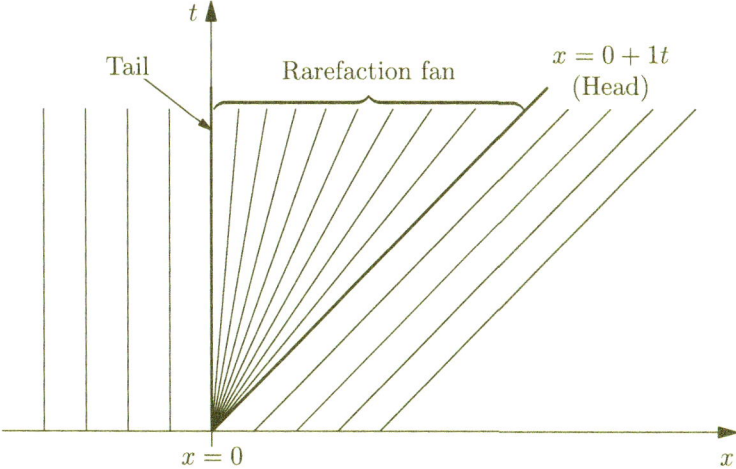

**Fig. 2.13** Illustration of the rarefaction solution (2.65) to initial-value problem (2.64)

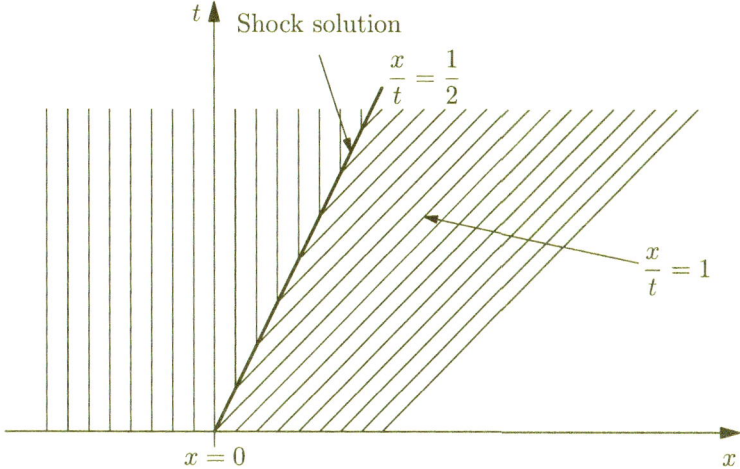

**Fig. 2.14** Illustration of the shock solution (2.66) to problem (2.64). Characteristics diverge from the shock path. This solution is called rarefaction shock

ing from *compression*. This *compressibility* condition is ensured by the *Lax entropy condition*:

$$\lambda(q_L) > s > \lambda(q_R) \ . \tag{2.67}$$

Here $s$ is the shock speed, while $\lambda(q_L)$ and $\lambda(q_R)$ are characteristic speeds. Note that characteristics *run into the shock*, which is *compressed* by the characteristics, see Fig. 2.15.

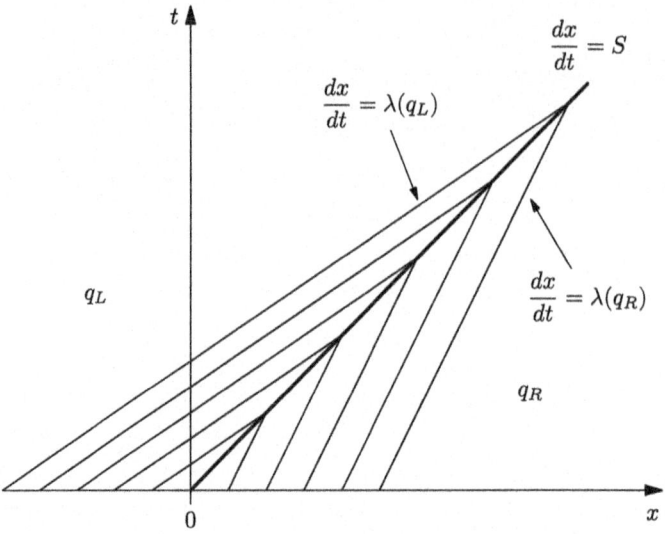

**Fig. 2.15** Picture of characteristics for an entropy-satisfying shock. Characteristic curves run into the shock path

### 2.3.6 The Riemann Problem for Burgers's Equation

The Riemann problem is defined as

$$\left.\begin{array}{l} \text{PDE:}\ \partial_t q + \partial_x f(q) = 0\,,\quad f(q) = \tfrac{1}{2}q^2\,, \\ \text{IC:}\ \ q(x,0) = \begin{cases} q_L & \text{if } x < 0\,, \\ q_R & \text{if } x > 0\,. \end{cases} \end{array}\right\} \quad (2.68)$$

The solution is given by the following two cases: (i) shock if $q_L > q_R$ and (ii) rarefaction otherwise, namely

$$\left.\begin{array}{l} q(x,t) = \begin{cases} q_L & \text{if } x - st < 0 \\ q_R & \text{if } x - st > 0 \end{cases} \Bigg\} \text{ if } q_L > q_R\,, \\ s = \tfrac{1}{2}(q_L + q_R) \\[1em] q(x,t) = \begin{cases} q_L & \text{if } \dfrac{x}{t} \le q_L \\ \dfrac{x}{t} & \text{if } q_L < \dfrac{x}{t} < q_R \\ q_R & \text{if } \dfrac{x}{t} \ge q_R \end{cases} \Bigg\} \text{ if } q_L \le q_R\,. \end{array}\right\} \quad (2.69)$$

Figure 2.16 illustrates the solution structure for the two cases. The bottom frame shows the shock case while the top frame shows the rarefaction case.

## 2.4 First-Order Non-linear Systems

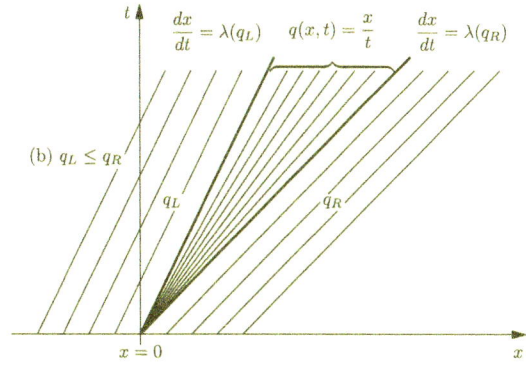
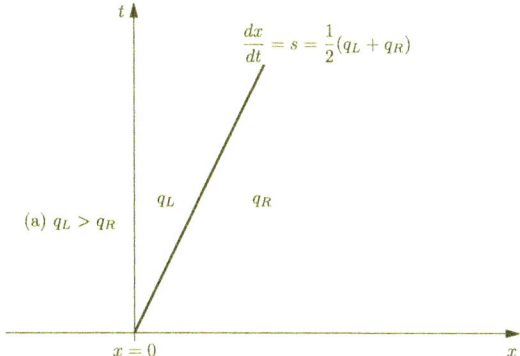

**Fig. 2.16** Solution of the Riemann problem for the Burgers equation. Frame **a**: shock wave if $q_L > q_R$. Frame **b**: rarefaction wave if $q_L \leq q_R$

## 2.4 First-Order Non-linear Systems

To end this Chapter we point out that the general setting is that of non-linear systems of $N$ hyperbolic balance laws in three space dimensions, which written in differential conservative form read

$$\partial_t \mathbf{Q} + \partial_x \mathbf{F}(\mathbf{Q}) + \partial_y \mathbf{G}(\mathbf{Q}) + \partial_z \mathbf{H}(\mathbf{Q}) = \mathbf{S}(\mathbf{Q}) \,, \tag{2.70}$$

where

$$\mathbf{Q} = \begin{bmatrix} q_1 \\ q_2 \\ \ldots \\ q_N \end{bmatrix} \;;\; \mathbf{F} = \begin{bmatrix} f_1 \\ f_2 \\ \ldots \\ f_N \end{bmatrix} \;;\; \mathbf{G} = \begin{bmatrix} g_1 \\ g_2 \\ \ldots \\ g_N \end{bmatrix} \;;\; \mathbf{H} = \begin{bmatrix} h_1 \\ h_2 \\ \ldots \\ h_N \end{bmatrix} \;;\; \mathbf{S} = \begin{bmatrix} s_1 \\ s_2 \\ \ldots \\ s_N \end{bmatrix} \,. \tag{2.71}$$

Here the independent variables are $x$, $y$, $z$ and $t$. $\mathbf{Q}(x, y, z, t)$ is the vector of dependent variables, called *conserved variables*; $\mathbf{F}(\mathbf{Q})$ is the flux vector in the $x$-direction; $\mathbf{G}(\mathbf{Q})$ is the flux vector in the $y$-direction and $\mathbf{H}(\mathbf{Q})$ is the flux vector in the $z$-direction;

$S(Q)$ is the vector of source terms. Fluxes and sources are prescribed functions of $Q(x, y, z, t)$. For background on non-linear hyperbolic systems see for example [1, 2] and [3].

In Chap. 3 we shall apply directly most of the concepts studied in this chapter via linearised versions of the shallow water equations. In Chaps. 4 and 5 we shall utilise some of the studied concepts to analyse some properties of the shallow water equations. In Chap. 9 we shall use the concepts just studied to construct and interpret numerical methods for solving hyperbolic equations.

This chapter is largely based on lectures given by the author for a short course in Levico Terme, Italy, 2016, sponsored by *Centro Internazionale per la Ricerca Matematica*. See [4].

# References

1. E.F. Toro, *Riemann Solvers and Numerical Methods for Fluid Dynamics. A Practical Introduction*, 3rd edn. (Springer-Verlag, 2009)
2. E. Godlewski, P.A. Raviart, *Numerical Approximation of Hyperbolic Systems of Conservation Laws*, 2nd edn. (Springer, 2021)
3. R.J. LeVeque, *Finite Volume Methods for Hyperbolic Problems* (Cambridge University Press, 2002)
4. A. Farina, A. Mikelić, G. Saccomandi, A. Sequeira, E.F. Toro, *Non-Newtonian Fluid Mechanics and Complex Flows*. Lecture Notes in Mathematics, vol. 2212 (Springer, 2018)

# Chapter 3
# Linear Shallow Water Equations

**Abstract** This Chapter introduces the reader to linear hyperbolic systems derived from the non-linear shallow water equations and to basic notions on wave propagation. The purpose is two fold. First to interpret the physical meaning of linearised models and second, practice on the application of salient mathematical notions on hyperbolic partial differential equations, such as eigenvalues, eigenvectors, hyperbolicity and the construction of analytical solutions to the general initial-value problem and its special case, the Riemann problem. The final section is a case study based on a more physically oriented, alternative approach to linearisation, and on which basic mathematical concepts on hyperbolic equations are consolidated. All contents are particularly useful in a teaching/self-studying setting. Useful background is found in Chaps. 1 and 2.

## 3.1 Linearised Models

Here we discuss simple models that result from linearising the basic shallow water equations presented in Chap. 1. Linearised models do have a practical value. However, our motivation for studying linearised models here is, in the main, pedagogical. The equations are simple enough to manipulate so that basic mathematical and physical notions of wave propagation can be made apparent without the intricacies of non-linear models. Analytical solutions are also possible; these are useful for the purpose of assessing the performance of numerical methods against known analytical solutions. In studying these simple models, use is made of mathematical methods which the reader can then apply to more complex situations. The important concept of Riemann problem is easily introduced and the exact solution is derived. The Riemann problem and mathematical techniques to find its solution are also relevant to subsequent chapters, where the Riemann problem for the non-linear equations is solved approximately using local linearisations.

We start by considering the full non-linear, time-dependent two-dimensional shallow water equations for a rectangular channel of variable bed elevation, written in non-conservative form, namely

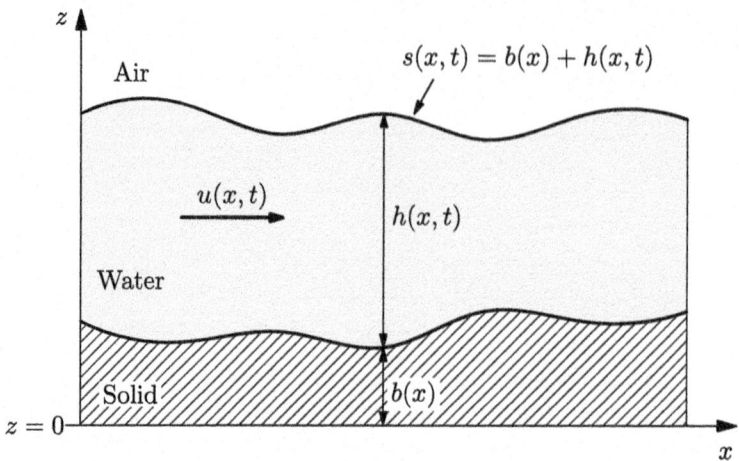

**Fig. 3.1** Side view of the physical set up for the shallow water equations. The water layer lies between the impermeable bottom $b(x)$ above a horizontal datum $z = 0$ and the free surface $s(x, t) = b(x) + h(x, t)$

$$\partial_t h + u\partial_x h + h\partial_x u + v\partial_y h + h\partial_y v = 0 \,, \tag{3.1}$$

$$\partial_t u + u\partial_x u + g\partial_x h + v\partial_y u = -g\partial_x b \,, \tag{3.2}$$

$$\partial_t v + u\partial_x v + v\partial_y v + g\partial_y h = -g\partial_y b \,. \tag{3.3}$$

Here $h(x, y, t)$, $u(x, y, t)$ and $v(x, y, t)$ are depth, $x$-velocity component and $y$-velocity component respectively. The bed elevation above a horizontal datum is given by the function $b(x, y)$, which is assumed to be prescribed; g denotes the acceleration due to gravity, taken as $g = 9.81$ m/s$^2$. Figure 3.1 depicts the physical set up. When written in *quasi-linear form*, Eqs. (3.1)–(3.3) become

$$\partial_t \mathbf{W} + \mathbf{A}(\mathbf{W})\partial_x \mathbf{W} + \mathbf{B}(\mathbf{W})\partial_y \mathbf{W} = \mathbf{S} \,, \tag{3.4}$$

where the vectors $\mathbf{W}$ and $\mathbf{S}$ and the coefficient matrices $\mathbf{A}(\mathbf{W})$ and $\mathbf{B}(\mathbf{W})$ are given as

$$\left.\begin{aligned}\mathbf{W} = \begin{bmatrix} h \\ u \\ v \end{bmatrix}, \quad \mathbf{A}(\mathbf{W}) = \begin{bmatrix} u & h & 0 \\ g & u & 0 \\ 0 & 0 & u \end{bmatrix}, \\ \mathbf{B}(\mathbf{W}) = \begin{bmatrix} v & 0 & h \\ 0 & v & 0 \\ g & 0 & v \end{bmatrix}, \quad \mathbf{S} = \begin{bmatrix} 0 \\ -g\partial_x b \\ -g\partial_y b \end{bmatrix} \,.\end{aligned}\right\} \tag{3.5}$$

## 3.1 Linearised Models

In this chapter we obtain linearised forms of these equations. There are essentially two ways of obtaining global, as opposed to local, linearised models from non-linear models. One approach consists of expressing the solution as the sum of a known solution, the uniform solution say, and a *small* perturbation to it. Linear equations are then derived for the perturbation. The second approach is more direct; one obtains linear models by simply *freezing* coefficients in all non-linear terms in the equations. In this chapter we utilise both, beginning with the latter approach.

### *A First Linear Model*

Consider the one-dimensional version of the non-linear shallow water Eqs. (3.1)–(3.3) for a rectangular channel with horizontal bed elevation. If we are interested in momentum alone, we have the equation

$$\partial_t u + u \partial_x u + g \partial_x h = 0 . \tag{3.6}$$

For flow of uniform depth $h(x, t)$ we obtain

$$\partial_t u + u \partial_x u = 0 . \tag{3.7}$$

This is the inviscid Burgers equation written in quasi-linear form, in which the coefficient of the non-linear term $uu_x$ is the *characteristic speed*, $\lambda(u) \equiv u$. By *freezing* this coefficient to $\hat{\lambda}$ say, we obtain

$$\partial_t u + \hat{\lambda} \partial_x u = 0 , \tag{3.8}$$

which is the *linear advection equation* with constant coefficient $\hat{\lambda}$, a wave propagation speed. See Chap. 2.

Let us now consider one-dimensional flow in a channel of rectangular cross-section with variable bed elevation $b(x)$. The relevant equations in non-conservative form are

$$\partial_t h + u \partial_x h + h \partial_x u = 0 \tag{3.9}$$

and

$$\partial_t u + u \partial_x u + g \partial_x h = -g \partial_x b . \tag{3.10}$$

Written in quasi-linear matrix form these equations read

$$\partial_t \mathbf{W} + \mathbf{A}(\mathbf{W}) \partial_x \mathbf{W} = \mathbf{S} , \tag{3.11}$$

where the vectors $\mathbf{W}$ and $\mathbf{S}$ and the coefficient matrix $\mathbf{A}(\mathbf{W})$ are given as

$$\mathbf{W} = \begin{bmatrix} h \\ u \end{bmatrix}, \quad \mathbf{A}(\mathbf{W}) = \begin{bmatrix} u & h \\ g & u \end{bmatrix}, \quad \mathbf{S} = \begin{bmatrix} 0 \\ -g\partial_x b \end{bmatrix}. \tag{3.12}$$

In the one-dimensional $2 \times 2$ system (3.11) the coefficient matrix $\mathbf{A}(\mathbf{W})$ is obviously a function of the vector of unknowns $\mathbf{W}$, which makes the problem non-linear. By *freezing* the coefficient matrix, or more specifically its entries, we obtain a linear system with constant coefficients, namely

$$\partial_t \mathbf{W} + \hat{\mathbf{A}} \partial_x \mathbf{W} = \mathbf{S}, \tag{3.13}$$

where the coefficient matrix $\hat{\mathbf{A}}$ is constant and given as

$$\hat{\mathbf{A}} = \begin{bmatrix} \hat{u} & \hat{h} \\ g & \hat{u} \end{bmatrix}. \tag{3.14}$$

The particular choice for the entries $\hat{u}$ and $\hat{h}$ depends on the purpose of the intended linearisation.

## 3.2 Eigenstructure and Characteristic Variables

Here we investigate the eigenstructure of the linear system (3.13) and its use in transforming the given set of equations to a new system in *characteristic variables*. The eigenvalues of (3.13) are the eigenvalues of the matrix $\hat{\mathbf{A}}$ given by Eq. (3.14), which in turn are the roots of the *characteristic polynomial*

$$|\hat{\mathbf{A}} - \lambda \mathbf{I}| = 0, \tag{3.15}$$

where $|\mathbf{X}|$ denotes *determinant of* $\mathbf{X}$, $\mathbf{I}$ is the identity matrix and $\lambda$ is a scalar parameter, the unknown quantity of the problem. Substitution of $\hat{\mathbf{A}}$ and $\mathbf{I}$ into (3.15) leads to

$$(\hat{u} - \lambda)^2 - g\hat{h} = 0. \tag{3.16}$$

The roots of this equation are the sought eigenvalues, and in this case they are both real and given by

$$\lambda_1 = \hat{u} - \hat{a}, \quad \lambda_2 = \hat{u} + \hat{a}. \tag{3.17}$$

Here $\hat{a}$ is the *celerity*, given by

$$\hat{a} = \sqrt{g\hat{h}}, \tag{3.18}$$

a constant here. System (3.13) is thus *striclty hyperbolic* for $\hat{h} > 0$. This is the *wet-bed* case, $\hat{h} > 0$ and $\hat{a} > 0$; the real eigenvalues are distinct.

The corresponding right eigenvectors are

## 3.2 Eigenstructure and Characteristic Variables

$$\mathbf{R}^{(1)} = \alpha_1 \begin{bmatrix} 1 \\ -\hat{a}/\hat{h} \end{bmatrix}, \quad \mathbf{R}^{(2)} = \alpha_2 \begin{bmatrix} 1 \\ \hat{a}/\hat{h} \end{bmatrix}, \tag{3.19}$$

where $\alpha_1$ and $\alpha_2$ are scaling factors. Choosing $\alpha_1 = \hat{h}$ and $\alpha_2 = \hat{h}$ gives

$$\mathbf{R}^{(1)} = \begin{bmatrix} \hat{h} \\ -\hat{a} \end{bmatrix}, \quad \mathbf{R}^{(2)} = \begin{bmatrix} \hat{h} \\ \hat{a} \end{bmatrix}. \tag{3.20}$$

Next we *diagonalise the system*, which is a key step in transforming the dependent variables $\mathbf{W}(x, t)$ to a new set of dependent variables $\mathbf{Q}(x, t)$, called *characteristic variables*. See Logan [1] for related concepts. Now consider the matrix

$$\mathbf{R} = \begin{bmatrix} \hat{h} & \hat{h} \\ -\hat{a} & \hat{a} \end{bmatrix}, \tag{3.21}$$

the columns of which are the right eigenvectors (3.20). The inverse of $\mathbf{R}$ is

$$\mathbf{R}^{-1} = \frac{1}{2\hat{a}\hat{h}} \begin{bmatrix} \hat{a} & -\hat{h} \\ \hat{a} & \hat{h} \end{bmatrix}. \tag{3.22}$$

It is then easy to verify that

$$\hat{\mathbf{A}} = \mathbf{R}\mathbf{\Lambda}\mathbf{R}^{-1} \text{ or } \mathbf{\Lambda} = \mathbf{R}^{-1}\hat{\mathbf{A}}\mathbf{R}, \tag{3.23}$$

where $\mathbf{\Lambda}$ is a diagonal matrix, the entries of which are the eigenvalues $\lambda_i$ of $\hat{\mathbf{A}}$, namely

$$\mathbf{\Lambda} = \begin{bmatrix} \lambda_1 & 0 \\ 0 & \lambda_2 \end{bmatrix}. \tag{3.24}$$

Note that the *diagonalisation* (3.23) is only possible if the matrix $\mathbf{R}^{-1}$ is defined, which in turn requires the right eigenvectors (3.20) to be *linearly independent*. For this reason a hyperbolic system is sometimes defined as a *diagonalisable system*.

Assuming that the inverse matrix $\mathbf{R}^{-1}$ exists, we then define a new set of dependent variables $\mathbf{Q} = [q_1, q_2]^T$ via the transformation

$$\mathbf{Q} = \mathbf{R}^{-1}\mathbf{W} \text{ or } \mathbf{W} = \mathbf{R}\mathbf{Q}. \tag{3.25}$$

The new variables $\mathbf{Q}$ are called *characteristic variables*. See Logan [1]. The importance of characteristic variables stems from the fact that in terms of these new variables the original set of partial differential equations (PDEs) becomes a simpler set of *decoupled equations*, as we shall see. Finding the solution to the decoupled system is then much easier than finding the solution to the original system directly.

We need the partial derivatives $\partial_t \mathbf{W}$ and $\partial_x \mathbf{W}$ in Eq. (3.13). From (3.25), since $\hat{\mathbf{A}}$ and thus $\mathbf{R}$ are constant, the sought derivatives are

$$\partial_t \mathbf{W} = \mathbf{R}\partial_t \mathbf{Q}, \quad \partial_x \mathbf{W} = \mathbf{R}\partial_x \mathbf{Q}.$$

Direct substitution of these expressions into Eq. (3.13) gives

$$\mathbf{R}\partial_t \mathbf{Q} + \hat{\mathbf{A}}\mathbf{R}\partial_x \mathbf{Q} = \mathbf{S}.$$

Multiplication of this equation from the left by $\mathbf{R}^{-1}$ and use of (3.25) gives

$$\partial_t \mathbf{Q} + \Lambda \partial_x \mathbf{Q} = \bar{\mathbf{S}}, \tag{3.26}$$

where the source term vector is now

$$\bar{\mathbf{S}} = \mathbf{R}^{-1}\mathbf{S}. \tag{3.27}$$

The form (3.26) of Eq. (3.13) is called the *canonical form* or *characteristic form*. When written in full this system becomes

$$\partial_t \begin{bmatrix} q_1 \\ q_2 \end{bmatrix} + \begin{bmatrix} \lambda_1 & 0 \\ 0 & \lambda_2 \end{bmatrix} \partial_x \begin{bmatrix} q_1 \\ q_2 \end{bmatrix}_x = \begin{bmatrix} \bar{s}_1 \\ \bar{s}_2 \end{bmatrix}. \tag{3.28}$$

Clearly, each PDE in this system is of the form

$$\partial_t q_i + \lambda_i \partial_x q_i = \bar{s}_i, \quad \text{for } i = 1, 2. \tag{3.29}$$

Let us now assume the homogeneous case in which the source term $\bar{\mathbf{S}}$ in (3.26) vanishes. Then each decoupled equation is a *linear advection equation* with constant coefficient $\lambda_i$, and involves the *single unknown* $q_i(x, t)$. The system is therefore *decoupled*; the characteristic speeds are $\lambda_i$ and associated with this decoupled system there are two pairs of decoupled ordinary differential equations (ODEs), namely

$$\frac{dq_i}{dt} = 0 \text{ on curves } \frac{dx}{dt} = \lambda_i \quad \text{for } i = 1, 2. \tag{3.30}$$

The meaning of this and its relevance for solving the general IVP for (3.26) will become more obvious in the next section.

**Remark 3.1** (*Characteristic form*) For a discussion on the notions of *characteristic equations* or *equations in characteristic form* for non-linear systems, see Chap. 5 of Logan [1].

## 3.3 The General Initial-Value Problem

This Section solves the general IVP for system (3.26). Due to the decoupling provided by transforming to characteristic variables, the problem is reduced to solving the general IVP for a scalar linear advection equation with constant coefficient.

### 3.3.1 Recalling the General IVP for the Scalar Case

As is well known, see Chap. 2, the solution of the general IVP for the linear advection equation

$$\left. \begin{array}{l} \text{PDE:} \ \partial_t q + \lambda \partial_x q = 0 \,, \\ \text{IC:} \quad q(x, 0) = q^{(0)}(x) \,, \end{array} \right\} \quad (3.31)$$

with constant wave propagation speed $\lambda$ is given by

$$q(x, t) = q^{(0)}(x - \lambda t) \,. \quad (3.32)$$

Here the initial condition $q^{(0)}(x)$ is an arbitrary function of a single variable $\hat{x}$. The reader can easily verify that (3.32) is the solution of (3.31) by calculating the partial derivatives $\partial_t q$ and $\partial_x q$ in (3.32), and substituting them in the PDE in (3.31). A geometric interpretation of the solution now follows.

Consider curves $x = x(t)$ in the $x$-$t$ plane and regard $q$ as a function of $t$. Restrict the curves to satisfy the ODE

$$\frac{dx}{dt} = \lambda \,. \quad (3.33)$$

Then, on such curves the PDE $\partial_t q + \lambda \partial_x q = 0$ can be written as an ODE, as we shall see. The rate of change of $q$ along $x = x(t)$ is

$$\frac{dq}{dt} = \frac{\partial q}{\partial t} + \frac{dx}{dt} \frac{\partial q}{\partial x} \,. \quad (3.34)$$

In view of the ODE (3.33), which is satisfied by the curves $x = x(t)$, and in view of the PDE in (3.31), we have

$$\frac{dq}{dt} = \frac{\partial q}{\partial t} + \lambda \frac{\partial q}{\partial x} = 0 \,. \quad (3.35)$$

Therefore $dq/dt = 0$, along the characteristic curves $x = x(t)$ satisfying (3.33), that is, the the rate of change of $q$ along the characteristic is zero; in other words, $q$ is *constant along such curves* $x = x(t)$. The speed $\lambda$ in (3.32) is called the *characteristic speed* and according to (3.33) it is the slope of the curve $x = x(t)$ in the $t$-$x$ plane.

**Fig. 3.2** Characteristics of the linear advection equation in the $x$-$t$ plane for positive characteristic speed $\lambda$. Initial condition at time $t = 0$ fixes the initial position $x_0$

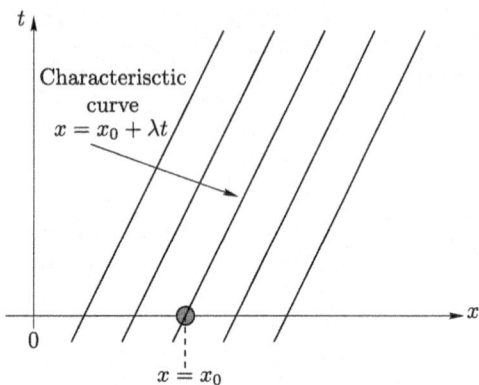

In practice it is more common to use the $x$-$t$ plane to sketch the characteristics, in which case the slope of the curves in question is $1/\lambda$.

Figure 3.2 illustrates the family of characteristic curves $x = x(t)$ given by the ODE (3.33) for $\lambda > 0$; it is a one-parameter family of curves. A particular member of this family is determined when an *initial condition* (IC) at time $t = 0$ for the ODE (3.33) is added, namely

$$x(0) = x_0 . \tag{3.36}$$

Then, the single characteristic curve that passes through the point $(x_0, 0)$, according to (3.33) is

$$x = x_0 + \lambda t . \tag{3.37}$$

This particular characteristic is illustrated in Fig. 3.2. Now we may regard the initial position $x_0$ as a parameter and in this way we reproduce the full one-parameter family of characteristics. The fact that the curves are *parallel* is typical of linear hyperbolic PDEs with constant coefficients. As $q$ remains constant along characteristics, given the initial value $q(x, 0) = q^{(0)}(x)$ at time $t = 0$ on the point $x_0$, along the whole of the characteristic curve $x(t) = x_0 + \lambda t$ that passes through the initial point $x_0$ on the $x$-axis, the solution is as given in (3.32). That this is so follows from the fact that $q(x, t)$ along the characteristic line through the point $x_0$ is simply $q(x, t) = q^{(0)}(x_0)$, but from (3.37) we have $x_0 = x - \lambda t$ and thus $q(x, t) = q^{(0)}(x_0) = q^{(0)}(x - \lambda t)$, which is (3.32).

The interpretation of the solution is this: *given an initial profile $q^{(0)}(x)$, the PDE will simply translate this profile with velocity $\lambda$ to the right if $\lambda > 0$ and to the left if $\lambda < 0$*. The shape of the initial profile *remains unchanged*. The model equation in (3.31) under study contains some of the basic features of wave propagation phenomena, where a wave is understood as some recognisable feature of a disturbance that travels at a finite speed; further reading on this topic may be found in [2–4]. See also Chap. 2.

### 3.3.2 The General IVP for the System Case

We now study the general IVP for the linearised homogeneous version of the system of PDEs (3.13), namely

$$\left. \begin{array}{l} \text{PDEs: } \partial_t \mathbf{W} + \hat{\mathbf{A}} \partial_x \mathbf{W} = \mathbf{0} \ , \\ \text{ICs: } \mathbf{W}(x, 0) \equiv \mathbf{W}^{(0)}(x) = \begin{bmatrix} w_1^{(0)}(x) \\ w_2^{(0)}(x) \end{bmatrix} . \end{array} \right\} \quad (3.38)$$

We find the general solution of the IVP (3.38) by first solving the IVP for the corresponding *canonical system* in terms of *characteristic variables*, namely

$$\left. \begin{array}{l} \text{PDEs: } \partial_t \mathbf{Q} + \mathbf{\Lambda} \partial_x \mathbf{Q} = \mathbf{0} \ , \\ \text{ICs: } \mathbf{Q}(x, 0) \equiv \mathbf{Q}^{(0)}(x) = \begin{bmatrix} q_1^{(0)}(x) \\ q_2^{(0)}(x) \end{bmatrix} , \end{array} \right\} \quad (3.39)$$

where the initial conditions satisfy (3.25), namely

$$\mathbf{Q}^{(0)} = \mathbf{R}^{-1} \mathbf{W}^{(0)} \text{ or } \mathbf{W}^{(0)} = \mathbf{R} \mathbf{Q}^{(0)} \ . \quad (3.40)$$

As the PDEs in (3.39) are decoupled and each component equation is a linear advection equation of the same form as that in (3.31), the solution of the IVP (3.39) is direct. Each of the unknowns $q_i(x, t)$ has initial condition $q_i^{(0)}(x)$ and is given by (3.32); the sought solution of (3.39) is therefore immediate, namely

$$\mathbf{Q}(x, t) = \begin{bmatrix} q_1(x, t) \\ q_2(x, t) \end{bmatrix} = \begin{bmatrix} q_1^{(0)}(x - \lambda_1 t) \\ q_2^{(0)}(x - \lambda_2 t) \end{bmatrix} . \quad (3.41)$$

The solution of the general IVP (3.38) in terms of the original variables $\mathbf{W}$ is now obtained by transforming back according to (3.25), namely $\mathbf{W} = \mathbf{R}\mathbf{Q}$, or in full

$$\mathbf{W}(x, t) = \begin{bmatrix} w_1(x, t) \\ w_2(x, t) \end{bmatrix} = \begin{bmatrix} \hat{h} & \hat{h} \\ -\hat{a} & \hat{a} \end{bmatrix} \begin{bmatrix} q_1^{(0)}(x - \lambda_1 t) \\ q_2^{(0)}(x - \lambda_2 t) \end{bmatrix} . \quad (3.42)$$

First we need to know $q_1^{(0)}(x)$ and $q_2^{(0)}(x)$ in terms of the initial data for the original variables. From (3.40) we have

$$\begin{bmatrix} q_1^{(0)}(x) \\ q_2^{(0)}(x) \end{bmatrix} = \frac{1}{2\hat{a}\hat{h}} \begin{bmatrix} \hat{a} & -\hat{h} \\ \hat{a} & \hat{h} \end{bmatrix} \begin{bmatrix} w_1^{(0)}(x) \\ w_2^{(0)}(x) \end{bmatrix} . \quad (3.43)$$

That is

$$q_1^{(0)}(x) = \frac{1}{2\hat{a}\hat{h}}\left[\hat{a}w_1^{(0)}(x) - \hat{h}w_2^{(0)}(x)\right] \tag{3.44}$$

and

$$q_2^{(0)}(x) = \frac{1}{2\hat{a}\hat{h}}\left[\hat{a}w_1^{(0)}(x) + \hat{h}w_2^{(0)}(x)\right]. \tag{3.45}$$

Therefore

$$q_1(x,t) = q_1^{(0)}(x - \lambda_1 t) = \frac{1}{2\hat{a}\hat{h}}\left[\hat{a}w_1^{(0)}(x - \lambda_1 t) - \hat{h}w_2^{(0)}(x - \lambda_1 t)\right] \tag{3.46}$$

and

$$q_2(x,t) = q_2^{(0)}(x - \lambda_2 t) = \frac{1}{2\hat{a}\hat{h}}\left[\hat{a}w_1^{(0)}(x - \lambda_2 t) + \hat{h}w_2^{(0)}(x - \lambda_2 t)\right]. \tag{3.47}$$

But from (3.42) we have

$$w_1(x,t) = \hat{h}q_1(x,t) + \hat{h}q_2(x,t) \tag{3.48}$$

and

$$w_2(x,t) = -\hat{a}q_1(x,t) + \hat{a}q_2(x,t). \tag{3.49}$$

Substituting (3.46) and (3.47) into (3.48) and (3.49) gives the final solution in terms of the original variables as

$$\left.\begin{aligned}w_1(x,t) &= \frac{1}{2}\left[w_1^{(0)}(x - \lambda_1 t) + w_1^{(0)}(x - \lambda_2 t)\right] \\ &+ \frac{1}{2}\frac{\hat{h}}{\hat{a}}\left[-w_2^{(0)}(x - \lambda_1 t) + w_2^{(0)}(x - \lambda_2 t)\right]\end{aligned}\right\} \tag{3.50}$$

and

$$\left.\begin{aligned}w_2(x,t) &= \frac{1}{2}\frac{\hat{a}}{\hat{h}}\left[-w_1^{(0)}(x - \lambda_1 t) + w_1^{(0)}(x - \lambda_2 t)\right] \\ &+ \frac{1}{2}\left[w_2^{(0)}(x - \lambda_1 t) + w_2^{(0)}(x - \lambda_2 t)\right].\end{aligned}\right\} \tag{3.51}$$

The above solution is general in that it is valid for any prescribed initial condition $\mathbf{W}(x, 0) = [w_1^{(0)}(x), w_2^{(0)}(x)]^T$; the components $w_1^{(0)}(x)$ and $w_2^{(0)}(x)$ are arbitrary functions of distance $x$. In the next section we solve the Riemann problem, an initial value problem with special initial condition.

## 3.4 The Riemann Problem

The Riemann problem for a set of PDEs is an initial value problem for such PDEs in which the initial condition has a special form. Conventionally, the initial data for the Riemann problem is regarded as being *piece-wise constant* but generalisations of the Riemann problem are also possible, in which case the initial data is, for instance, *piece-wise linear* or *piece-wise quadratic* [3]. For a one-dimensional time-dependent PDE the conventional Riemann problem has initial data consisting of two constant states $\mathbf{W}_L$ (Left) and $\mathbf{W}_R$ (Right) separated by a discontinuity at a point $x = x_0$. Here we first recall the Riemann problem for the scalar linear advection equation and then for a $2 \times 2$ linear system. The main purpose is to introduce the reader to this important topic via simple equations.

### 3.4.1 Recalling the Scalar Case

The Riemann problem for the scalar linear advection equation with constant wave propagation speed $\lambda$ is the IVP

$$\left.\begin{array}{ll} \text{PDE:} & \partial_t q + \lambda \partial_x q = 0 \,, \\ \text{IC:} & q(x, 0) = q^{(0)}(x) = \begin{cases} q_L & \text{if } x < 0 \,, \\ q_R & \text{if } x > 0 \,, \end{cases} \end{array}\right\} \quad (3.52)$$

where $q_L$ (left) and $q_R$ (right) are two constant values, as depicted in Fig. 3.3. The initial data has a discontinuity at $x = 0$. The IVP (3.52) is the simplest initial-value problem one can pose; the trivial case would result when $q_L = q_R$. From the previous discussion on the solution (3.32) of the general IVP (3.31) we expect any point on the initial profile to propagate a distance $d = \lambda t$ in time $t$. In particular, we expect the initial discontinuity at $x = 0$ to propagate a distance $d = \lambda t$ in time $t$. The particular characteristic curve $x = \lambda t$ will then separate those characteristic curves to the left,

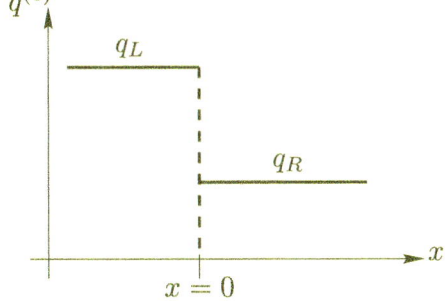

**Fig. 3.3** Initial data for the Riemann problem. At the initial time the data consists of two constant states separated by a discontinuity at $x = 0$

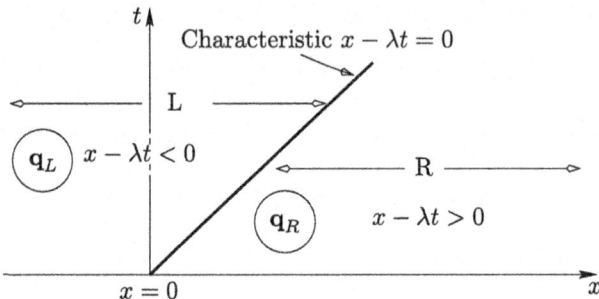

**Fig. 3.4** Solution of the Riemann problem in the *x-t* plane for the linear advection equation with positive characteristic speed $\lambda$

on which the solution takes the value $q_L$, from those curves to the right, on which the solution takes the value $q_R$; see Fig. 3.4. Thus the solution of the Riemann problem (3.52) is simply

$$q(x,t) = q^{(0)}(x - \lambda t) = \begin{cases} q_L & \text{if } x - \lambda t < 0, \\ q_R & \text{if } x - \lambda t > 0. \end{cases} \qquad (3.53)$$

Obviously, the solution (3.53) follows directly from the general solution (3.32), namely $q(x,t) = q^{(0)}(x - \lambda t)$. Note that from (3.52) $q^{(0)}(x - \lambda t) = q_L$ in the **L** region, see Fig. 3.4, because the argument satisfies $x - \lambda t < 0$, and $q^{(0)}(x - \lambda t) = q_R$ in the **R** region because the argument satisfies $x - \lambda t > 0$. The solution of the Riemann problem can be represented in the *x-t* plane, as shown in Fig. 3.4. Through any point $x_0$ on the *x*-axis one can draw a characteristic; as $\lambda$ is constant these characteristics are all parallel to each other. For the solution of the Riemann problem the characteristic that passes through $x = 0$ is significant. This is the only one across which the solution changes discontinuously; it is a singularity.

### 3.4.2 The System Case

Here we study the solution of the Riemann problem for the $2 \times 2$ linear shallow water system, namely

### 3.4 The Riemann Problem

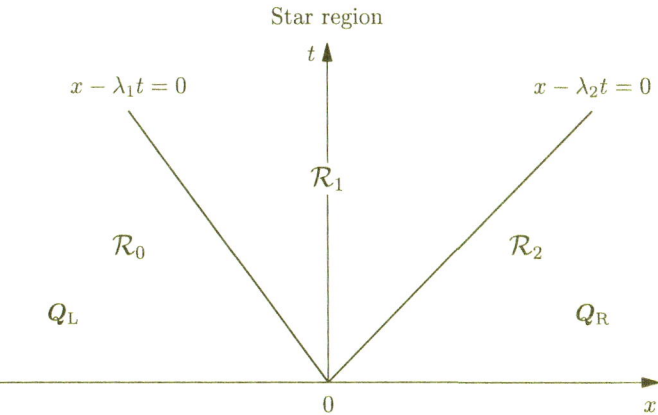

**Fig. 3.5** Structure of the solution of the Riemann problem for the 2 × 2 linear shallow water system. The solution in the $x$-$t$ half plane involves three regions $\mathcal{R}_0$, $\mathcal{R}_1$, $\mathcal{R}_2$ separated by the two waves determined by the two eigenvalues $\lambda_1$ and $\lambda_2$

$$\text{PDEs: } \partial_t \mathbf{W} + \hat{\mathbf{A}} \partial_x \mathbf{W} = \mathbf{0} \,,$$

$$\text{ICs: } \mathbf{W}^{(0)}(x) = \begin{bmatrix} w_1(x) \\ w_2(x) \end{bmatrix} = \begin{cases} \mathbf{W}_L = \begin{bmatrix} w_{1L} \\ w_{2L} \end{bmatrix} & \text{if } x < 0 \,, \\ \\ \mathbf{W}_R = \begin{bmatrix} w_{1R} \\ w_{2R} \end{bmatrix} & \text{if } x > 0 \,. \end{cases} \tag{3.54}$$

The structure of the solution of this Riemann problem is illustrated in Fig. 3.5 in the $x$-$t$ plane. There are three regions $\mathcal{R}_0$, $\mathcal{R}_1$, $\mathcal{R}_2$ separated by the two waves determined by the two eigenvalues $\lambda_1$ and $\lambda_2$. Note that

$$\left.\begin{aligned} &\text{Region } \mathcal{R}_0 = \{(x,t) \text{ such that } x - \lambda_1 t < 0\} \,, \\ &\text{Region } \mathcal{R}_1 = \{(x,t) \text{ such that } x - \lambda_1 t > 0 \text{ and } x - \lambda_2 t < 0\} \,, \\ &\text{Region } \mathcal{R}_2 = \{(x,t) \text{ such that } x - \lambda_2 t > 0\} \,. \end{aligned}\right\} \tag{3.55}$$

The complete solution to the Riemann problem (3.54) can now be directly obtained from the solution of the general IVP given by (3.50)–(3.51). Note for example that in evaluating the solution for $w_1(x,t)$ given by (3.50) in region $\mathcal{R}_0$ we have

$$\left.\begin{aligned} w_1^{(0)}(x - \lambda_1 t) = w_{1L} \,, \quad w_1^{(0)}(x - \lambda_2 t) = w_{1L} \,, \\ w_2^{(0)}(x - \lambda_1 t) = w_{2L} \,, \quad w_2^{(0)}(x - \lambda_2 t) = w_{2L} \,, \end{aligned}\right\} \tag{3.56}$$

as their arguments satisfy the corresponding conditions in (3.55), namely $x - \lambda_1 t < 0$ and $x - \lambda_2 t < 0$. Similarly for $w_2(x,t)$. Not surprisingly, the solution for $w_1(x,t)$

and $w_2(x,t)$ in region $\mathcal{R}_0$ is simply the initial data to the left of the origin $x = 0$, namely $w_1(x,t) = w_{1L}$, $w_2(x,t) = w_{2L}$. Similarly, the reader can verify that the solution for $w_1(x,t)$ and $w_2(x,t)$ in the region $\mathcal{R}_2$ is simply the initial data to the right of the origin $x = 0$, namely $w_1(x,t) = w_{1R}$, $w_2(x,t) = w_{2R}$. The interesting region is $\mathcal{R}_1$, which lies between the two waves and is usually called the *Star Region*; see Fig. 3.5. Here we have

$$\left.\begin{array}{l} w_1^{(0)}(x - \lambda_1 t) = w_{1R} \text{ because } x - \lambda_1 t > 0, \\ w_2^{(0)}(x - \lambda_2 t) = w_{2L} \text{ because } x - \lambda_2 t < 0. \end{array}\right\} \quad (3.57)$$

Thus the final solution in the *Star Region* $\mathcal{R}_1$ for $w_1 \equiv h$ (depth) and $w_2 \equiv u$ (velocity) is

$$\left.\begin{array}{l} h_* = \dfrac{1}{2}(h_L + h_R) + \dfrac{1}{2}\dfrac{\hat{h}}{\hat{a}}(u_L - u_R), \\ u_* = \dfrac{1}{2}(u_L + u_R) + \dfrac{1}{2}\dfrac{\hat{a}}{\hat{h}}(h_L - h_R). \end{array}\right\} \quad (3.58)$$

Note that $h_*$ and $u_*$ for depth and particle velocity in the *Star Region* involve arithmetic means of their initial data as the first term.

### 3.4.3 Example

As an example we solve the Riemann problem for the linearised shallow water equations for a rectangular horizontal channel, with specific initial data. We assume the initial discontinuity is placed at the position $x = 0$ and we choose the following initial conditions: to the left of $x = 0$ we take $h_L = 1$ m and $u_L = 0$ m/s; to the right of $x = 0$ we take $h_R = \frac{1}{2}$ m and $u_R = 0$ m/s. For the entries of the linearised matrix we choose $\hat{h} = 9/g$ and $\hat{u} = 1$ m/s, with $g = 9.8$ m/s². The resulting celerity is $\hat{a} = 3$ m/s and the eigenvalues are $\lambda_1 = -2$ and $\lambda_2 = 4$. The solution in the *Star Region*, from (3.58), is $h_* = \frac{3}{4}$ and $u_* = g/12 \approx 0.8167$.

Figure 3.6 shows the solution for free-surface position $h(x,t)$ and particle velocity $u(x,t)$ at time $t = 1$ s. Frame (a) shows the initial profile for water depth. The initial velocity profile is identically zero and is not shown. Frame (b) shows the free-surface profile at time $t = 1$ s. Frame (c) shows the particle velocity profile at the same time. There are two waves, one propagating to the left at speed $\lambda_1 = -2$, and the other propagating to the right at speed $\lambda_2 = 4$.

This problem is analogous to the *dam-break problem* in which the initial discontinuity comes about by imagining a wall or gate at $x = 0$ that is suddenly removed at *very high speed*. The initial velocities in the dam-break problem are always $u_L = 0$ and $u_R = 0$; the dam-break problem is thus a special case of the Riemann problem. See Chaps. 4, 6 and 7 for related discussions.

## 3.5 A Linear Model with Source Terms

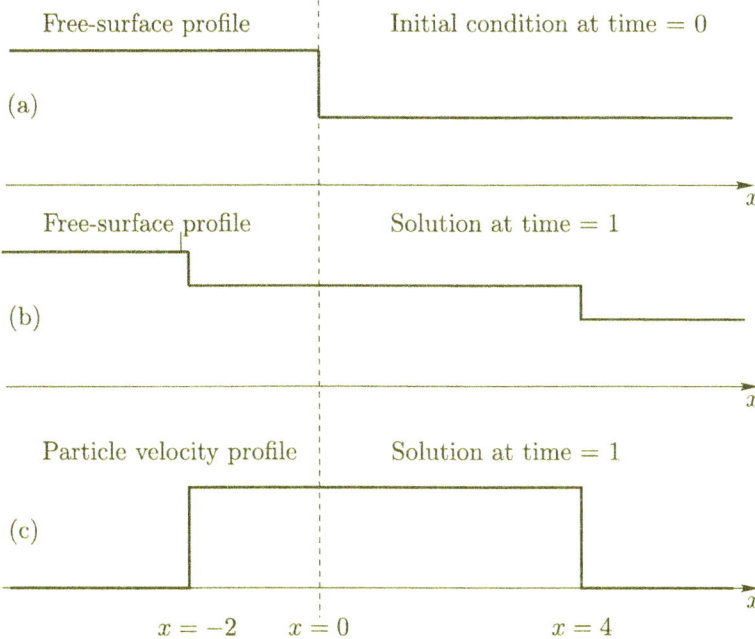

**Fig. 3.6** Solution of a Riemann problem for the linearised shallow water equations: **a** initial condition for free-surface position; **b** solution for free-surface position at time $t = 1$ s; **c** solution for particle velocity at time $t = 1$ s

**Remark 3.2** For linear hyperbolic systems with constant coefficients, the wave speeds in the Riemann problem solution are identical to the characteristic speeds given by the eigenvalues of the system. This is not so for non-linear systems.

**Remark 3.3** For linear hyperbolic systems with constant coefficients all waves present in the Riemann problem solution are discontinuous, or trivial of course. This is not so for non-linear systems, which allow for continuous waves, such as rarefaction waves; see Chap. 4.

### 3.5 A Linear Model with Source Terms

As an example we study the general IVP for the linearised shallow water equations for a rectangular channel of *variable bed elevation* $b(x)$, namely

$$\left. \begin{array}{l} \text{PDEs: } \partial_t \mathbf{W} + \hat{\mathbf{A}} \partial_x \mathbf{W} = \mathbf{S} \,, \\[6pt] \text{ICs: } \mathbf{W}(x, 0) \equiv \mathbf{W}^{(0)}(x) = \begin{bmatrix} w_1^{(0)}(x) \\ w_2^{(0)}(x) \end{bmatrix} , \end{array} \right\} \quad (3.59)$$

where

$$\mathbf{S} = \begin{bmatrix} s_1 \\ s_2 \end{bmatrix} = \begin{bmatrix} 0 \\ -gb'(x) \end{bmatrix} \quad (3.60)$$

is the vector of source terms due to variable bed elevation. As for the homogeneous case studied in Sect. 3.3.2, we rely on the solution of the inhomogeneous scalar case, namely the IVP

$$\left. \begin{array}{l} \text{PDE:} \ \partial_t q + \lambda \partial_x q = s \, , \\ \text{IC:} \ \ q(x,0) = q^{(0)}(x) + r(x) \, , \end{array} \right\} \quad (3.61)$$

with constant wave propagation speed $\lambda$. We look for solutions to (3.61) that are of the form

$$q(x,t) = q^{(0)}(x - \lambda t) + r(x) \, , \quad (3.62)$$

where $r(x)$ is an arbitrary function of distance $x$. It is trivial to verify that the sought form of the source term $s$ in (3.61) must be

$$s(x) = \lambda r'(x) \, , \quad (3.63)$$

which is precisely the form of a source term due to a variable bed elevation. After transforming the equations in (3.59) to characteristic variables, we obtain the IVP

$$\left. \begin{array}{l} \text{PDEs:} \ \mathbf{Q}_t + \mathbf{\Lambda} \mathbf{Q}_x = \tilde{\mathbf{S}} \, , \\ \text{ICs:} \ \ \mathbf{Q}(x,0) \equiv \mathbf{Q}^{(0)}(x) = \begin{bmatrix} q_1^{(0)}(x) \\ q_2^{(0)}(x) \end{bmatrix} \, , \end{array} \right\} \quad (3.64)$$

where the transformed source term vector satisfies (3.27), for which the case (3.60) yields

$$\tilde{s}_1 = \frac{1}{2}\frac{g}{\hat{a}}b'(x) \, , \quad \tilde{s}_2 = -\frac{1}{2}\frac{g}{\hat{a}}b'(x) \, . \quad (3.65)$$

The general solution of (3.64) is

$$q_1(x,t) = q_1^{(0)}(x - \lambda_1 t) + \frac{1}{2}\frac{g}{\hat{a}\lambda_1}b(x) \quad (3.66)$$

and

$$q_2(x,t) = q_2^{(0)}(x - \lambda_2 t) - \frac{1}{2}\frac{g}{\hat{a}\lambda_2}b(x) \, . \quad (3.67)$$

In terms of the original variables, the solution is

$$w_1(x,t) = \frac{1}{2}\left[w_1^{(0)}(x-\lambda_1 t) + w_1^{(0)}(x-\lambda_2 t)\right]$$
$$+ \frac{1}{2}\frac{\hat{h}}{\hat{a}}\left[-w_2^{(0)}(x-\lambda_1 t) + w_2^{(0)}(x-\lambda_2 t)\right] + \frac{\hat{a}^2}{\lambda_1\lambda_2}b(x) \quad (3.68)$$

and

$$w_2(x,t) = \frac{1}{2}\frac{\hat{a}}{\hat{h}}\left[-w_1^{(0)}(x-\lambda_1 t) + w_1^{(0)}(x-\lambda_2 t)\right]$$
$$+ \frac{1}{2}\left[-w_2^{(0)}(x-\lambda_1 t) + w_2^{(0)}(x-\lambda_2 t)\right] - \frac{g\hat{u}}{\lambda_1\lambda_2}b(x) \, . \quad (3.69)$$

Note that solution (3.68)–(3.69) is a simple generalisation of (3.50)–(3.51). For the shallow water equations $w_1(x,t) = h(x,t)$ and $w_2(x,t) = u(x,t)$, of course. This exact solution can be very useful for testing numerical methods intended for the shallow water equations with variable bed elevation.

## 3.6 Case Study: Alternative Linearisation

The purpose of this section is to study an alternative, more physically-based linearisation approach as applied to the one-dimensional non-linear shallow water equations. The model is then used to consolidate salient concepts on hyperbolic systems, such as eigenvalues, eigenvectors, hyperbolicity and construction of analytical solutions to the general initial-value problem, including the Riemann problem, as a special case.

Figure 3.1 recalls the physical set up. The region of interest lies above a horizontal datum at $z = 0$. The fluid lies above the channel bottom elevation $b(x) \geq 0$; the slope $b'(x)$ of the bottom elevation gives rise to a source term in the equations; $h(x,t)$ is the flow depth; $u(x,t)$ is the particle velocity; the free surface, positioned at $s(x,t) = b(x) + h(x,t)$, is under the influence of the acceleration due to gravity $g$.

We are interested in an alternative linearisation approach, as applied to the one-dimensional non-linear shallow water Eqs. (3.9)–(3.10), recalled here

$$\left.\begin{array}{l}\partial_t h + u\partial_x h + h\partial_x u = 0, \\ \partial_t u + u\partial_x u + g\partial_x h = -gb'(x).\end{array}\right\} \quad (3.70)$$

In what follows we neglect the source term, $b'(x) = 0$, though strictly speaking, this is not necessary for deriving the linearisation.

## 3.6.1 The Linear Equations

To derive a linearised form of Eq. (3.70) we consider small perturbations in surface elevation and in particle velocity, denoted respectively by $\eta(x, t)$ and $v(x, t)$, given as follows

$$h(x, t) = H + \eta(x, t), \quad u(x, t) = 0 + v(x, t). \tag{3.71}$$

$H$ is the, constant, unperturbed water depth and $0 + v(x, t)$ means a small perturbation $v(x, t)$ to stationary fluid. It is assumed that $\eta$ is small compared with $H$, that $v$ is small and that derivatives of the perturbations are also small. Then by substituting $h$ and $u$ from Eq. (3.71) into (3.70) and neglecting second-order terms we obtain a set of linearised shallow water equations

$$\left. \begin{array}{l} \partial_t \eta + H \partial_x v = 0, \\ \partial_t v + g \partial_x \eta = 0. \end{array} \right\} \tag{3.72}$$

The reader is invited to verify the above derivation by performing all the steps. In matrix form, Eq. (3.72) read

$$\partial_t \mathbf{Q} + \mathbf{A} \partial_x \mathbf{Q} = \mathbf{0}, \tag{3.73}$$

where the vector of unknowns $\mathbf{Q}(x, t)$ and the coefficient matrix $\mathbf{A}$ are respectively given as

$$\mathbf{Q} = \begin{bmatrix} \eta \\ v \end{bmatrix}, \quad \mathbf{A} = \begin{bmatrix} 0 & H \\ g & 0 \end{bmatrix}. \tag{3.74}$$

We note in passing that Eq. (3.72) reproduce the well known linear second-order wave equation, for both $\eta$ and $v$, namely

$$\partial_t^{(2)} \eta = gH \partial_x^{(2)} \eta \quad \text{and} \quad \partial_t^{(2)} v = gH \partial_x^{(2)} v. \tag{3.75}$$

The reader is invited to the verify the above. We remark that the second-order linear wave equation, either form in (3.75) for $\eta$ or $v$, is a very popular hyperbolic model for the study of wave propagation phenomena in classical textbooks. Here we adopt the first-order form (3.73).

## 3.6.2 The Eigenstructure

The eigenvalues of the matrix $\mathbf{A}$ in (3.73) are obtained from the characteristic polynomial, as in (3.15). They are

$$\lambda_1 = -a, \quad \lambda_2 = a, \tag{3.76}$$

## 3.6 Case Study: Alternative Linearisation

with

$$a = \sqrt{gH} \qquad (3.77)$$

denoting the celerity, the speed of propagation of vanishingly small-amplitude surface waves.

A left eigenvector $\mathbf{L}_1 = [l_1, l_2]$ of $\mathbf{A}$ corresponding to an eigenvalue $\lambda$ is found from the equation

$$[l_1, l_2] \begin{bmatrix} 0 & H \\ g & 0 \end{bmatrix} = [\lambda l_1, \lambda l_2], \qquad (3.78)$$

which leads to the following algebraic system for the components $l_1, l_2$,

$$l_2 g = \lambda l_1, \qquad l_1 H = \lambda l_2. \qquad (3.79)$$

These two equations are not independent (verify) and thus we effectively have a single equation, which has a one-parameter family of solutions. Using the second equation we may write

$$l_2 = \frac{H}{\lambda} l_1. \qquad (3.80)$$

Now, we fix the parameter $\alpha_1$ and set $l_1 = \alpha_1$. Therefore the eigenvalue $\lambda = \lambda_1 = -a$ in (3.80) gives the two components of the left eigenvector $\mathbf{L}_1$ corresponding to the eigenvalue $\lambda_1 = -a$, as

$$l_1 = \alpha_1, \qquad l_2 = -\frac{H}{a}\alpha_1. \qquad (3.81)$$

Recall that $\alpha_1$ is a *scaling parameter* and is open to choice. To find the left eigenvector $\mathbf{L}_2$ corresponding to $\lambda = \lambda_2 = a$ we set $l_1 = \alpha_2$ and $\lambda = \lambda_2 = a$ in (3.80) to obtain

$$l_1 = \alpha_2, \qquad l_2 = \frac{H}{a}\alpha_2, \qquad (3.82)$$

where $\alpha_2$ is again a scaling parameter open to choice. Then the two left eigenvectors corresponding to the eigenvalues $\lambda_1 = -a$ and $\lambda_2 = a$ are respectively given by

$$\mathbf{L}_1 = \alpha_1 \left[ 1, -\frac{H}{a} \right], \qquad \mathbf{L}_2 = \alpha_2 \left[ 1, \frac{H}{a} \right]. \qquad (3.83)$$

Analogously, a right eigenvector $\mathbf{R}_1 = [r_1, r_2]^T$ corresponding to an eigenvalue $\lambda$ satisfies

$$\begin{bmatrix} 0 & H \\ g & 0 \end{bmatrix} \begin{bmatrix} r_1 \\ r_2 \end{bmatrix} = \begin{bmatrix} \lambda r_1 \\ \lambda r_2 \end{bmatrix}, \qquad (3.84)$$

from which we obtain the relation

$$r_2 = \frac{\lambda}{H} r_1. \tag{3.85}$$

By fixing a parameter $\beta_1$ and setting $r_1 = \beta_1$ and $\lambda = \lambda_1 = -a$ in (3.85) one obtains

$$r_1 = \beta_1, \qquad r_2 = -\frac{a}{H}\beta_1. \tag{3.86}$$

For the eigenvalue $\lambda = \lambda_2 = a$, setting $r_1 = \beta_2$ and $\lambda = \lambda_2 = a$ in (3.85) gives

$$r_1 = \beta_2, \qquad r_2 = \frac{a}{H}\beta_2. \tag{3.87}$$

Then the right eigenvectors corresponding to the eigenvalues $\alpha_1 = -a$ and $\alpha_2 = a$ are

$$\mathbf{R}_1 = \beta_1 \begin{bmatrix} 1 \\ -a/H \end{bmatrix}, \qquad \mathbf{R}_2 = \beta_2 \begin{bmatrix} 1 \\ a/H \end{bmatrix}. \tag{3.88}$$

If we want the left and right eigenvectors to be *orthonormal*, then the scaling parameters must be chosen to satisfy

$$\alpha_1 \beta_1 = 1/2, \qquad \alpha_2 \beta_2 = 1/2. \tag{3.89}$$

Choosing $\alpha_1 = \alpha_2 = 1$ gives $\beta_1 = \beta_2 = \frac{1}{2}$. Then the matrices $\mathbf{L}$ and $\mathbf{R}$ formed by the left and right eigenvectors are given by

$$\mathbf{L} = \begin{bmatrix} 1 & -H/a \\ 1 & H/a \end{bmatrix}, \qquad \mathbf{R} = \frac{1}{2}\begin{bmatrix} 1 & 1 \\ -a/H & a/H \end{bmatrix}. \tag{3.90}$$

It is easy to show that $\mathbf{R}^{-1} = \mathbf{L}$ where the matrix $\mathbf{R}^{-1}$ denotes the inverse matrix of $\mathbf{R}$ (verify).

### 3.6.3 Equations in Characteristic Variables

First we define the characteristic variables $\mathbf{C} = [c_1, c_2]^T$ by the transformation

$$\mathbf{C} = \mathbf{LQ}. \tag{3.91}$$

Note that we could also use $\mathbf{C} = \mathbf{R}^{-1}\mathbf{Q}$, but (3.91) is more direct. For our system, written in full, we have

$$\begin{bmatrix} c_1 \\ c_2 \end{bmatrix} = \begin{bmatrix} 1 & -H/a \\ 1 & H/a \end{bmatrix} \begin{bmatrix} \eta \\ v \end{bmatrix}, \tag{3.92}$$

## 3.6 Case Study: Alternative Linearisation

which gives the characteristic variables as linear functions of the original variables, namely

$$c_1 = \eta - \frac{H}{a}v, \qquad c_2 = \eta + \frac{H}{a}v. \tag{3.93}$$

Now, the governing equations can be expressed in terms of the new variables $\mathbf{C}$, the characteristic variables. From (3.91), as $\mathbf{L}$ is constant, we obtain

$$\partial_t \mathbf{Q} = \mathbf{L}^{-1} \partial_t \mathbf{C}, \qquad \partial_x \mathbf{Q} = \mathbf{L}^{-1} \partial_x \mathbf{C}. \tag{3.94}$$

Then the orginal system (3.73) becomes

$$\mathbf{L}^{-1} \partial_t \mathbf{C} + \mathbf{A} \mathbf{L}^{-1} \partial_x \mathbf{C} = \mathbf{0}. \tag{3.95}$$

Multiplying (3.95) from the left by $\mathbf{L}$ gives

$$\partial_t \mathbf{C} + (\mathbf{L}\mathbf{A}\mathbf{L}^{-1}) \partial_x \mathbf{C} = \mathbf{0}. \tag{3.96}$$

It is easily verified that

$$\mathbf{L}\mathbf{A}\mathbf{L}^{-1} = \Lambda = \begin{bmatrix} \lambda_1 & 0 \\ 0 & \lambda_2 \end{bmatrix} = \begin{bmatrix} -a & 0 \\ 0 & a \end{bmatrix} \tag{3.97}$$

and thus the equations in characteristic variables become

$$\partial_t \mathbf{C} + \Lambda \partial_x \mathbf{C} = \mathbf{0}. \tag{3.98}$$

The equations have become completely decoupled, that is

$$\left.\begin{array}{l} \partial_t c_1 + \lambda_1 \partial_x c_1 = 0, \\ \partial_t c_2 + \lambda_2 \partial_x c_2 = 0. \end{array}\right\} \quad \rightarrow \quad \left.\begin{array}{l} \partial_t c_1 - a \partial_x c_1 = 0, \\ \partial_t c_2 + a \partial_x c_2 = 0. \end{array}\right\} \tag{3.99}$$

Each equation involves a single unknown advected with a constant characteristic speed, just as in the linear advection equation studied in Sect. 3.3.1. See also Chap. 2.

### 3.6.4 The General Initial-Value Problem

We want to solve the initial-value problem

$$\left.\begin{array}{ll} \text{PDEs:} & \partial_t \mathbf{Q} + \mathbf{A} \partial_x \mathbf{Q} = \mathbf{0}, \\ \text{ICs:} & \mathbf{Q}(x, 0) = \mathbf{Q}^{(0)}(x). \end{array}\right\} \tag{3.100}$$

Here, the initial condition $\mathbf{Q}^{(0)}(x)$ at time $t = 0$ is an arbitrary function of $x$ alone. This IVP is most easily solved in terms of characteristic variables $\mathbf{C}$ by replacing the IVP (3.100) by the equivalent IVP given in terms of characteristic variables $\mathbf{C}$, namely

$$\left.\begin{array}{rl} \text{PDEs:} & \partial_t \mathbf{C} + \Lambda \partial_x \mathbf{C} = \mathbf{0}, \\ \text{ICs:} & \mathbf{C}(x, 0) = \mathbf{C}^{(0)}(x) = \mathbf{L}\mathbf{Q}^{(0)}(x). \end{array}\right\} \quad (3.101)$$

From (3.93) the initial conditions $\mathbf{C}^{(0)}(x)$ are

$$\left.\begin{array}{l} c_1^{(0)}(x) = \eta^{(0)}(x) - \dfrac{H}{a} v^{(0)}(x), \\ c_2^{(0)}(x) = \eta^{(0)}(x) + \dfrac{H}{a} v^{(0)}(x). \end{array}\right\} \quad (3.102)$$

Hence, the solution of (3.101), in terms of characteristic variables, is

$$\left.\begin{array}{l} c_1(x, t) = c_1^{(0)}(x - \lambda_1 t) = \eta^{(0)}(x + at) - \dfrac{H}{a} v^{(0)}(x + at), \\ c_2(x, t) = c_2^{(0)}(x - \lambda_2 t) = \eta^{(0)}(x - at) + \dfrac{H}{a} v^{(0)}(x - at). \end{array}\right\} \quad (3.103)$$

In order to obtain the solution in terms of the original variables $\mathbf{Q}$ we transform back in (3.91), that is $\mathbf{Q} = \mathbf{RC}$. Then the complete solution to the original IVP (3.100) is

$$\left.\begin{array}{l} \eta(x, t) = \dfrac{1}{2}\left[\eta^{(0)}(x + at) + \eta^{(0)}(x - at) + \dfrac{H}{a}[-v^{(0)}(x + at) + v^{(0)}(x - at)]\right], \\ v(x, t) = \dfrac{1}{2}\left[v^{(0)}(x + at) + v^{(0)}(x - at) + \dfrac{a}{H}[-\eta^{(0)}(x + at) + \eta^{(0)}(x - at)]\right]. \end{array}\right\} \quad (3.104)$$

The reader is invited to verify all the above calculations. Obviously, solution (3.104) is valid for arbitrary initial conditions $\mathbf{Q}^{(0)}(x)$ in (3.100).

### 3.6.5 The Riemann Problem

The Riemann problem for system (3.73) is the initial-value problem

$$\left.\begin{array}{rl} \text{PDEs:} & \partial_t \mathbf{Q} + \mathbf{A}\partial_x \mathbf{Q} = \mathbf{0}, \\ \text{ICs:} & \mathbf{Q}(x, 0) = \mathbf{Q}^{(0)}(x) = \begin{cases} \mathbf{Q}_L & \text{if } x < 0, \\ \mathbf{Q}_R & \text{if } x > 0. \end{cases} \end{array}\right\} \quad (3.105)$$

Here the initial conditions $\mathbf{Q}_L$ and $\mathbf{Q}_R$ are very special, they are constant. Figure 3.5 illustrates the structure of the solution of (3.105) in the $x$-$t$ half plane. There are three constant regions $\mathcal{R}_0$, $\mathcal{R}_1$, $\mathcal{R}_2$ separated by the two waves determined by the

## 3.6 Case Study: Alternative Linearisation

two eigenvalues $\lambda_1$ and $\lambda_2$. The vector $\mathbf{Q}$ is known in regions $\mathcal{R}_0$ and $\mathcal{R}_2$; it is given by the initial conditions. The problem is to find the solution $\mathbf{Q}^*(x, t)$ in the intermediate region $\mathcal{R}_1$, called the *Star Region*. As seen in Chap. 2 the solution at any point $(x, t)$ for any $N \times N$ linear system can be written as a linear combination of the right eigenvectors, namely

$$\mathbf{Q}(x, t) = \mathbf{Q}_L + \sum_{i=1}^{I} \delta_i \mathbf{R}_i , \qquad (3.106)$$

where the positive integer $I(x, t)$ is such that for

$$\lambda_I < \frac{x}{t} < \lambda_{I+1} . \qquad (3.107)$$

The coefficients $\delta_i$ emerge as the solution of the following $N \times N$ linear algebraic system

$$\sum_{i=1}^{N} \delta_i \mathbf{R}_i = \mathbf{Q}_R - \mathbf{Q}_L = \Delta \mathbf{Q} . \qquad (3.108)$$

For our particular equations $N = 2$ and the linear algebraic system is

$$\delta_1 \begin{bmatrix} \frac{1}{2} \\ -\frac{1}{2}a/H \end{bmatrix} + \delta_2 \begin{bmatrix} \frac{1}{2} \\ \frac{1}{2}a/H \end{bmatrix} = \begin{bmatrix} \Delta \eta \\ \Delta v \end{bmatrix} = \begin{bmatrix} \eta_R - \eta_L \\ v_R - v_L \end{bmatrix} . \qquad (3.109)$$

That is

$$\left. \begin{array}{c} \delta_1 + \delta_2 = 2 \Delta \eta , \\ -\delta_1 + \delta_2 = 2 \dfrac{H}{a} \Delta v . \end{array} \right\} \qquad (3.110)$$

The solution of (3.110) is

$$\delta_1 = \Delta \eta - \frac{H}{a} \Delta v , \quad \delta_2 = \Delta \eta + \frac{H}{a} \Delta v . \qquad (3.111)$$

Here we are interested in the solution $\mathbf{Q}^* = [\eta^*, v^*]^T$ in the *Star Region* $\mathcal{R}_1$ in Fig. 3.5, which is given by points $(x, t)$ such that

$$\lambda_1 = -a < \frac{x}{t} < \lambda_2 = a . \qquad (3.112)$$

Finally, with $I = 1$ in (3.106) the solution is given by

$$\mathbf{Q}^* = \mathbf{Q}_L + \delta_1 \mathbf{R}_1 . \qquad (3.113)$$

That is

$$\left.\begin{aligned} \eta^* &= \tfrac{1}{2}(\eta_L + \eta_R) - \frac{1}{2}\frac{H}{a}(v_R - v_L)\,, \\ v^* &= \tfrac{1}{2}(v_L + v_R) - \frac{1}{2}\frac{a}{H}(\eta_R - \eta_L)\,. \end{aligned}\right\} \quad (3.114)$$

The reader is invited to verify the calculations.

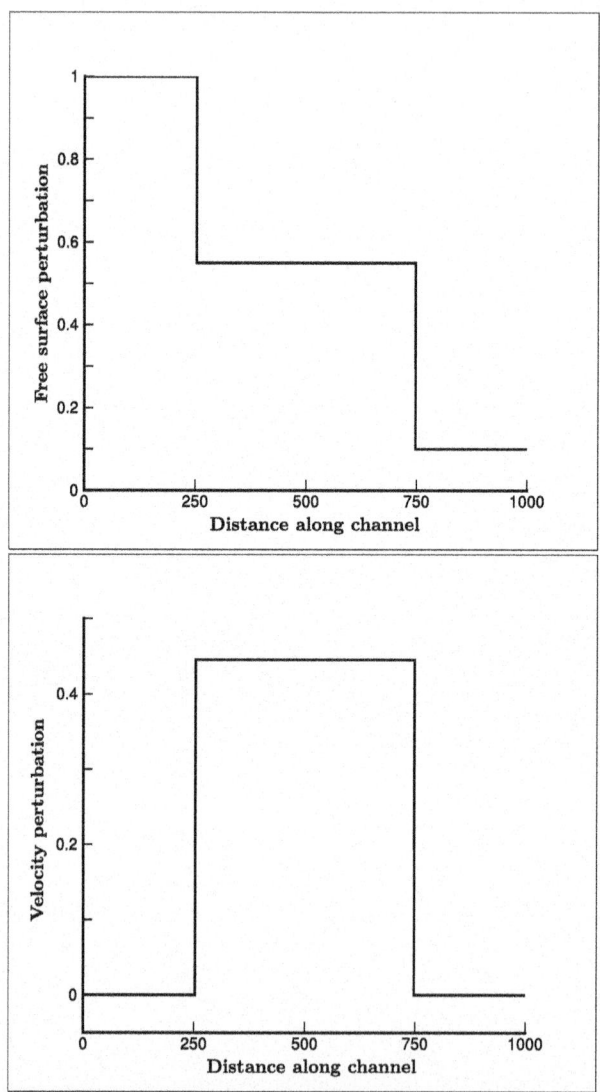

**Fig. 3.7** Solution of the Riemann problem for the linearised shallow water Eq. (3.73) with data as given. Free-surface elevation and velocity at time $T_{out} = 25\,\text{s}$

## A Riemann Problem Example

An example of a Riemann problem for the linearised shallow water equations of this section is defined by the following conditions:

1. **Domain**: the spatial domain is defined by the interval $x \in [0, 1000]$ m.
2. **Parameters**: as parameters of the problem we choose $H = 10$ m, height of the undisturbed water level; $g = 9.8$ m/s$^2$, so that the celerity is $a = 9.9$ m/s.
3. **Initial conditions**: $\eta_L = 1.0$ m, $\eta_R = 0.1$ m, $v_L = 0.0$ m/s, $v_R = 0.0$ m/s.
4. **Position of the initial discontinuity at time** $t = 0$ is $x_0 = 500$ m.

The exact solution in the *Star Region* between the two waves is $\eta^* = 0.55$ m; $v^* = 0.4455$ m/s. Solution profiles are shown in Fig. 3.7 at time $T_{out} = 25$ s.

# References

1. J.D. Logan, *An Introduction to Nonlinear Partial Differential Equations* (John Wiley and Sons, 1994)
2. R.J. LeVeque, *Finite Volume Methods for Hyperbolic Problems* (Cambridge University Press, 2002)
3. E.F. Toro, *Riemann Solvers and Numerical Methods for Fluid Dynamics. A Practical Introduction*, 3rd edn. (Springer-Verlag, 2009)
4. E.C. Zachmanoglou, D.W. Thoe, *Introduction to Partial Differential Equations* (Dover Publications, Inc., New York, 1986)

# Chapter 4
# Properties of the Nonlinear Equations

**Abstract** This Chapter is devoted to the study of mathematical properties of the two-dimensional non-linear shallow water equations, starting from the eigenstructure of the equations, which is discussed in terms of both conserved variables and primitive variables. Hyperbolicity of the equations in one and two space dimensions is proved and the nature of characteristic fields is established as being either linearly degenerate or genuinely non-linear. The rotational invariance of the two-dimensional equations is proved; this property is of much value when designing numerical methods for unstructured meshes. Finally, the two-dimensional steady shallow water equations are analysed, proving their hyperbolicity in the supercritical regime. A list of suggested exercises is given at the end of the Chapter. Useful background is found in Chaps. 1 and 2.

## 4.1 Recalling the Equations

In this Chapter we study some elementary properties of the shallow water equations and prepare the ground for studying elementary waves in Chap. 5 and for solving the Riemann problem exactly in Chaps. 6 and 7. Most of this Chapter is devoted to the time-dependent case but also included is the eigenstructure of the steady supercritical two-dimensional shallow water equations. Use is made of some basic mathematical concepts related to the theory of hyperbolic conservation laws.

In order to make the Chapter reasonably self-contained, we recall appropriate definitions as they are required, with reference to previous chapters and textbooks. Useful background is found in Chaps. 1 and 2 of this book. For general background on partial differential equations (PDEs) I recommend the book by Zachmanoglou and Thoe [1]; Chap. 10 of this contains a very useful introduction to hyperbolic PDEs. See also Chap. 2 of [2] for a self-contained summary of basic properties of hyperbolic conservation laws. More comprehensive and rigorous treatments are found in the books by Jeffrey [3], Logan [4], Chorin and Marsden [5], Smoller [6], LeVeque [7], Godlewski and Raviart [8], Hörmander [9], Tveito and Winther [10] and Dafermos [11]. The discretisation techniques studied in this book are strongly based on the underlying physical and mathematical properties of the equations and thus the need for a basic understanding of these.

The two-dimensional shallow water equations were derived in Chap. 1. When written in differential conservation-law form with source terms, the equations read

$$\partial_t \mathbf{Q} + \partial_x \mathbf{F}(\mathbf{Q}) + \partial_y \mathbf{G}(\mathbf{Q}) = \mathbf{S}(\mathbf{Q}), \tag{4.1}$$

where $\mathbf{Q}$, $\mathbf{F}(\mathbf{Q})$ and $\mathbf{G}(\mathbf{Q})$ are the vectors of conserved variables, fluxes in the $x$- and $y$-directions, and sources, given by

$$\left. \begin{array}{c} \mathbf{Q} = \begin{bmatrix} q_1 \\ q_2 \\ q_3 \end{bmatrix} = \begin{bmatrix} h \\ hu \\ hv \end{bmatrix}, \quad \mathbf{F}(\mathbf{Q}) = \begin{bmatrix} f_1 \\ f_2 \\ f_3 \end{bmatrix} = \begin{bmatrix} hu \\ hu^2 + \frac{1}{2}gh^2 \\ huv \end{bmatrix}, \\[1em] \mathbf{G}(\mathbf{Q}) = \begin{bmatrix} g_1 \\ g_2 \\ g_3 \end{bmatrix} = \begin{bmatrix} hv \\ hvu \\ hv^2 + \frac{1}{2}gh^2 \end{bmatrix}, \quad \mathbf{S}(\mathbf{Q}) = \begin{bmatrix} s_1 \\ s_2 \\ s_3 \end{bmatrix} = \begin{bmatrix} 0 \\ -gh\partial_x b \\ -gh\partial_y b \end{bmatrix}. \end{array} \right\} \tag{4.2}$$

Here $u(x, y, t)$ and $v(x, y, t)$ are respectively the $x$- and $y$-components of velocity, $h(x, y, t)$ is the *depth*, related to the *free surface* $s(x, y, t)$ and the *bed* $b(x, y)$ via

$$s(x, y, t) = b(x, y) + h(x, y, t). \tag{4.3}$$

The function $b(x, y)$ defines the bed profile and for most problems of interest here it is prescribed and does not depend on time $t$; g is the acceleration due to gravity taken as $g = 9.81$ m/s², a constant. There are two distinct situations of practical interest, namely the *wet-bed case* in which the water depth $h$ is greater than zero and the *dry-bed case*, in which portions of the bed are dry, that is $h = 0$. $\mathbf{S}(\mathbf{Q})$ is a source term vector that accounts for various physical and geometric effects. As given, the source term is accounting for variable bathymetry. Note that $b(x, y)$ is a parameter of the problem, a prescribed function, not an unknown of the problem. For many practical applications there will be additional terms in the vector $\mathbf{S}(\mathbf{Q})$ to account for Coriolis forces, wind forces, bottom friction, etc.

In the next Section we study the eigenstructure of these equations, that is we find the eigenvalues and eigenvectors of the system. These concepts play a key role in understanding the mathematical character of the governing equations as well as in studying the physical character of shallow water free-surface waves.

## 4.2 Eigenstructure in Terms of Conserved Variables

We study the eigenstructure of the governing Eqs. (4.1) and (4.2) in terms of the *conserved variables*. To this end one only considers the *principal part of the PDEs*, that is the left-hand side in (4.1). By virtue of the chain rule, Eq. (4.1) may be written in *quasi-linear form* as

## 4.2 Eigenstructure in Terms of Conserved Variables

$$\partial_t \mathbf{Q} + \mathbf{A}(\mathbf{Q})\partial_x \mathbf{Q} + \mathbf{B}(\mathbf{Q})\partial_y \mathbf{Q} = \mathbf{0}, \tag{4.4}$$

where $\mathbf{A}(\mathbf{Q})$ and $\mathbf{B}(\mathbf{Q})$ are the matrices

$$\left.\begin{aligned}\mathbf{A}(\mathbf{Q}) = \partial \mathbf{F}/\partial \mathbf{Q} &= \begin{bmatrix} \partial f_1/\partial q_1 & \partial f_1/\partial q_2 & \partial f_1/\partial q_3 \\ \partial f_2/\partial q_1 & \partial f_2/\partial q_2 & \partial f_2/\partial q_3 \\ \partial f_3/\partial q_1 & \partial f_3/\partial q_2 & \partial f_3/\partial q_3 \end{bmatrix}, \\[4pt] \mathbf{B}(\mathbf{Q}) = \partial \mathbf{G}/\partial \mathbf{Q} &= \begin{bmatrix} \partial g_1/\partial q_1 & \partial g_1/\partial q_2 & \partial g_1/\partial q_3 \\ \partial g_2/\partial q_1 & \partial g_2/\partial q_2 & \partial g_2/\partial q_3 \\ \partial g_3/\partial q_1 & \partial g_3/\partial q_2 & \partial g_3/\partial q_3 \end{bmatrix},\end{aligned}\right\} \tag{4.5}$$

called the *Jacobian matrices* corresponding to the fluxes $\mathbf{F}(\mathbf{Q})$ and $\mathbf{G}(\mathbf{Q})$, respectively. The entries of these matrices are the partial derivatives of the components $f_i$ of $\mathbf{F}(\mathbf{Q})$ and $g_i$ of $\mathbf{G}(\mathbf{Q})$ with respect to the components $q_j$ of the vector of conserved variables $\mathbf{Q}$.

It is important to note that the arguments of the flux functions, actually their components, are the components of the vector of conserved variables. To make this clear when calculating the entries of the matrices in (4.5), we explicitly express the flux functions purely in terms of the components $q_1, q_2, q_3$ of $\mathbf{Q}$, namely

$$\left.\begin{aligned}\mathbf{Q} &= \begin{bmatrix} q_1 \\ q_2 \\ q_3 \end{bmatrix} = \begin{bmatrix} h \\ hu \\ hv \end{bmatrix}, \\[4pt] \mathbf{F}(\mathbf{Q}) &= \begin{bmatrix} f_1 \\ f_2 \\ f_3 \end{bmatrix} = \begin{bmatrix} hu \\ hu^2 + \tfrac{1}{2}gh^2 \\ huv \end{bmatrix} = \begin{bmatrix} q_2 \\ q_2^2/q_1 + \tfrac{1}{2}gq_1^2 \\ q_2 q_3/q_1 \end{bmatrix}, \\[4pt] \mathbf{G}(\mathbf{Q}) &= \begin{bmatrix} g_1 \\ g_2 \\ g_3 \end{bmatrix} = \begin{bmatrix} hv \\ hvu \\ hv^2 + \tfrac{1}{2}gh^2 \end{bmatrix} = \begin{bmatrix} q_3 \\ q_2 q_3/q_1 \\ q_3^2/q_1 + \tfrac{1}{2}gq_1^2 \end{bmatrix}.\end{aligned}\right\} \tag{4.6}$$

**Remark 4.1** (*Non-conservative formulation*) Note that although the variables in the quasi-linear form (4.4) are the conserved variables, the formulation of the equations is not conservative.

Next we calculate the entries of the Jacobian matrices and express them in terms of the non-conservative (or primitive) variables $u, v, a$.

**Proposition 4.2** *The Jacobian matrices $\mathbf{A}$ and $\mathbf{B}$ for the conservative formulation (4.4), (4.6) are given by*

$$\mathbf{A}(\mathbf{Q}) = \begin{bmatrix} 0 & 1 & 0 \\ a^2 - u^2 & 2u & 0 \\ -uv & v & u \end{bmatrix}, \quad \mathbf{B}(\mathbf{Q}) = \begin{bmatrix} 0 & 0 & 1 \\ -uv & v & u \\ a^2 - v^2 & 0 & 2v \end{bmatrix}. \quad (4.7)$$

*Proof* As pointed out earlier, it should be borne in mind that the components of the fluxes are to be thought of in terms of the components of the vector of conserved variables. Clearly, the first row of **A** is

$$\partial f_1/\partial q_1 = 0, \quad \partial f_1/\partial q_2 = 1, \quad \partial f_1/\partial q_3 = 0.$$

Also,

$$\partial f_2/\partial q_1 = -q_2^2/q_1^2 + gq_1 = a^2 - u^2,$$

where the last equality follows from the definitions for conserved variables and of *celerity*

$$a = \sqrt{gh}. \quad (4.8)$$

The calculation of the remaining entries of **A** follows in the same way and is left to the reader as a simple exercise. The entries of **B** are found in a similar manner. □

**Proposition 4.3** *The eigenvalues of* **A** *and* **B** *are respectively given by*

$$\lambda_1 = u - a, \quad \lambda_2 = u, \quad \lambda_3 = u + a, \quad (4.9)$$

*and*

$$\lambda_1 = v - a, \quad \lambda_2 = v, \quad \lambda_3 = v + a, \quad (4.10)$$

*where a is the celerity, as defined in Eq. (4.8).*

*Proof* By definition the eigenvalues of the matrix **A** are given by the roots of the *characteristic polynomial* equation

$$|\mathbf{A} - \lambda \mathbf{I}| = 0, \quad (4.11)$$

where $|\mathbf{X}|$ denotes the determinant of a matrix **X**, $\lambda$ is a scalar parameter and **I** is the identity matrix. Expanding (4.11) gives the cubic algebraic equation

$$(u - \lambda)\left[-\lambda(2u - \lambda) - (a^2 - u^2)\right] = 0,$$

the roots of which are easily found to be those given in (4.9). Note that the eigenvalues are all real; they are also distinct under all circumstances, except for the case of dry bed $h = 0$, in which case $a = 0$ and $\lambda_1 = \lambda_2 = \lambda_3 = u$. Similarly, the eigenvalues of **B** are found by solving the characteristic polynomial equation $|\mathbf{B} - \lambda \mathbf{I}| = 0$. The details are left to the reader. □

## 4.2 Eigenstructure in Terms of Conserved Variables

We next give expressions for the left and right eigenvectors of the Jacobian matrices.

**Proposition 4.4** *The* right eigenvectors *of* $\mathbf{A}$ *in (4.7) are given by*

$$\mathbf{R}^{(1)} = \alpha_1 \begin{bmatrix} 1 \\ u-a \\ v \end{bmatrix}, \quad \mathbf{R}^{(2)} = \alpha_2 \begin{bmatrix} 0 \\ 0 \\ 1 \end{bmatrix}, \quad \mathbf{R}^{(3)} = \alpha_3 \begin{bmatrix} 1 \\ u+a \\ v \end{bmatrix} \quad (4.12)$$

*and the* left *eigenvectors of* $\mathbf{A}$ *are given by*

$$\left. \begin{aligned} \mathbf{L}^{(1)} &= \hat{\alpha}_1 \begin{bmatrix} u+a, & -1, & 0 \end{bmatrix}, \\ \mathbf{L}^{(2)} &= \hat{\alpha}_2 \begin{bmatrix} -v, & 0, & 1 \end{bmatrix}, \\ \mathbf{L}^{(3)} &= \hat{\alpha}_3 \begin{bmatrix} u-a, & -1, & 0 \end{bmatrix}, \end{aligned} \right\} \quad (4.13)$$

*where the coefficients* $\alpha_1, \alpha_2, \alpha_3, \hat{\alpha}_1, \hat{\alpha}_2$ *and* $\hat{\alpha}_3$ *are* scaling factors.

**Proof** Recall that a *right eigenvector* of the matrix $\mathbf{A}$ corresponding to an eigenvalue $\lambda$ is a column vector $\mathbf{R} = [r_1, r_2, r_3]^{\mathrm{T}}$ such that

$$\mathbf{AR} = \lambda \mathbf{R}, \quad (4.14)$$

or in full

$$\left. \begin{aligned} r_2 &= \lambda r_1, \\ (a^2 - u^2) r_1 + 2u r_2 &= \lambda r_2, \\ -uv r_1 + v r_2 + u r_3 &= \lambda r_3. \end{aligned} \right\} \quad (4.15)$$

To find the eigenvector $\mathbf{R}^{(1)}$ corresponding to the eigenvalue $\lambda = \lambda_1 = u - a$ we substitute this into Eq. (4.15) to obtain

$$r_2 = (u-a) r_1, \quad (4.16)$$

$$(a^2 - u^2) r_1 + 2u r_2 = (u-a) r_2, \quad (4.17)$$

$$-uv r_1 + v r_2 + u r_3 = (u-a) r_3. \quad (4.18)$$

Note that Eqs. (4.16) and (4.17) are actually equivalent, and thus we only have two independent equations for the three unknowns $r_1, r_2, r_3$. We adopt Eqs. (4.16) and (4.18), which give a one-parameter family of solutions. Prescribing the parameter as $\alpha_1$ and setting $r_1 = \alpha_1$ it follows that

$$\mathbf{R}^{(1)} = \alpha_1 \begin{bmatrix} 1 \\ u-a \\ v \end{bmatrix},$$

as claimed. The calculations for $\mathbf{R}^{(2)}$ and $\mathbf{R}^{(3)}$ are left to the reader as an exercise. In order to compute a *left eigenvector* $\mathbf{L} = [l_1, l_2, l_3]$ corresponding to an eigenvalue $\lambda$, we solve the equations

$$\mathbf{LA} = \lambda \mathbf{L}. \tag{4.19}$$

The details to verify (4.13) are left to the reader as an exercise. □

**Proposition 4.5** *The* right eigenvectors *of the matrix* $\mathbf{B}$ *are*

$$\mathbf{R}^{(1)} = \beta_1 \begin{bmatrix} 1 \\ u \\ v-a \end{bmatrix}, \quad \mathbf{R}^{(2)} = \beta_2 \begin{bmatrix} 0 \\ 1 \\ 0 \end{bmatrix}, \quad \mathbf{R}^{(3)} = \beta_3 \begin{bmatrix} 1 \\ u \\ v+a \end{bmatrix} \tag{4.20}$$

*and the* left eigenvectors *of* $\mathbf{B}$ *are given by*

$$\left. \begin{aligned} \mathbf{L}^{(1)} &= \hat{\beta}_1 \left[ v+a,\, 0,\, -1 \right], \\ \mathbf{L}^{(2)} &= \hat{\beta}_2 \left[ -u,\, 1,\, 0 \right], \\ \mathbf{L}^{(3)} &= \hat{\beta}_3 \left[ v-a,\, 0,\, -1 \right], \end{aligned} \right\} \tag{4.21}$$

*where the coefficients* $\beta_1$, $\beta_2$, $\beta_3$, $\hat{\beta}_1$, $\hat{\beta}_2$ *and* $\hat{\beta}_3$ *are scaling factors.*

**Proof** Left to the reader as an exercise. □

**Exercise 4.6** (*Bi-orthonormality of left and right eigenvectors of* $\mathbf{A}$) The reader is encouraged to verify that the left and right eigenvectors (4.12) and (4.13) of the Jacobian matrix $\mathbf{A}$ are *bi-orthonormal*, that is they satisfy the relations

$$\mathbf{L}^{(i)} \cdot \mathbf{R}^{(j)} = \begin{cases} 1 & \text{if } i = j, \\ 0 & \text{otherwise.} \end{cases} \tag{4.22}$$

Note that for this to be satisfied the scaling factors must be chosen thus:

$$\hat{\alpha}_1 = \frac{1}{2a\alpha_1}, \quad \hat{\alpha}_2 = \frac{1}{\alpha_2}, \quad \hat{\alpha}_3 = -\frac{1}{2a\alpha_3}. \tag{4.23}$$

**Exercise 4.7** (*Bi-orthonormality of left and right eigenvectors of* $\mathbf{B}$) Verify that the left and right eigenvectors of $\mathbf{B}$ are bi-orthonormal and derive the corresponding relationships between the scaling factors $\beta_i$ and $\hat{\beta}_i$.

In the next Section we analyse the eigenstructure of the shallow water equations in terms of a different set of variables.

## 4.3 Eigenstructure in Terms of Primitive Variables

One may formulate the governing equations in terms of variables other than the conserved variables $q_1 = h$, $q_2 = hu$ and $q_3 = hv$. Alternative variables are $h$, $u$ and $v$, called *primitive variables* or *physical variables*. It can be easily verified that the conservative Eq. (4.1) with variable bed elevation may be expressed as

$$\partial_t h + u\partial_x h + h\partial_x u + v\partial_y h + h\partial_y v = 0, \tag{4.24}$$

$$\partial_t u + u\partial_x u + g\partial_x h + v\partial_y u = -g\partial_x b, \tag{4.25}$$

$$\partial_t v + u\partial_x v + v\partial_y v + g\partial_y h = -g\partial_y b. \tag{4.26}$$

Clearly the non-conservative form (4.24) of the mass equation is obtained from the first of the vector Eq. (4.1) by simply expanding $x$ and $y$ derivatives. The non-conservative form (4.25) of the $x$-momentum equation is obtained by expanding all partial derivatives in the $x$-momentum equation in (4.1), followed by use of (4.24) and algebraic manipulations. An analogous procedure is followed to obtain the non-conservative form (4.26) of the $y$-momentum equation. When written in quasi-linear form, Eqs. (4.24)–(4.26) become

$$\partial_t \mathbf{W} + \mathbf{A}(\mathbf{W})\partial_x \mathbf{W} + \mathbf{B}(\mathbf{W})\partial_y \mathbf{W} = \mathbf{S}, \tag{4.27}$$

with appropriate definitions for the vectors $\mathbf{W}$, $\mathbf{S}$ and the coefficient matrices $\mathbf{A}(\mathbf{W})$ and $\mathbf{B}(\mathbf{W})$, namely

$$\left.\begin{array}{c} \mathbf{W} = \begin{bmatrix} h \\ u \\ v \end{bmatrix}, \quad \mathbf{A}(\mathbf{W}) = \begin{bmatrix} u & h & 0 \\ g & u & 0 \\ 0 & 0 & u \end{bmatrix}, \\[1em] \mathbf{B}(\mathbf{W}) = \begin{bmatrix} v & 0 & h \\ 0 & v & 0 \\ g & 0 & v \end{bmatrix}, \quad \mathbf{S} = \begin{bmatrix} 0 \\ -g\partial_x b \\ -g\partial_y b \end{bmatrix}. \end{array}\right\} \tag{4.28}$$

**Proposition 4.8** *The eigenvalues of* $\mathbf{A}$ *and* $\mathbf{B}$ *in (4.28) are respectively given by*

$$\lambda_1 = u - a, \quad \lambda_2 = u, \quad \lambda_3 = u + a \tag{4.29}$$

*and*

$$\lambda_1 = v - a, \quad \lambda_2 = v, \quad \lambda_3 = v + a. \tag{4.30}$$

**Proof** The proof follows the same steps as for the conserved variable case studied in the previous Section and is left to the reader as an exercise. □

**Proposition 4.9** *The* right eigenvectors *of* **A** *in (4.28) are*

$$\mathbf{R}^{(1)} = \alpha_1 \begin{bmatrix} h \\ -a \\ 0 \end{bmatrix}, \quad \mathbf{R}^{(2)} = \alpha_2 \begin{bmatrix} 0 \\ 0 \\ 1 \end{bmatrix}, \quad \mathbf{R}^{(3)} = \alpha_3 \begin{bmatrix} h \\ a \\ 0 \end{bmatrix}, \qquad (4.31)$$

*where $\alpha_1$, $\alpha_2$ and $\alpha_3$ are scaling factors. The* left eigenvectors *of* **A** *in (4.28) are*

$$\left. \begin{aligned} \mathbf{L}^{(1)} &= \hat{\alpha}_1 \left[ a, -h, 0 \right], \\ \mathbf{L}^{(2)} &= \hat{\alpha}_2 \left[ 0, 0, 1 \right], \\ \mathbf{L}^{(3)} &= \hat{\alpha}_3 \left[ a, h, 0 \right], \end{aligned} \right\} \qquad (4.32)$$

*where $\hat{\alpha}_1$, $\hat{\alpha}_2$ and $\hat{\alpha}_3$ are scaling factors.*

**Proof** The proof is left to the reader as an exercise. See the previous Section. □

**Proposition 4.10** *The right eigenvectors of* **B** *in (4.28) are*

$$\mathbf{R}^{(1)} = \beta_1 \begin{bmatrix} h \\ 0 \\ -a \end{bmatrix}, \quad \mathbf{R}^{(2)} = \beta_2 \begin{bmatrix} 0 \\ 1 \\ 0 \end{bmatrix}, \quad \mathbf{R}^{(3)} = \beta_3 \begin{bmatrix} h \\ 0 \\ a \end{bmatrix}, \qquad (4.33)$$

*where $\beta_1$, $\beta_2$ and $\beta_3$ are scaling factors. The* left eigenvectors *of* **B** *in (4.28) are*

$$\left. \begin{aligned} \mathbf{L}^{(1)} &= \hat{\beta}_1 \left[ a, 0, -h \right], \\ \mathbf{L}^{(2)} &= \hat{\beta}_2 \left[ 0, 1, 0 \right], \\ \mathbf{L}^{(3)} &= \hat{\beta}_3 \left[ a, 0, h \right], \end{aligned} \right\} \qquad (4.34)$$

*where $\hat{\beta}_1$, $\hat{\beta}_2$ and $\hat{\beta}_3$ are scaling factors.*

**Proof** The proof is left to the reader as an exercise. □

**Exercise 4.11** Verify the bi-orthonormality property for the left and right eigenvectors of **A**, and for the left and right eigenvectors of **B**. Also, find the relationships between the scaling factors for the property to be satisfied.

## 4.4 Hyperbolic Character of the 2D Equations

Consider the two-dimensional (2D) equations in conserved variables but written in quasi-linear form (4.4) with coefficient matrices **A** and **B** given by (4.7). Further, consider a matrix **C** that is a *linear combination* of **A** and **B**, namely

$$\mathbf{C} = \omega_1 \mathbf{A} + \omega_2 \mathbf{B}, \qquad (4.35)$$

## 4.4 Hyperbolic Character of the 2D Equations

where the coefficients $\omega_1$ and $\omega_2$ are two real parameters that define a non-zero vector $\boldsymbol{\omega} = [\omega_1, \omega_2]$, such that

$$|\boldsymbol{\omega}| = \sqrt{\omega_1^2 + \omega_2^2} > 0. \tag{4.36}$$

Then the matrix **C** is given by

$$\mathbf{C}(\mathbf{Q}) = \begin{bmatrix} 0 & \omega_1 & \omega_2 \\ (a^2 - u^2)\omega_1 - uv\omega_2 & 2u\omega_1 + v\omega_2 & u\omega_2 \\ -uv\omega_1 + (a^2 - v^2)\omega_2 & v\omega_1 & u\omega_1 + 2v\omega_2 \end{bmatrix}. \tag{4.37}$$

**Proposition 4.12** *The eigenvalues of* **C** *are given by*

$$\lambda_1 = u\omega_1 + v\omega_2 - a|\boldsymbol{\omega}|, \quad \lambda_2 = u\omega_1 + v\omega_2, \quad \lambda_3 = u\omega_1 + v\omega_2 + a|\boldsymbol{\omega}|. \tag{4.38}$$

*The corresponding* right eigenvectors *of* **C** *are*

$$\mathbf{R}^{(1)} = \begin{bmatrix} 1 \\ u - \dfrac{a\omega_1}{|\boldsymbol{\omega}|} \\ v - \dfrac{a\omega_2}{|\boldsymbol{\omega}|} \end{bmatrix}, \quad \mathbf{R}^{(2)} = \begin{bmatrix} 0 \\ \omega_2 \\ -\omega_1 \end{bmatrix}, \quad \mathbf{R}^{(3)} = \begin{bmatrix} 1 \\ u + \dfrac{a\omega_1}{|\boldsymbol{\omega}|} \\ v + \dfrac{a\omega_2}{|\boldsymbol{\omega}|} \end{bmatrix} \tag{4.39}$$

*and the* left eigenvectors *of* **C** *are given by*

$$\left. \begin{aligned} \mathbf{L}^{(1)} &= \left[ \dfrac{u\omega_1 + v\omega_2}{2a|\boldsymbol{\omega}|} + \dfrac{1}{2}, -\dfrac{\omega_1}{2a|\boldsymbol{\omega}|}, -\dfrac{\omega_2}{2a|\boldsymbol{\omega}|} \right], \\ \mathbf{L}^{(2)} &= \left[ -\dfrac{(u\omega_2 - v\omega_1)}{|\boldsymbol{\omega}|^2}, -\dfrac{\omega_2}{|\boldsymbol{\omega}|^2}, -\dfrac{\omega_1}{|\boldsymbol{\omega}|^2} \right], \\ \mathbf{L}^{(3)} &= \left[ -\dfrac{u\omega_1 + v\omega_2}{2a|\boldsymbol{\omega}|} + \dfrac{1}{2}, \dfrac{\omega_1}{2a|\boldsymbol{\omega}|}, \dfrac{\omega_2}{2a|\boldsymbol{\omega}|} \right]. \end{aligned} \right\} \tag{4.40}$$

*Proof* Left to the reader as an exercise. □

We next recall the definition of *hyperbolicity* and show that the time-dependent two-dimensional shallow water equations

$$\partial_t \mathbf{Q} + \partial_x \mathbf{F}(\mathbf{Q}) + \partial_y \mathbf{G}(\mathbf{Q}) = \mathbf{0} \tag{4.41}$$

are hyperbolic.

**Definition 4.13** (*Hyperbolicity*) A system of $N$ *conservation laws* (4.41) with Jacobian matrices $\mathbf{A}(\mathbf{Q})$ and $\mathbf{B}(\mathbf{Q})$ is said to be *hyperbolic* if the matrix $\mathbf{C}$ formed by the linear combination of the Jacobian matrices $\mathbf{A}(\mathbf{Q})$ and $\mathbf{B}(\mathbf{Q})$,

$$\mathbf{C} = \omega_1 \mathbf{A} + \omega_2 \mathbf{B}, \tag{4.42}$$

has $N$ real eigenvalues for any vector $\mathbf{Q}$ of conserved variables and any vector $\boldsymbol{\omega} = [\omega_1, \omega_2]$, such that $\boldsymbol{\omega} \neq \mathbf{0}$. The system is called *strictly hyperbolic* if in addition the eigenvalues are all distinct. The equations are *weakly hyperbolic* if they are *non-strictly hyperbolic* (real eigenvalues but not distinct) and there is no complete set of $N$ corresponding linear independent eigenvectors. The equations are *elliptic* if the eigenvalues are complex.

**Remark 4.14** The above definition extends to systems of conservation laws in any number of space dimensions and for any number $N$ of equations. Also, the time variable may be a *time like variable*. For instance, the two-dimensional steady supercritical shallow water equations are also hyperbolic, where the time-like variable is the direction of flow, and not time. See Sect. 4.7.

**Corollary 4.15** *The time-dependent two-dimensional shallow water Eqs. (4.1) and (4.2) are hyperbolic. For a wet bed ($h > 0$) they are strictly hyperbolic.*

**Proof** This follows immediately from the fact that the eigenvalues (4.38) are all real. They are also distinct whenever $a \neq 0$, that is whenever $h > 0$. □

By *eigenstructure* of a hyperbolic system it is meant the eigenvalues and the corresponding eigenvectors. Eigenvalues are associated with wave speeds and *characteristic fields*. In the next Section we study the nature of the characteristic fields in the shallow water equations.

## 4.5 Nature of Characteristic Fields

Consider a one-dimensional (1D) hyperbolic system of $N$ conservation laws

$$\partial_t \mathbf{Q} + \partial_x \mathbf{F}(\mathbf{Q}) = \mathbf{0}, \tag{4.43}$$

with real eigenvalues $\lambda_i(\mathbf{Q})$ and corresponding right eigenvectors $\mathbf{R}^{(i)}(\mathbf{Q})$. The *characteristic speed* $\lambda_i(\mathbf{Q})$ defines a *characteristic field*, the $\lambda_i$-field; we also speak of the $\mathbf{R}^{(i)}$-field or simply the $i$-field. Before introducing some relevant definitions, we recall that the *gradient* of an eigenvalue $\lambda_i(\mathbf{Q})$ is given by

$$\nabla \lambda_i(\mathbf{Q}) = \left[ \frac{\partial}{\partial q_1} \lambda_i, \frac{\partial}{\partial q_2} \lambda_i, \ldots, \frac{\partial}{\partial q_N} \lambda_i \right]^{\mathrm{T}}. \tag{4.44}$$

## 4.5 Nature of Characteristic Fields

**Definition 4.16** (*Linearly degenerate fields*) A $\lambda_i$-characteristic field is said to be *linearly degenerate* if

$$\nabla \lambda_i(\mathbf{Q}) \cdot \mathbf{R}^{(i)}(\mathbf{Q}) = 0, \forall \mathbf{Q} \in \Re^N, \quad (4.45)$$

where $\Re^N$ is the set of real-valued vectors of $N$ components, called *phase space*.

Note that for a linear system with constant coefficients the eigenvalues $\lambda_i$ are constant and therefore $\nabla \lambda_i(\mathbf{Q}) = \mathbf{0}$. Hence all characteristic fields of a linear hyperbolic system with constant coefficients are linearly degenerate. See Chap. 2.

**Definition 4.17** (*Phase space*) The *phase space* is the space of vectors $\mathbf{Q} = [q_1, \ldots, q_N]^T$; for a $2 \times 2$ system we speak of the *phase plane* $q_1$–$q_2$.

**Definition 4.18** (*Genuinely non-linear fields*) A $\lambda_i$-characteristic field is said to be *genuinely non-linear* if

$$\nabla \lambda_i(\mathbf{Q}) \cdot \mathbf{R}^{(i)}(\mathbf{Q}) \neq 0, \ \forall \mathbf{Q} \in \Re^N. \quad (4.46)$$

We now apply these definitions to the analysis of the characteristic fields of the $x$-split two-dimensional shallow water equations.

**Proposition 4.19** (Nature of the characteristic fields) *For the $x$-split two-dimensional shallow water equations*

$$\partial_t \mathbf{Q} + \partial_x \mathbf{F}(\mathbf{Q}) = \mathbf{0}, \quad (4.47)$$

*with* $\mathbf{Q}$ *and* $\mathbf{F}(\mathbf{Q})$ *as defined in (4.2), the $\lambda_2(\mathbf{Q})$-characteristic field is* linearly degenerate *and the $\lambda_1(\mathbf{Q})$ and $\lambda_3(\mathbf{Q})$-characteristic fields are* genuinely non-linear.

*Proof* We first show that the $\lambda_2(\mathbf{Q})$-characteristic field is linearly degenerate, that is

$$\nabla \lambda_2(\mathbf{Q}) \cdot \mathbf{R}^{(2)}(\mathbf{Q}) = 0$$

for all vectors $\mathbf{Q}$. First we express the eigenvalue $\lambda_2(\mathbf{Q})$ in terms of the conserved variables $q_1, q_2, q_3$, as defined in (4.6). We have

$$\lambda_2 = u = \frac{hu}{h} = \frac{q_2}{q_1}$$

and thus

$$\nabla \lambda_2 = \left[ \frac{\partial}{\partial q_1} \lambda_2, \ \frac{\partial}{\partial q_2} \lambda_2, \ \frac{\partial}{\partial q_3} \lambda_2 \right]^T = \left[ -\frac{u}{h}, \ \frac{1}{h}, \ 0 \right]^T.$$

As $\mathbf{R}^{(2)}(\mathbf{Q}) = \beta_2 [0, \ 0, \ 1]^T$, see (4.12), clearly the dot product $\nabla \lambda_2(\mathbf{Q}) \cdot \mathbf{R}^{(2)}(\mathbf{Q})$ vanishes for all vectors $\mathbf{Q}$; hence the $\lambda_2(\mathbf{Q})$-characteristic field is linearly degenerate as claimed and the first part of the proposition is thus proved. For the other two characteristic fields, simple calculations give

$$\nabla \lambda_1 \cdot \mathbf{R}^{(1)}(\mathbf{Q}) = -\frac{3}{2a} \neq 0 \text{ and } \nabla \lambda_3 \cdot \mathbf{R}^{(3)} = \frac{3}{2a} \neq 0. \quad (4.48)$$

Therefore the $\lambda_1(\mathbf{Q})$ and $\lambda_3(\mathbf{Q})$ characteristic fields are genuinely non-linear and the proof of the proposition is complete. □

## 4.6 Integral Forms and Rotational Invariance

Consider the homogeneous (no source terms) time-dependent two-dimensional shallow water equations

$$\partial_t \mathbf{Q} + \partial_x \mathbf{F}(\mathbf{Q}) + \partial_y \mathbf{G}(\mathbf{Q}) = \mathbf{0}, \quad (4.49)$$

with $\mathbf{Q}$ and $\mathbf{F}(\mathbf{Q})$ as defined in (4.1) and (4.2). A corresponding integral form of these conservation laws is

$$\frac{\partial}{\partial t} \int_V \mathbf{Q} \, dV + \int_\Omega [\cos\theta \mathbf{F}(\mathbf{Q}) + \sin\theta \mathbf{G}(\mathbf{Q})] \, d\Omega = \mathbf{0}, \quad (4.50)$$

where $V$ is any control volume in the $x$–$y$ plane, $\Omega$ is its boundary and

$$\mathbf{n} \equiv [n_1, n_2] \equiv [\cos\theta, \sin\theta] \quad (4.51)$$

is the *outward unit normal vector* to the boundary $\Omega$. See Chap. 2. Figure 4.1 depicts the situation. The outward unit normal vector $\mathbf{n}$ is expressed in terms of the angle $\theta$ between the positive $x$-direction (reference direction) and the vector $\mathbf{n}$. We next prove an important property, called the *rotational invariance* of the 2D shallow water equations. We remark that the property allows the proof of *hyperbolicity in time* for the two-dimensional Eqs. (4.1) and (4.2). This is omitted here; for details see [2].

The rotational invariance property can also be used for computational purposes when dealing with domains whose boundaries are not aligned with the Cartesian coordinate directions. The finite volume method then allows direct discretisation of the spatial domain in physical space, without having to explicitly transform coordi-

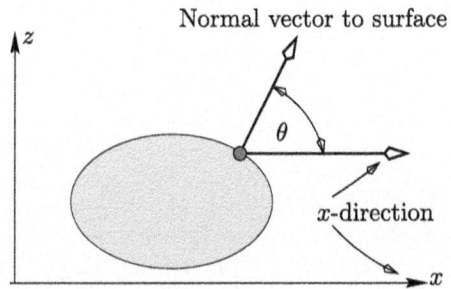

**Fig. 4.1** Control volume in $x$–$y$ plane; $\theta$ is angle between the outward unit vector $\mathbf{n}$ normal to the surface $\Omega$ of the control volume and the $x$ direction (the reference direction)

## 4.6 Integral Forms and Rotational Invariance

nates, allowing for both structured and unstructured meshes to be used. See Chaps. 3 and 16 of the textbook by Toro [2, 12, 13]. See also Chaps. 9 and 13 of this book.

We now define the *normal flux component* through a surface $\Omega$ as

$$\mathbf{H} \equiv \mathbf{n} \cdot [\mathbf{F(Q)}, \mathbf{G(Q)}] = \cos\theta \mathbf{F(Q)} + \sin\theta \mathbf{G(Q)}, \quad (4.52)$$

and then show that the following property is valid.

**Proposition 4.20** (Rotational invariance) *The two-dimensional homogeneous shallow water Eqs. (4.1) and (4.2) satisfy*

$$\mathbf{H} \equiv \mathbf{n} \cdot [\mathbf{F(Q)}, \mathbf{G(Q)}] = \mathbf{T}^{-1}\mathbf{F}(\mathbf{T(Q)}), \quad (4.53)$$

*for all vectors* $\mathbf{Q}$ *and for all real angles* $\theta$, *or equivalently, normal directions of the surface* $\Omega$. *Here* $\mathbf{T} = \mathbf{T}(\theta)$ *is a rotation matrix and* $\mathbf{T}^{-1}(\theta)$ *is its inverse, given respectively as*

$$\mathbf{T} = \begin{bmatrix} 1 & 0 & 0 \\ 0 & \cos\theta & \sin\theta \\ 0 & -\sin\theta & \cos\theta \end{bmatrix}, \quad \mathbf{T}^{-1} = \begin{bmatrix} 1 & 0 & 0 \\ 0 & \cos\theta & -\sin\theta \\ 0 & \sin\theta & \cos\theta \end{bmatrix}. \quad (4.54)$$

***Proof*** First we calculate $\mathbf{T(Q)}$. The result is

$$\mathbf{T(Q)} = \begin{bmatrix} h, h\hat{u}, h\hat{v} \end{bmatrix}^T,$$

with velocity components $\hat{u}$ and $\hat{v}$ given by

$$\hat{u} = u\cos\theta + v\sin\theta, \quad \hat{v} = -u\sin\theta + v\cos\theta.$$

Next we compute $\mathbf{F(T(Q))}$ and obtain

$$\mathbf{F(T(Q))} = \begin{bmatrix} h\hat{u}, h\hat{u}^2 + \tfrac{1}{2}gh^2, h\hat{u}\hat{v} \end{bmatrix}^T.$$

Now we apply the inverse rotation $\mathbf{T}^{-1}$ to $\mathbf{F(T(Q))}$:

$$\mathbf{T}^{-1}(\mathbf{F(T(Q))}) = \begin{bmatrix} h\hat{u} \\ \cos\theta\left(h\hat{u}^2 + \tfrac{1}{2}gh^2\right) - \sin\theta\left(h\hat{u}\hat{v}\right) \\ \sin\theta\left(h\hat{u}^2 + \tfrac{1}{2}gh^2\right) + \cos\theta\left(h\hat{u}\hat{v}\right) \end{bmatrix}.$$

The vector equality

$$\mathbf{T}^{-1}(\mathbf{F(T(Q))}) = \cos\theta\mathbf{F} + \sin\theta\mathbf{G}$$

is easily verified. For the first component the identity is immediate. Elementary algebraic manipulations involving trigonometric identities verify the result for the second and third components and the proof of the proposition is thus complete. □

In the next Section we study the two-dimensional steady shallow water equations and show that these are hyperbolic if the flow is supercritical.

## 4.7 Steady Supercritical Flow

We study the eigenstructure of the two-dimensional steady supercritical shallow water equations. We begin with the primitive variable formulation of Sect. 4.3, namely

$$\partial_t \mathbf{W} + \mathbf{A}(\mathbf{W})\partial_x \mathbf{W} + \mathbf{B}(\mathbf{W})\partial_y \mathbf{W} = \mathbf{0}, \tag{4.55}$$

with

$$\mathbf{W} = \begin{bmatrix} h \\ u \\ v \end{bmatrix}, \quad \mathbf{A}(\mathbf{W}) = \begin{bmatrix} u & h & 0 \\ g & u & 0 \\ 0 & 0 & u \end{bmatrix}, \quad \mathbf{B}(\mathbf{W}) = \begin{bmatrix} v & 0 & h \\ 0 & v & 0 \\ g & 0 & v \end{bmatrix}. \tag{4.56}$$

The steady (or time-independent) equations result from (4.55) by neglecting the time derivative, namely

$$\mathbf{A}(\mathbf{W})\mathbf{W}_x + \mathbf{B}(\mathbf{W})\mathbf{W}_y = \mathbf{0}. \tag{4.57}$$

By premultiplying (4.57) from the left by the inverse $\mathbf{A}^{-1}$ of matrix $\mathbf{A}$ we obtain

$$\mathbf{W}_x + \mathbf{C}(\mathbf{W})\mathbf{W}_y = \mathbf{0}, \tag{4.58}$$

where

$$\mathbf{C} = \mathbf{A}^{-1}\mathbf{B} = \frac{1}{u(u^2 - a^2)} \begin{bmatrix} u^2v & -huv & u^2h \\ -guv & u^2v & -a^2u \\ g(u^2 - a^2) & 0 & v(u^2 - a^2) \end{bmatrix}. \tag{4.59}$$

We now define the Froude number $\mathcal{F}$ as

$$\mathcal{F} = \sqrt{\frac{u^2 + v^2}{a^2}}. \tag{4.60}$$

**Proposition 4.21** *The eigenvalues of* $\mathbf{C}$ *are given by*

$$\lambda_1 = \frac{uv + a^2\sqrt{\mathcal{F}^2 - 1}}{u^2 - a^2}, \quad \lambda_2 = \frac{v}{u}, \quad \lambda_3 = \frac{uv - a^2\sqrt{\mathcal{F}^2 - 1}}{u^2 - a^2}. \tag{4.61}$$

## 4.7 Steady Supercritical Flow

***Proof*** The proof follows the same steps as for the time-dependent case, and is left to the reader as an exercise. □

**Remark 4.22** (*Steady supercritical flow*) The steady, two-dimensional shallow water equations are *hyperbolic* if the flow is *supercritical*, that is if

$$\mathcal{F} = \sqrt{\frac{u^2 + v^2}{a^2}} > 1, \qquad (4.62)$$

in which case the eigenvalues (4.61) are all real. Otherwise, the equations are *mixed elliptic-hyperbolic*.

We next state a proposition concerning the right and left eigenvectors of **C**.

**Proposition 4.23** *The left eigenvectors of* **C** *in (4.58) corresponding to the eigenvalues (4.61) are*

$$\left.\begin{array}{l} \mathbf{L}^{(1)} = \hat{\alpha}_1 \left[ a^2 \sqrt{\mathcal{F}^2 - 1}, \; hv, \; -hu \right], \\[6pt] \mathbf{L}^{(2)} = \hat{\alpha}_2 \left[ a^2, \; hu, \; hv \right], \\[6pt] \mathbf{L}^{(3)} = \hat{\alpha}_3 \left[ -a^2 \sqrt{\mathcal{F}^2 - 1}, \; hv, \; -hu \right], \end{array}\right\} \qquad (4.63)$$

*where* $\hat{\alpha}_1$, $\hat{\alpha}_2$ *and* $\hat{\alpha}_3$ *are scaling factors.*

*The right eigenvectors of* **C** *in (4.58) corresponding to the eigenvalues (4.61) are*

$$\left.\begin{array}{l} \mathbf{R}^{(1)} = \alpha_1 \begin{bmatrix} -h(u\sqrt{\mathcal{F}^2 - 1} - v) \\ a^2\sqrt{\mathcal{F}^2 - 1} - uv \\ u^2 - a^2 \end{bmatrix}, \\[20pt] \mathbf{R}^{(2)} = \alpha_2 \begin{bmatrix} 0 \\ u \\ v \end{bmatrix}, \\[20pt] \mathbf{R}^{(3)} = \alpha_3 \begin{bmatrix} h(u\sqrt{\mathcal{F}^2 - 1} + v) \\ -a^2\sqrt{\mathcal{F}^2 - 1} - uv \\ u^2 - a^2 \end{bmatrix}, \end{array}\right\} \qquad (4.64)$$

*where* $\alpha_1$, $\alpha_2$ *and* $\alpha_3$ *are scaling factors.*

***Proof*** The proof is left to the reader as an exercise. □

As the steady supercritical equations are hyperbolic in the flow direction $x$, one can use *space-marching* methods to solve the steady Eqs. (4.57) and (4.58), just as done for the time-dependent equations using *time-marching* methods. As the name suggests, in the space-marching approach one advances the solution in the direction $s$ ($x$ here), in which the relevant system is hyperbolic. For an example of a space-marching scheme, as applied to the steady 2D supersonic Euler equations, used in conjunction with a Godunov type method see for instance Toro and Chakraborty [14].

## 4.8 Concluding Remarks

We have studied salient mathematical properties of the two-dimensional non-linear shallow water equations including, hyperbolicity, the nature of characteristic fields, the rotational invariance property of the two-dimensional equations and the two-dimensional steady supercritical shallow water equations. Prerequisite background is found in Chaps. 1 and 2. The contents of this Chapter constitute essential background material for Chap. 5 on elementary waves and Chaps. 6 and 7 on the exact solution of the Riemann problem. A list of suggested exercises is given at the end of the Chapter.

## 4.9 Suggested Exercises

This Section gives a few problems for the reader to practice in applying basic mathematical concepts to the analysis of hyperbolic conservation laws. As a matter of fact, the solutions to these problems might be relevant to some applications.

**4.1 Homogeneity Property.** This exercise concerns the so-called homogeneity property, as defined below. Consider the $x$-split shallow water equations

$$\partial_t \mathbf{Q} + \partial_x \mathbf{F}(\mathbf{Q}) = \mathbf{0}, \tag{4.65}$$

where $\mathbf{Q}$ and $\mathbf{F}(\mathbf{Q})$ are the vectors of conserved variables and fluxes, given respectively by

$$\mathbf{Q} = \begin{bmatrix} h \\ hu \\ hv \end{bmatrix}, \ \mathbf{F}(\mathbf{Q}) = \begin{bmatrix} hu \\ hu^2 + \tfrac{1}{2}gh^2 \\ huv \end{bmatrix}. \tag{4.66}$$

A system such as (4.65) and (4.66) is said to satisfy the *homogeneity property* if the following identity holds:

$$\mathbf{F}(\mathbf{Q}) = \mathbf{A}(\mathbf{Q})\mathbf{Q}. \tag{4.67}$$

## 4.9 Suggested Exercises

Note that this property is certainly satisfied for linear systems with constant coefficients, that is with Jacobian matrix **A** constant. It is also well-known that the time-dependent Euler equations of gas dynamics do satisfy this property for certain equations of state; see [2] for details. We note that the satisfaction of the *homogeneity property* is useful for constructing certain numerical methods of the flux-vector splitting type.

1. Show that the shallow water Eqs. (4.65) and (4.66) *do not* satisfy the homogeneity property, that is
$$\mathbf{F}(\mathbf{Q}) \neq \mathbf{A}(\mathbf{Q})\mathbf{Q}.$$

2. Look for an approximation $\hat{\mathbf{A}}(\mathbf{Q}_L, \mathbf{Q}_R)$ to the Jacobian matrix **A** for the shallow water equations, given in terms of the data $\mathbf{Q}_L$ and $\mathbf{Q}_R$ from a Riemann problem, such that the homogeneity property is satisfied.

**4.2 Flux Evolution Equation.** An interesting property of conservation laws concerns an evolution equation for the flux [15].

1. Verify that system (4.65) and (4.66) satisfies the following evolution equation
$$\partial_t \mathbf{F}(\mathbf{Q}) + \mathbf{A}(\mathbf{Q})\partial_x \mathbf{F} = \mathbf{0}. \tag{4.68}$$

2. Show that (4.68) is valid for any hyperbolic system (4.65). State clearly the assumptions made for the proof to be valid.

**4.3 Saint Venant Equations.** The homogeneous (no sources) Saint Venant equations are
$$\partial_t A + \partial_x (Au) = 0 \tag{4.69}$$

and
$$\partial_t (Au) + \partial_x (Au^2) + gA\partial_x h = 0. \tag{4.70}$$

See Chap. 1 for details.

1. Express these equations in quasi-linear form
$$\partial_t \mathbf{W} + \mathbf{A}\partial_x \mathbf{W} = \mathbf{0}$$
   in terms of the variables $A$ and $u$.
2. Find the eigenvalues of the coefficient matrix **A**.
3. Find the corresponding right eigenvectors of **A**.
4. Find the corresponding left eigenvectors of **A**.
5. Show that the left and right eigenvectors of **A** are *bi-orthonormal*.
6. Determine the nature of the characteristic fields corresponding to the eigenvalues.

**4.4 The Two-Dimensional Steady Supercritical Equations.** Recall that the steady (or time-independent) shallow water equations are

$$\mathbf{A}(\mathbf{Q})\partial_x\mathbf{Q} + \mathbf{B}(\mathbf{Q})\partial_y\mathbf{Q} = \mathbf{0}, \tag{4.71}$$

where $\mathbf{Q} = [h, hu, hv]^T$ is the vector of conserved variables and

$$\mathbf{A}(\mathbf{Q}) = \begin{bmatrix} 0 & 1 & 0 \\ a^2 - u^2 & 2u & 0 \\ -uv & v & u \end{bmatrix}, \quad \mathbf{B}(\mathbf{Q}) = \begin{bmatrix} 0 & 0 & 1 \\ -uv & v & u \\ a^2 - v^2 & 0 & 2v \end{bmatrix} \tag{4.72}$$

are the Jacobian matrices.

1. Express (4.71) and (4.72) as

$$\partial_x\mathbf{Q} + \mathbf{C}(\mathbf{Q})\partial_y\mathbf{Q} = \mathbf{0},$$

where $\mathbf{C}(\mathbf{Q}) = \mathbf{A}^{-1}(\mathbf{Q})\mathbf{B}(\mathbf{Q})$; see (4.58) and (4.59).
2. Find the eigenvalues of the coefficient matrix $\mathbf{C}$.
3. Find the corresponding right eigenvectors of $\mathbf{C}$.
4. Find the corresponding left eigenvectors of $\mathbf{C}$.
5. Show that the left and right eigenvectors of $\mathbf{C}$ are *bi-orthonormal*.
6. Determine the nature of the characteristic fields corresponding to the eigenvalues.

# References

1. E.C. Zachmanoglou, D.W. Thoe, *Introduction to Partial Differential Equations* (Dover Publications, Inc., New York, 1986)
2. E.F. Toro, *Riemann Solvers and Numerical Methods for Fluid Dynamics. A Practical Introduction*, 3rd edn. (Springer-Verlag, 2009)
3. A. Jeffrey, *Quasilinear Hyperbolic Systems and Waves* (Pitman, 1976)
4. J.D. Logan, *An Introduction to Nonlinear Partial Differential Equations* (John Wiley and Sons, 1994)
5. A.J. Chorin, J.E. Marsden, *A Mathematical Introduction to Fluid Mechanics* (Springer–Verlag, 1993)
6. J. Smoller, *Shock Waves and Reaction–Diffusion Equations* (Springer–Verlag, 1994)
7. R.J. LeVeque, *Finite Volume Methods for Hyperbolic Problems* (Cambridge University Press, 2002)
8. E. Godlewski, P.A. Raviart, *Numerical Approximation of Hyperbolic Systems of Conservation Laws*, 2nd edn. (Springer, 2021)
9. L. Hörmander, *Lectures on Nonlinear Hyperbolic Differential Equations*. Mathématiques et Applications, vol. 26 (Springer–Verlag, 1997)
10. A. Tveito, R. Winther, *Introduction to Partial Differential Equations* (Springer–Verlag, 1998)
11. C.M. Dafermos, *Hyperbolic Conservation Laws in Continuum Physics*. Grundlehren der Mathematischen Wissenschaften, vol. 325 (Springer, 2000)

12. E.F. Toro, *Riemann Solvers and Numerical Methods for Fluid Dynamics* (Springer–Verlag, 1997)
13. E.F. Toro, *Riemann Solvers and Numerical Methods for Fluid Dynamics*, 2nd edn. (Springer–Verlag, 1999)
14. E.F. Toro, A. Chakraborty, Development of an approximate Riemann solver for the steady supersonic Euler equations. Aeronaut. J. **98**, 325–339 (1994)
15. J.M. Ghidaglia, G. LeCoq, I. Toumi, Two flux schemes for computing two phases flows throught multidimensional finite volume methods, in *Proceedings of the NURETH–9 Conference* (American Nuclear Society, 1999)

# Chapter 5
# Elementary Waves in Shallow Water

**Abstract** This Chapter is devoted to the study of elementary waves emerging from the solution of the Riemann problem for the augmented one-dimensional and for the split two-dimensional shallow water equations. The dam-break problem is introduced as a physical, motivating example of a special case of a Riemann problem. Four possible wave patters in the solution of the Riemann problem are identified, each comprising rarefactions, shocks and contact discontinuities or shear waves. For each wave type, mathematical relations across the wave structure are established. These include generalised Riemann invariants and Rankine-Hugoniot jump conditions. Useful shock relations are established for left and right shocks, connecting shock speed with shock Froude number. These relations can be readily used to setup shock test problems with exact solutions, to test numerical methods. Finally we study shock waves from conservative formulations of the shallow water equations in terms of physical variables, rather than conserved variables. It is found that such shocks are slower than those from the conserved formulation using the conserved variables. The Chapter is concluded with a list of suggested exercises. Useful background is found in Chaps. 1, 2 and 4. The contents of this Chapter will be used to solve the complete Riemann problem exactly in Chaps. 6 and 7.

## 5.1 Elementary Waves in the Riemann Problem

The solution of the Riemann problem for the shallow water equations generates a wave system comprising a combination of waves, including rarefaction waves and shock waves. There may also be contact discontinuities or shear waves, depending on whether the shallow water equations are suitably augmented, as done here in fact. To facilitate the study of the Riemann problem, here we consider the simpler case in which the Riemann problem contains just one isolated wave. The full problem is studied in Chaps. 6 and 7.

## 5.1.1 Motivation: The Dam-Break Problem

A problem of considerable physical interest is the so-called dam-break problem. Assume a horizontal channel of uniform, rectangular cross-section. Suppose further that the channel has two uniform water levels, both at rest, separated by a wall at a position $x = 0$, as depicted in Fig. 5.1.

If the wall separating the two uniform levels of water collapses, two dominant features emerge from the process in the form of waves. A right-facing wave travels into the shallow portion of the fluid, raising the depth abruptly. The left-facing wave travels into the deep water region and has the effect of reducing the free-surface height. The details of the physical processes occurring in the vicinity of the wall shortly after the collapse of the wall are indeed very complex and are not correctly modelled (not even remotely) by the shallow water equations.

If the wall collapses in a sufficiently short time, the wave pattern emerging is almost that of a *centered wave system* with a left *rarefaction wave* and a right-facing *shock wave*, as depicted in Fig. 5.2d. Such wave systems may be approximated by the shallow water equations. In the language of free-surface water flows, a rarefaction is also called a *depression* and a shock is also called a *bore*. If one assumes that the wave phenomenon is correctly governed by the shallow water equations and that the wall vanishes instantaneously at time $t = 0$, then the situation may be described quite accurately by Fig. 5.2. The initial conditions for the water depth at time $t = 0$ are represented by Fig. 5.2a, where the wall is replaced by a discontinuity in water depth at the position $x = 0$. Figure 5.2b shows the water depth profile at a later time $t_*$ after the wall collapses and Fig. 5.2c shows the corresponding particle velocity distribution at the same time. Figure 5.2d summarises the wave process as a function of space and time. The right wave is a shock, a discontinuous wave, and the left wave is a rarefaction, a smooth wave.

In the next Section we study elementary waves arising from the *Riemann problem*, a generalization of the dam-break problem just discussed.

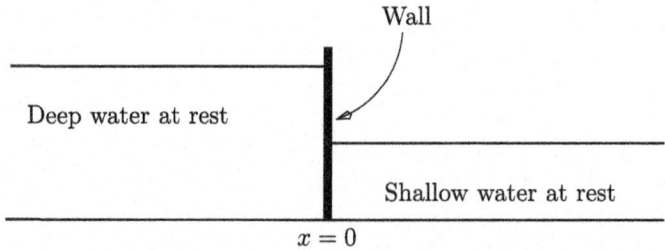

**Fig. 5.1** Side view of the initial conditions for a dam-break problem. A wall at $x = 0$ separates two uniform levels of water at rest. The collapse of the dam results in a wave system, as illustrated in Fig. 5.2

**Fig. 5.2** Time evolution of the dam-break problem for the case in which the water depth on the left is larger than that on the right: **a** initial depth, **b** depth at a later time and **c** corresponding velocity distribution; **d** wave diagram of the full process in the $x$-$t$ half plane

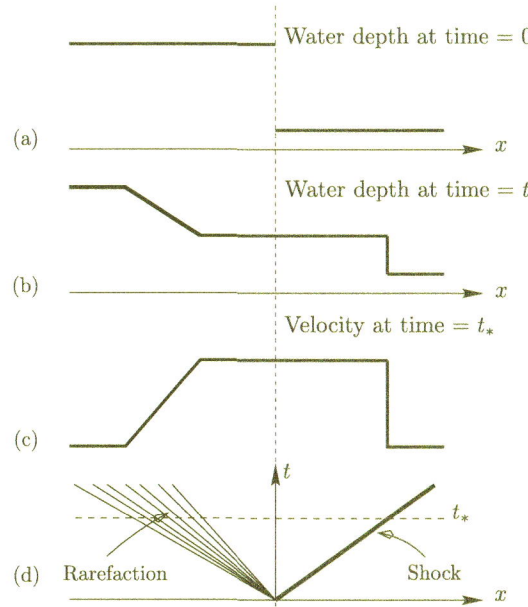

## 5.1.2 The Riemann Problem and Wave Patterns

The Riemann problem for the shallow water equations is a generalisation of the dam-break problem, and is formally defined as the initial-value problem (IVP)

$$\text{PDEs: } \partial_t \mathbf{Q} + \partial_x \mathbf{F}(\mathbf{Q}) = \mathbf{0},$$
$$\text{ICs: } \mathbf{Q}(x,0) = \begin{cases} \mathbf{Q}_L : \text{constant if } x < 0, \\ \mathbf{Q}_R : \text{constant if } x > 0. \end{cases} \quad (5.1)$$

The vectors $\mathbf{Q}$ and $\mathbf{F}(\mathbf{Q})$ in (5.1) are

$$\mathbf{Q} = \begin{bmatrix} h \\ hu \\ h\psi \end{bmatrix}, \quad \mathbf{F}(\mathbf{Q}) = \begin{bmatrix} h \\ hu^2 + \tfrac{1}{2}gh^2 \\ hu\psi \end{bmatrix} \quad (5.2)$$

and the constant initial states $\mathbf{Q}_L$ and $\mathbf{Q}_R$ at time $t = 0$, to the left of $x = 0$ and to the right of $x = 0$, respectively are

$$\mathbf{Q}_L = \begin{bmatrix} h_L \\ h_L u_L \\ h_L \psi_L \end{bmatrix}, \quad \mathbf{Q}_R = \begin{bmatrix} h_R \\ h_R u_R \\ h_R \psi_R \end{bmatrix}. \quad (5.3)$$

Here we consider the augmented conservation laws (5.1) for two cases, namely

1. The third equation is a transport equation for a passive scalar $\psi(x, t)$ representing the concentration of a chemical species.
2. The third equation arises from the split two-dimensional shallow water equations involving the tangential velocity component $v(x, t)$.

Mathematically, in both cases either $\psi(x, t)$ or $v(x, t)$, behave identically, as a passive scalar. Therefore in what follows we often use $\psi(x, t)$ to mean a concentration or the tangential velocity. The third equation then gives rise to a middle wave, which is either a contact discontinuity (species transport) or a shear wave (split 2D case). In the Riemann problem (5.1) the particle velocity components $u_L$ and $u_R$ are allowed to be distinct from zero, while in the dam-break problem $u_L = u_R = 0$. There are four possible wave patterns that may occur in the solution of the Riemann problem (5.1). These are illustrated in Fig. 5.3. Note that in each case there are three wave families; the left and right wave families correspond to the purely one-dimensional shallow water equations. The middle wave arises from the presence of the $y$-momentum equation in (5.2) or the inclusion of a chemical species transport equation. Case (a) in Fig. 5.3 is when both left and right waves are rarefactions; case (b) is when the left wave is a rarefaction and the right wave is a shock; case (c) is when the left is a shock and right wave is a rarefaction, and case (d) is when both left and right waves are shock waves. In general, the left and right waves are shocks or rarefactions, while the middle wave is always a *shear wave*, across which the tangential velocity component $v$ changes discontinuously or a contact discontinuity across which the concentration $\psi$ changes discontinuously.

Hence the structure of the solution of the Riemann problem (5.1) looks as depicted in Fig. 5.4. There are three wave families, which are associated with the eigenvalues $\lambda_1$, $\lambda_2$, $\lambda_3$ separating four constant regions $\mathcal{R}_0$, $\mathcal{R}_1$, $\mathcal{R}_2$ and $\mathcal{R}_3$. The corresponding four constant states are denoted, from left to right, by $\mathbf{Q}_L$, $\mathbf{Q}_{*L}$, $\mathbf{Q}_{*R}$, $\mathbf{Q}_R$. The region between the left and right waves is called the *Star Region* and is subdivided into two subregions $\mathcal{R}_1$ and $\mathcal{R}_2$, in which the solution is unknown. In regions $\mathcal{R}_0$ and $\mathcal{R}_3$ $\mathbf{Q}(x, t)$ is determined by the initial conditions. We note that the solution $\mathbf{Q}(x, t)$ of the Riemann problem (5.1) is a *similarity solution*, that is $\mathbf{Q}$ depends on the ratio $x/t$.

Next we study the much simpler case in which the initial data states for the Riemann problem are connected by a single wave, that is, the solution of the Riemann problem consists of a single non-trivial wave; the remaining waves are assumed to have zero strength. This assumption is entirely justified as we can always solve the Riemann problem with general data and then select the constant states on either side of a particular wave as the initial data for a Riemann problem. In Chaps. 6–8 we consider the general case in which all types of waves might simultaneously be present in the solution of the Riemann problem. First we recall some useful mathematical wave relations.

## 5.1 Elementary Waves in the Riemann Problem

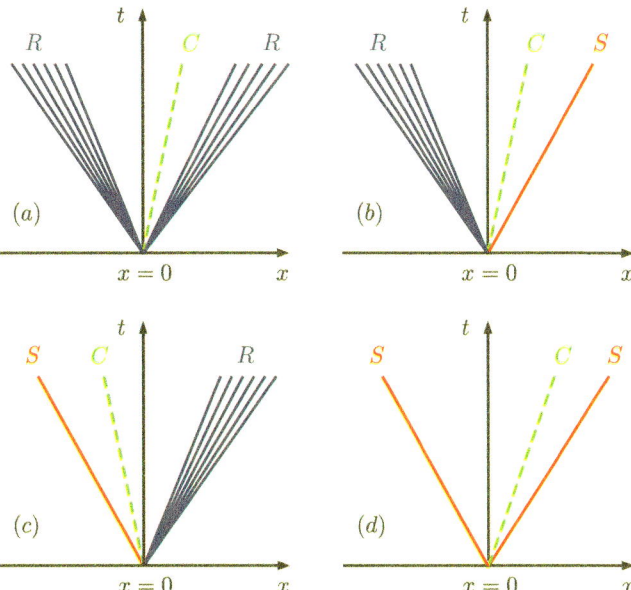

**Fig. 5.3** Four wave patterns arising from the solution of the Riemann problem in the $x$–$t$ plane: **a** left rarefaction, shear/contact and right rarefaction; **b** left rarefaction, shear/contact and right shock; **c** left shock, shear/contact and right rarefaction; **d** left shock, shear/contact and right shock. There are variations of these basic four patterns, including transcritical and supercritical cases. See Fig. 5.4

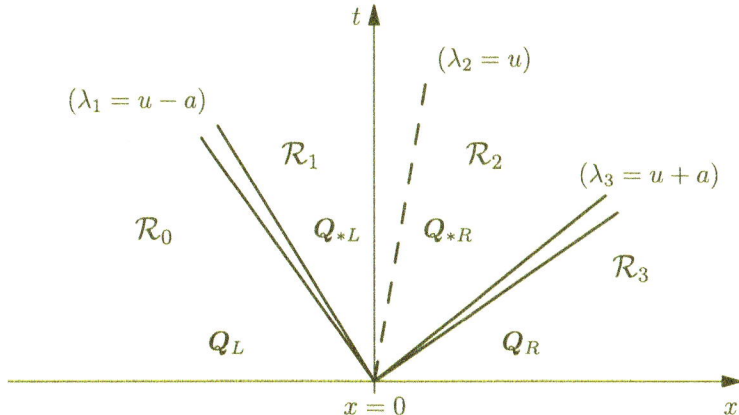

**Fig. 5.4** Structure of the general solution of the Riemann problem for the augmented shallow water equations. The left and right waves associated with the eigenvalues $\lambda_1 = u - a$ and $\lambda_3 = u + a$ respectively are either shocks or rarefactions. The middle wave associated with the eigenvalue $\lambda_2 = u$ is a shear wave, across which the tangential velocity component $v$ changes discontinuously, or a contact discontinuity across which the concentration $\psi$ changes discontinuously. The solution is first sought in regions $\mathcal{R}_1$ and $\mathcal{R}_2$, which comprise the *Star Region*. See Fig. 5.3

## 5.1.3 Relations Across the Wave Structure

Here we recall some mathematical relations that are valid across the structure of the various types of waves arising from the Riemann problem. We consider relations for rarefactions, contacts/shear waves and shock waves. For background on the eigenstructure of the shallow water equations see Chap. 4.

**Generalised Riemann Invariants**

Before proceeding to the detailed study of rarefaction waves, we recall the important notion of *generalised Riemann invariants* for a general $N \times N$ hyperbolic system, which we may express in quasi-linear form as

$$\partial_t \mathbf{W} + \mathbf{A}(\mathbf{W})\partial_x \mathbf{W} = \mathbf{0} . \tag{5.4}$$

Here the vector of unknowns

$$\mathbf{W} = [w_1, w_2, \ldots, w_N]^\mathrm{T} \tag{5.5}$$

represents either the conserved variables or some other suitable set of variables. Recall that any system of conservation laws may always be expressed in quasi-linear form via the Jacobian matrix.

For the wave associated with the $\lambda_i(\mathbf{W})$-characteristic field, with eigenvalue $\lambda_i(\mathbf{W})$, the corresponding right eigenvector is denoted as

$$\mathbf{R}^{(i)} = \left[ r_1^{(i)}, r_2^{(i)}, \cdots, r_N^{(i)} \right]^\mathrm{T} . \tag{5.6}$$

**Definition 5.1** The *generalised Riemann invariants* are relations that hold true *across* the wave structure of *simple waves*, which lead to the following $N - 1$ ordinary differential equations in phase space:

$$\frac{dw_1}{r_1^{(i)}} = \frac{dw_2}{r_2^{(i)}} = \frac{dw_3}{r_3^{(i)}} = \cdots = \frac{dw_N}{r_N^{(i)}} . \tag{5.7}$$

These equations relate ratios of changes $dw_s$ of a quantity $w_s$ to the respective component $r_s^{(i)}$ of the right eigenvector $\mathbf{R}^{(i)}$ corresponding to the $\lambda_i(\mathbf{W})$ wave family.

In Eq. (5.7) we emphasize that the ratios are to be interpreted as meaning proportionality, that is

$$dw_s \propto r_s^{(i)} . \tag{5.8}$$

5.1 Elementary Waves in the Riemann Problem                                      91

If $r_s^{(i)} = 0$ then $\mathrm{d}w_s = 0$ and therefore $w_s$ does not change across the respective wave. For a detailed discussion on Riemann invariants, see for example the book by Jeffrey [1]; see also LeVeque [2] and Godlewski and Raviart [3]. We shall apply wave relations (5.7) to study a particular class of waves.

**Rankine-Hugoniot Conditions**

*Rankine-Hugoniot jump conditions* or simply Rankine-Hugoniot conditions constitute an important concept that is applicable to hyperbolic conservation laws

$$\partial_t \mathbf{Q} + \partial_x \mathbf{F}(\mathbf{Q}) = \mathbf{0} . \tag{5.9}$$

They apply to a discontinuous wave travelling with speed $S$, which is related to jumps in conserved variables $\mathbf{Q}$ and fluxes $\mathbf{F}(\mathbf{Q})$ across the wave as follows:

$$\mathbf{F}_{ahead} - \mathbf{F}_{behind} = S \left( \mathbf{Q}_{ahead} - \mathbf{Q}_{behind} \right) . \tag{5.10}$$

Here subscript *ahead* denotes the state immediately ahead of the discontinuity and *behind* denotes the state immediately behind the discontinuity.

A simple illustration of Rankine-Hugoniot conditions is given via the inviscid Burgers' equation

$$\partial_t q + \partial_x f(q) = 0 , \quad \text{with flux function } f(q) = \tfrac{1}{2} q^2 . \tag{5.11}$$

See Chap. 2. Assume a shock wave solution with states $q_{ahead}, q_{behind}$ on either side of the wave, travelling with speed $S$. Application of the Rankine-Hugoniot conditions (5.10) gives

$$\frac{1}{2} q_{ahead}^2 - \frac{1}{2} q_{behind}^2 = S(q_{ahead} - q_{behind}).$$

The expression for the shock speed $S$ is therefore explicit

$$S = \frac{1}{2} \frac{\left( q_{ahead}^2 - q_{behind}^2 \right)}{(q_{ahead} - q_{behind})} = \frac{1}{2} \left( q_{ahead} + q_{behind} \right) . \tag{5.12}$$

**Remark 5.2** Such an explicit solution for the shock speed is only possible for the *scalar* conservation law (5.11); the fact that this is elegantly given by an arithmetic mean of the states ahead and behind is determined entirely by the particular expression for the flux function $f(q)$ in (5.11).

Next we apply wave relations, generalised Riemann invariants or Rankine-Hugoniot conditions, to specific waves in the shallow water equations

## 5.2 Single Rarefaction Wave

Here we apply generalised Riemann invariants to isolated rarefaction waves arising from the Riemann problem for the shallow water equations.

### 5.2.1 Left Rarefaction Wave

Assume a left rarefaction wave associated with the eigenvalue $\lambda_1 = u - a$ connecting the two constant states $\mathbf{Q}_L$ (left) and $\mathbf{Q}_{*L}$ (right), as depicted in Fig. 5.5. The rarefaction wave occupies a wedge $\mathcal{R}_L$ defined as

$$\mathcal{R}_L = \left\{ (x,t) / \ u_L - a_L \leq \frac{x}{t} \leq u_{*L} - a_{*L} \right\}, \tag{5.13}$$

where the characteristic line $x/t = u_L - a_L$ defines the **head of the rarefaction** and the characteristic line $x/t = u_{*L} - a_{*L}$ defines the **tail of the rarefaction**. The eigenvalue $\lambda_1(\mathbf{Q})$ increases monotonically across the wave from head to tail, as seen from the inclination of the characteristic lines in Fig. 5.5.

We now apply generalised Riemann invariants (5.7) across the $\lambda_1$-wave. We adopt the conserved variables $\mathbf{Q} = [h, hu, h\psi]^T$. For $\lambda_1 = u - a$ the right eigenvector, suitably scaled, is $\mathbf{R}^{(1)} = [1, u - a, \psi]^T$. Therefore

$$\frac{dh}{1} = \frac{d(hu)}{u - a} = \frac{d(h\psi)}{\psi}. \tag{5.14}$$

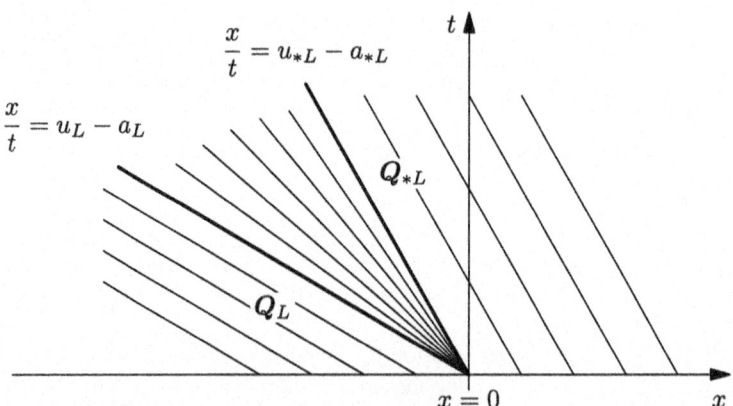

**Fig. 5.5** Left rarefaction wave associated with the eigenvalue $\lambda_1 = u - a$ connecting states $\mathbf{Q}_L$ and $\mathbf{Q}_{*L}$. The characteristic line $x/t = u_L - a_L$ defines the **head** and the characteristic line $x/t = u_{*L} - a_{*L}$ defines the **tail**

## 5.2 Single Rarefaction Wave

From the first and third ratios we write

$$\psi dh = d(\psi h) = h d\psi + \psi dh.$$

Therefore $d\psi = 0$ and therefore across the $\lambda_1$-wave we have

$$\psi : \text{constant} \rightarrow \psi_{*L} = \psi_L \text{ or } v_{*L} = v_L . \tag{5.15}$$

This means that the passive scalar $\psi$, or the tangential velocity component, is constant across the left rarefaction wave.

Analogously, from equating the first and second ratios in (5.14), followed by integration in phase space, we obtain

$$u + 2a = \text{constant} . \tag{5.16}$$

From here we establish

$$u_{*L} + 2a_{*L} = u_L + 2a_L . \tag{5.17}$$

For solving the complete Riemann problem in Chap. 6, it is convenient to introduce a wave function $f_L$ so that (5.17) can be expressed as

$$u_{*L} = u_L - f_L ; \quad f_L = 2(a_{*L} - a_L) . \tag{5.18}$$

**Solution inside a left rarefaction**

Consider a left rarefaction wave and a point $\hat{P} = (\hat{x}, \hat{t}) \in \mathcal{R}_L$, inside the wave, as depicted in Fig. 5.6. Consider now a characteristic line through $\hat{P} = (\hat{x}, \hat{t})$ and the origin $(0, 0)$, of slope (known)

$$\frac{\hat{x}}{\hat{t}} = \hat{u}_L - \hat{a}_L . \tag{5.19}$$

The unknowns of the problem are $\hat{u}_L = u(\hat{x}, \hat{t})$ and $\hat{a}_L = a(\hat{x}, \hat{t})$, noting that $h$ follows from $a = \sqrt{gh}$. Application of the left generalised Riemann invariant (5.16) to connect the point $\hat{P}$ to the left initial condition gives

$$\hat{u}_L + 2\hat{a}_L = u_L + 2a_L . \tag{5.20}$$

Equations (5.19) and (5.20) are two equations for the two unknowns $\hat{a}$ and $\hat{u}$, whose solution is

$$\hat{a}_L = a(\hat{x}, \hat{t}) = \frac{1}{3}(u_L + 2a_L - \frac{\hat{x}}{\hat{t}}) , \quad \hat{u}_L = u(\hat{x}, \hat{t}) = \frac{1}{3}(u_L + 2a_L + \frac{2\hat{x}}{\hat{t}}) . \tag{5.21}$$

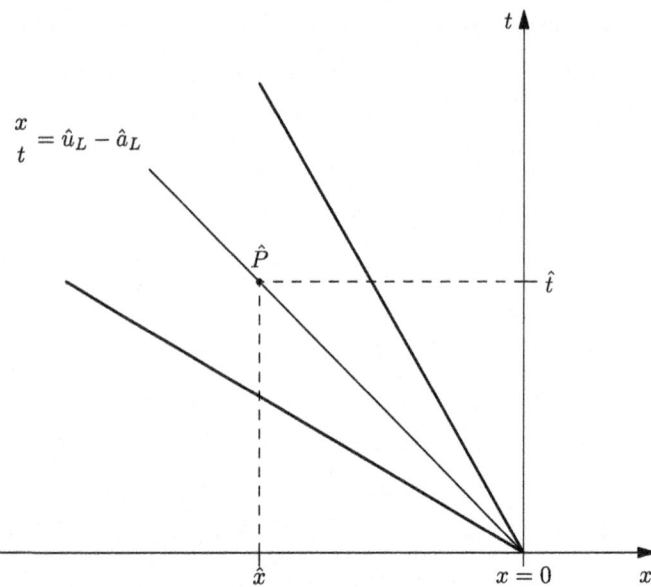

**Fig. 5.6** Point $\hat{P} = (\hat{x}, \hat{t})$ inside left rarefaction wave. We seek the solution $\hat{a}_L$ for the celerity $a$ and $\hat{u}_L$ for the particle velocity $u$ at the point $\hat{P} = (\hat{x}, \hat{t})$ in terms of its prescribed coordinates $(\hat{x}, \hat{t})$

### 5.2.2 Right Rarefaction Wave

Consider a right rarefaction wave associated with the eigenvalue $\lambda_3 = u + a$, connecting the constant states $\mathbf{Q}_{*R}$ (left) and $\mathbf{Q}_R$ (right), as depicted in Fig. 5.7. The wave occupies a wedge $\mathcal{R}_R$ defined as

$$\mathcal{R}_R = \left\{ (x, t) / \ u_{*R} + a_{*R} \leq \frac{x}{t} \leq u_R + a_R \right\} . \tag{5.22}$$

Note that the eigenvalue $\lambda_3 = u + a$ increases monotonically from left to right, as seen from the inclination of the characteristic lines. For $\lambda_1 = u + a$ the right eigenvector is $\mathbf{R}^{(3)} = [1, u + a, \psi]^T$. Application of the generalised Riemann invariants (5.7) across the $\lambda_3$-wave gives

$$\frac{dh}{1} = \frac{d(hu)}{u + a} = \frac{d(h\psi)}{\psi} . \tag{5.23}$$

From comparing the first and third ratios we obtain $d\psi = 0$ and so across the $\lambda_3$-wave the concentration variable and the tangential velocity component are constant. That is

$$\psi : \text{constant} \to \psi_{*R} = \psi_R \text{ or } v : \text{constant} \to v_{*R} = v_R . \tag{5.24}$$

## 5.3 Single Shock Wave

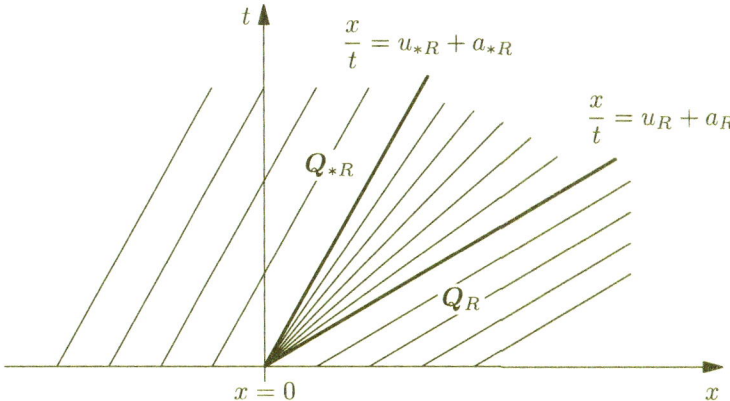

**Fig. 5.7** Right rarefaction wave connecting states $\mathbf{Q}_{*R}$ and $\mathbf{Q}_R$. The characteristic line $x/t = u_R + a_R$ defines the head while $x/t = u_{*R} + a_{*R}$ defines the tail

Also, from (5.23) we obtain

$$u - 2a = \text{constant},  \tag{5.25}$$

which gives

$$u_{*R} - 2a_{*R} = u_R - 2a_R.  \tag{5.26}$$

In readiness for solving the complete Riemann problem in Chap. 6, it is convenient to introduce a wave function $f_R$ so that (5.26) can be expressed as

$$u_{*R} = u_R + f_R ; \quad f_R = 2(a_{*R} - a_R).  \tag{5.27}$$

As for the left rarefaction wave case, the solution at $\hat{P} = (\hat{x}, \hat{t}) \in \mathcal{R}_R$ inside the right rarefaction wave is easily found to be

$$\hat{a}_R = \frac{1}{3}(-u_R + 2a_R + \frac{\hat{x}}{\hat{t}}) , \quad \hat{u}_R = \frac{1}{3}(u_R - 2a_R + \frac{2\hat{x}}{\hat{t}}).  \tag{5.28}$$

Next we deal with isolated shock waves.

## 5.3 Single Shock Wave

We seek relations across shock waves in the shallow water equations from applying Rankine-Hugoniot conditions (5.10), starting with a right-facing shock.

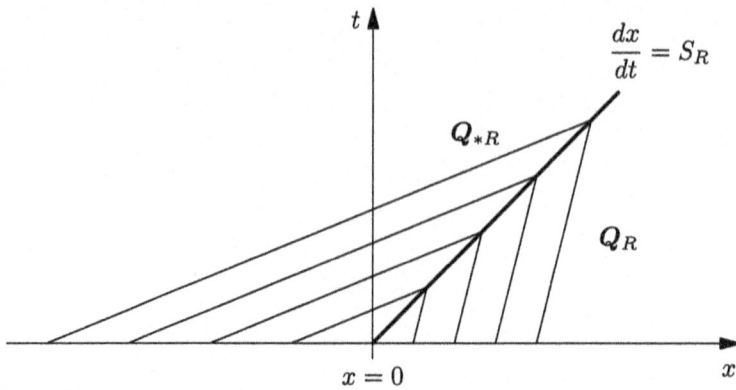

**Fig. 5.8** Right-facing shock wave of speed $S_R$ connecting constant states $\mathbf{Q}_R$ (ahead) and $\mathbf{Q}_{*R}$ (behind)

### 5.3.1 Right-Facing Shock Wave

Consider an isolated right-facing shock wave of speed $S_R$ associated with the $\lambda_3 = u + a$ characteristic field. Figure 5.8 shows the wave in the $x$–$t$ half plane. Characteristics $x/t = \lambda_3 = u + a$ converge into the shock path described by $x/t = S_R$, from left and right. This is a fundamental feature of a compressive shock wave that obeys the Lax entropy condition

$$\lambda_3(\mathbf{Q}_{*R}) > S_R > \lambda_3(\mathbf{Q}_R) \ . \tag{5.29}$$

Characteristics run into the shock path, as illustrated in Fig. 5.8.

The **Rankine-Hugoniot Conditions** applied across the right shock wave read

$$S_R(\mathbf{Q}_R - \mathbf{Q}_{*R}) = \mathbf{F}(\mathbf{Q}_R) - \mathbf{F}(\mathbf{Q}_{*R}) \ . \tag{5.30}$$

In order to facilitate the analysis we apply the transformation of frames of reference

$$\hat{u}_{*R} = u_{*R} - S_R \ , \quad \hat{u}_R = u_R - S_R \ , \tag{5.31}$$

which is illustrated in Fig. 5.9. In the new frame, Fig. 5.9b, the shock propagation speed is 0 and the vectors of conserved variables and fluxes ahead of the shock are

$$\hat{\mathbf{Q}}_R = \begin{bmatrix} h_R \\ h_R \hat{u}_R \\ h_R \psi_R \end{bmatrix} \ , \quad \mathbf{F}(\hat{\mathbf{Q}}_R) = \begin{bmatrix} h_R \hat{u}_R \\ h_R \hat{u}_R^2 + \frac{1}{2} g h_R^2 \\ h_R \hat{u}_R \psi_R \end{bmatrix} \ , \tag{5.32}$$

## 5.3 Single Shock Wave

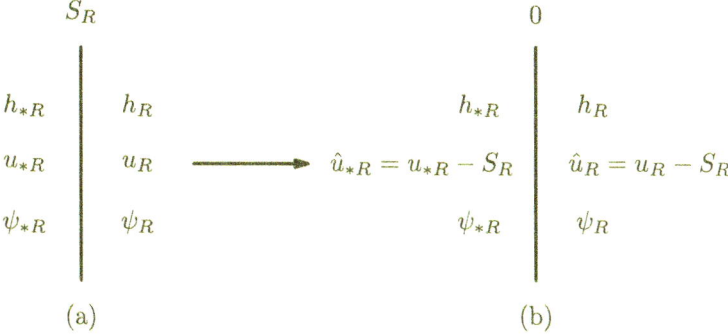

**Fig. 5.9** Right shock wave in two frames of reference. Frame **a** is the original frame of reference and frame **b** is the moving frame of references in which the shock is stationary

while those behind the shock are

$$\hat{\mathbf{Q}}_{*R} = \begin{bmatrix} h_{*R} \\ h_{*R}\hat{u}_{*R} \\ h_{*R}\psi_{*R} \end{bmatrix}, \quad \mathbf{F}(\hat{\mathbf{Q}}_{*R}) = \begin{bmatrix} h_{*R}\hat{u}_{*R} \\ h_{*R}\hat{u}_{*R}^2 + \frac{1}{2}gh_{*R}^2 \\ h_{*R}\hat{u}_{*R}\psi_{*R} \end{bmatrix}. \quad (5.33)$$

The Rankine-Hugoniot conditions in the frame Fig. 5.9b moving with the shock are

$$\mathbf{F}(\hat{\mathbf{Q}}_{*R}) - \mathbf{F}(\hat{\mathbf{Q}}_R) = 0 \times (\hat{\mathbf{Q}}_{*R} - \hat{\mathbf{Q}}_R), \quad (5.34)$$

which give

$$\mathbf{F}(\hat{\mathbf{Q}}_{*R}) = \mathbf{F}(\hat{\mathbf{Q}}_R).$$

Written in full, the above flux equality gives

$$\left.\begin{array}{rcl} h_{*R}\hat{u}_{*R} &=& h_R\hat{u}_R, \\ h_{*R}\hat{u}_{*R}^2 + \frac{1}{2}gh_{*R}^2 &=& h_R\hat{u}_R^2 + \frac{1}{2}gh_R^2, \\ h_{*R}\hat{u}_{*R}\psi_{*R} &=& h_R\hat{u}_R\psi_R. \end{array}\right\} \quad (5.35)$$

The first equation in (5.35) says that the mass flux is constant across the shock, that is

$$- M_R \equiv h_{*R}\hat{u}_{*R} = h_R\hat{u}_R. \quad (5.36)$$

Using this into the third of Eq. (5.35) reveals that $\psi$ is constant across the shock wave. It also says that the tangential velocity component $v$ is constant across the shock wave. That is

$$\psi_{*R} = \psi_R, \quad v_{*R} = v_R. \quad (5.37)$$

Compare with (5.24). We only need to work with the first two equations in (5.35); the second one gives

$$(h_{*R}\hat{u}_{*R})\hat{u}_{*R} - (h_R\hat{u}_R)\hat{u}_R = \frac{1}{2}g(h_R^2 - h_{*R}^2) . \tag{5.38}$$

Use of (5.36) into (5.38) gives

$$M_R = \frac{\frac{1}{2}g(h_R^2 - h_{*R}^2)}{\hat{u}_R - \hat{u}_{*R}} . \tag{5.39}$$

But from (5.36) we may write

$$\hat{u}_{*R} = -\frac{M_R}{h_{*R}} , \quad \hat{u}_R = -\frac{M_R}{h_R} . \tag{5.40}$$

Use of (5.40) into (5.39) followed by some manipulations yields

$$M_R = \sqrt{\frac{1}{2}gh_Rh_{*R}(h_R + h_{*R})} . \tag{5.41}$$

From (5.31)

$$u_{*R} = u_R + (\hat{u}_{*R} - \hat{u}_R) . \tag{5.42}$$

Inserting (5.40) into (5.42) followed by some algebraic manipulations gives

$$u_{*R} = u_R + f_R ; \quad f_R = (h_{*R} - h_R)\sqrt{\frac{1}{2}g\frac{(h_{*R} + h_R)}{h_Rh_{*R}}} . \tag{5.43}$$

As for previous cases we have introduced a shock wave function $f_R$ that relates quantities across the wave. This will prove useful when solving the full Riemann problem in Chap. 6. From (5.31) the speed of the right-facing shock wave is

$$S_R = u_R - \hat{u}_R . \tag{5.44}$$

Use of (5.40) into (5.44) followed by manipulations gives

$$S_R = u_R + q_R a_R , \quad q_R = \sqrt{\frac{1}{2}\frac{(h_R + h_{*R})h_{*R}}{h_R^2}} . \tag{5.45}$$

This expression relates the shock speed $S_R$ to the unknown depth $h_{*R}$ behind the shock. Note that for the limiting case $h_{*R}/h_R = 1$ the shock speed coincides with the characteristic speed, that is $S_R = u + a$, as expected.

## 5.3 Single Shock Wave

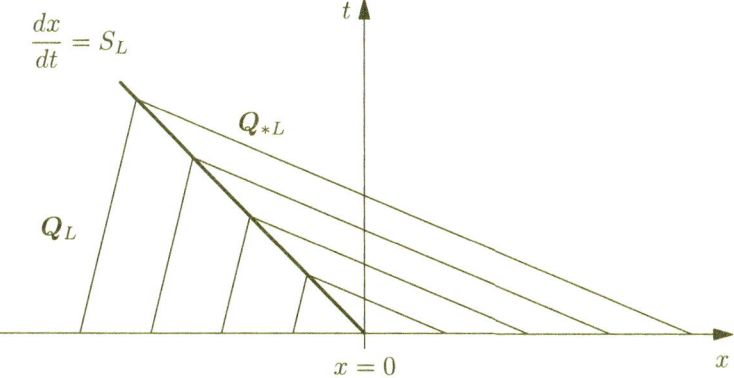

**Fig. 5.10** Left-facing shock wave of speed $S_L$ connecting states $\mathbf{Q}_L$ (ahead) and $\mathbf{Q}_{*L}$ (behind)

### 5.3.2 Left-Facing Shock Wave

For a left-facing shock of speed $S_L$ associated with the eigenvalue $\lambda_1 = u - a$ the analysis is similar to that of a right shock. Figure 5.10 depicts a left-facing shock wave with $\lambda_1$-characteristics $x/t = \lambda_1 = u - a$ from left and right converging into the shock path $x/t = S_L$, consistently with the Lax entropy condition

$$\lambda_1(\mathbf{Q}_L) > S_L > \lambda_3(\mathbf{Q}_{*L}) . \tag{5.46}$$

The analysis for this case is entirely analogous to the right-facing shock case and therefore the presentation will be succinct. The transformation of frames of reference is

$$\hat{u}_{*L} = u_{*L} - S_L ; \quad \hat{u}_L = u_L - S_L . \tag{5.47}$$

Then the Rankine-Hugoniot conditions applied to the equations on the transformed frame of reference give

$$\left. \begin{array}{rcl} h_{*L}\hat{u}_{*L} & = & h_L\hat{u}_L , \\ h_{*L}\hat{u}_{*L}^2 + \frac{1}{2}gh_{*L}^2 & = & h_L\hat{u}_L^2 + \frac{1}{2}gh_L^2 , \\ h_{*L}\hat{u}_{*L}\psi_{*L} & = & h_L\hat{u}_L\psi_L . \end{array} \right\} \tag{5.48}$$

The first of Eq. (5.48) says that the mass flux

$$M_L \equiv h_{*L}\hat{u}_{*L} = h_L\hat{u}_L \tag{5.49}$$

is constant across the shock wave. Using this condition into the third of Eq. (5.48) reveals that $\psi$, and hence the tangential velocity component $v$, are constant across the left shock. That is

$$\psi_{*L} = \psi_L, \quad v_{*L} = v_L. \tag{5.50}$$

Compare with (5.37). Analogous manipulations to those for a right-facing shock yield

$$M_L = \sqrt{\frac{1}{2}gh_L h_{*L}(h_L + h_{*L})} \tag{5.51}$$

and

$$u_{*L} = u_L - f_L; \quad f_L = (h_{*L} - h_L)\sqrt{\frac{1}{2}g\frac{(h_{*L} + h_L)}{h_L h_{*L}}}. \tag{5.52}$$

This relates $u_{*L}$ to $h_{*L}$ via the left shock wave function $f_L$. Also, from (5.47) the shock speed is

$$S_L = u_L - \hat{u}_L. \tag{5.53}$$

Use of (5.49) into (5.53) followed by manipulations gives the shock speed as

$$S_L = u_L - q_L a_L; \quad q_L = \sqrt{\frac{1}{2}\frac{(h_L + h_{*L})h_{*L}}{h_L^2}}. \tag{5.54}$$

This expression relates the shock speed $S_L$ to the water depth $h_{*L}$ behind the shock. Again, in the limiting case $h_{*L}/h_L = 1$ we have $S_L = u_L - a_L$, as would be expected.

## 5.4 Contact Discontinuity and Shear Wave

An isolated contact discontinuity associated with the characteristic field $\lambda_2 = u$, connecting the (constant) states $\mathbf{Q}_{*L}$ and $\mathbf{Q}_{*R}$ is depicted in Fig. 5.11. The wave is a single discontinuity travelling with speed $u_*$ and characteristics either side of the discontinuity run parallel to it, namely

$$\lambda_2(\mathbf{Q}_{*L}) = u_* = \lambda_2(\mathbf{Q}_{*R}). \tag{5.55}$$

An eigenvector analysis using generalised Riemann invariants (5.7) provides the sought jump conditions across the contact discontinuity. The right eigenvector corresponding to $\lambda_2 = u$ is $\mathbf{R}^{(2)} = [0, 0, 1]^T$, from which we have

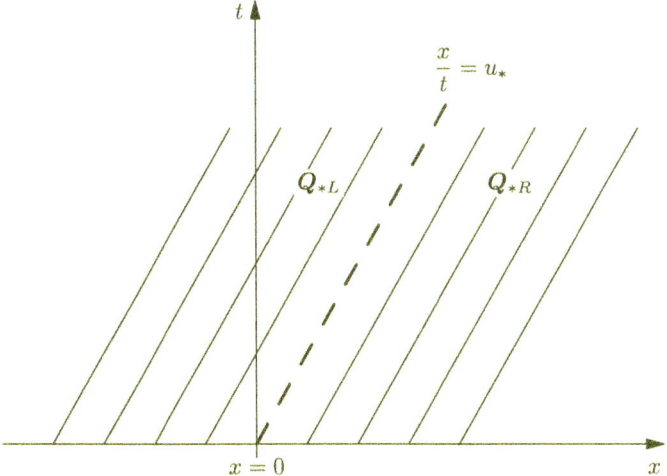

**Fig. 5.11** Contact wave/shear wave associated with the linearly degenerate field $\lambda_2$, connecting states $\mathbf{Q}_{*L}$ and $\mathbf{Q}_{*R}$. Characteristics either side of the wave are parallel to the wave, just as in the linear advection equation

$$\left.\begin{array}{l} u_{*L} = u_{*R} = u_* \,, \\ h_{*L} = h_{*R} = h_* \,, \\ \psi_{*L} \neq \psi_{*R} \,. \end{array}\right\} \tag{5.56}$$

These wave relations across the contact/shear wave say that:

1. The particle velocity $u$ (normal velocity) is constant across the wave and so is constant in the entire *Star Region* in Fig. 5.11.
2. The water depth $h$ is constant across the wave and so is constant in the entire *Star Region* in Fig. 5.11. See also Fig. 5.4.
3. The only quantity that changes across the wave (discontinuously) is $\psi$ and the tangential velocity component $v$.

**Exercise 5.3** Verify relations (5.56) by applying the generalised Riemann invariants across the $\lambda_2$ wave.

**Exercise 5.4** Show that the Rankine-Hugoniot conditions across the contact/shear wave (5.56) are identically satisfied if $u_{*L} = u_{*R} = u_*$ and $h_{*L} = h_{*R} = h_*$.

## 5.5 The Full Wave System

Figure 5.12 depicts the full structure of the solution of the Riemann problem in the $x$–$t$ plane. The left and right waves can be shocks or rarefactions. The velocity and depth are constant in the *Star Region*; $\psi$ is also constant in $\mathcal{R}_1 \cup \mathcal{R}_0$ and in $\mathcal{R}_2 \cup \mathcal{R}_3$ but with a discontinuous jump across the contact/shear wave. See (5.56).

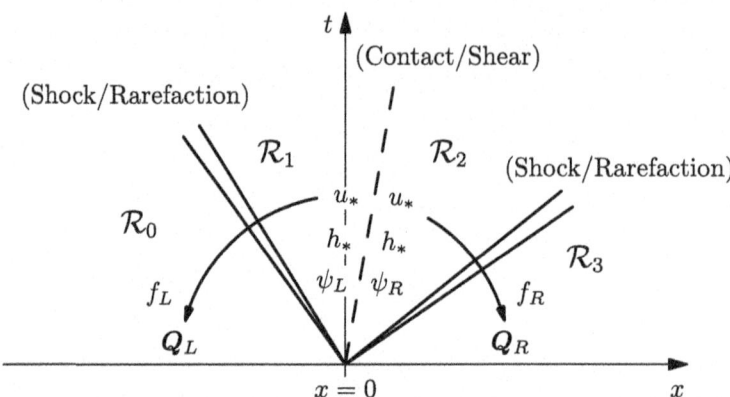

**Fig. 5.12** Structure of the full wave system emerging from the Riemann problem for the augmented shallow water equations. Compare with Figs. 5.3 and 5.4

The main step in solving the Riemann problem is the *Star Problem*, namely that of determining the velocity $u_*$ and the depth $h_*$ in the *Star Region*. Once $u_*$ and $h_*$ are known, the rest of the solution through the wave structure follows.

To find the velocity $u_*$ and the depth $h_*$ we first assemble together all the wave relations derived for each elementary wave in isolation. Note that the velocity $u_*$ is connected to $\mathbf{Q}_L$ via a function $f_L$ and that the velocity $u_*$ is connected to $\mathbf{Q}_R$ via a function $f_R$. The functions $f_L$ and $f_R$ are given in (5.18) and (5.27) for rarefactions and in (5.43) and (5.52) for shocks. The functions $f_L$ and $f_R$ depend on the unknown depth $h_*$, the *wave type* (shock or rarefaction) and, parametrically, on the initial conditions $\mathbf{Q}_L$ and $\mathbf{Q}_R$, that is

$$f_L = f_L(h_*, w_L; \mathbf{Q}_L) \; ; \quad f_R = f_R(h_*, w_R; \mathbf{Q}_R) \; . \tag{5.57}$$

Here $w_L$ and $w_R$ denote logical variables that identify the wave type; $w_K$ denotes either a shock or a rarefaction, for $K = L$ and $K = R$. The complete solution procedure for the *Star Problem* is then summarised in the following theorem.

**Theorem 5.5** *The solution $h_*$ for the Riemann problem (5.1) is the root of the non-linear algebraic equation*

$$f(h) \equiv f_L(h, w_L; h_L) + f_R(h, w_R; h_R) + \Delta u = 0 \; , \quad \Delta u \equiv u_R - u_L \; , \tag{5.58}$$

*with*

$$f_L(h, w_L; h_L) = \begin{cases} 2(\sqrt{gh} - \sqrt{gh_L}) & \text{if } h \leq h_L \ (w_L: \text{rarefaction}), \\ (h - h_L)\sqrt{\frac{1}{2}g\frac{(h + h_L)}{hh_L}} & \text{if } h > h_L \ (w_L: \text{shock}) \end{cases} \quad (5.59)$$

*and*

$$f_R(h, w_R; h_R) = \begin{cases} 2(\sqrt{gh} - \sqrt{gh_R}) & \text{if } h \leq h_R \ (w_R: \text{rarefaction}), \\ (h - h_R)\sqrt{\frac{1}{2}g\frac{(h + h_R)}{hh_R}} & \text{if } h > h_R \ (w_R: \text{shock}). \end{cases} \quad (5.60)$$

Once the depth $h_*$ has been found the solution for the velocity $u_*$ follows as

$$u_* = \tfrac{1}{2}(u_L + u_R) + \tfrac{1}{2}[f_R(h_*, w_R; h_R) - f_L(h_*, w_L; h_L)]. \quad (5.61)$$

**Proof** (*Sketch of the Proof*). First note that the particle velocity $u_*$ and depth $h_*$ are constant across the contact discontinuity according to (5.56). In fact $u_*$ and $h_*$ are constant in the entire *Star region*. Then, the function $f_L$ is used to relate $u_*$ to the left initial condition $\mathbf{Q}_L$ across the left wave. In case the left wave is a shock we have the relation (5.52) and if it is a rarefaction we use (5.18). Analogously, the function $f_R$ is used to relate $u_*$ to the right initial condition $\mathbf{Q}_R$ across the right wave. If the right wave is a shock we have the relation (5.43) and if it is a rarefaction we use (5.27). As $u_* = u_{*L} = u_{*R}$, see (5.56), we can eliminate $u_*$ resulting in Eq. (5.58). Then the particle velocity could be written in terms of the function $f_L$, for both the shock and rarefaction cases. See (5.52) and (5.18). So we could compute $u_*$ directly from $f_L$ once $h_*$ is known. Alternatively, we could compute $u_*$ directly from $f_R$ using (5.43) or (5.27). Solution (5.61) results from a mean of the two possible solutions. This concludes the proof. □

The practical implementation of the full Riemann solver is presented in Chap. 6. This involves the iterative solution of the non-linear algebraic Eq. (5.58) to find the value $h_*$ in the *Star Region*, the evaluation of $u_*$ from (5.61) and the computation of all quantities across the full wave structure that includes all types of waves. Full details are given in Chap. 6 and a computer programme is included in Chap. 8.

## 5.6 Useful Shock Relations

Useful shock relations may be written in terms of shock speed and shock Froude number. The relations can be readily used to set up test problems for an isolated shock, with exact solution, to test numerical methods, for example.

### 5.6.1 Left Shock

Consider a left-facing shock wave associated with the eigenvalues $\lambda_1 = u - a$. From (5.54) we may write the left shock speed as

$$S_L = u_L - a_L q_L , \\ q_L = \sqrt{\left[\frac{1}{2} \frac{(h_* + h_L) h_*}{h_L^2}\right]} . \qquad (5.62)$$

We now derive some useful shock relations in terms of the initial data and the shock *Froude number*. We introduce the two Froude numbers

$$\mathcal{F}_L = \frac{u_L}{a_L}, \quad \mathcal{F}_S = \frac{S_L}{a_L} . \qquad (5.63)$$

The quantity $\mathcal{F}_S$ is called the *shock Froude number*. Clearly, since for a left shock we require $h_* \geq h_L$, then $q_L \geq 1$ and thus

$$q_L = \mathcal{F}_L - \mathcal{F}_S \geq 1 . \qquad (5.64)$$

The quantity $\mathcal{F}_L$ is a function of the initial data state, which is prescribed, and $\mathcal{F}_S$ is an unknown parameter of the problem; in fact $\mathcal{F}_S$ is a function of the solution itself. Thus, for prescribed initial data ahead of the shock and an *assumed* shock Froude number $\mathcal{F}_S$, we can completely determine the state behind the shock.

Equation (5.64), along (5.62), may be regarded as a quadratic in $h_*$, which if solved in terms of the shock Froude number $\mathcal{F}_S$ and the data ahead of the shock gives

$$h_* = h_L \left[\frac{-1 + \sqrt{1 + 8(\mathcal{F}_L - \mathcal{F}_S)^2}}{2}\right] . \qquad (5.65)$$

Then the particle velocity $u_*$ behind the shock follows from (5.52) as

$$u_* = u_L - (h_* - h_L) \sqrt{\frac{1}{2} g \frac{(h_* + h_L)}{h_L h_*}} . \qquad (5.66)$$

**Summary**. For given initial conditions $h_L, u_L, v_L$ ahead of the *left* shock, and an *assumed* shock Froude number $\mathcal{F}_S$, the state behind the shock is completely determined and is given by (5.65)–(5.66) for $h_*$ and $u_*$. The speed of the shock is given by (5.62) or (5.63). Note that this is a **one-parameter family of solutions**, the free parameter open to choice is the shock Froude number $\mathcal{F}_S$, or equivalently the shock speed $S_L$. The tangential velocity component (or the concentration) is constant across the shock, as seen in (5.50). This exact shock solution may be very useful when setting up test problems for the purpose of assessing numerical methods.

## 5.6.2 Right Shock

Consider a right-facing shock wave associated with the eigenvalues $\lambda_3 = u + a$, of shock speed $S_R$

$$\left. \begin{array}{l} S_R = u_R + a_R q_R \, , \\[4pt] q_R = \sqrt{\dfrac{1}{2}\left[\dfrac{(h_* + h_R)h_*}{h_R^2}\right]} \, . \end{array} \right\} \quad (5.67)$$

We now introduce the Froude numbers

$$\mathcal{F}_S = \frac{S_R}{a_R} \, , \quad \mathcal{F}_R = \frac{u_R}{a_R} \, , \quad (5.68)$$

where $\mathcal{F}_S$ is the *shock Froude number*. Then, from Eq. (5.67) we write

$$q_R = \mathcal{F}_S - \mathcal{F}_R \geq 1 \, , \quad (5.69)$$

where $\mathcal{F}_R$ is a function of the initial data state, which is prescribed, and $\mathcal{F}_S$ is effectively the shock speed. Therefore, for prescribed initial data ahead of the right shock and an *assumed* shock Froude number $\mathcal{F}_S$, or assumed shock speed, we can completely determine the state behind the shock. Equation (5.69) is as a quadratic in $h_*$, the solution of which gives the water depth behind the shock as

$$h_* = h_R \left[\frac{-1 + \sqrt{1 + 8(\mathcal{F}_S - \mathcal{F}_R)^2}}{2}\right] \, . \quad (5.70)$$

The particle velocity $u_*$ behind the shock follows from (5.43) as

$$u_* = u_R + (h_* - h_R)\sqrt{\frac{1}{2}g\frac{(h_* + h_R)}{h_R h_*}} \, . \quad (5.71)$$

**Summary.** For given initial conditions $h_R$, $u_R$, $v_R$ ahead of the *right* shock, and *assuming* a shock Froude number $\mathcal{F}_S$, the state behind the shock is *completely determined* and is given by (5.70) and (5.71) for $h_*$ and $u_*$. The speed of the shock is given by (5.67) or (5.68). The tangential velocity component and the concentration variable are constant across the shock, as seen in (5.37). Note that this is a **one-parameter family of solutions**; the free parameter open to choice is the shock Froude number $\mathcal{F}_S$, or equivalently, the shock speed $S_R$.

**Example 5.6** Assume a right state with data $h_R = 1.0\ m$, $u_R = 0\ m/s$, and acceleration due to gravity as $g = 9.81\ m/s^2$. Figure 5.13 shows water depth $h_*$ (top) and particle velocity $u_*$ (bottom) behind the shock wave versus shock Froude number $\mathcal{F}_S$, for $1 \leq \mathcal{F}_S \leq 10$.

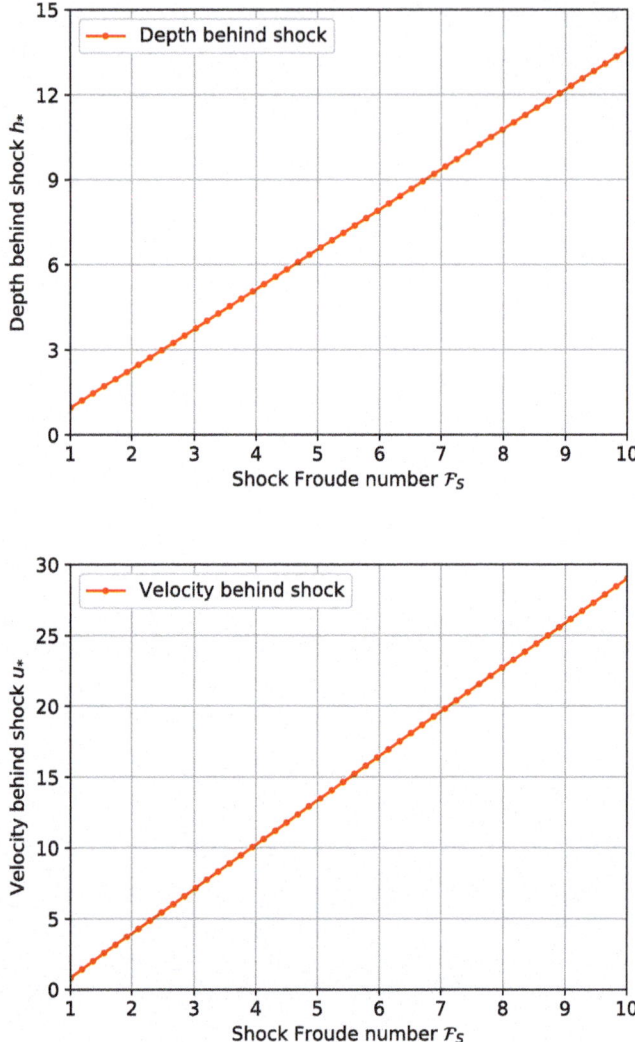

**Fig. 5.13** Water depth $h_*$ (top) and particle velocity $u_*$ (bottom) behind the shock wave versus shock Froude number $\mathcal{F}_S$. Here $h_R = 1\ m$ and $u_R = 0\ m/s$

We have studied all possible situations involving a single non-trivial wave arising as solution of the Riemann problem for the augmented shallow water equations. We shall apply this knowledge in Chaps. 6 and 7 to study the general Riemann problem in which all type of waves may be present simultaneously.

But before doing so, in the next Section we address the question of formulation of the equations in the presence of shock waves.

## 5.7 Non-conservative Formulation and Shocks

This Section illustrates the importance of correctly formulating the equations in the presence of shock waves.

### 5.7.1 Conservative System in Non-conserved Variables

Consider the one-dimensional shallow water equations in (5.1), namely

$$\partial_t \begin{bmatrix} h \\ hu \end{bmatrix} + \partial_x \begin{bmatrix} hu \\ hu^2 + \frac{1}{2}gh^2 \end{bmatrix} = \mathbf{0} \ . \tag{5.72}$$

See Chap. 1. These express the physical laws of *conservation of mass and momentum*. On the other hand, *under the assumption of smooth solutions*, we can expand derivatives so as to write the equations in primitive variable formulation as follows:

$$\partial_t h + u \partial_x h + h \partial_x u = 0 \ , \tag{5.73}$$

$$\partial_t u + u \partial_x u + g \partial_x h = 0 \ . \tag{5.74}$$

It is tempting to derive *new conservation-law forms* of the shallow water equations starting from Eqs. (5.73)–(5.74). One possibility is to keep the mass equation as in (5.72) and rewrite the momentum Eq. (5.74) as

$$\partial_t u + \partial_x \left( \tfrac{1}{2} u^2 + gh \right) = 0 \ , \tag{5.75}$$

where under the assumption of smoothness we have replaced the term $uu_x$ by $(\frac{1}{2}u^2)_x$, as for example in the Burgers equation. Now we have an *alternative conservative form* of the shallow water equations, namely

$$\partial_t \begin{bmatrix} h \\ u \end{bmatrix} + \partial_x \begin{bmatrix} hu \\ \frac{1}{2}u^2 + gh \end{bmatrix} = \mathbf{0} \ . \tag{5.76}$$

Mathematically, this is a system of conservation laws. It expresses conservation of mass, which is physically correct of course, and *conservation of particle speed u*; this second (mathematical) conservation law is however, physically meaningless. System (5.76) is mathematically conservative but physically non-conservative. We shall see that we can still use (5.76) for the shallow water equations, provided solutions are smooth. In the presence of shock waves, formulations (5.72) and (5.76) lead to different solutions, as we now demonstrate.

## 5.7.2 Shock Waves

Without loss of generality we consider a right-facing shock wave associated with the eigenvalue $\lambda_2 = u + a$, in which the state ahead of the shock is given by the constants $h_R, u_R$.

**Theorem 5.7** *A right-facing shock wave solution of (5.72) associated with the eigenvalue $\lambda_2 = u + a$ has shock speed*

$$S_{cons} = u_R + \sqrt{\frac{1}{2}g\left[\frac{(h_* + h_R)h_*}{h_R}\right]}, \qquad (5.77)$$

*while a right-facing shock wave solution of (5.76) associated with the eigenvalue $\lambda_2 = u + a$ has speed*

$$S_{nonc} = u_R + \sqrt{\frac{2gh_*^2}{h_* + h_R}}, \qquad (5.78)$$

*where $h_*$ is the water depth behind the shock. These shock speeds satisfy*

$$S_{nonc} \leq S_{cons}, \quad \text{with } S_{nonc} = S_{cons} \text{ if and only if } h_* = h_R. \qquad (5.79)$$

**Proof** The derivation of the shock speed for the conservative form (5.72) was carried out in Sect. 5.3.1. The derivation of the shock speed for the *alternative conservative form* (5.76) is carried out using the same methodology as for (5.72). The Rankine-Hugoniot conditions applied to Eq. (5.76) in the transformed frame of reference give

$$\left.\begin{array}{r}h_* \hat{u}_* = h_R \hat{u}_R, \\ \frac{1}{2}\hat{u}_*^2 + gh_* = \frac{1}{2}\hat{u}_R^2 + gh_R.\end{array}\right\} \qquad (5.80)$$

Similar manipulations of these equations to those of Sect. 5.3.1 produce

$$M_R = h_* h_R \sqrt{\frac{2g}{h_* + h_R}} \qquad (5.81)$$

and the shock speed $S_{nonc}$ corresponding to the physically non-conservative system (5.76) is then found to be

$$S_{nonc} = u_R + \sqrt{\frac{2gh_*^2}{h_* + h_R}},$$

as claimed. By comparing the two shock speeds $S_{nonc}$ and $S_{cons}$ it is found that

$$S_{nonc} \leq S_{cons}.$$

## 5.7 Non-conservative Formulation and Shocks

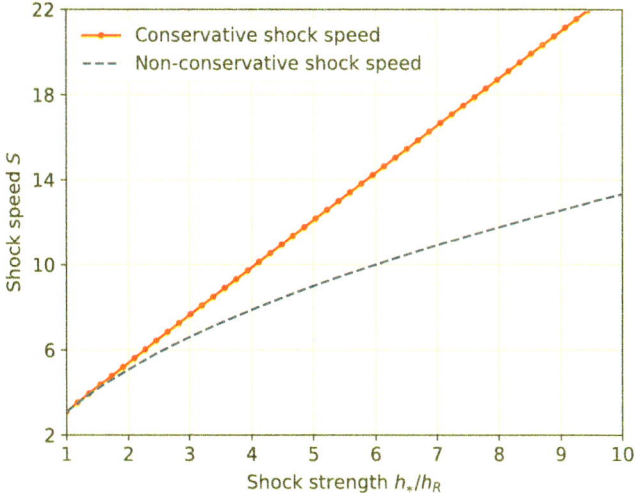

**Fig. 5.14** Shock-wave speeds $S_{cons}$ (higher) and $S_{ncon}$ (lower) versus shock strength $h_*/h_R$, for conservative and non-conservative formulations of the equations. Here $h_R = 1\ m, u_R = 0\ m/s$ and $1 \leq h_*/h_R \leq 10$

This results from the inequality

$$0 \leq (h_* - h_R)^2 \ .$$

Equality of the shock speeds holds only if the shock is trivial, $h_* = h_R$, that is, if the solution is smooth. □

Figures 5.14 and 5.15 show comparisons between the shock speeds $S_{cons}$ obtained from the physically conservative formulation (5.72) and $S_{nonc}$ obtained from the physically non-conservative formulation (5.76). Data ahead of the shock are prescribed values for depth $h_R$ and velocity $u_R$. In Fig. 5.14 the comparison is carried out for a large range of assumed values of the *shock strength* $h_*/h_R$. Figure 5.15 is a zoom of Fig. 5.14 for a small range of the *shock strength* $h_*/h_R$ near the trivial case of no shock $h_*/h_R \approx 1$.

**Remark 5.8** (*Abandoning conservation*). Note that in system (5.76) conservation of mass is preserved. It is only conservation of momentum that has been abandoned and yet this is sufficient to give completely wrong solutions for shocks. See Figs. 5.14 and 5.15. Numerical solution procedures based on (5.76) will also be expected to produce shocks with the wrong propagation speed, naturally, even when using a *conservative* numerical method.

**Remark 5.9** (*Weak shocks*). Weak shocks are those with shock strength given as $h_*/h_R = 1 + \epsilon$, where $\epsilon$ is a small positive quantity and $h_R$ is the depth ahead of the shock. From the results of Fig. 5.15, for $0 \leq \epsilon \leq 1$, it is seen that for week shocks,

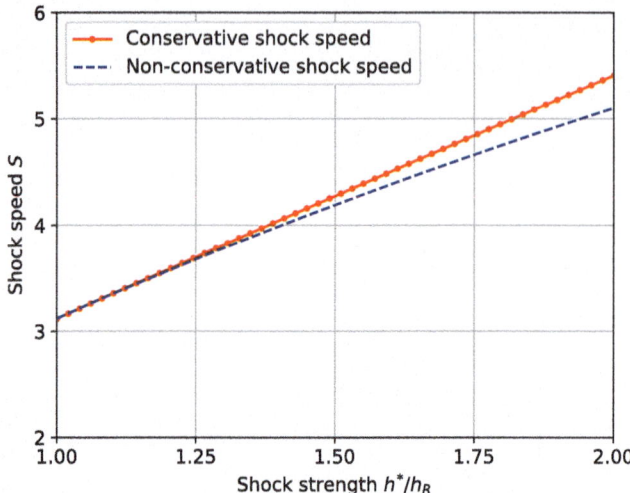

**Fig. 5.15** Shock-wave speeds $S_{cons}$ (higher) and $S_{ncon}$ (lower) versus shock strength $h_*/h_R$, for conservative and non-conservative formulations of the equations. Here $h_R = 1\ m, u_R = 0\ m/s$ and $1 \leq h_*/h_R \leq 2$. Errors around $h_*/h_R \approx 1$ are small

the non-conservative error in shock speed is relatively small and one could therefore contemplate using the non-conservative formulation of the shallow water equations, if weak shock waves could be anticipated reliably.

## 5.8 Concluding Remarks

In this Chapter we have studied all the necessary properties of the shallow water equations needed to solve the Riemann problem exactly. A good working knowledge of these properties is also useful to understand the bases for constructing approximate Riemann solvers, done in Chap. 11, and for interpreting numerical results correctly. Having studied all possible isolated-wave configurations emerging from the Riemann problem, in Chap. 6 we solve exactly the general Riemann problem for the shallow water equations for the case of positive depth throughout the domain. In Chap. 7 we solve the Riemann problem exactly for conditions admitting vacuum, or incipient vacuum, where the water depth $h$ is allowed to be zero. This Chapter is concluded with a list of suggested exercises.

## 5.9 Suggested Exercises

**5.1 Saint Venant Equations.** The homogeneous (no sources) Saint Venant equations are

$$\partial_t A + \partial_x (Au) = 0 \tag{5.82}$$

and

$$\partial_t (Au) + \partial_x (Au^2) + gA\partial_x h = 0 . \tag{5.83}$$

See Chap. 1 for details.

1. Express these equations in quasi-linear form

$$\partial_t \mathbf{W} + \mathbf{A}\partial_x \mathbf{W} = \mathbf{0}$$

   in terms of the variables $A$ and $u$.
2. Find the eigenvalues of the coefficient matrix $\mathbf{A}$ and corresponding right eigenvectors
3. Determine the nature of the characteristic fields corresponding to the eigenvalues.
4. Derive the Riemann invariants associated with isolated rarefaction wave solutions of the Riemann problem for (5.82)–(5.83).

**5.2** (*Equations with variable bed elevation*). Consider the time-dependent one-dimensional shallow water equations with variable bed elevation

$$\partial_t h + u\partial_x h + h\partial_x u = 0 , \tag{5.84}$$

$$\partial_t u + u\partial_x u + g\partial_x h = -gb'(x) , \tag{5.85}$$

where $b = b(x)$ defines the bed elevation above a horizontal datum. See Chap. 1. Here $b(x)$ is independent of time and thus it is legitimate to write

$$\partial_t b = 0 . \tag{5.86}$$

1. Regard Eqs. (5.84)–(5.86) as a $3 \times 3$ system for the three variables $h, u, b$. Write the system in quasi-linear form, with matrix $\mathbf{A}$.
2. Find the three eigenvalues of the system.
3. Find the corresponding right eigenvectors of $\mathbf{A}$.
4. Determine the nature of the three characteristic fields corresponding to the eigenvalues.
5. Discuss isolated elementary wave solutions of the Riemann problem for (5.84)–(5.86).
6. By applying Riemann invariants to each wave family give a detailed discussion on jumps of the quantities $h, u, b$ across the three wave families. For instance, a question to answer is this: does the bed (b) jump across the genuinely non-linear fields?

# References

1. A. Jeffrey, *Quasilinear Hyperbolic Systems and Waves* (Pitman, 1976)
2. R.J. LeVeque, *Finite Volume Methods for Hyperbolic Problems* (Cambridge University Press, 2002)
3. E. Godlewski, P.A. Raviart, *Numerical Approximation of Hyperbolic Systems of Conservation Laws*, 2nd edn. (Springer, Berlin, 2021)

# Chapter 6
# Exact Riemann Solver: Wet Bed

**Abstract** This Chapter presents a solver for computing the exact solution through the entire wave structure of the Riemann problem for the non-linear shallow water equations for the case of a wet bed, that is for $h(x, t) > 0$ (no vacuum). The computing algorithm is based on a detailed study of elementary waves in the Riemann problem performed in Chap. 5. There are two steps in computing the complete solution. First, the water depth $h_*$ and velocity $u_*$ are found in the Star Region; the solution $h_*$ is the root of a non-linear algebraic equation $f(h) = 0$, which is solved iteratively with a Newton-Raphson method. A detailed study of the behaviour of the depth function $f(h)$ is carried out to determine existence and uniqueness of the solution, and for selecting the best solution method. The solution $u_*$ for velocity follows directly from $h_*$. The second step is concerned with finding the solution through the entire wave system, which includes identifying the type of waves present and a solution sampling procedure to determine the vector $Q(x, t)$ at any point $(x, t)$ in the $x$-$t$ half plane. The solution will be used in successive chapters for constructing upwind numerical methods for the general initial-boundary value problem, for applying boundary conditions and for assessing the accuracy of numerical approximations. Useful background is found in Chaps. 1–5. The exact solution for the Riemann problem in the presence of wet/dry fronts is presented in Chap. 7. The listing of a computer programme for the exact Riemann solver, including both wet and dry-bed conditions, is given in Chap. 8.

## 6.1 Introduction

In this Chapter we pose and solve exactly the Riemann problem for the shallow water equations for the case of a wet bed. The Riemann problem is a generalisation of the so-called *dam-break problem*, the solution of which was given by Stoker [1]. As to the solution of the Riemann problem, it appears as if Marshall and Méndez [2] were the first to report a general procedure to solve the Riemann problem exactly for the case of a wet bed. They applied the methodology developed by Godunov [3] for the Euler equations of gas dynamics, which nowadays is known to be computationally expensive and thus alternative schemes are desirable. An efficient solver was presented by Toro [4], which is based on his earlier work on compressible gas dynamics [5]; see also [6]. This method reduces the problem to solving a *single algebraic non-linear equation* for the water depth by an iterative technique, such as the Newton-Raphson method. The remaining flow variables follow directly through the complete structure of the solution of the Riemann problem. There are several reasons for studying the exact solution of the Riemann problem. First, it is the simplest initial value problem for the full set of time-dependent non-linear equations, the solution of which may include simultaneously both smooth solutions, such as rarefactions, as well as discontinuous solutions, such as shocks, contact discontinuities and shear waves. This Chapter builds upon Chap. 5, which gives a detailed characterisation of elementary waves present in the Riemann problem.

The information provided by the Riemann problem solution is fundamental to the understanding of basic features of wave propagation in shallow water models and for the understanding of the more general initial-boundary value problem. The exact solution can also be used locally in the Random Choice Method (RCM) of Glimm [7, 8] for solving the general initial-boundary value problem numerically. For the same purpose the exact solution can be used locally in the Godunov method. The exact Riemann solver presented in this Chapter is very efficient and leads to Godunov methods that are only marginally more expensive than those based on approximate Riemann solvers. Another important reason for wishing to find the exact solution of the Riemann problem is that it can be used to test numerical methods in the early stages of development. All too often, workers in the field do not pay sufficient attention to the use of exact solutions for carefully assessing the performance of their methods intended for more complex applications. A good understanding of the solution process can be a useful exercise for the understanding of physical and mathematical concepts associated with the shallow water equations and for the development of approximate Riemann solvers. Finally, it is worth pointing out the important role of the Riemann problem in implementing boundary conditions to solve the general initial-boundary value problem numerically. Even in the case in which the numerical methodology being used in the interior of the computational domain is not of the upwind type, at the boundaries one cannot do without using the Riemann problem, particularly at reflecting, fixed or moving boundaries.

In this Chapter we deal with the case in which the water depth is everywhere positive (wet bed). In the following Chap. 7 we deal with the case in which the depth of water is zero, either in one of the initial data states or as a consequence of the interaction of two *wet bed* states satisfying a special relationship. The particular Riemann problem in which portions of the bed are dry plays an important role in dealing with the so-called *wet fronts* or *dry fronts* or *wet/dry fronts*, which are very challenging to the numerical modeller. In Chap. 11 we study approximate solutions to the Riemann problem, which are intended for local use in Godunov-type methods.

The reader is strongly advised to revise Chap. 5 before proceeding with the detailed study of this Chapter.

## 6.2 The Riemann Problem and Solution Strategy

We are concerned with the exact solution of the Riemann problem for the **augmented** shallow water equations for the case of a wet bed, namely

$$\begin{aligned}
\text{PDEs:} \quad & \partial_t \mathbf{Q} + \partial_x \mathbf{F}(\mathbf{Q}) = \mathbf{0}\,, \\
\text{ICs:} \quad & \mathbf{Q}(x,0) = \begin{cases} \mathbf{Q}_L & \text{if } x < 0\,, \\ \mathbf{Q}_R & \text{if } x > 0\,. \end{cases}
\end{aligned} \qquad (6.1)$$

For the case of the $x$-split two-dimensional shallow water equations the vectors $\mathbf{Q}$ and $\mathbf{F}(\mathbf{Q})$ are

$$\mathbf{Q} = \begin{bmatrix} h \\ hu \\ hv \end{bmatrix}, \quad \mathbf{F}(\mathbf{Q}) = \begin{bmatrix} h \\ hu^2 + \tfrac{1}{2}gh^2 \\ huv \end{bmatrix}\,. \qquad (6.2)$$

For the case in which the augmented equations include the transport of a passive scalar $\psi(x,t)$ the vectors $\mathbf{Q}$ and $\mathbf{F}(\mathbf{Q})$ are

$$\mathbf{Q} = \begin{bmatrix} h \\ hu \\ h\psi \end{bmatrix}, \quad \mathbf{F}(\mathbf{Q}) = \begin{bmatrix} h \\ hu^2 + \tfrac{1}{2}gh^2 \\ hu\psi \end{bmatrix}\,. \qquad (6.3)$$

Here $\psi(x,t)$ is meant to represent the concentration of a chemical species transported with the fluid velocity $u(x,t)$. As seen in Chap. 5, the mathematical behaviour of the tangential velocity component $v(x,t)$ in (6.2) is identical to that of $\psi(x,t)$ in (6.3).

As anticipated in Chap. 5, the structure of the general solution looks as depicted in Fig. 6.1, which shows three wave families associated with the three eigenvalues $\lambda_k$ that separate four constant regions $\mathcal{R}_k$. The unknown *Star Region* is formed by $\mathcal{R}_1$ and $\mathcal{R}_2$. In regions $\mathcal{R}_0$ and $\mathcal{R}_3$ the vector $\mathbf{Q}$ is given by the initial conditions. The middle wave in Fig. 6.1 is a shear wave for $v$ and a contact discontinuity for $\psi(x,t)$. We shall often use $\psi(x,t)$ to represent either case. The entire $x$-$t$ half plane is divided by the contact/shear wave associated with $\lambda_2 = u$ into two large regions, as shown in Fig. 6.4, a left portion formed by $\mathcal{R}_0$ and $\mathcal{R}_1$ and a right portion formed by $\mathcal{R}_2$ and $\mathcal{R}_3$. In each of these portions there is a non-linear wave, the nature of which is determined as follows

1. **Left non-linear wave**. The nature of the wave is determined by comparing $h_*$ to $h_L$, namely

$$\left.\begin{aligned} h_* > h_L &: \text{ the left wave is a shock wave}\,, \\ h_* \leq h_L &: \text{ the left wave is a rarefaction wave}\,, \end{aligned}\right\} \qquad (6.4)$$

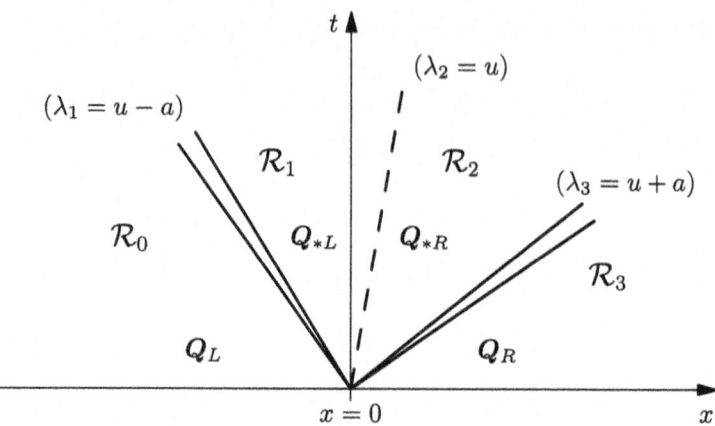

**Fig. 6.1** Structure of the general solution of the Riemann problem for the augmented shallow water equations for the case of a wet bed

2. **Right non-linear wave**. The nature of the wave is determined by comparing $h_*$ to $h_R$, namely

$$\left. \begin{array}{l} h_* > h_R \; : \; \text{the right wave is a shock wave}, \\ h_* \leq h_R \; : \; \text{the right wave is a rarefaction wave}. \end{array} \right\} \quad (6.5)$$

## 6.3 Solution in the Star Region

As anticipated in Chap. 5, the solution in the *Star Region* is determined by the solution for the water depth $h_*$. Once $h_*$ has been computed, all remaining unknowns follow directly. Therefore, the crucial step is performed by deriving a depth function $f(h) = f_L(h, h_L) + f_R(h, h_R) + \Delta u$; see Fig. 6.2. The function $f_L(h, h_L)$ is derived by connecting the velocity $u_*$ to the left data state $\mathbf{Q}_L$ using appropriate wave relations. Analogously, the function $f_R(h, h_R)$ is derived by connecting the velocity $u_*$ to the right data state $\mathbf{Q}_R$. See Fig. 6.2. The wave functions $f_L$ and $f_R$ depend on the unknown depth $h_*$, the *wave type* (shock or rarefaction) and, parametrically, on the initial conditions $\mathbf{Q}_L$ and $\mathbf{Q}_R$, so we may write

$$f_L = f_L(h_*, w_L; \mathbf{Q}_L) \; ; \quad f_R = f_R(h_*, w_R; \mathbf{Q}_R) \; . \quad (6.6)$$

Here $w_L$ and $w_R$ denote logical variables that identify the wave type; $w_K$ denotes either a shock or a rarefaction, for $K = L$ and $K = R$. In what follows we drop the arguments $w_L$, $w_R$ and often also $h_L$ and $h_R$.

## 6.3 Solution in the Star Region

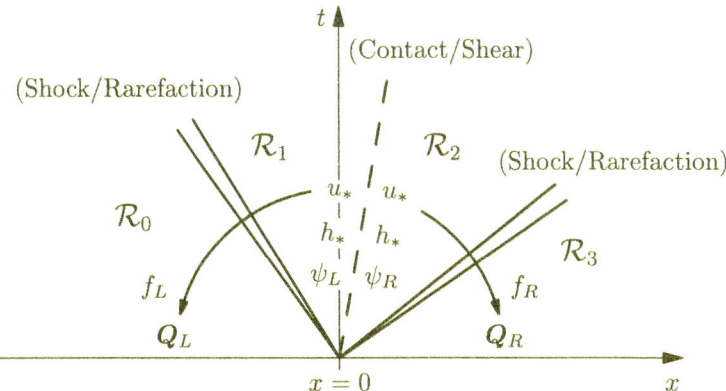

**Fig. 6.2** Connecting $u_*$ to data states $\mathbf{Q}_L$ and $\mathbf{Q}_R$ via the functions $f_L$ and $f_R$

### 6.3.1 Non-linear Equation for Water Depth $h_*$

As demonstrated in Chap. 5, the complete solution procedure for the water depth $h_*$ in the *Star Region* is summarised in the following theorem.

**Theorem 6.1** *The solution $h_*$ for the Riemann problem (6.1) is the root of the non-linear algebraic equation*

$$f(h) := f_L(h; h_L) + f_R(h; h_R) + \Delta u = 0, \quad \Delta u \equiv u_R - u_L, \tag{6.7}$$

*with*

$$f_L(h; h_L) = \begin{cases} 2(\sqrt{gh} - \sqrt{gh_L}) & \text{if } h \leq h_L \text{ (rarefaction)}, \\ (h - h_L)\sqrt{\dfrac{1}{2}g\dfrac{(h + h_L)}{hh_L}} & \text{if } h > h_L \text{ (shock)}, \end{cases} \tag{6.8}$$

*and*

$$f_R(h; h_R) = \begin{cases} 2(\sqrt{gh} - \sqrt{gh_R}) & \text{if } h \leq h_R \text{ (rarefaction)}, \\ (h - h_R)\sqrt{\dfrac{1}{2}g\dfrac{(h + h_R)}{hh_R}} & \text{if } h > h_R \text{ (shock)}. \end{cases} \tag{6.9}$$

*Once the depth $h_*$ has been found the solution for the velocity $u_*$ follows as*

$$u_* = \tfrac{1}{2}(u_L + u_R) + \tfrac{1}{2}[f_R(h_*, w_R; h_R) - f_L(h_*, w_L; h_L)]. \tag{6.10}$$

**Proof** See Chap. 5 for details. □

Finding the root $h_*$ in the *Star Region* involves an iterative numerical procedure applied to non-linear algebraic equation (6.7). The solution for $u_*$ follows directly from (6.10). There appears to be no general closed-form solution for the non-linear algebraic equation (6.7) for the depth in the Star Region; exact solutions are possible for some special cases. Although the solution for $h_*$ must be found *iteratively*, we still use the expression *exact Riemann solver*, which must be understood as an iterative solution to the accuracy provided by a computer.

### 6.3.2 Analysis of the Depth Function $f(h)$

Before devising an iterative scheme for solving $f(h) = 0$ in (6.7) to find $h_*$, we analyse the behaviour of the *depth function* $f(h)$, depicted in Fig. 6.3 for four situations. The root is the intersection of the curve $f(h)$ with the $h$-axis. The analysis is useful in that the behaviour of $f(h)$ determines three important issues of interest:

1. whether the solution for $h_*$ exists,
2. whether it is unique, and
3. what is the best numerical approach to find the root(s).

To begin with, the first derivative of $f = f_L + f_R$ is determined by the first derivatives of $f_L$ and $f_R$; these are found to be given by

$$f'_K(h, h_K) = \begin{cases} \dfrac{g}{a_K} & \text{if } h \leq h_K, \\ g_K(h) - \dfrac{g(h - h_K)}{4h^2 g_K(h)} & \text{if } h > h_K, \end{cases} \quad (6.11)$$

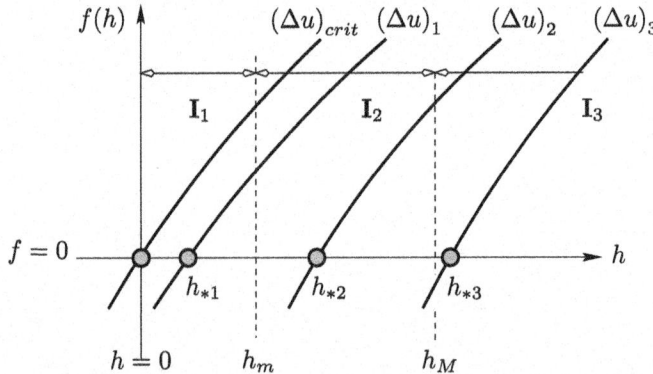

**Fig. 6.3** Behaviour of depth function $f(h)$ against depth $h$; $f(h)$ is a monotone increasing function of $h$ and is concave down. The sought root is the intersection of the curve $f(h)$ with the $h$-axis. Three $h$ intervals have physical relevance, in which the solution exists and is unique

## 6.3 Solution in the Star Region

where

$$g_K(h) = \sqrt{\frac{1}{2}g\left(\frac{h+h_K}{hh_K}\right)}. \tag{6.12}$$

Expressions for the second derivatives are also easily found but they are omitted here; we note, however, that the depth function $f(h)$ is *concave down*, that is $f'' < 0$, as illustrated in Fig. 6.3.

For a given function $f_K(h, h_K)$, there are two possible branches, namely the shock branch $f_{Ksho}$ for $h > h_K$, and the rarefaction branch $f_{Krar}$ for $h \leq h_K$; see (6.8) and (6.9). An interesting observation is that at a given data point $h_K$, we have

$$f_{Krar}^{(l)}(h_K, h_K) = f_{Ksho}^{(l)}(h_K, h_K), \tag{6.13}$$

for $l = 0, 1, 2$ and $K = L, R$. That is, the shock and rarefaction branches coincide at $h = h_K$ and so do their first and second derivatives. As $f' = f'_L + f'_R$ and $f'' = f''_L + f''_R$, it follows that the depth function $f(h)$ is a monotone increasing function of $h$ and concave down, as illustrated in Fig. 6.3. Thus for appropriate combinations of data, the equation $f(h) = 0$ has a unique solution.

For given data $h_L, h_R, u_L, u_R$ the solution is crucially dependent on the data velocity difference $\Delta u = u_R - u_L$. The shape of $f(h)$ is determined by the data $h_L, h_R$ but the height, and thus the root, is determined by $\Delta u$. As seen in Fig. 6.3, there are three $h$ intervals of physical interest ($h_* > 0$). Within each of these intervals along the $h$-axis the complete wave pattern in the Riemann problem solution is also determined. The intervals are

$$\left.\begin{array}{l} I_1 = (0, h_m] \quad : \text{two rarefactions}, \\ I_2 = (h_m, h_M) : \text{one rarefaction and one shock}, \\ I_3 = [h_M, \infty) \: : \text{two shocks}. \end{array}\right\} \tag{6.14}$$

Here

$$h_m = \min(h_L, h_R), \quad h_M = \max(h_L, h_R). \tag{6.15}$$

For fixed $h_L, h_R$ the solution depends entirely on $\Delta u = u_R - u_L$, as we now illustrate through some examples. For a choice such as $(\Delta u)_3$ in Fig. 6.3 the solution is $h_* = h_{*3}$ and lies in $I_3$; note that $h_L < h_*$ and $h_R < h_*$ and thus the solution in this case consists of two shock waves. For a choice such as $(\Delta u)_2$, $h_* = h_{*2} \in I_2$, and one wave is a rarefaction and the other one is a shock. If $\Delta u = (\Delta u)_1, h_* = h_{*1} \in I_1$ and both waves are rarefactions. For choices of $\Delta u$ such that $\Delta u \geq (\Delta u)_{crit}, h_* \leq 0$; here $(\Delta u)_{crit}$ is a critical value of the velocity difference $\Delta u$, defined in Eq. (6.17). The negative value for $h$ is obviously unphysical, as the depth cannot be negative. The case $h_* = 0$ is, however, admissible. We call this *incipient dry bed*, or *incipient cavitation*, and is dealt with in Chap. 7.

We remark that all previous possible situations can be determined *a-priori* in terms of the functions $f_L$ and $f_R$ evaluated on the data values. That is

$$\left. \begin{array}{ll} h_* \in I_3 \text{ (shock/shock)} & \text{if } f(h_M) < 0, \\ h_* \in I_2 \text{ (shock/rarefaction)} & \text{if } f(h_m) \leq 0 \leq f(h_M), \\ h_* \in I_1 \text{ (rarefaction/rarefaction)} & \text{if } f(h_m) \geq 0. \end{array} \right\} \quad (6.16)$$

The critical case of zero depth, $h_* = 0$, corresponds to the condition

$$f(0) = -2(a_L + a_R) + u_R - u_L = 0.$$

Thus, in order for the exact Riemann solver for a wet bed to be applicable, the following **depth positivity condition** on the data must be fulfilled:

$$(\Delta u)_{crit} \equiv 2(a_L + a_R) > u_R - u_L. \quad (6.17)$$

We finally note that if $f(h_m) \geq 0$, that is the two non-linear waves are rarefaction waves, then the rarefaction branches in (6.8) and (6.9) must be used when evaluating $f(h)$ in (6.7). In this case we obtain

$$2(a - a_L) + 2(a - a_R) + u_R - u_L = 0,$$

where we have replaced $\sqrt{gh}$ by the celerity $a$. The solution for $a = a_*$ (and thus for $h_*$) is immediate

$$a_* = \tfrac{1}{2}(a_L + a_R) - \tfrac{1}{4}(u_R - u_L). \quad (6.18)$$

The corresponding particle velocity is given by

$$u_* = \tfrac{1}{2}(u_L + u_R) + a_L - a_R. \quad (6.19)$$

**Two-rarefaction approximation.** We call (6.18)–(6.19) the two-rarefaction approximation. It is exact in the case in which both non-linear waves are rarefactions. For cases in which $f(h_m) < 0$, the solutions (6.18)–(6.19) are approximations to the exact solution. The two-rarefaction approximation is found to be quite accurate also for cases in which a shock wave is present. To some extent this is justified in view of the property (6.13). We note that the properties of the rarefaction branches associated with the two-rarefaction approximation turn out to be very useful when seeking theoretical bounds for the wave speeds emerging from the Riemann problem; see Toro et al. [9]. This property will be used in Chap. 11 when establishing wave speed estimates for HLL-type approximate Riemann solvers.

### 6.3.3 Iterative Solution for $h_*$

Here we solve *numerically* the algebraic non-linear Eq. (6.7), namely

$$f(h) \equiv f_L(h, h_L) + f_R(h, h_R) + \Delta u = 0 \tag{6.20}$$

for the unknown $h$ in the Star Region. There appears to be no general closed-form solution available to this equation and we therefore apply an iterative numerical method. Given the particularly simple behaviour of $f(h)$, the availability of its derivative $f'(h)$ and the availability of a reliable guess value that can be obtained from the two-rarefaction approximation (6.18), we suggest the use of a Newton-Raphson iteration scheme [10]

$$h^{(k+1)} = h^{(k)} - \frac{f(h^{(k)})}{f'(h^{(k)})}, \tag{6.21}$$

for $k = 0, 1, \ldots, K$. To start the iteration we use the two-rarefaction approximation (6.18) to the celerity $a = \sqrt{gh}$ to provide a guess value $h^{(0)}$ as follows

$$h^{(0)} = \frac{1}{g}\left[\frac{1}{2}(a_L + a_R) - \frac{1}{4}(u_R - u_L)\right]^2. \tag{6.22}$$

The iteration (6.21) is stopped whenever the change in $h$ is smaller than a prescribed tolerance $TOL$, that is when

$$\Delta h = \frac{|h^{(k+1)} - h^{(k)}|}{(h^{(k+1)} + h^{(k)})/2} < TOL. \tag{6.23}$$

In practice we choose $TOL = 10^{-6}$. For use in the Godunov method, the choice $TOL = 10^{-4}$ is sufficiently reliable.

Having formulated and solved numerically Eq. (6.20) for $h_*$, the solution for $u_*$ follows directly from (6.10) in terms of $f_L(h_*, h_L)$ and $f_R(h_*, h_R)$. The rest of the solution through the complete wave structure then follows by applying standard wave relations introduced in Chap. 5. This is the subject of the next section.

## 6.4 Sampling the Complete Solution

So far we have an algorithm for finding $h_*$ and $u_*$ in the Star Region of Figs. 6.1 and 6.2, whenever the depth positivity condition (6.17) holds. The aim here is to find the solution through the complete wave structure in the $x$-$t$ half plane. To this end it is convenient to divide the half plane by the contact/shear line $x/t = u_*$, into the two subregions $\mathcal{R}_L = \mathcal{R}_0 \cup \mathcal{R}_1$ and $\mathcal{R}_R = \mathcal{R}_2 \cup \mathcal{R}_3$, as depicted in Fig. 6.4.

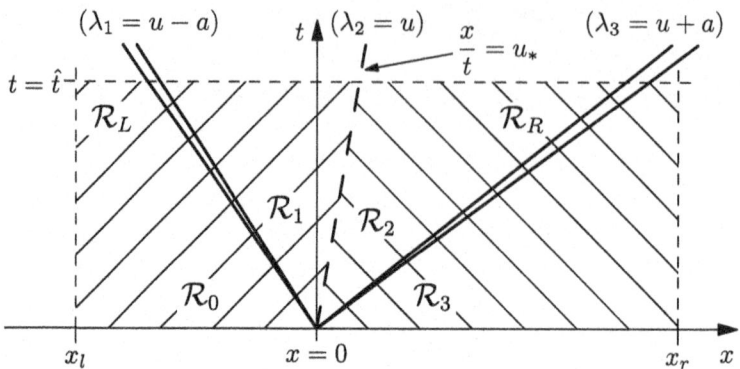

**Fig. 6.4** The contact/shear wave line $x/t = u_*$ divides the $x$-$t$ half plane into two subregions $\mathcal{R}_L = \mathcal{R}_0 \cup \mathcal{R}_1$ and $\mathcal{R}_R = \mathcal{R}_2 \cup \mathcal{R}_3$

### 6.4.1 Passive Scalars

We represent the solution in terms of the vector of physical variables $\mathbf{W} = [h, u, v]^T$ for the $x$-split two-dimensional shallow water equation, or $\mathbf{W} = [h, u, \psi]^T$ for transport of a passive scalar $\psi$. In the former case the middle wave is a *shear wave* and in the latter case it is a *contact discontinuity*. As established in Chap. 5 the solution for the tangential velocity component $v$ and the passive scalar $\psi$ in the complete Riemann problem is particularly simple and essentially decoupled from the rest of the solution. In Chap. 5 it was established that $v$ and $\psi$ are constant across the non-linear waves but change discontinuously across the middle wave defined by the characteristic line $x/t_* = u_*$. That is, across the full wave structure of Fig. 6.2 we have

$$v(x/t_*) = \begin{cases} v_L & \text{if } x/t_* < u_* \\ v_R & \text{if } x/t_* > u_* \end{cases} \quad \psi(x/t_*) = \begin{cases} \psi_L & \text{if } x/t_* < u_*, \\ \psi_R & \text{if } x/t_* > u_*. \end{cases} \quad (6.24)$$

See also Fig. 6.4.

Now we focus on obtaining the solution for the unknowns $h(x, t)$ and $u(x, t)$ in the full wave structure. For a given spatial domain $[x_l, x_r]$ with $x_l < 0 < x_r$, around the origin $x = 0$ of the Riemann problem, and an arbitrary time $t_* > 0$ we seek the solution value of $\mathbf{W}(x, t_*)$, for any $x \in [x_l, x_r]$. As the solution is a *similarity solution*, we perform the sampling procedure by defining the *speed*

$$S = x/t_*. \quad (6.25)$$

Then, by comparing $h_*$ with $h_L$ and $h_R$, see Eqs. (6.4)–(6.5), we determine whether the left and right non linear waves, respectively in $\mathcal{R}_L$ and $\mathcal{R}_R$ in Fig. 6.4, are shocks or rarefactions. We analyse each case separately.

## 6.4.2 Left of Contact/shear: $S = x/t_* \leq u_*$

For a wave to the left of the contact/shear in Fig. 6.4 we need to determine whether it is a shock or a rarefaction. See Fig. 6.5.

1. **Left shock**. If $h_* > h_L$, then the left wave is a shock wave. From the detailed analysis reported in Chap. 5, the shock speed is

$$\left. \begin{aligned} S_L &= u_L - a_L q_L \,, \\ q_L &= \sqrt{\frac{1}{2}\left[\frac{(h_* + h_L)h_*}{h_L^2}\right]} \,. \end{aligned} \right\} \tag{6.26}$$

See Fig. 6.5a. The complete solution on the left side of the contact/shear wave is

$$\mathbf{W}(x, t_*) \equiv \begin{cases} \mathbf{W}_{*L} = [h_*, u_*, v_L]^T & \text{if } S_L \leq x/t_* \leq u_* \,, \\ \mathbf{W}_L = [h_L, u_L, v_L]^T & \text{if } x/t_* \leq S_L \,. \end{cases} \tag{6.27}$$

See solution (6.24) for passive scalars.

2. **Left rarefaction**. If $h_* \leq h_L$ then the left wave in Fig. 6.4 is a rarefaction. The speeds of the *head* and *tail* are respectively

$$S_{HL} = u_L - a_L \,, \quad S_{TL} = u_* - a_* \,. \tag{6.28}$$

See Fig. 6.5b. The solution inside left rarefaction is

$$\mathbf{W}_{Lfan} \equiv \begin{cases} a = \dfrac{1}{3}\left(u_L + 2a_L - \dfrac{x}{t_*}\right), \\ u = \dfrac{1}{3}\left(u_L + 2a_L + \dfrac{2x}{t_*}\right), \\ v = v_L. \end{cases} \tag{6.29}$$

Therefore, for a left rarefaction wave the solution to the left of the contact/ shear wave is

$$\mathbf{W}(x, t_*) = \begin{cases} \mathbf{W}_L & \text{if } x/t_* \leq S_{HL} \,, \\ \mathbf{W}_{Lfan} & \text{if } S_{HL} \leq x/t_* \leq S_{TL} \,, \\ \mathbf{W}_{*L} & \text{if } S_{TL} \leq x/t_* \leq u_* \,, \end{cases} \tag{6.30}$$

where the characteristic speeds $S_{HL}$ and $S_{TL}$ are given by (6.28) and $\mathbf{W}_{Lfan}$ is given by (6.29); see Fig. 6.5b. See solution (6.24) for passive scalars.

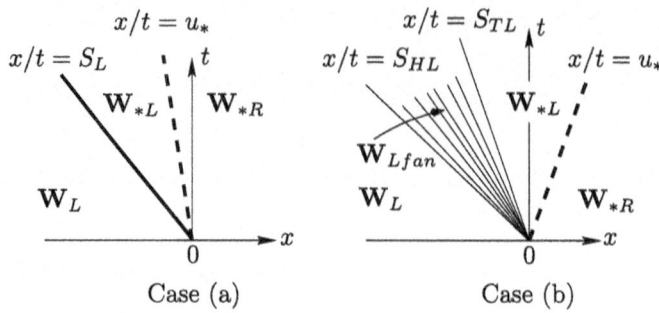

**Fig. 6.5** Wave configurations to the left of the contact shear wave: **a** left wave is a shock wave, **b** left wave is a rarefaction wave

### 6.4.3 Right of Contact/shear: $S = x/t_* \geq u_*$

For a wave to the right of the contact/shear in Fig. 6.4 we need to determine whether it is a shock or a rarefaction. See Fig. 6.6.

1. **Right shock.** If $h_* > h_R$ the right wave is a shock wave of speed

$$\left. \begin{array}{l} S_R = u_R + a_R q_R \,, \\[6pt] q_R = \sqrt{\dfrac{1}{2}\left[\dfrac{(h_* + h_R)h_*}{h_R^2}\right]} \,. \end{array} \right\} \quad (6.31)$$

The complete solution on the right of the contact/shear is

$$\mathbf{W}(x, t_*) \equiv \begin{cases} \mathbf{W}_{*R} = [h_*, u_*, v_R]^T & \text{if } u_* \leq x/t_* \leq S_R \,, \\ \mathbf{W}_L = [h_R, u_R, v_R]^T & \text{if } S_R \leq x/t_* \,, \end{cases} \quad (6.32)$$

where $S_R$ is the shock speed given by (6.31); see Fig. 6.6a. See solution (6.24) for passive scalars.

2. **Right rarefaction.** If $h_* \leq h_R$ the right wave is a rarefaction. The speeds of the *tail* and *head* are

$$S_{TR} = u_* + a_* \,, \quad S_{HR} = u_R + a_R \,. \quad (6.33)$$

See Fig. 6.6b. The solution for $u$, $a$ and $v$ inside the right rarefaction fan is

$$\mathbf{W}_{Rfan} \equiv \begin{cases} a = \dfrac{1}{3}\left(-u_R + 2a_R + \dfrac{\hat{x}}{\hat{t}}\right), \\[6pt] u = \dfrac{1}{3}\left(u_R - 2a_R + \dfrac{2\hat{x}}{\hat{t}}\right), \\[6pt] v = v_R \,. \end{cases} \quad (6.34)$$

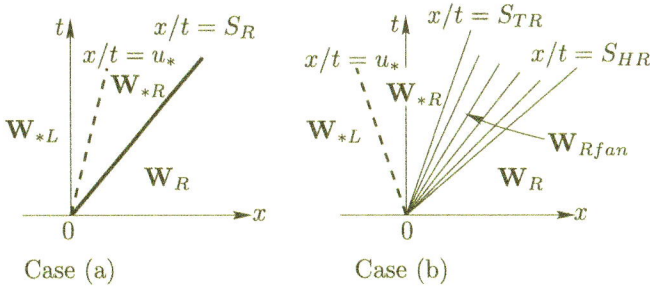

**Fig. 6.6** Wave configurations to the right of the contact/shear wave: **a** right wave is a shock wave, **b** right wave is a rarefaction wave

Therefore, for a right rarefaction wave the solution to the right of the contact/shear wave is

$$\mathbf{W}(x, t_*) = \begin{cases} \mathbf{W}_{*R} & \text{if } u_* \leq x/t_* \leq S_{TR}, \\ \mathbf{W}_{Rfan} & \text{if } S_{TR} \leq x/t_* \leq S_{HR}, \\ \mathbf{W}_R & \text{if } S_{HR} \leq x/t_*. \end{cases} \quad (6.35)$$

The characteristic speeds $S_{TR}$ and $S_{HR}$ are given by (6.33) and $\mathbf{W}_{Rfan}$ is given by (6.34); see Fig. 6.6b. See solution (6.24) for passive scalars.

We have a method to compute the exact solution of the one-dimensional Riemann problem for the variables $h$ and $u$ through the complete wave structure. For the $x$-split two-dimensional shallow water equations we add the tangential velocity component $v$ and for the transport of a passive scalar we add the concentration variable $\psi$. Both $v$ and $\psi$ change only across the middle wave (discontinuously). See solution (6.24) for passive scalars.

## 6.5 Conclusions

We have presented an exact solver for computing the solution through the entire wave structure of the Riemann problem for the non-linear shallow water equations for the case of a wet bed. The computing algorithm builds upon the detailed study of isolated elementary waves in the Riemann problem conducted in Chap. 5. Two steps are needed for computing the complete solution. In the first, the water depth $h_*$ and particle velocity $u_*$ in the *Star Region* are determined. The computation of $h_*$ requires the iterative, numerical solution of a non-linear algebraic equation $f(h) = 0$; to this end a detailed study of the behaviour of the depth function $f(h)$ proved useful in establishing existence and uniqueness of the solution and for choosing the best numerical solution method to compute it. The solution $u_*$ for velocity follows directly from $h_*$. The second step is concerned with finding the solution for the entire wave system, which includes identifying the type of waves present and a solution sampling procedure to determine the vector $Q(x, t)$ at any point $(x, t)$ in the x-t half plane. The solution will be used in Chaps. 9 and 10 for constructing upwind numerical methods for the general initial-boundary value problem and for assessing the accuracy of numerical approximations. The Riemann problem is also useful when applying boundary conditions; even if the numerical solution in the interior of the domain is found through a centred, non-upwind method, the application of boundary conditions cannot do without resorting to the notion of Riemann problem, especially in the case of reflecting solid and moving boundaries. A sound knowledge of the exact solution is of fundamental value in the designing of approximate Riemann solvers, as done in Chap. 11, and for judging the correctness of newly proposed solvers. Useful background is found in Chaps. 2, 4 and 5. The exact solution for the Riemann problem in the presence of wet/dry fronts is presented in Chap. 7. The listing of a computer programme for the exact Riemann solver, including both wet and dry-bed conditions, is given in Chap. 8.

## 6.6 Exercises

Consider the Riemann problem for the augmented sallow water equations

$$\text{PDEs: } \partial_t \mathbf{Q} + \partial_x \mathbf{F}(\mathbf{Q}) = \mathbf{0} \,,$$
$$\text{ICs: } \mathbf{Q}(x, 0) = \begin{cases} \mathbf{Q}_L \text{ if } x < 0 \,, \\ \mathbf{Q}_R \text{ if } x > 0 \,. \end{cases} \quad (6.36)$$

For the case of the x-split two-dimensional shallow water equations the vectors $\mathbf{Q}$ and $\mathbf{F}(\mathbf{Q})$ are

$$\mathbf{Q} = \begin{bmatrix} h \\ hu \\ hv \end{bmatrix}, \quad \mathbf{F}(\mathbf{Q}) = \begin{bmatrix} h \\ hu^2 + \frac{1}{2}gh^2 \\ huv \end{bmatrix}. \quad (6.37)$$

**6.1** For initial conditions

$$h_L = h_R, \quad u_L = -u_R, \quad v_L \neq v_R \,. \quad (6.38)$$

1. Assuming $u_L < 0$ find the complete solution of the Riemann problem (6.36).
2. Assuming $u_L > 0$ find the complete solution of the Riemann problem (6.36).

**6.2** For initial conditions

$$h_L = h_R, \quad u_L = -u_R + 2V_{wall}, \quad v_L \neq v_R, \quad (6.39)$$

where $V_{wall} > 0$ denotes the velocity of the boundary.

1. Assuming $u_L < 0$ find the complete solution of the Riemann problem (6.36).
2. Assuming $u_L > 0$ find the complete solution of the Riemann problem (6.36).

**Boundary Riemann problems**. The problems above arise in the application of boundary conditions for the case of reflective boundaries, for the fixed boundary case $V_{wall} = 0$ and for the moving boundary case $V_{wall} \neq 0$. See Chaps. 10 and 11 on the subject of boundary conditions.

# References

1. J.J. Stoker, *Water Waves: The Mathematical Theory with Applications* (Wiley, 1992)
2. E. Marshall, R. Méndez, Computational aspects of the random choice method for shallow water equations. J. Comput. Phys. **39**, 1–21 (1981)
3. S.K. Godunov, Finite difference methods for the computation of discontinuous solutions of the equations of fluid dynamics. Mat. Sb. **47**, 271–306 (1959)
4. E.F. Toro, Riemann problems and the WAF method for solving two-dimensional shallow water equations. Phil. Trans. Roy. Soc. London **A338**, 43–68 (1992)
5. E.F. Toro, A fast Riemann solver with constant covolume applied to the random choice method. Int. J. Numer. Meth. Fluids **9**, 1145–1164 (1989)
6. E.F. Toro, *Riemann Solvers and Numerical Methods for Fluid Dynamics. A Practical Introduction*, 3rd ed. (Springer, 2009)
7. J. Glimm, Solution in the large for nonlinear hyperbolic systems of equations. Comm. Pure. Appl. Math. **18**, 697–715 (1965)
8. A.J. Chorin, Random choice solutions of hyperbolic systems. J. Comput. Phys. **22**, 517–533 (1976)
9. E.F. Toro, L.O. Müller, A. Siviglia, Bounds for wave speeds in the Riemann problem: direct theoretical estimates. Comput. Fluids **209**(104640) (2020)
10. J.H. Mathews, *Numerical Methods* (Prentice–Hall International, Inc., 1987)

# Chapter 7
# Exact Riemann Solver: Dry Bed

**Abstract** In this Chapter we solve the Riemann problem for the split one-dimensional shallow water equations for the case in which the solution is adjacent to *dry regions*, or *vacuum regions*, that is when $h(x, y, t) = 0$. The boundary separating regions of water and no water, called the wet/dry front, emerges from the exact solution of the Riemann problem as the edge (tail) of a strong rarefaction wave, which is the only wave present in the solution structure. This wet/dry front is a very fast wave of speed $S = u_L + 2a_L$, for the case of a dry right data state, and is the source of considerable challenges in computational practice. We solve the Riemann problem exactly under two conditions: (i) vacuum is present in one of the initial states at time $t = 0$ and (ii) vacuum appears from the interaction of two non-vacuum states, generating two wet/dry fronts. The solution strategy is then extended to the split two-dimensional shallow water equations and for the case of pollutant transport models. A computer-programme listing is given in Chap. 8, for the complete exact Riemann solver that can deal with both wet and dry bed conditions. Useful background is found in Chaps. 1–6.

## 7.1 Introduction

In the previous Chap. 6 we solved the Riemann problem exactly for the case in which the water depth is strictly positive everywhere. In dry regions the water depth is obviously zero, a perfectly acceptable physical situation, in which the shallow water equations or any other equations based on the continuum assumption are obviously not applicable. Now we consider *wet* regions that are adjacent to dry regions, which effectively means solving the shallow water equations in the wet regions, *right up to the boundary between wet and dry regions*. The simplest case is a horizontal dam adjacent to a dry horizontal region, see Fig. 7.1. This is indeed a special case of the Riemann problem.

In this Chapter we solve the Riemann problem for cases in which a dry-bed region is either present at the initial time or appears as the result of the interaction of two *wet-bed* states; see Toro [1]. Formally, the Riemann problem for the one-dimensional shallow water equations is the initial value problem (IVP)

# 7 Exact Riemann Solver: Dry Bed

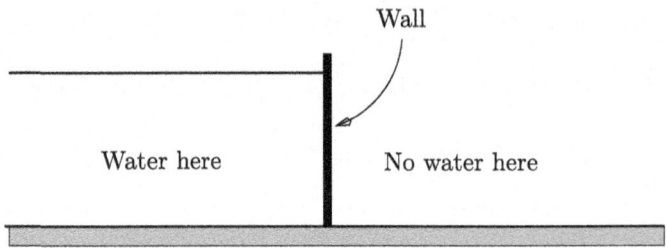

**Fig. 7.1** The Riemann problem for which one data state has no water (dry bed)

$$\text{PDEs:} \quad \partial_t \mathbf{Q} + \partial_x \mathbf{F}(\mathbf{Q}) = \mathbf{0} \, ,$$
$$\text{ICs:} \quad \mathbf{Q}(x, 0) = \begin{cases} \mathbf{Q}_L & \text{if } x < 0 \, , \\ \mathbf{Q}_R & \text{if } x > 0 \, , \end{cases} \quad (7.1)$$

for which the vectors $\mathbf{Q}$ and $\mathbf{F}(\mathbf{Q})$ are

$$\mathbf{Q} = \begin{bmatrix} h \\ hu \end{bmatrix}, \quad \mathbf{F}(\mathbf{Q}) = \begin{bmatrix} hu \\ hu^2 + \frac{1}{2} gh^2 \end{bmatrix} . \quad (7.2)$$

The structure of the solution of Riemann problems in which dry-bed regions are present is *distinct* from that of the wet-bed Riemann problem studied in Chap. 6. In the wet-bed case, for the purely one-dimensional equations, there are always two wave families and four possible wave patterns. Any attempts at using the wet-bed Riemann problem structure will naturally fail, simply because the structure of the solution is different. For instance, in the wet-bed Riemann problem one first finds the value of the depth $h_*$ in the **Star Region**. In the dry-bed Riemann problem the star region does not exist. When using approximate Riemann solvers in methods of the Godunov type, it is tempting to deal with dry-bed Riemann problems by replacing the dry states by wet states with *very shallow water*. But then the solution is different to that of the original problem. This issue is discussed at some length at the end of Chap. 11.

## 7.2 Admissible Wet/Dry Interface Waves

In this section we prove an important result concerning the admissible waves connecting a wet-bed region with a dry-bed region [1].

**Proposition 7.1** *A shock wave (bore) cannot be adjacent to a region of dry bed.*

**Proof** Consider the Riemann problem (7.1) for the one-dimensional shallow water equations with data expressed in terms of physical variables

$$\mathbf{W}_L = \begin{bmatrix} h_L \\ u_L \end{bmatrix}, \quad \mathbf{W}_R \equiv \mathbf{W}_0 = \begin{bmatrix} h_0 \\ u_0 \end{bmatrix}, \tag{7.3}$$

such that $\mathbf{W}_L$ is the data for a wet-bed portion ($h_L > 0$) and $\mathbf{W}_0$ is the data for the dry-bed portion; $h_0 = 0$ and $u_0$ is *arbitrary*.

Let us now suppose that $\mathbf{W}_L$ and $\mathbf{W}_0$ are connected by a shock wave of speed $S$. Then, application of the Rankine-Hugoniot conditions, in terms of conserved variables, gives

$$\left. \begin{aligned} h_L u_L &= h_0 u_0 + S(h_L - h_0) , \\ h_L u_L^2 + \tfrac{1}{2} g h_L^2 &= h_0 u_0^2 + \tfrac{1}{2} g h_0^2 + S(h_L u_L - h_0 u_0) . \end{aligned} \right\} \tag{7.4}$$

As $h_0 = 0$, the first equation gives $S = u_L$, which means that the speed of the assumed shock wave is identical to the particle velocity behind the shock. Substitution of $S = u_L$ in the second equation gives $h_L = 0$, which contradicts our assumption that the left state is wet ($h_L > 0$) and the proposition is thus proved. □

**Remark 7.2** From the result of the previous proposition it follows that a *contact discontinuity* can be adjacent to a region of dry bed; this makes perfect physical sense, as the wave separates a region of water from a region of no water, or vacuum.

**Remark 7.3** The particle velocity in the dry bed region $u_0$ is arbitrary and can be set to $u_0 = S$. In practice, however, one sets $u_0 = 0$ in the understanding that *the particle velocity of nothing should be zero*. This choice causes a discontinuity jump in particle velocity right at the wet/dry front, which might contribute to numerical difficulties in capturing this feature by numerical means.

## 7.3 Dry Bed: Three Possible Cases

There are three cases involving a dry bed, or vacuum. These are illustrated in Fig. 7.2. Case (a) is where the dry bed is on the initial right-hand side. The solution consists of a *single* left rarefaction wave, that is a rarefaction wave associated with the left eigenvalue $\lambda_1 = u - a$. Case (b) is that where the dry bed is on the initial left-hand side and the solution consists of a single right rarefaction wave associated with the right eigenvalue $\lambda_2 = u + a$. Case (c) is where a dry bed is not present at $t = 0$ but is generated in the interaction of the data states $\mathbf{W}_L$ and $\mathbf{W}_R$, if these do not satisfy the **depth-positivity condition**

$$u_R - u_L < (\Delta u)_{crit} := 2(a_L + a_R) . \tag{7.5}$$

Next we consider each case separately.

**Fig. 7.2** Three cases in which the solution of the Riemann problem involves a dry-bed state; **a** dry-bed region is on the right, **b** dry-bed region is on the left and **c** dry-bed region appears in the interaction of two wet-bed states violating the depth positivity condition (7.5)

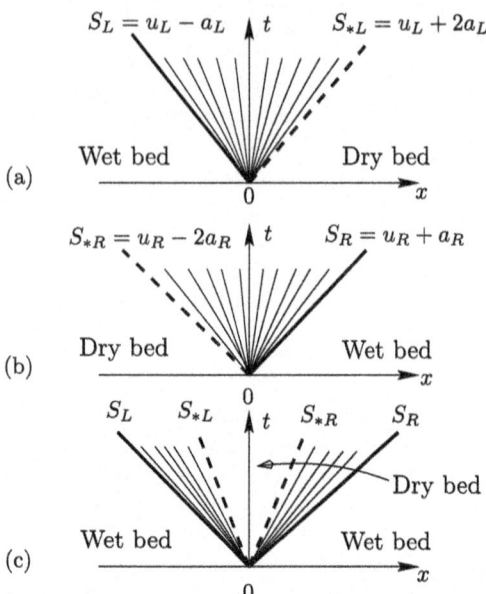

## 7.3.1 The Dry Bed Is on the Right Side

This Riemann problem has initial data

$$\mathbf{W}(x, 0) = \begin{cases} \mathbf{W}_L \ne \mathbf{W}_0 & \text{if } x < 0, \\ \mathbf{W}_0 \text{ (dry bed)} & \text{if } x > 0. \end{cases} \tag{7.6}$$

The structure of the general solution is shown in Fig. 7.2a. The expected right shock is missing from the wave pattern. The absence of a wave corresponding to the wave family $\lambda_2 = u + a$ is physically correct as there is no medium (water) for any wave of this family to propagate through. The solution contains a left rarefaction wave associated with $\lambda_1 = u - a$, and a *contact discontinuity* of speed $S_{*L}$; see the dashed line on the right-hand side of Fig. 7.2a that coincides with the *tail* of the rarefaction. The exact solution follows from the methods developed in Chaps. 4, 5 and 6. Consider a point $P_0$ along the wet/dry front (contact discontinuity) and denote respectively by $a_c$ and $u_c$ the celerity and the particle velocity along the characteristic of speed $S_{*L}$ that coincides with the front

$$\frac{dx}{dt} = S_{*L} = u_c - a_c. \tag{7.7}$$

By connecting $a_c$ and $u_c$ to the data on the left-hand side via the left Riemann invariant we obtain

$$u_c + 2a_c = u_L + 2a_L. \tag{7.8}$$

## 7.3 Dry Bed: Three Possible Cases

See Chap. 5. But along the contact discontinuity $h_c = 0$, and thus $a_c = 0$. Hence the speed of the contact discontinuity is

$$S_{*L} = u_c = u_L + 2a_L .\tag{7.9}$$

Therefore the complete solution is

$$\mathbf{W}_{LO}(x,t) = \begin{cases} \mathbf{W}_L & \text{if } x/t \leq u_L - a_L , \\ \mathbf{W}_{Lfan}(x,t) & \text{if } u_L - a_L \leq x/t \leq S_{*L} , \\ \mathbf{W}_0 & \text{if } S_{*L} \leq x/t , \end{cases}\tag{7.10}$$

where $\mathbf{W}_{Lfan}(x,t)$ is given by

$$\mathbf{W}_{Lfan}(x,t) = \begin{cases} a = \dfrac{1}{3}\left(u_L + 2a_L - \dfrac{x}{t}\right), \\ u = \dfrac{1}{3}\left(u_L + 2a_L + \dfrac{2x}{t}\right). \end{cases}\tag{7.11}$$

This is the solution inside the rarefaction fan.

### Example

Figure 7.3 shows the typical solution profile for depth and particle velocity at a specified time. Note that the solution for depth is continuous, with a discontinuity in the $x$ derivative right at the front, where the particle velocity jumps discontinuously from the maximum value at the peak to $u_0 = 0$ in the dry region. As discussed earlier, $u_0$ could be given a different value, such as $u_0 = u_c$. Numerically, it is very difficult

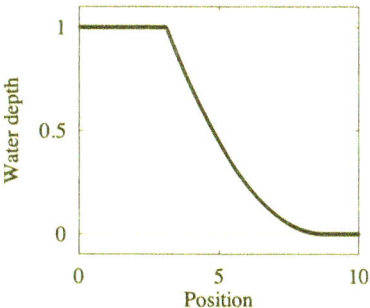
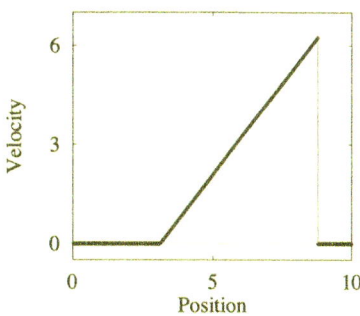

**Fig. 7.3** Exact solution for a dry-bed Riemann problem, with initial conditions $h_L = 1$, $h_R = 0$, $u_L = 0$, $u_R = 0$. The spatial domain is the interval $[0, 10]$ in meters; the initial discontinuity is positioned at $x_0 = 5.0\,m$ and the solution is displayed at time $T_{out} = 0.6$ s

to compute correctly the position of the wet/dry front, which is most evident when the velocity profile is plotted.

### 7.3.2 The Dry Bed Is on the Left Side

This Riemann problem has initial data of the form

$$\mathbf{W}(x, 0) = \begin{cases} \mathbf{W}_0 & \text{if } x < 0, \\ \mathbf{W}_R \neq \mathbf{W}_0 & \text{if } x > 0. \end{cases} \tag{7.12}$$

The structure of the exact solution is shown in Fig. 6.2b. The wave associated with the family $\lambda_1 = u - a$ is missing. The solution consists of a right rarefaction wave associated with $\lambda_2 = u + a$, whose *tail* coincides with the wet/dry front, the speed of which is given by

$$S_{*R} = u_R - 2a_R. \tag{7.13}$$

The complete solution is

$$\mathbf{W}_{RO}(x,t) = \begin{cases} \mathbf{W}_0 & \text{if } x/t \leq S_{*R}, \\ \mathbf{W}_{Rfan}(x,t) & \text{if } S_{*R} \leq x/t \leq u_R + a_R, \\ \mathbf{W}_R & \text{if } u_R + a_R \leq x/t, \end{cases} \tag{7.14}$$

where $\mathbf{W}_{Rfan}(x, t)$ is given by

$$\mathbf{W}_{Rfan}(x, t) \equiv \begin{cases} a = \dfrac{1}{3}\left(-u_R + 2a_R + \dfrac{x}{t}\right), \\ u = \dfrac{1}{3}\left(u_R - 2a_R + \dfrac{2x}{t}\right). \end{cases} \tag{7.15}$$

The reader is encouraged to verify the above calculations.

### 7.3.3 Generation of Vacuum from Wet-Bed States

For general wet-bed data $\mathbf{W}_L \neq \mathbf{W}_0$ and $\mathbf{W}_R \neq \mathbf{W}_0$, there can be special combinations of particle velocities and celerities via the *depth positivity condition* (7.5), such that a dry bed is created in the interaction of the two data states. For such a situation the solution looks like Fig. 7.2c. There are two rarefaction waves associated with the eigenvalues $\lambda_1 = u - a$ and $\lambda_2 = u + a$, with two wet/dry fronts attached to

## 7.4 Passive Scalars

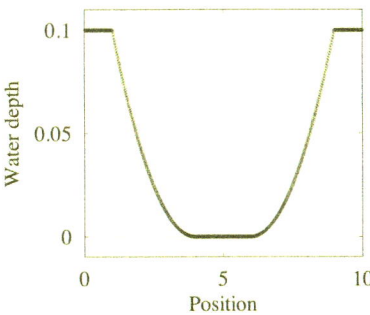
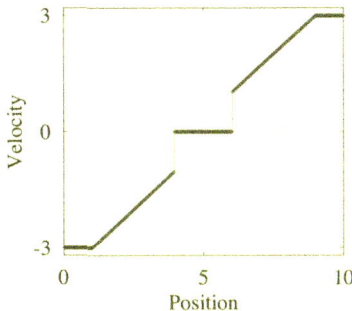

**Fig. 7.4** Exact solution for a Riemann problem in which a dry-bed region between two rarefaction waves results from the interaction of two wet-bed states satisfying condition (7.17). Initial conditions are $h_L = 0.1\ m, h_R = 0.1\ m, u_L = -3\ m/s, u_R = 3\ m/s$. The spatial domain is the interval [0, 10] in meters; the initial discontinuity is positioned at $x_0 = 5.0\ m$ and the solution is displayed at time $T_{out} = 1$ s. This example corresponds to the wave structure shown in Fig. 7.2c

their tails and a *new region of dry bed* between the rarefaction waves. The complete solution can be written in terms of the previous two cases as

$$\mathbf{W}(x,t) = \begin{cases} \mathbf{W}_{L0}(x,t) & \text{if } x/t \leq S_{*L}, \\ \mathbf{W}_0 \text{ (dry bed)} & \text{if } S_{*L} \leq x/t \leq S_{*R}, \\ \mathbf{W}_{R0}(x,t) & \text{if } S_{*R} \leq x/t, \end{cases} \quad (7.16)$$

where $\mathbf{W}_{L0}$ and $\mathbf{W}_{R0}$ are given by (7.10)–(7.11) and (7.14)–(7.15), respectively. For this case to apply we require the condition $S_{*L} \leq S_{*R}$, that is

$$(\Delta u)_{crit} \equiv 2(a_L + a_R) \leq u_R - u_L. \quad (7.17)$$

Here, the *depth positivity condition* (7.5) is violated. The reader is encouraged to verify the above calculations.

**Example**. Figure 7.4 shows typical solution profiles for depth and particle velocity for a Riemann problem in which a dry-bed region results from the interaction of two wet-bed states satisfying the condition (7.17). Note the discontinuity in particle velocity at the two wet/dry fronts. See Fig. 7.2c.

## 7.4 Passive Scalars

As discussed in Chap. 6, the one-dimensional shallow water equations in (7.1)–(7.2) may be extended to include the split two-dimensional case or the transport of pollutants via additional advection equations for their concentrations. In either case there will be an additional advection equation for a passive scalar $\psi(x, y, t)$ to represent either the tangential velocity component, say $v(x, y, t)$, or the concentration of a chemical species. It is also possible to have more advection equations to include both the tangential velocity and concentrations for species equations. Here we assume a single advection equation for $\psi(x, t)$

$$\partial_t \psi + u \partial_x \psi = 0 . \tag{7.18}$$

When (7.18) is used in conjunction with the continuity equation in (7.1), it can be re-written in conservation-law form as follows

$$\partial_t (h\psi) + \partial_x (uh\psi) = 0 . \tag{7.19}$$

For a Riemann problem with dry right-hand side the initial data for the passive scalar is

$$\psi(x, 0) = \begin{cases} \psi_L & \text{if } x < 0 , \\ \psi_R = 0 & \text{if } x > 0 . \end{cases} \tag{7.20}$$

The exact solution is

$$\psi(x/t) = \begin{cases} \psi_L & \text{if } x/t < S_{*L} , \\ \psi_R & \text{if } x/t > S_{*L} , \end{cases} \tag{7.21}$$

where $S_{*L}$ is the speed of the right-facing wet/dry front. See Fig. 7.2a. The distribution of $\psi(x/t)$ consists of two uniform states $\psi_L$ and $\psi_R = 0$, joined by a travelling discontinuity of speed $S_{*L}$. When $\psi(x, t)$ represents the tangential velocity component $v(x, y, t)$ this wave is strictly speaking a *shear wave*. When $\psi(x, t)$ is the concentration of a chemical species, then this wave is a contact discontinuity travelling with the wet/dry front speed $S_{*L}$.

For a Riemann problem with dry left-hand side, the initial data for the passive scalar is

$$\psi(x, 0) = \begin{cases} \psi_L = 0 & \text{if } x < 0 , \\ \psi_R & \text{if } x > 0 , \end{cases} \tag{7.22}$$

and the solution is

$$\psi(x/t) = \begin{cases} \psi_L = 0 & \text{if } x/t < S_{*R} , \\ \psi_R & \text{if } x/t > S_{*R} . \end{cases} \tag{7.23}$$

The distribution of $\psi(x/t)$ consists of two uniform states $\psi_L = 0$ and $\psi_R$ joined by a travelling *contact discontinuity* of speed $S_{*R}$. See Fig. 7.2b.

For a Riemann problem in which a dry bed state is created from the interaction of water states not satisfying the depth positivity condition, the initial data for the passive scalar is of the form

## 7.5 Conclusions

$$\psi(x,0) = \begin{cases} \psi_L & \text{if } x < 0, \\ \psi_R & \text{if } x > 0, \end{cases} \qquad (7.24)$$

and the exact solution is

$$\psi(x/t) = \begin{cases} \psi_L & \text{if} \quad x/t < S_{*L}, \\ 0 & \text{if } S_{*L} < x/t < S_{*R}, \\ \psi_R & \text{if} \quad x/t > S_{*R}. \end{cases} \qquad (7.25)$$

The distribution of $\psi(x/t)$ consists of three uniform states, namely $\psi_L$, $\psi = 0$ and $\psi_R$, joined by two travelling contact discontinuities of speeds $S_{*R}$ and $S_{*L}$. See Fig. 7.2c.

**Remark 7.4** A general Riemann solver must include a check on the initial data to verify the depth positivity condition (7.5) in order to deal with either the wet-bed Riemann problem of Chap. 6 or with the dry-bed Riemann problem of this Chapter. Approximate Riemann solvers used in Godunov-type methods should also include a check on a depth positivity condition, which in all probability will be solver dependent. For example, in the field of gas dynamics, linearised Riemann solvers will predict incipient cavitation well before it is predicted by the exact *pressure positivity condition*. To my knowledge this problem has not yet been studied in sufficient detail, for both gas dynamics and shallow water flows.

**Remark 7.5** The speed of a wet/dry front is of the form $S_{*R} = u_R - 2a_R$ for a left dry state and $S_{*L} = u_L + 2a_L$ for a right dry state. These speeds can be larger than those associated with the eigenvalues $\lambda_1 = u - a$ and $\lambda_2 = u + a$. This has a bearing on the stability condition implemented in numerical methods. One usually chooses the time step based on the possible maximum of the eigenvalues in absolute value. In [2] it is argued that this practice is not entirely correct. In the presence of wet/dry fronts this practice will most certainly lead to numerical instabilities and for which the numerical method itself may not be wholly responsible.

## 7.5 Conclusions

The complete solution to the dry-bed Riemann problem has been presented, which includes three cases. These exact solutions can be used to assess the performance of numerical methods to solve problems including wet/dry fronts and also to assist the construction of Godunov-type methods, in which Riemann problem solutions are used locally. Useful background is found in Chaps. 2, 4, 5 and 6. Chapter 8 gives the programme listing for a complete exact Riemann solver that can deal with both wet and dry bed conditions.

# References

1. E.F. Toro, The dry-bed problem in shallow-water flows. Technical report, College of Aeronautics, Cranfield Institute of Technology, UK. CoA 9007 (1990)
2. E.F. Toro, L.O. Müller, A. Siviglia, Bounds for wave speeds in the Riemann problem: direct theoretical estimates. Comput. Fluids **209**(104640) (2020)

# Chapter 8
# Tests with Exact Solution

**Abstract** In this Chapter we present a suite of carefully selected test problems with exact solution for the non-linear shallow water equations on a horizontal bottom, admitting both wet-bed and dry-bed conditions. The purpose of these test problems is to rigorously assess numerical methods for the general initial-boundary value problem, against exact solutions for the kind of physical situations that are bound to be challenging to the numerical modeller. Each of the test problems has been carefully designed and offers a specific challenge to numerical methods. The exact solutions will be used to asses the suitability of approximate Riemann solvers and the accuracy of approximate numerical solutions in successive chapters. A computer programme listing for the Riemann solvers is given at the end of the chapter, for both wet and dry-bed conditions. Useful background is found in Chaps. 6 and 7, as well as in Refs. [1, 2] for the wet-bed case and [3] for the dry-bed case.

## 8.1 Introduction

This chapter presents a suite of carefully selected test problems for the shallow water equations with exact solution in the hope that they will be of use to numerical practitioners involved in the testing of numerical methods or developing code from scratch, or both. Exact solutions do not abound but when available they should be used to systematically assess methods and codes. Tests similar to the ones presented here are chosen by practitioners but often the exact solution is excluded from the comparisons. Each of the test problems has been carefully designed and offers a specific challenge to numerical methods. They have not been chosen to *show off* the power of particular methods but to *challenge* the suitability of numerical methods intended for application to the shallow water equations.

Five test problems for a straight channel of constant rectangular cross-section and horizontal bed elevation are presented in Table 8.1. In all cases we take a channel of length 50 m. The initial data in Table 8.1 is given for water depth $h$, water particle velocity $u$; $x_0$ is the position of the initial discontinuity, which simulates a *gate* withdrawn at high speed, and $T_{out}$ is the output time, at which the solution is to be displayed.

**Table 8.1** Data for five test problems with exact solution. The length of the channel is 50 m, $x_0$ is the position of the initial discontinuity and $T_{out}$ is the output time in seconds

| Test | $h_L(m)$ | $u_L(m/s)$ | $h_R(m)$ | $u_R(m/s)$ | $x_0(m)$ | $T_{out}(s)$ |
|---|---|---|---|---|---|---|
| 1 | 1.0 | 2.5 | 0.1 | 0.0 | 10.0 | 7.0 |
| 2 | 1.0 | −5.0 | 1.0 | 5.0 | 25.0 | 2.5 |
| 3 | 1.0 | 0.0 | 0.0 | 0.0 | 20.0 | 4.0 |
| 4 | 0.0 | 0.0 | 1.0 | 0.0 | 30.0 | 4.0 |
| 5 | 0.1 | −3.0 | 0.1 | 3.0 | 25.0 | 5.0 |

**Table 8.2** Exact solution in the *Start Region* for Tests 1 and 2. For Test 1 the guess value for $h_*$ comes from the two-shock approximation and for Test 2 comes from the two-rarefaction approximation, which is exact in this case

| Test | $h_*$ | $u_*$ | Approximation | $h_{guess}$ | Iterations |
|---|---|---|---|---|---|
| 1 | 0.611753 | 3.86398 | Two-shock | 0.7357571 | 2 |
| 2 | 0.040564 | 0.0 | Two-rarefaction | 0.0405637 | 1 |

All five test problems are solved exactly using the exact Riemann solvers presented in Chap. 6 for the wet-bed case and in Chap. 7 for the dry-bed case, and embodied in the computer programme given in Sect. 8.9. The exact solutions are obtained from the following values for the computational parameters:

1. The acceleration due to gravity is $g = 9.8$ m/s$^2$,
2. The tolerance $TOL$ in the solver is $TOL = 10^{-6}$,
3. The maximum number of iterations allowed in the exact solver is $NITER = 50$,
4. The solution is plotted with $MCELLS = 500$ points.

Table 8.2 shows the values of the exact solution in the *Star Region* for Tests 1 and 2 for the water depth $h_*$ and the particle velocity $u_*$, see Chap. 6. These numbers can also be useful to test numerical methods. Also displayed are the approximations used as the initial guess for the depth $h$ in the iterative procedure and the number of iterations required for convergence. For Test 1 the program uses the *two-shock approximation*, see Chap. 11, to provide a guess value for $h_*$ and the solver takes two iterations for convergence. For Test 2 the guess value is taken from the *two-rarefaction approximation*, which for this test is actually the exact solution and thus the solver goes through the iteration procedure just once.

In Sect. 8.7 we give details of a transformation whereby the inhomogeneous equations resulting from the inclusion of variable bed elevation with *constant slope* become homogeneous, and thus exact solutions are again possible. In Sect. 8.9 we give the listing of a FORTRAN program to solve the general Riemann problem for the shallow water equations for both wet and dry bed conditions.

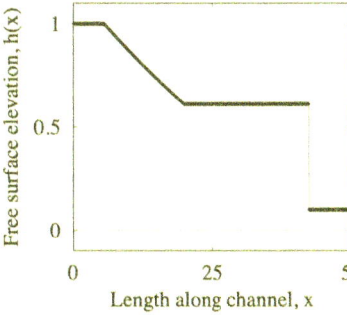
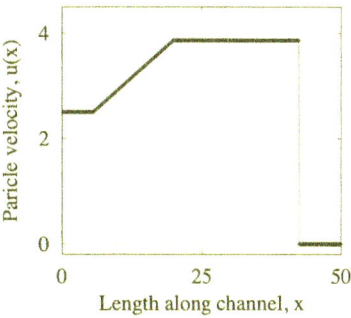

**Fig. 8.1** Exact solution to Test 1 at time $T_{out} = 7.0$ s; initial discontinuity is at $x_0 = 10.0$ m. See Table 8.1

## 8.2 Test 1: Left Critical Rarefaction and Right Shock

The initial data for this test have been chosen so as to produce a strong right propagating shock wave and a *critical* or *transcritical* left rarefaction wave. Figure 8.1 shows profiles of water depth and particle velocity at the output time $T_{out} = 7.0$ s.

When using this test problem to assess a numerical method one expects the correct numerical resolution of the shock wave. Items to watch include (i) correct speed of propagation, (ii) correct strength of the jump, (iii) width of the shock layer and (iv) absence (or otherwise) of *spurious oscillations* in the vicinity of the shock. The other important feature of this test is the *critical, or transcritical* left rarefaction wave; that is the left eigenvalue $\lambda_1 = u - a$ changes from a negative to a positive value as the wave is crossed from left to right. Then, obviously $\lambda_1$ goes through the value $\lambda_1 = u - a = 0$, that is $u = a$, hence the particle velocity equals the celerity. This is called critical flow.

Numerical methods are bound to experience difficulties at *critical points*, even those methods that are theoretically entropy satisfying, such as the first-order Godunov method used in conjunction with the exact Riemann solver. See Chaps. 10 and 11. Entropy-violating methods will produce an unphysical jump at the critical point inside the rarefaction wave. Such an unphysical jump is known as a *rarefaction shock*, which does not satisfy the Lax entropy condition; see Chap. 2.

Table 8.2 gives the exact values for depth $h_*$ and particle velocity $u_*$ in the **Star Region**. These may be helpful to the reader to check her/his own computations.

## 8.3 Test 2: Two Rarefactions and Nearly Dry Bed

Test 2 has chosen initial data such that the solution consists of two strong rarefaction waves travelling in opposite directions. The water in the *Star Region* between these waves is very shallow, small $h$. There are good numerical reasons for choosing this

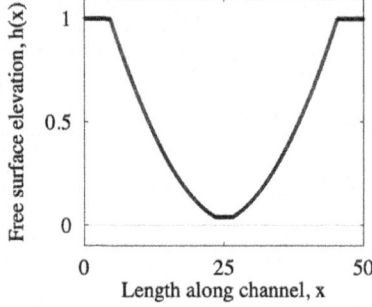 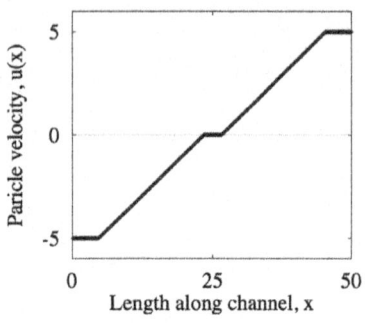

**Fig. 8.2** Exact solution to Test 2 at at time $T_{out} = 2.5$ s; initial discontinuity is at $x_0 = 25.0$ m. See Table 8.1

test. It is well known that a large class of methods will compute a negative depth $h$ in the vicinity of very shallow water produced by strong rarefactions; such situation may lead codes to crash when attempting to compute the celerity. Figure 8.2 shows solution profiles for water depth and particle velocity at the output time $T_{out} = 2.5$ s. See Table 8.2 for solution values $h_*$ and $u_*$ in the Star Region.

## 8.4 Test 3: Right Dry-Bed Riemann Problem

This test problem has data containing a dry-bed right state. As seen in Chap. 7, the solution consists of a single left rarefaction wave, associated with the left eigenvalue $\lambda_1 = u - a$, with the wet/dry front attached to the tail of the rarefaction wave. Numerical methods will experience difficulties in capturing the wet/dry front. One major difficulty is to propagate the front at the correct speed. My experience is that the resulting position error grows with time. Thus in a real application in which such fronts are to be propagated by 10–20 km, say, the propagation speed and thus the predicted wave arrival time will suffer from considerable errors, rendering the predictions unreliable or even useless. Methods based on Riemann problems have another difficulty; strictly speaking, they require a special modification of the appropriate Riemann solvers in order to account correctly for the wave structure that emerges; see Chap. 7. Riemann solvers derived on the assumption that the bed is wet throughout are incorrect for this problem. As the velocity experiences a jump at the front, numerical methods tend to produce spurious oscillations in the vicinity of the front, even some good high-resolution non-linear methods. Figure 8.3 shows exact profiles of depth and particle velocity at time $T_{out} = 4.0$ s. Numerical practitioners occasionally display results for momentum $hu$, rather than velocity $u$. In this case, numerical results, as compared with the exact solution, may look quite satisfactory [1].

## 8.6 Test 5: Generation of a Dry-Bed Region

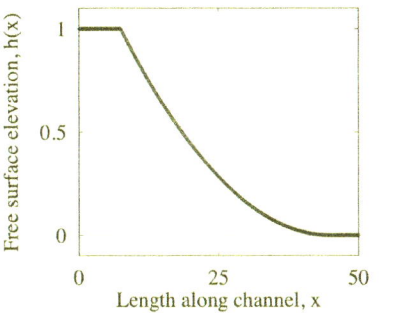
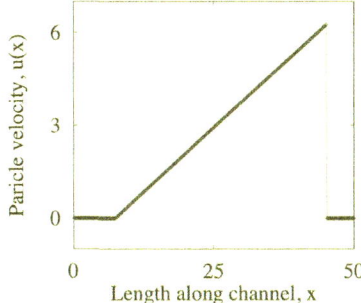

**Fig. 8.3** Exact solution to Test 3 at at time $T_{out} = 4.0$ s; initial discontinuity is at $x_0 = 20.0$ m. See Table 8.1

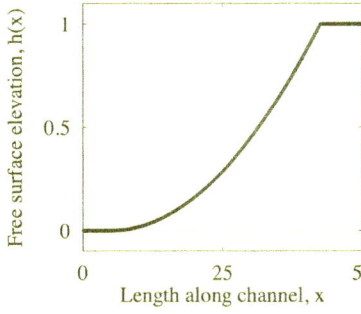
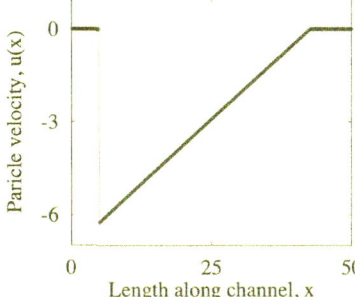

**Fig. 8.4** Exact solution to Test 4 at at time $T_{out} = 4.0$ s; initial discontinuity is at $x_0 = 30.0$ m. See Table 8.1

## 8.5 Test 4: Left Dry-Bed Riemann Problem

This test problem is the mirror image of Test 3 above, see Table 8.1. It is sometimes advisable to run codes on problems with mirror-image data to see if the expected symmetries are reproduced. Figure 8.4 shows profiles of depth and particle velocity at time $T_{out} = 4.0$ s.

## 8.6 Test 5: Generation of a Dry-Bed Region

The data states for this test have been chosen in such a way as to produce a solution consisting of two rarefaction waves with a portion of dry bed between them; see Table 8.1. The data *does not* satisfy the *depth-positivity condition*

$$2(a_L + a_R) > u_R - u_L, \tag{8.1}$$

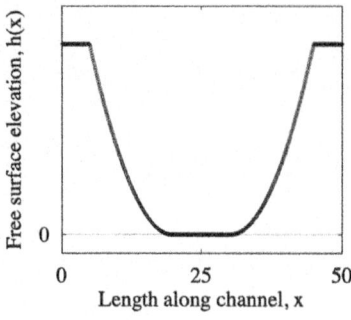 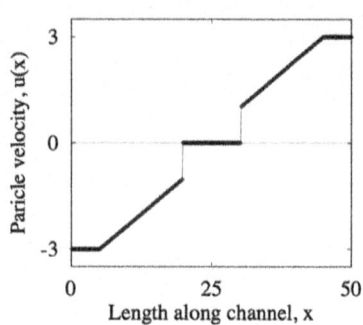

**Fig. 8.5** Exact solution to Test 5 at at time $T_{out} = 5.0$ s; initial discontinuity is at $x_0 = 25.0$ m. See Table 8.1

discussed in Chaps. 6 and 7. For this example $2(a_L + a_R) < u_R - u_L$, as the reader can easily verify from Table 8.1. Figure 8.5 shows profiles of depth and particle velocity at time $T_{out} = 5.0$ s. Based on the discussion for Tests 3 and 4, numerical methods are bound to experience serious difficulties in resolving the flow, particularly the two wet/dry fronts travelling in opposite directions.

## 8.7 Test Problems with Constant Slope

This section devises test problems with exact solution for the case in which a rectangular channel of constant cross-sectional area has variable bed elevation but with *constant slope*. To this end, we follow the work of Watson, Peregrine and Toro [4]. Adopting subscripts for partial derivatives, the governing equations are

$$h_t + (hu)_x = 0, \tag{8.2}$$

$$(hu)_t + (hu^2 + \tfrac{1}{2}gh^2)_x = -ghb_x. \tag{8.3}$$

First define the negative of the bed slope as

$$\alpha = -b_x. \tag{8.4}$$

In non-conservative form, the governing equations transform to

$$h_t + uh_x + hu_x = 0 \tag{8.5}$$

and

$$u_t + uu_x + gh_x = g\alpha. \tag{8.6}$$

## 8.7 Test Problems with Constant Slope

See Chap. 4. Following Watson et al. [4] we make the change of variables

$$\left.\begin{array}{l}\chi = x - \frac{1}{2}g\alpha t^2, \\ \tau = t, \\ \hat{u} = u - \alpha g t, \\ \hat{h} = h.\end{array}\right\} \quad (8.7)$$

Assuming that the slope $\alpha = constant$, it is easy to check that for any quantity $\psi$ one has

$$\left.\begin{array}{l}\psi_t = \psi_\tau - g\alpha t \psi_\chi, \\ \psi_x = \psi_\chi.\end{array}\right\} \quad (8.8)$$

From Eq. (8.7) we have

$$\left.\begin{array}{l}h_t = \hat{h}_t, \\ h_x = \hat{h}_x, \\ u_t = \hat{u}_t + g\alpha, \\ u_x = \hat{u}_x.\end{array}\right\} \quad (8.9)$$

Application of (8.8) and (8.9) to $h$ gives

$$\left.\begin{array}{l}h_t = \hat{h}_\tau - g\alpha t \hat{h}_\chi, \\ h_x = \hat{h}_\chi.\end{array}\right\} \quad (8.10)$$

Application of (8.8) and (8.9) to $u$ gives

$$\left.\begin{array}{l}u_t = \hat{u}_\tau - g\alpha t \hat{u}_\chi + g\alpha, \\ u_x = \hat{u}_\chi.\end{array}\right\} \quad (8.11)$$

Use of (8.10)–(8.11) in the original *inhomogeneous* equations (8.6)–(8.7) leads to *homogeneous equations* (no source term for the slope):

$$\hat{h}_\tau + \hat{u}\hat{h}_\chi + \hat{h}\hat{u}_\chi = 0 \quad (8.12)$$

and

$$\hat{u}_\tau + \hat{u}\hat{u}_\chi + g\hat{h}_\chi = 0. \quad (8.13)$$

This new system (8.12)–(8.13) is *exactly as the original shallow water equations* (8.6)–(8.7), but now *without* explicit presence of the source term.

One can now design test problems with exact solutions, including rarefactions, shocks and a constant slope, by utilising the exact Riemann solvers of Chaps. 6 and 7 for the homogenous equations and then reinterpret the solution according to (8.7) to account for the effect of the slope.

## 8.8 Closing Remarks

A suite of carefully selected test problems with exact solution, admitting both wet-bed and dry-bed conditions, has been presented. The purpose of these test problems is to rigorously assess numerical methods for the general initial-boundary value problem, against exact solutions for the kind of physical situations that are bound to be challenging to the numerical modeller. Each of the test problems has been carefully designed and offers a specific challenge to numerical methods. The exact solutions will be used to asses the suitability of approximate Riemann solvers and the accuracy of approximate numerical solutions in successive chapters.

Not included in these test problems is the case that contains linearly degenerate characteristic fields. These are associated with intermediate contact discontinuity waves in the transport of chemical species in one or two-dimensional shallow water models. These characteristic fields are also associated with vortical flows, in which the tangential velocity component jumps discontinuously across the shear wave. In both cases numerical methods experience serious difficulties in resolving these slowly moving discontinuous features. Two difficulties may be present: excessive numerical dissipation and spurious oscillations. In Chap. 12 we add an intermediate characteristic field to Tests 1 and 2 of Table 8.1.

A computer programme listing for the Riemann solvers is given in the Sect. 8.9, for both wet and dry-bed conditions. Useful background for this chapter is found in Chaps. 6 and 7, as well as in Refs. [1, 2] for the wet-bed case and [3] for the dry-bed case.

## 8.9 Computer Program for the Exact Riemann Solver

The listing of computer program to compute the exact solution of the Riemann problem for the non-linear one-dimensional time-dependent shallow water equations is now given. The input data file is called datain.ini and the output is written to result.out. Figure 8.6 shows a flow chart of the complete program. The main program calls either the routine: DRYBED or WETBED, depending on whether dry bed conditions are present in the solution or not. In the case of a dry bed portion being present, DRYBED will call DRYLEF if the left state is dry at time $t = 0$, or DRYMID if the middle state is dry at time $t > 0$ when the initial data does not obey the *depth-positivity condition*, or DRYRIG if the right state is dry at time $t = 0$. If the bed is wet throughout in space and time then the subroutine WETBED is called. This will in turn call STARTE, which provides a starting value for the depth $h_*$ in the Star Region. Then the solution for $h_*$ is computed iteratively, after which the solution for $u_*$ is evaluated. The complete solution is then sampled by calling the routine SAMWET. This last routine can also be used to sample the solution when using the exact Riemann solver locally in the Random Choice Method (RCM) of Glimm [5, 6], or the Godunov method [7]. For Glimm's method one calls SAMWET

## 8.9 Computer Program for the Exact Riemann Solver

**Fig. 8.6** Flow chart for exact Riemann solver

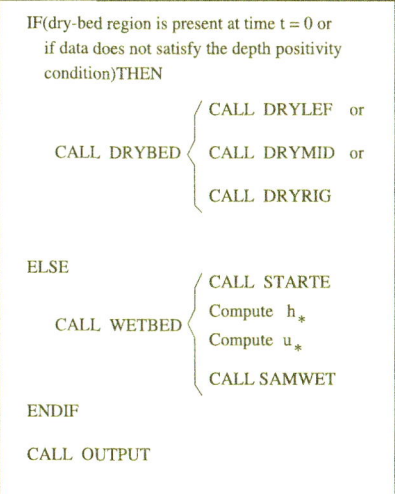

with $S = \theta x/t$, where $\theta$ is a random number. See Chaps. 9 and 10 for details on these numerical methods.

```
C-----------------------------------------------------------C
C                                                           C
C              EXACT RIEMANN SOLVER FOR                     C
C              THE SHALLOW WATER EQUATIONS                  C
C                                                           C
C     Name of program: HW-SWRPEXACT                         C
C     Purpose: to solve exactly the general Riemann         C
C              problem for the one-dimensional shallow      C
C              water equations for wet and dry bed          C
C              conditions                                   C
C     Programer: E. F. Toro                                 C
C     Last revision: May 21st 2000                          C
C                                                           C
C     Theory is found in Chapters 6 and 7                   C
C                                                           C
C      This program is part of HYPER-WAT,                   C
C      a sub-library of                                     C
C                                                           C
C      NUMERICA                                             C
C      A Library of Source Codes for                        C
C      Teaching, Research and Applications                  C
C      by E. F. Toro                                        C
C                                                           C
C-----------------------------------------------------------C
```

```
*
      IMPLICIT NONE
*
C     Declaration of variables
*
      REAL    CHALEN,CL,CR,DCRIT,DL,DR,GATE,GRAVIT,TIMOUT,
     &        TOL,UL,UR
*
      INTEGER MCELLS, NITER
*
      COMMON /STATES/ CL, DL, UL, CR, DR, UR
      COMMON /ACCELE/ GRAVIT
      COMMON /TOLERA/ NITER, TOL
      COMMON /DOMAIN/ CHALEN, GATE, MCELLS, TIMOUT
*
C     Initial data and computational parameters are read in
*
      OPEN(UNIT=1,FILE='datain.ini',STATUS='UNKNOWN')
*
      READ(1,*)CHALEN    ! length of channel
      READ(1,*)GATE      ! position of gate
      READ(1,*)GRAVIT    ! acceleration due to gravity
      READ(1,*)MCELLS    ! number of cells in profile
      READ(1,*)TOL       ! tolerance for convergence test
      READ(1,*)NITER     ! iterations in exact solver
      READ(1,*)TIMOUT    ! output time
      READ(1,*)DL        ! depth on left reservoir
      READ(1,*)UL        ! velocity in left reservoir
      READ(1,*)DR        ! depth in right reservoir
      READ(1,*)UR        ! velocity in right reservoir
*
      CLOSE(1)
*
C     Compute celerity on left and right states
*
      CL = SQRT(GRAVIT*DL)
      CR = SQRT(GRAVIT*DR)
*
C     Use the "depth positivity condition" to identify
C     type of data and thus of solution and to call
C     appropriate exact solver
*
      DCRIT = (UR-UL) - 2.0*(CL+CR)
*
      IF(DL.LE.0.0.OR.DR.LE.0.0.OR.DCRIT.GE.0.0)THEN
```

## 8.9 Computer Program for the Exact Riemann Solver

```
*
C       Dry bed cases
*
        CALL DRYBED
*
      ELSE
*
C       Wet bed case
*
        CALL WETBED
*
      ENDIF
*
C     Results are printed out
*
      CALL OUTPUT
*
      END
*
*-----------------------------------------------------------*
*
      SUBROUTINE OUTPUT
*
C     Purpose: to output exact solution at chosen
C              output time TIMOUT
*
      IMPLICIT NONE
*
C     Declaration of variables
*
      INTEGER  MX, I, MCELLS
*
      REAL     D, U,CHALEN,GATE,TIMOUT,XCOORD
*
      PARAMETER (MX = 3000)
*
      DIMENSION D(MX), U(MX)
*
      COMMON /SOLUTI/ D, U
      COMMON /DOMAIN/ CHALEN, GATE, MCELLS, TIMOUT
*
      OPEN(UNIT=1,FILE='result.out',STATUS='UNKNOWN')
*
      DO 10 I = 1, MCELLS
         XCOORD = REAL(I)*CHALEN/REAL(MCELLS)
```

```
            WRITE(1,20)XCOORD, D(I), U(I)
 10    CONTINUE
*
 20    FORMAT(3(F10.5,4X))
*
       CLOSE(1)
*
       END
*
*-----------------------------------------------------------------*
*
       SUBROUTINE WETBED
*
C      Purpose: to solve the Riemann problem exactly for
C               the wet-bed case
*
       IMPLICIT NONE
*
C      Declaration of variables
*
       INTEGER  I,IT,MCELLS,MX,NITER
*
       REAL     CHA,CHALEN,CL,CR,CS,D,D0,DL,DR,DS,DSAM,FL,
      &         FLD,FR,FRD,GATE,GRAVIT,S,TIMOUT,TOL,U,UL,
      &         UR,US,USAM,XCOORD
*
       PARAMETER (MX = 3000)
*
       DIMENSION D(MX), U(MX)
*
       COMMON /SOLUTI/ D, U
       COMMON /STATES/ CL, DL, UL, CR, DR, UR
       COMMON /STARSO/ CS, DS, US
       COMMON /ACCELE/ GRAVIT
       COMMON /TOLERA/ NITER, TOL
       COMMON /DOMAIN/ CHALEN, GATE, MCELLS, TIMOUT
*
C      Find starting value for iteration
*
       WRITE(6,*)
       WRITE(6,*)'Exact Solution in Star Region'
       WRITE(6,*)'============================='
       WRITE(6,*)
*
       CALL STARTE
```

### 8.9 Computer Program for the Exact Riemann Solver

```
*
C       Store starting value in D0
*
        D0 = DS
*
C       Start iteration
*
        WRITE(6,*)'    IT   ','   DS           ', '   CHA '
        WRITE(6,*)
        DO 10 IT = 1, NITER
*
            CALL GEOFUN(FL,FLD,DS,DL,CL)
            CALL GEOFUN(FR,FRD,DS,DR,CR)
            DS  = DS - (FL + FR + UR-UL)/(FLD + FRD)
            CHA = ABS(DS-D0)/(0.5*(DS+D0))
            WRITE(6,30)IT,DS,CHA
            IF(CHA.LE.TOL)GOTO 20
            IF(DS.LT.0.0)DS = TOL
            D0 = DS
*
 10     CONTINUE
*
        WRITE(6,*)'Number of NITER iterations exceeded,
     &             STOP'
*
        STOP
*
 20     CONTINUE
 30     FORMAT(I6,2X,2(F12.7,2X))
*
C       Converged solution for depth DS in Star Region.
C       Compute velocity US in Star Region
*
        US = 0.5*(UL + UR) + 0.5*(FR - FL)
*
        WRITE(6,*)
        WRITE(6,*)'Depth in Star Region    h* =',DS
        WRITE(6,*)'Velocity in Star Region u* =',US
        WRITE(6,*)
*
        CS = SQRT(GRAVIT*DS)
*
C       Evaluate exact solution at time TIMOUT
*
        DO 40 I = 1, MCELLS
```

```
*
              XCOORD = REAL(I)*CHALEN/REAL(MCELLS) - GATE
              S      = XCOORD/TIMOUT
*
C             Sample solution throughout wave structure at
C             time TIMOUT
*
              CALL SAMWET(DSAM,USAM,S)
*
C             Store solution
*
              D(I) = DSAM
              U(I) = USAM
*
 40      CONTINUE
*
         END
*
*----------------------------------------------------------------*
*
         SUBROUTINE GEOFUN(F,FD,D,DK,CK)
*
C        Purpose: to evaluate functions FL, FR and their
C                 derivatives in iterative Riemann solver,
C                 for wet-bed case.
*
         IMPLICIT NONE
*
C        Declaration of variables
*
         REAL    C,CK,D,DK,F,FD,GES,GRAVIT
*
         COMMON /ACCELE/ GRAVIT
*
         IF(D.LE.DK)THEN
*
C           Wave is rarefaction wave (or depression)
*
            C  = SQRT(GRAVIT*D)
            F  = 2.0*(C-CK)
            FD = GRAVIT/C
         ELSE
*
C           Wave is shock wave (or bore)
*
```

8.9 Computer Program for the Exact Riemann Solver        153

```
            GES = SQRT(0.5*GRAVIT*(D+DK)/(D*DK))
            F   = (D-DK)*GES
            FD  = GES - 0.25*GRAVIT*(D-DK)/(GES*D*D)
         ENDIF
*
         END
*
*-------------------------------------------------------------*
*
         SUBROUTINE STARTE
*
C        Purpose: to provide starting value for Newton-Raphson
C                 iteration. The Two-Rarefaction Riemann
C                 Solver (TRRS) and Two-Shock Riemann Solver
C                 (TSRS) are used adaptively
*
         IMPLICIT NONE
*
C        Declaration of variables
*
         REAL    CL,CR,CS,DL,DMIN,DR,DS,GEL,GER, GRAVIT,
       &         UL,UR,US
*
         COMMON /STATES/ CL, DL, UL, CR, DR, UR
         COMMON /STARSO/ CS, DS, US
         COMMON /ACCELE/ GRAVIT
*
         DMIN = MIN(DL,DR)
*
C        Use Two-Rarefaction (TRRS) solution as starting value
*
         DS = (1.0/GRAVIT)*(0.5*(CL+CR)-0.25*(UR-UL))**2
*
         IF(DS.LE.DMIN)THEN
*
C           Use Two-Rarefaction (TSRS) approximation as
C           starting value
*
            WRITE(6,*)'TR approximation, h* =',DS
         ELSE
*
C           Use two-shock (TSRS) solution as starting value
C           with DS as computed from TRRS as estimate
*
            WRITE(6,*)'TS approximation, h* =',DS
```

```
              GEL  = SQRT(0.5*GRAVIT*(DS+DL)/(DS*DL))
              GER  = SQRT(0.5*GRAVIT*(DS+DR)/(DS*DR))
              DS   = (GEL*DL + GER*DR - (UR-UL))/(GEL + GER)
*
        ENDIF
        WRITE(6,*)
*
        END
*
*-----------------------------------------------------------------*
*
        SUBROUTINE SAMWET(D,U,S)
*
C       Purpose: to sample solution through wave structure at
C                TIMOUT for wet-bed case
*
        IMPLICIT NONE
*
C       Declaration of variables
*
        REAL     C,CL,CR,CS,D,DL,DR,DS,GRAVIT,QL,QR,S,SHL,
       &         SHR,SL,SR,STL,STR,U,UL,UR,US
*
        COMMON /STATES/ CL, DL, UL, CR, DR, UR
        COMMON /STARSO/ CS, DS, US
        COMMON /ACCELE/ GRAVIT
*
        IF(S.LE.US)THEN
**********************************************
C       Sample left wave
**********************************************
           IF(DS.GE.DL)THEN
*
C          Left shock
*
              QL = SQRT((DS + DL)*DS/(2.0*DL*DL))
              SL = UL - CL*QL
*
              IF(S.LE.SL)THEN
*
C             Sample point lies to the left of the shock
*
                 D = DL
                 U = UL
```

## 8.9 Computer Program for the Exact Riemann Solver 155

```
            ELSE
*
C             Sample point lies to the right of the shock
*
              D = DS
              U = US
            ENDIF
         ELSE
*
C          Left rarefaction
*
           SHL = UL - CL
*
           IF(S.LE.SHL)THEN
*
C             Sample point lies to the right of the
C             rarefaction
*
              D = DL
              U = UL
           ELSE
*
              STL = US - CS
*
              IF(S.LE.STL)THEN
*
C                Sample point lies inside the rarefaction
*
                 U = (UL + 2.0*CL + 2.0*S)/3.0
                 C = (UL + 2.0*CL - S)/3.0
                 D = C*C/GRAVIT
              ELSE
*
C                Sample point lies in the STAR region
*
                 D = DS
                 U = US
              ENDIF
           ENDIF
         ENDIF
*
      ELSE
*******************************************
C        Sample right wave
*******************************************
```

```
*
            IF(DS.GE.DR)THEN
*
C           Right shock
*
            QR = SQRT((DS + DR)*DS/(2.0*DR*DR))
            SR = UR + CR*QR
*
            IF(S.GE.SR)THEN
*
C              Sample point lies to the right of the shock
*
               D = DR
               U = UR
            ELSE
*
C              Sample point lies to the left of the shock
*
               D = DS
               U = US
            ENDIF
*
         ELSE
*
C           Right rarefaction
*
            SHR = UR + CR
*
            IF(S.GE.SHR)THEN
*
C              Sample point lies to the right of the
C              rarefaction
*
               D = DR
               U = UR
            ELSE
*
               STR = US + CS
*
               IF(S.GE.STR)THEN
*
C                 Sample point lies inside the rarefaction
*
                  U = (UR  - 2.0*CR + 2.0*S)/3.0
                  C = (-UR + 2.0*CR + S)/3.0
```

8.9 Computer Program for the Exact Riemann Solver    157

```
                  D = C*C/GRAVIT
               ELSE
*
C                 Sample point lies in the STAR region
*
                  D = DS
                  U = US
               ENDIF
            ENDIF
         ENDIF
      ENDIF
*
      END
*
*----------------------------------------------------------------*
*
      SUBROUTINE DRYBED
*
C     Pupose: to compute the exact solution in the case
C             in which a portion of dry bed is present
*
      IMPLICIT NONE
*
C     Declaration of variables
*
      INTEGER I,MCELLS,MX
*
      REAL    CHALEN,CL,CR,D,DL,DR,DSAM,GATE,S,TIMOUT,
     &        U,UL,UR,USAM,XCOORD
*
      PARAMETER (MX = 3000)
*
      DIMENSION D(MX), U(MX)
*
      COMMON /SOLUTI/ D, U
      COMMON /STATES/ CL, DL, UL, CR, DR, UR
      COMMON /DOMAIN/ CHALEN, GATE, MCELLS, TIMOUT
*
      DO 10 I = 1, MCELLS
*
         XCOORD = REAL(I)*CHALEN/REAL(MCELLS) - GATE
         S      = XCOORD/TIMOUT
C
         IF(DL.LE.0.0)THEN
*
```

```
C              Left state is dry
*
               CALL SAMLEF(DSAM,USAM,S)
            ELSE
               IF(DR.LE.0.0)THEN
*
C                 Right state is dry
*
                  CALL SAMRIG(DSAM,USAM,S)
               ELSE
*
C                 Middle state is dry
*
                  CALL SAMMID(DSAM,USAM,S)
               ENDIF
            ENDIF
*
            D(I) = DSAM
            U(I) = USAM
*
 10      CONTINUE
*
         END
*
*-----------------------------------------------------------------*
*
         SUBROUTINE SAMLEF(D,U,S)
*
C        Purpose: to sample the solution through the wave
C                 structure at time TIMOUT, for the case in
C                 which the left state is dry. Solution
C                 consists of single right rarefaction
*
         IMPLICIT NONE
*
C        Declaration of variables
*
         REAL    C,CL,CR,D,DL,DR,GRAVIT,S,SHR,STR,U,UL,UR
*
         COMMON /STATES/ CL, DL, UL, CR, DR, UR
         COMMON /ACCELE/ GRAVIT
*
         SHR = UR + CR
*
         IF(S.GE.SHR)THEN
```

8.9 Computer Program for the Exact Riemann Solver 159

```
*
C          Sampling point lies to the right of the
C          rarefaction
*
           D = DR
           U = UR
      ELSE
*
           STR = UR-2.0*CR
*
           IF(S.GE.STR)THEN
*
C             Sampling point lies inside the rarefaction
*
              U = ( UR - 2.0*CR + 2.0*S)/3.0
              C = (-UR + 2.0*CR + S)/3.0
              D = C*C/GRAVIT
           ELSE
*
C             Sampling point lies in dry-bed state
*
              D = DL
              U = UL
           ENDIF
      ENDIF
*
      END
*
*-----------------------------------------------------------*
*
      SUBROUTINE SAMMID(D,U,S)
*
C     Purpose: to sample the solution through the wave
C              structure at time TIMOUT, for the case in
C              which the middle state is dry. Solution
C              consists of a left and a right rarefaction
C              with a dry portion in the the middle
*
      IMPLICIT NONE
*
C     Declaration of variables
*
      REAL    C,CL,CR,D,DL,DR,GRAVIT,S,SHL,SHR,SSL,SSR,
     &        U,UL,UR
*
```

```
      COMMON /STATES/ CL, DL, UL, CR, DR, UR
      COMMON /ACCELE/ GRAVIT
*
C     Compute wave speeds
*
      SHL = UL - CL
      SSL = UL + 2.0*CL
      SSR = UR - 2.0*CR
      SHR = UR + CR
*
      IF(S.LE.SHL)THEN
*
C        Sampling point lies to the left of the left
C        rarefaction
*
         D = DL
         U = UL
      ENDIF
*
      IF(S.GT.SHL.AND.S.LE.SSL)THEN
*
C        Sampling point lies inside the left rarefaction
*
         U = (UL + 2.0*CL + 2.0*S)/3.0
         C = (UL + 2.0*CL - S)/3.0
         D = C*C/GRAVIT
      ENDIF
*
      IF(S.GT.SSL.AND.S.LE.SSR)THEN
*
C        Sampling point lies inside the middle dry bed region
*
         D = 0.0
         U = 0.0
      ENDIF
*
      IF(S.GT.SSR.AND.S.LE.SHR)THEN
*
C        Sampling point lies inside the right rarefaction
*
         U = ( UR - 2.0*CR + 2.0*S)/3.0
         C = (-UR + 2.0*CR + S)/3.0
         D = C*C/GRAVIT
      ENDIF
*
```

## 8.9 Computer Program for the Exact Riemann Solver

```
      IF(S.GT.SHR)THEN
*
C        Sampling point lies to the right of the right
C        rarefaction
*
         D = DR
         U = UR
      ENDIF
*
      END
*
*----------------------------------------------------------------*
*
      SUBROUTINE SAMRIG(D,U,S)
*
C     Purpose: to sample the solution through the wave
C              structure at time TIMOUT, for the case in
C              which the right state is dry. Solution
C              consists of single left rarefaction
*
      IMPLICIT NONE
*
C     Declaration of variables
*
      REAL    C,CL,CR,D,DL,DR,GRAVIT,S,SHL,STL,U,UL,UR
*
      COMMON /STATES/ CL, DL, UL, CR, DR, UR
      COMMON /ACCELE/ GRAVIT
*
      SHL = UL - CL
*
      IF(S.LE.SHL)THEN
*
C        Sampling point lies to the left of the rarefaction
*
         D = DL
         U = UL
      ELSE
*
         STL = UL + 2.0*CL
*
         IF(S.LE.STL)THEN
*
C           Sampling point lies inside the rarefaction
*
```

```
              U = (UL + 2.0*CL + 2.0*S)/3.0
              C = (UL + 2.0*CL - S)/3.0
              D = C*C/GRAVIT
          ELSE
*
C         Sampling point lies in right dry-bed state
*
              D = DR
              U = UR
          ENDIF
      ENDIF
*
      END
*
*------------------------------------------------------------*
*
```

# References

1. E.F. Toro, A fast Riemann solver with constant covolume applied to the random choice method. Int. J. Numer. Meth. Fluids **9**, 1145–1164 (1989)
2. E.F. Toro, Riemann problems and the WAF method for solving two-dimensional shallow water equations. Phil. Trans. Roy. Soc. London **A338**, 43–68 (1992)
3. E.F. Toro, The dry-bed problem in shallow-water flows. Technical report, College of Aeronautics, Cranfield Institute of Technology, UK. CoA 9007 (1990)
4. G. Watson, D. H. Peregrine, and E. F. Toro. Numerical solution of the shallow water equations on a beach using the weighted average flux method, in *Computational Fluid Dynamics*, vol. 1 (Elsevier, 1992), pp. 495–502
5. J. Glimm, Solution in the large for nonlinear hyperbolic systems of equations. Comm. Pure. Appl. Math. **18**, 697–715 (1965)
6. A.J. Chorin, Random choice solutions of hyperbolic systems. J. Comput. Phys. **22**, 517–533 (1976)
7. S.K. Godunov, Finite difference methods for the computation of discontinuous solutions of the equations of fluid dynamics. Mat. Sb. **47**, 271–306 (1959)

# Chapter 9
# Notions on Numerical Methods

**Abstract** This chapter is a succinct, largely self-contained presentation of some very basic notions on numerical methods for solving hyperbolic equations. The subject is dealt with entirely in terms of the simplest partial differential equation (PDE), namely the linear advection equation with constant wave propagation speed. We begin with the simplest numerical approximation method, namely the finite difference method. Most well-known schemes, as finite difference methods, are presented, including the Lax-Friedrichs method, the Lax-Wendroff method, the FORCE method, the Godunov centred method and the Godunov upwind method. The main properties of these schemes are also studied, including truncation error, accuracy, monotonicity and linear stability. Some theoretical notions are also included, notably the Godunov theorem, which is stated and proved. This result sets the theoretical bases for the construction of non-linear methods for hyperbolic equations. Sample numerical results are presented. Useful background is found in Chap. 2.

## 9.1 Numerical Approximation of Hyperbolic Equations

This chapter is concerned with some fundamentals on numerical methods for partial differential equations of hyperbolic type, such as the shallow water equations, the system of specific interest in this book. Here we introduce some basic concepts on numerical discretisation methods for hyperbolic equations, all based on the simplest equation, namely the linear advection equation (LAE). To this end we first consider the initial-boundary value problem (IBVP) for LAE

$$\left.\begin{array}{ll} \text{PDE:} & \partial_t q + \lambda \partial_x q = 0, \quad x \in [a, b], \ t > 0, \\ \text{IC:} & q(x, 0) = h(x), \quad x \in [a, b], \ t = 0, \\ \text{BCs:} & q(a, t) = b_L(t); \quad q(b, t) = b_R(t), \ t \geq 0. \end{array}\right\} \quad (9.1)$$

Here $[a, b]$ defines the spatial domain; $h(x)$ is the initial condition (IC) at the initial time $t = 0$, a prescribed function of $x$; $b_L(t)$ and $b_R(t)$ are prescribed functions of time and define boundary conditions (BCs) at $x = a$ (left) and at $x = b$ (right).

### 9.1.1 Finite Difference Approximation to PDEs

One approach to solve problem (9.1) is by the method of finite differences, which requires the following steps:

1. *Partition of the spatial domain* $[a, b]$ *into* $M + 2$ *equidistant points*

$$x_i = a + i\Delta x, \quad i = 0, \ldots, M+1, \quad \Delta x = \frac{b-a}{M+1}. \tag{9.2}$$

$M$ is a chosen positive integer and $\Delta x$ is the mesh spacing. Figure 9.1 illustrates a finite difference mesh. There are $M$ *interior points*: $x_1, x_2, \ldots, x_M$; and two *boundary points*: $x_0 = a$ and $x_{M+1} = b$.

2. *Partition of the temporal domain* $[0, T_{out}]$ *into a set of time points, or time levels,*

$$t_n = n\Delta t, \quad n = 0, \ldots, N_{out}, \ldots. \tag{9.3}$$

See Fig. 9.1. Here $t_0 = 0$: initial time; $T_{out} = \Delta t N_{out}$; $\Delta t$: time step. We assume a fixed relationship between $\Delta t$ and $\Delta x$ of the form

$$\Delta x = \Delta t \times K, \quad K > 0 : \text{constant}. \tag{9.4}$$

The spatial mesh parameter $\Delta x$ is chosen through the choice of $M$, that is, the number of interior points. There are no particular constraints in choosing $M$. The choice of the time step $\Delta t$ is constrained by accuracy or stability considerations [1–3].

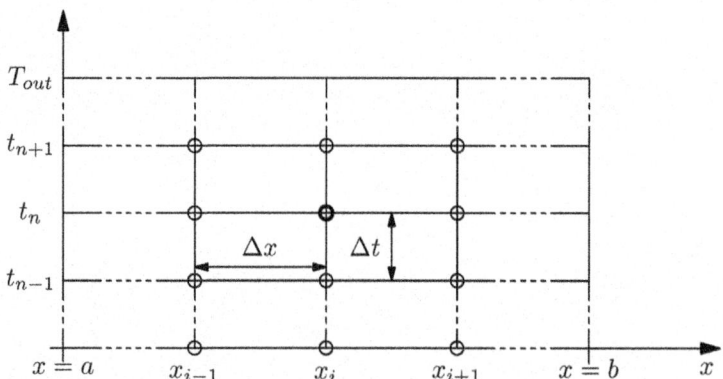

**Fig. 9.1** Finite difference mesh defining a set of discrete points $(x_i, t_n)$ resulting from partitions of the spatial $x$ and temporal $t$ domains

## 9.1 Numerical Approximation of Hyperbolic Equations

The continuous domain $[a, b] \times [0, \infty)$ has been replaced by a mesh made up of a finite number of points $(x_i, t_n)$. We now need to replace the continuous distribution of the function $q(x, t)$ by a finite number of discrete values $q(x_i, t_n)$ associated with these points. Then in order to solve the differential equation in this discrete setting we also need to represent in discrete form the partial derivatives $\partial_t q(x, t)$ and $\partial_x q(x, t)$ in (9.1). Here we do so by finite difference approximations. In this manner the partial differential equation is represented by *difference equations*, expressions that relate approximate discrete values of the solution at neighbouring points. The *differential operator* is replaced by a *numerical operator*, as we shall see.

Consider the generic point $(x_i, t_n)$ of the mesh, as shown in Fig. 9.1. We seek an approximation to $q(x_i, t_n)$ and this will be denoted by $q_i^n$, that is

$$q_i^n \approx q(x_i, t_n) \,. \tag{9.5}$$

The temporal partial derivative $\partial_t q(x, t)$ can be approximated in a variety of ways, such as

$$\partial_t q(x_i, t_n) = \begin{cases} \dfrac{q(x_i, t_{n+1}) - q(x_i, t_n)}{\Delta t} + \mathcal{O}(\Delta t) \,, & \text{Forward}, \\ \dfrac{q(x_i, t_n) - q(x_i, t_{n-1})}{\Delta t} + \mathcal{O}(\Delta t) \,, & \text{Backward}, \\ \dfrac{q(x_i, t_{n+1}) - q(x_i, t_{n-1})}{2\Delta t} + \mathcal{O}(\Delta t^2) \,, & \text{Centred}. \end{cases} \tag{9.6}$$

Analogously, for the spatial partial derivative $\partial_x q(x, t)$ in (9.1) at the point $(x_i, t_n)$ we write

$$\partial_x q(x_i, t_n) = \begin{cases} \dfrac{q(x_{i+1}, t_n) - q(x_i, t_n)}{\Delta x} + \mathcal{O}(\Delta x) \,, & \text{Forward}, \\ \dfrac{q(x_i, t_n) - q(x_{i-1}, t_n)}{\Delta x} + \mathcal{O}(\Delta x) \,, & \text{Backward}, \\ \dfrac{q(x_{i+1}, t_n) - q(x_{i-1}, t_n)}{2\Delta x} + \mathcal{O}(\Delta x^2) \,, & \text{Centred}. \end{cases} \tag{9.7}$$

Various combinations of these finite-difference approximations will lead to various well-known methods.

## 9.1.2 Well-Known Finite Difference Methods

Next we recall some well-known finite difference methods in the literature.

**The Upwind Method of Godunov: Finite Difference Version**

This method uses the following approximations to partial derivatives

$$\left.\begin{aligned}\partial_t q(x_i, t_n) &= \frac{q(x_i, t_{n+1}) - q(x_i, t_n)}{\Delta t} + \mathcal{O}(\Delta t) \,, \\ \partial_x q(x_i, t_n) &= \begin{cases} \dfrac{q(x_i, t_n) - q(x_{i-1}, t_n)}{\Delta x} + \mathcal{O}(\Delta x) \text{ if } \lambda > 0 \,, \\ \dfrac{q(x_{i+1}, t_n) - q(x_i, t_n)}{\Delta x} + \mathcal{O}(\Delta x) \text{ if } \lambda < 0 \,. \end{cases}\end{aligned}\right\} \quad (9.8)$$

**Remarks:**
1. The time derivative is approximated by a *forward-in-time* formula.
2. The space derivative is approximated by a one-sided, **upwind**, spatial derivative discretisation, according to the sign of the wave propagation speed.
3. For linear equations the method was first proposed by Courant et al. [4].
4. Godunov [5] extended the upwind method in **conservation form** to solve non-linear systems of hyperbolic equations.

The differential operator in (9.1) is

$$L_e(q) \equiv \partial_t q(x, t) + \lambda \partial_x q(x, t) = 0 \,, \quad (9.9)$$

which when applied to the function $q(x, t)$ at the point $(x_i, t_n)$ of the mesh, for $\lambda > 0$, becomes

$$\left.\begin{aligned} L_e(q(x_i, t_n)) &= \partial_t q(x_i, t_n) + \lambda \partial_x q(x_i, t_n) \\ &= \frac{q(x_i, t_{n+1}) - q(x_i, t_n)}{\Delta t} + \mathcal{O}(\Delta t) \\ &\quad + \lambda [\frac{q(x_i, t_n) - q(x_{i-1}, t_n)}{\Delta x}] + \mathcal{O}(\Delta x) \\ &= 0 \,. \end{aligned}\right\} \quad (9.10)$$

By suppressing truncation errors $\mathcal{O}(\Delta t) + \mathcal{O}(\Delta x)$ and replacing $q(x_i, t_n)$ by $q_i^n$ we obtain

$$\frac{q_i^{n+1} - q_i^n}{\Delta t} + \lambda \left( \frac{q_i^n - q_{i-1}^n}{\Delta x} \right) = 0 \,. \quad (9.11)$$

## 9.1 Numerical Approximation of Hyperbolic Equations

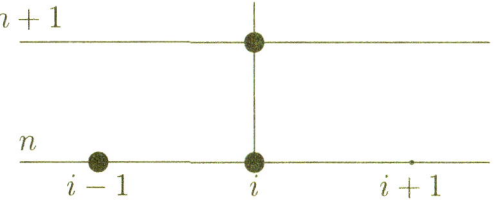

**Fig. 9.2** Stencil for Godunov's upwind method for positive characteristic speed $\lambda > 0$. Note the one-sided (upwind) character of the stencil

Solving for $q_i^{n+1}$ we obtain the Godunov upwind numerical scheme

$$q_i^{n+1} = q_i^n - \frac{\lambda \Delta t}{\Delta x} \left( q_i^n - q_{i-1}^n \right) . \tag{9.12}$$

In this *explicit method* the unknown at the new time level n+1 is $q_i^{n+1}$, on the left hand side. All terms on the right-hand side at time level $n$ are known.

It is convenient to introduce the **Courant-Friedrichs-Lewy number**, or **CFL number**, or simply **Courant number**, defined as

$$c = \frac{\lambda \Delta t}{\Delta x} = \frac{\lambda}{\Delta x / \Delta t} . \tag{9.13}$$

This is a *dimensionless quantity*, it is the ratio of two speeds, the speed $\lambda$ in the PDE in (9.1) and the *mesh speed* $\Delta x / \Delta t$. Then the Godunov upwind scheme (9.12) for $\lambda > 0$ becomes

$$q_i^{n+1} = q_i^n - c \left( q_i^n - q_{i-1}^n \right) . \tag{9.14}$$

Figure 9.2 displays the stencil of scheme (9.14), which is the set of points of the mesh that contribute to the scheme. The reader can easily verify that for $\lambda < 0$ the Godunov upwind scheme reads

$$q_i^{n+1} = q_i^n - c \left( q_{i+1}^n - q_i^n \right) , \tag{9.15}$$

with an appropriate stencil configuration.

### The FTCS Method

The FTCS method (Forward-in-Time Centred-in-Space) results from the following approximations to the partial derivatives

$$\left. \begin{aligned} \partial_t q(x_i, t_n) &= \frac{q(x_i, t_{n+1}) - q(x_i, t_n)}{\Delta t} + \mathcal{O}(\Delta t) , \\ \partial_x q(x_i, t_n) &= \frac{q(x_{i+1}, t_n) - q(x_{i-1}, t_n)}{2\Delta x} + \mathcal{O}(\Delta x^2) . \end{aligned} \right\} \tag{9.16}$$

**Fig. 9.3** Stencil for the FTCS method. Note the symmetric, or centred, character of the stencil

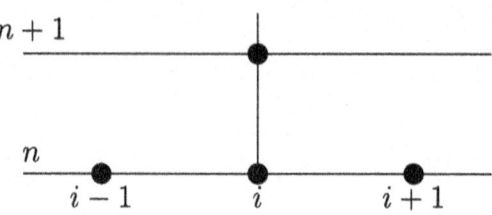

Substitution of these into the PDE, suppressing error terms and replacing exact values by approximate values, yields

$$\frac{q_i^{n+1} - q_i^n}{\Delta t} + \lambda \left( \frac{q_{i+1}^n - q_{i-1}^n}{2\Delta x} \right) = 0 . \tag{9.17}$$

By solving for $q_i^{n+1}$ and using the CFL number we obtain the FTCS numerical scheme

$$q_i^{n+1} = q_i^n - \frac{1}{2} c (q_{i+1}^n - q_{i-1}^n) . \tag{9.18}$$

Figure 9.3 shows the stencil for the FTCS numerical scheme. Unfortunately, FTCS is useless; it is *unconditionally unstable*. For the definition of stability of a scheme we recommend any book on numerical methods for evolutionary PDEs; see for example [1–3].

FTCS uses the same approximation to the time derivative as the Godunov method, but the spatial derivative is approximated via a centred, second-order accurate, discretization. Naively, one would have expected a better method than Godunov's upwind method. There are two ways to rescue FTCS. One modification results in the explicit Lax-Friedrichs scheme. The other way is to resort to an implicit version.

### The Lax-Friedrichs Method

The Lax-Friedrichs method results from replacing $q_i^n$ in the approximation to the time derivative of FTCS in (9.18) by a mean value of its closest neighbours, that is

$$q_i^n \longrightarrow \frac{1}{2}(q_{i-1}^n + q_{i+1}^n) .$$

Then

$$\frac{q_i^{n+1} - \frac{1}{2}(q_{i-1}^n + q_{i+1}^n)}{\Delta t} + \lambda \left( \frac{q_{i+1}^n - q_{i-1}^n}{2\Delta x} \right) = 0 , \tag{9.19}$$

## 9.1 Numerical Approximation of Hyperbolic Equations

**Fig. 9.4** Stencil for the Lax-Friedrichs method. Note the symmetry of the stencil and the missing point $(x_i, t_n)$

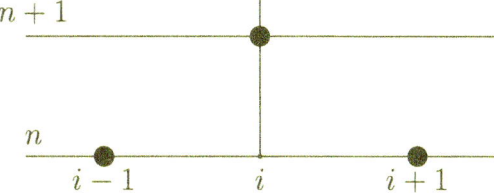

yielding the Lax-Friedrichs scheme

$$q_i^{n+1} = \frac{1}{2}(1+c)q_{i-1}^n + \frac{1}{2}(1-c)q_{i+1}^n . \tag{9.20}$$

Figure 9.4 shows the stencil of the Lax-Friedrichs scheme.

### The Lax-Wendroff Method

The Lax-Wendroff method [6] is constructed through a different approach, via the following steps:

1. The solution at $(x_i, t_{n+1})$ is expressed as a Taylor series in time

$$q(x_i, t_{n+1}) = q(x_i, t_n) + \Delta t \partial_t q(x_i, t_n) + \frac{1}{2}\Delta t^2 \partial_t^{(2)} q(x_i, t_n) + \mathcal{O}(\Delta t^3) . \tag{9.21}$$

2. By means of the Cauchy-Kovalevskaya procedure (or Lax-Wendroff procedure, as is sometimes called) one uses the PDE in (9.1) to replace time derivatives by space derivatives

$$\partial_t q(x,t) = -\lambda \partial_x q(x,t) , \quad \partial_t^{(2)} q(x,t) = \lambda^2 \partial_x^{(2)} q(x,t) . \tag{9.22}$$

In fact, for any order $k$, one can prove

$$\partial_t^{(k)} q(x,t) = (-\lambda)^k \partial_x^{(k)} q(x,t) . \tag{9.23}$$

3. By substituting (9.22) into (9.21) one obtains

$$q(x_i, t_{n+1}) = q(x_i, t_n) - \Delta t \lambda \partial_x q(x_i, t_n) + \frac{1}{2}\Delta t^2 \lambda^2 \partial_x^{(2)} q(x_i, t_n) + \mathcal{O}(\Delta t^3) \tag{9.24}$$

**Fig. 9.5** Stencil for the Lax-Wendroff method. Note the symmetry of the stencil

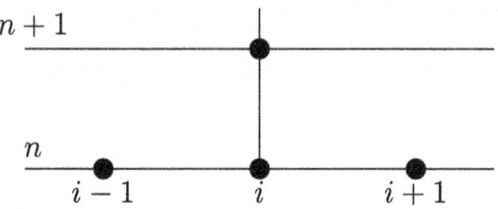

4. The spatial derivatives in (9.24) are approximated by centred finite differences

$$\left.\begin{array}{l}\partial_x q(x_i, t_n) = \dfrac{q(x_{i+1}, t_n) - q(x_{i-1}, t_n)}{2\Delta x} + \mathcal{O}(\Delta x^2) \,, \\[2mm] \partial_x^{(2)} q(x_i, t_n) = \dfrac{q(x_{i+1}, t_n) - 2q(x_i, t_n) + q(x_{i-1}, t_n)}{\Delta x^2} + \mathcal{O}(\Delta x^2) \,.\end{array}\right\} \quad (9.25)$$

5. Finally, by substituting (9.25) into (9.24), neglecting truncation errors and replacing exact values $q(x_i, t_n)$ by approximate values $q_i^n$ one obtains the Lax-Wendroff scheme

$$q_i^{n+1} = \frac{1}{2}c(1+c)q_{i-1}^n + (1-c^2)q_i^n - \frac{1}{2}c(1-c)q_{i+1}^n \,. \quad (9.26)$$

Figure 9.5 shows the stencil for the Lax-Wendroff method.

## 9.2 Basic Properties of Numerical Methods

Here we briefly review some basic properties of numerical methods in the terms of the model linear advection equation.

### 9.2.1 Forms of Expressing a Numerical Scheme

**General form of a scheme and examples.** All explicit schemes studied so far can be written in the general form

$$q_i^{n+1} = H(q_{i-l}^n, \ldots, q_i^n, \ldots, q_{i+r}^n) \,, \quad (9.27)$$

with $l, r$ two non-negative integers and $H(\ldots)$ a real-valued function of $l + r + 1$ arguments and

$$q_i^n \approx q(x_i, t_n) \,, \quad q_i^n \to 0 \text{ as } |i| \to \infty \quad (9.28)$$

## 9.2 Basic Properties of Numerical Methods

is a point-wise value that approximates the true solution $q(x, t)$ at the mesh point $(x_i, t_n)$, with $x_i = i\Delta x$, $t_n = n\Delta t$.

**Example 9.1** The Godunov upwind method in finite difference form. When the Godunov scheme (9.14)–(9.15) is written as in (9.27), the function $H$ takes the following forms:

$$\left. \begin{array}{l} \text{For } \lambda > 0 : H = cq_{i-1}^n + (1-c)q_i^n \,, \\ \text{For } \lambda < 0 : H = (1+c)q_i^n - cq_{i+1}^n \,. \end{array} \right\} \quad (9.29)$$

### Linear Schemes

Linear schemes are a special class of schemes (9.27) for the linear advection equation in (9.1), defined by the form

$$q_i^{n+1} = \sum_{k=-l}^{k=r} b_k q_{i+k}^n \,, \quad (9.30)$$

in which the coefficients $b_k$ are constant, that is, they do not depend on the solution. For example, for the Godunov finite difference method we have two cases:

1. For $\lambda > 0$: $l = 1$, $r = 0$, $b_{-1} = c$ and $b_0 = 1 - c$.
2. For $\lambda < 0$: we have $l = 0$, $r = 1$, $b_0 = 1 + c$, $b_1 = -c$.

A special class of schemes are three-point schemes of the form

$$q_i^{n+1} = b_{-1} q_{i-1}^n + b_0 q_i^n + b_1 q_{i+1}^n \,. \quad (9.31)$$

Table 9.1 lists a selection of well-known three-point schemes, their coefficients and some of their properties. For more details on these see [2].

**Table 9.1** Three-point schemes. Coefficients $b_k$, monotonicity and linear stability limit $C_{lim}$. Symbol * denotes unconditional instability. For background on all the methods listed see [2]

| Scheme | $b_{-1}$ | $b_0$ | $b_1$ | Monotone | $C_{lim}$ |
|---|---|---|---|---|---|
| Lax-Friedrichs | $\frac{1}{2}(1+c)$ | $0$ | $\frac{1}{2}(1-c)$ | Yes | 1 |
| Lax-Wendroff | $\frac{1}{2}(1+c)c$ | $1-c^2$ | $-\frac{1}{2}(1-c)c$ | No | 1 |
| FORCE | $\frac{1}{4}(1+c)^2$ | $\frac{1}{2}(1+c^2)$ | $\frac{1}{4}(1-c)^2$ | Yes | 1 |
| Godunov upwind | $c$ | $1-c$ | $0$ | Yes | 1 |
| Godunov centred | $\frac{1}{2}(1+2c)c$ | $1-2c^2$ | $-\frac{1}{2}(1-2c)c$ | No | $\frac{1}{2}\sqrt{2}$ |
| FTCS | $\frac{1}{2}$ | $1$ | $-\frac{1}{2}$ | No | * |

### 9.2.2 Monotonicity, Accuracy and Godunov's Theorem

**Definition 9.2** (*Monotone scheme*) A numerical scheme of the form (9.27) is called monotone if $H$ satisfies

$$\frac{\partial}{\partial q_k^n} H(q_{i-l}^n, q_{i-l+1}^n, \ldots, q_i^n, \ldots, q_{i+r}^n) \geq 0, \quad i-l \leq k \leq i+r. \tag{9.32}$$

**Remark** a linear scheme is monotone if and only if all its coefficients are non-negative. This follows from the definitions of linear schemes and monotonicity.

**Theorem 9.3** (A shortcut to accuracy) *A linear scheme of the form (9.30) is p-th order accurate in space and time ($p \geq 0$) in the sense of local truncation error, if and only if*

$$\sum_{k=-l}^{r} k^\eta b_k = (-c)^\eta, \quad \eta = 0, 1, \ldots, p, \tag{9.33}$$

where $c$ is the Courant number. For notational convenience we introduce $0^0 = 1$.

**Proof** For the proof and extensions to two and three dimensions see [2, 7]. □

**Example 9.4** (*The Godunov upwind finite difference method*) For $\lambda > 0$ the scheme is

$$q_i^{n+1} = H(q_{i-l}^n, q_i^n) = c q_{i-l}^n + (1-c) q_i^n. \tag{9.34}$$

$l = 1$, $r = 0$, $b_{-1} = c$, $b_0 = 1 - c$. Then we need to verify identity (9.33) for all possible non-negative integer values of $\eta$.

$$\eta = 0: \quad (-1)^0 \times c + 0^0 \times (1-c) = c + 1 - c = 1 = (-c)^0.$$

This merely says that the sum of the coefficients of the scheme is unity.

$$\eta = 1: \quad (-1)^1 \times c + 0^1 \times (1-c) = -c = (-c)^1.$$

Therefore the Godunov upwind scheme is first-order accurate. But just for curiosity let us try:

$$\eta = 2: \quad (-1)^2 \times c + 0^2 \times (1-c) = c \neq (-c)^2.$$

Hence the Godunov scheme is **not** second-order accurate, except for the trivial cases $c = 0$ and $c = 1$.

**Theorem 9.5** (Godunov's Theorem [5]) *There are no monotone, linear schemes (9.30) for the linear advection equation in (9.1) with constant $\lambda$, of accuracy two or higher.*

## 9.2 Basic Properties of Numerical Methods

***Proof*** It is sufficient to prove that there is no second order, linear, monotone method for LAE. Proceed by contradiction and assume there is a second order, linear, monotone method for LAE. From the accuracy theorem (9.33) we must have:

$$s_\eta = \sum_{k=-l}^{r} k^\eta b_k = \begin{cases} s_0 = 1, \eta = 0, \\ s_1 = -c, \eta = 1, \\ s_2 = c^2, \eta = 2. \end{cases} \quad (9.35)$$

But, in particular, from (9.35) plus some algebraic manipulations one obtains

$$\begin{aligned} s_2 &= \sum_{k=-l}^{r} k^2 b_k \\ &= \sum_{k=-l}^{r} (k+c)^2 b_k - 2c \sum_{k=-l}^{r} k b_k - c^2 \sum_{k=-l}^{r} b_k \\ &= \left[ \sum_{k=-l}^{r} (k+c)^2 b_k \right] - 2cs_1 - c^2 s_0. \end{aligned} \quad (9.36)$$

Use of (9.35) into (9.36) gives

$$c^2 = \left[ \sum_{k=-l}^{r} (k+c)^2 b_k \right] + c^2. \quad (9.37)$$

This implies a contradiction; for a monotone scheme all coefficients $b_k$ are non-negative but not simultaneously zero. Thus Godunov's theorem has been proved. □

### Consequences of Godunov's Theorem

From the theorem, linear monotone schemes are at most first-order accurate. But first-order methods are too inaccurate to be of practical use and therefore one must search for other classes of schemes. This is down to finding ways of circumventing Godunov's theorem. The key to this lies on the assumption of linear schemes. Thus a necessary condition for a numerical scheme to be oscillation-free (without new extrema) and of high-order of accuracy (for smooth solutions) is to be non-linear. In simple terms: *Schemes must be non-linear, even when applied to linear equations.*

Recall that schemes can be expressed in **the general form** (9.27). In what follows we introduce other forms.

## 9.2.3 Viscous Form of a Scheme

**Definition 9.6** (*The viscous form of a three-point scheme*) A three-point scheme is said to be written in viscous form if expressed as

$$q_i^{n+1} = q_i^n - \frac{1}{2}\frac{\Delta t}{\Delta x}[f(q_{i+1}^n) - f(q_{i-1}^n)] + \frac{1}{2}(d_{i+\frac{1}{2}}\Delta q_{i+\frac{1}{2}} - d_{i-\frac{1}{2}}\Delta q_{i-\frac{1}{2}}), \quad (9.38)$$

where $d_{i+\frac{1}{2}}$ is a function of $2k$ variables

$$d_{i+\frac{1}{2}} = d_{i+\frac{1}{2}}(q_{i-k+1}^n, q_{i-k+1}^n, \ldots, q_i^n, \ldots, q_{i+k}^n). \quad (9.39)$$

The function $d_{i+\frac{1}{2}}$ is called the *coefficient of numerical viscosity*.

**Viscous Form of a Three-Point Linear Scheme**

We study the viscous form a three-point linear scheme of the form

$$q_i^{n+1} = b_{-1}q_{i-1}^n + b_0 q_i^n + b_1 q_{i+1}^n. \quad (9.40)$$

The coefficients $b_{-1}$, $b_0$ and $b_1$ are constant. Assume the scheme to be at least first-order. Then from the accuracy theorem, see (9.33), we have

$$b_{-1} + b_0 + b_1 = 1, \quad b_{-1} - b_1 = c. \quad (9.41)$$

System (9.41) gives a one-parameter family of solutions. From the first equation we introduce $d = b_{-1} + b_1 = 1 - b_0$ and thus

$$b_{-1} = \frac{1}{2}(d+c), \quad b_0 = 1 - d, \quad b_1 = \frac{1}{2}(d-c). \quad (9.42)$$

Now in terms of $d$ scheme (9.40) becomes

$$q_i^{n+1} = q_i^n - \frac{1}{2}c(q_{i+1}^n - q_{i-1}^n) + \frac{1}{2}d(q_{i+1}^n - 2q_i^n + q_{i-1}^n). \quad (9.43)$$

This is the *viscous form* of scheme (9.40) and $d$ is the *coefficient of numerical viscosity* of the scheme.

**Remarks on the Viscous Form**

1. Particular values of $d$ give particular schemes, as we shall see.
2. The stability condition for the scheme becomes

## 9.2 Basic Properties of Numerical Methods

$$c^2 \leq d \leq 1 . \tag{9.44}$$

3. The monotonicity condition is

$$c \leq d \leq 1 . \tag{9.45}$$

4. A truncation error analysis gives the *coefficient of numerical viscosity*, in the sense of truncation error

$$\alpha_{visc} = \frac{1}{2} \Delta x \lambda \left( \frac{d - c^2}{c} \right) . \tag{9.46}$$

Thus effectively the coefficient $d$ measures the truncation error of the scheme. Further use of the viscous form (9.43) reveals that well-known schemes are obtained by an appropriate choice of the coefficient of numerical viscosity $d$. In particular, we can reproduce the Lax-Friedrichs, FORCE, Godunov upwind, Godunov centred and Lax-Wendroff schemes by appropriately choosing $d$, namely

$$d = \begin{cases} d_{LF} = 1 & \rightarrow \text{Lax-Friedrichs} , \\ d_{FORCE} = \frac{1}{2}(1 + c^2) & \rightarrow \text{FORCE} , \\ d_{GodU} = |c| & \rightarrow \text{Godunov upwind} , \\ d_{GodC} = 2c^2 & \rightarrow \text{Godunov centred} , \\ d_{LW} = c^2 & \rightarrow \text{Lax-Wendroff} . \end{cases} \tag{9.47}$$

Figure 9.6 shows a $d$-$c$ plot of the coefficient of numerical viscosity $d$ against the Courant number $c$, for the five schemes in (9.47) derived from the viscous form. For all linearly stable methods $d$ is bounded below by $d_{LW} = c^2$ (Lax-Wendroff) and bounded above by $d_{LF} = 1$ (Lax-Friedrichs). For all monotone schemes the coefficient of numerical viscosity $d = d(c)$ lies inside the triangle formed by the thick lines, with the Lax-Friedrichs scheme ($d_{LF} = 1$) above and the Godunov upwind method ($d_{GodU} = |c|$) below. The Lax-Friedrichs scheme is the least accurate of all linearly stable schemes, is has the largest $d$, the maximum value of the coefficient of numerical viscosity for stability. The Lax-Wendroff scheme is the most accurate linearly stable scheme, it has the smallest $d_{LW} = c^2$ for a fixed $c$, with second-order viscosity equal to zero. The Lax-Wendroff is not monotone for any value of $c$, except for $c = 0$ and $c = 1$. The Godunov centred scheme is monotone for $\frac{1}{2} \leq c \leq \frac{1}{2}\sqrt{2} = C_{lim}$ and non-monotone for $0 \leq c \leq \frac{1}{2}$. The most accurate scheme in the class of monotone methods is the Godunov upwind method, as stated below.

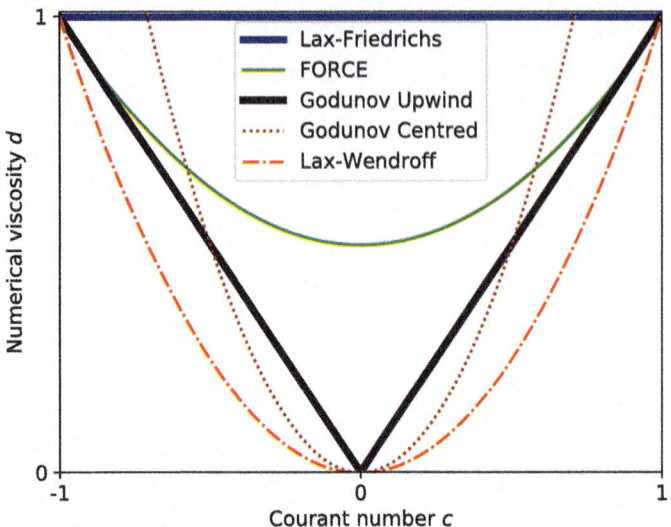

**Fig. 9.6** Numerical schemes represented in terms of their coefficient of numerical viscosity $d = d(c)$ in (9.47), as function of Courant number $c$. Monotone schemes lie within the triangle defined by the thick lines

**Theorem 9.7** *The Godunov upwind scheme for the linear advection equation is the monotone scheme with the smallest truncation error.*

***Proof*** The proof is left as an exercise. □

### 9.2.4 Conservative Form of a Scheme

Consider the scalar conservation law

$$\partial_t q + \partial_x f(q) = 0 . \tag{9.48}$$

**Definition 9.8** (*Conservative form of a scheme*) A numerical method for solving the scalar PDE (9.48) is said to be in conservative form if expressed as

$$q_i^{n+1} = q_i^n - \frac{\Delta t}{\Delta x} \left( f_{i+\frac{1}{2}} - f_{i-\frac{1}{2}} \right) , \tag{9.49}$$

where $f_{i+\frac{1}{2}}$ is the *numerical flux*.

Most well-known finite difference methods can be expressed in conservative form (9.49), with an appropriate numerical flux. In what follows we give the expression for the numerical flux for most well-known finite difference methods. In doing so, for some cases we assume the linear advection equation, with $f(q) = \lambda q$ in (9.48).

## 9.2 Basic Properties of Numerical Methods

For other schemes we shall keep $f(q)$ as the flux for a general scalar conservation law (9.48).

1. **The Godunov upwind method** [5]. For LAE the Godunov upwind flux is

$$f_{i+\frac{1}{2}}^{GodU} = \frac{1}{2}[1 + sign(c)](\lambda q_i^n) + \frac{1}{2}[1 - sign(c)](\lambda q_{i+1}^n) . \quad (9.50)$$

This formula is only valid for the linear advection equation; it contains both the $\lambda > 0$ and $\lambda < 0$ cases. For $\lambda > 0$ $f_{i+\frac{1}{2}} = \lambda q_i^n$ and $f_{i-\frac{1}{2}} = \lambda q_{i-1}^n$.
Verify that by inserting these fluxes into the conservative formula (9.49), the Godunov upwind method in finite difference form is reproduced, for both positive and negative wave speeds.

2. **The Godunov centred method** [8]. For the conservation law (9.48) with flux function $f(q)$, the numerical flux for the Godunov centred method is

$$f_{i+\frac{1}{2}}^{GodC} = f(q_{i+\frac{1}{2}}^{GodC}) ; \quad q_{i+\frac{1}{2}}^{GodC} = \frac{1}{2}(q_i^n + q_{i+1}^n) - \frac{\Delta t}{\Delta x}\left[f(q_{i+1}^n) - f(q_i^n)\right] . \quad (9.51)$$

Here $f$ is the flux function in (9.48) evaluated at various mesh points. This method, also due to Godunov [8], is a centred method, not upwind; it does not depend on a wave speed. Here the scheme is written in a way that is applicable to any conservation law (9.48), or in fact, any system of conservation laws.

3. **The Lax-Friedrichs method (or Lax method)** [9]. For the conservation law (9.48) with flux function $f(q)$, the numerical flux is

$$f_{i+\frac{1}{2}}^{LF} = \frac{1}{2}\left[f(q_i^n) + f(q_{i+1}^n)\right] - \frac{1}{2}\frac{\Delta x}{\Delta t}(q_{i+1}^n - q_i^n) . \quad (9.52)$$

This is also a centred flux (not upwind) and is easily implemented for solving any conservation law or system of conservation laws.

4. **The Lax-Wendroff method** [6]. For the conservation law (9.48) with flux function $f(q)$, the numerical flux for the Lax-Wendroff method is

$$f_{i+\frac{1}{2}}^{LW} = f(q_{i+\frac{1}{2}}^{LW}) ; \quad q_{i+\frac{1}{2}}^{LW} = \frac{1}{2}(q_i^n + q_{i+1}^n) - \frac{1}{2}\frac{\Delta t}{\Delta x}\left[f(q_{i+1}^n) - f(q_i^n)\right] . \quad (9.53)$$

This is also a centred flux which is easily implemented for solving any conservation law or system of conservation laws, leading to a method of second-order accuracy.

5. **The FORCE method** [10, 11]. For the conservation law (9.48) with flux function $f(q)$, the numerical flux for the FORCE method is

$$f_{i+\frac{1}{2}}^{FORCE} = \frac{1}{2}\left[f_{i+\frac{1}{2}}^{LF} + f_{i+\frac{1}{2}}^{LW}\right] . \quad (9.54)$$

**Table 9.2** Properties of seven finite difference methods, written as conservative schemes. Note that not all linearly stable methods have Courant number stability limit $C_{lim} = 1$ and not all schemes are monotone

| Scheme | Type | Accuracy | Monotonicity | Stability condition |
|---|---|---|---|---|
| Godunov upwind | Upwind | 1st order | Monotone | $\|c\| \leq C_{lim} = 1$ |
| Godunov centred | Centred | 1st order | Non-monotone | $\|c\| \leq C_{lim} = \frac{1}{2}\sqrt{2}$ |
| Lax-Friedrichs | Centred | 1st order | Monotone | $\|c\| \leq C_{lim} = 1$ |
| FORCE | Centred | 1st order | Monotone | $\|c\| \leq C_{lim} = 1$ |
| Lax-Wendroff | Centred | 2nd order | Non-monotone | $\|c\| \leq C_{lim} = 1$ |
| Warming-Beam | Upwind | 2nd order | Non-monotone | $\|c\| \leq C_{lim} = 2$ |
| Fromm | Centred | 2nd order | Non-monotone | $\|c\| \leq C_{lim} = 1$ |

The FORCE flux is precisely an arithmetic average of the Lax-Friedrichs flux and Lax-Wendroff flux. FORCE is also a centred method, which is easily implemented for solving any conservation law or system of conservation laws.

6. **The Warming-Beam method** [12]. For LAE the numerical flux for the Warming-Beam method is

$$f_{i+\frac{1}{2}}^{WB} = \begin{cases} -\frac{1}{2}(1-c)(\lambda q_{i-1}^n) + \frac{1}{2}(3-c)(\lambda q_i^n) & \text{if } \lambda > 0, \\ \frac{1}{2}(3+c)(\lambda q_{i+1}^n) - \frac{1}{2}(1+c)(\lambda q_{i+2}^n) & \text{if } \lambda < 0. \end{cases} \quad (9.55)$$

This is an upwind flux, written here specifically for the linear advection equation.

7. **The Fromm method** [13]. For LAE the numerical flux for the Fromm method is

$$f_{i+\frac{1}{2}}^{FR} = \begin{cases} -\frac{1}{4}(1-c)(\lambda q_{i-1}^n) + (\lambda q_i^n) + \frac{1}{4}(1-c)(\lambda q_{i+1}^n) & \text{if } \lambda > 0, \\ \frac{1}{4}(1+c)(\lambda q_i^n) + (\lambda q_{i+1}^n) - \frac{1}{4}(1+c)(\lambda q_{i+2}^n) & \text{if } \lambda < 0. \end{cases} \quad (9.56)$$

This is a centred method, written here specifically for the linear advection equation.

Table 9.2 summarises the main properties of the listed numerical methods (9.50)-(9.56). The reader is encouraged to verify these properties by using previous definitions given in this chapter. In Chap. 10 we shall extend some of these methods to non-linear systems, such as the shallow water equations.

## 9.3 Computational Results

In this section we show computational results for the linear advection equation from six methods, as applied to two test problems, one with smooth solution and another with discontinuous solution. Numerical results are compared with the exact solution.

### 9.3.1 Test Problems, Methods and Parameters

We solve the following initial-boundary value problem for the linear advection equation

$$
\begin{aligned}
&\text{PDE:} && \partial_t q + \lambda \partial_x q = 0 \,, \quad x \in [a, b] \,, \quad t > 0 \,, \\
&\text{IC:} && q(x, 0) = h(x) \,, \quad x \in [a, b] \,, t = 0 \,, \\
&\text{BCs (periodic):} && q(a, t) = q(b, t) \,, \quad \forall t > 0 \,,
\end{aligned}
\tag{9.57}
$$

with $\lambda = 1$, $a = -5$ and $b = 5$. We use 3 first-order methods: Lax-Friedrichs, FORCE and Godunov upwind; and 3 second-order methods: Lax-Wendroff, Warming-Beam and Fromm. All six methods have stability limit $C_{lim} = 1$ and therefore $C_{cfl} = S_c \times C_{lim} = S_c$, where $S_c$ is the safety coefficient. We consider two tests problems given by two types of initial conditions $h(x)$:

**Test 1: The Initial Condition Is Smooth**

In this test problem the initial condition $h(x)$ is given by a smooth Gaussian profile

$$q(x, 0) = h(x) = \alpha e^{-\beta x^2} \,, \quad \alpha = 1 \,, \quad \beta = 2 \,, \quad x \in [-5, 5] \,. \tag{9.58}$$

**Test 2: The Initial Condition Is Discontinuous**

In this test problem the initial condition $h(x)$ is a (discontinuous) square wave given as

$$q(x, 0) = h(x) = \begin{cases} 0 & \text{if } a = -5 \leq x < -2 \,, \\ 1 & \text{if } -2 \leq x \leq 2 \,, \\ 0 & \text{if } 2 < x \leq b = 5 \,. \end{cases} \tag{9.59}$$

**Computational Parameters**

We use two meshes: $M = 25$ cells and $M = 200$ cells. The time step $\Delta t$ results from choosing the CFL coefficient set to $C_{cfl} = 0.9$. The output time $T_{out} = 100$

is chosen so as to complete 10 cycles within the domain, at which time the exact solution coincides with the initial condition. In both test problems we impose periodic boundary conditions.

### 9.3.2 Results and Discussion

Results are displayed in the following four figures. Figure 9.7 shows results for Test 1 for three first-order methods; Fig. 9.8 shows results for Test 1 for three second-order methods; Fig. 9.9 shows results for Test 2 for three first-order methods; Fig. 9.10 shows results for Test 2 for three second-order methods.

**First-Order Methods for Test 1**

Results are shown in Fig. 9.7. The exact solution is a very smooth function but has large derivatives and a pronounced peak $q_{max} = 1$ located at $x = 0$ at the initial time and at the output time of $T_{out} = 100$ units. Subject to periodic boundary conditions, the initial profile has travelled towards the right, has left the domain through the right boundary and has re-entered the domain through the left boundary, completing a total of 10 cycles in 100 time units.

Results from the Lax-Friedrichs method (top) are shown for two meshes: 25 cells (left) and 200 cells (right). The coarser mesh result is too inaccurate; this is more evident at $x = 0$; the numerical error is very large. The finer mesh on the right, has improved the results, but not very appreciably, even after a mesh refinement by a factor 8. In theory, if the mesh continues to be refined the solution should get closer and closer to the exact solution, but clearly convergence is very slow. The slow convergence is to do with the first-order of accuracy of the scheme.

The results from the FORCE scheme (middle) have improved upon those of Lax-Friedrichs, but not significantly. Again we see an improvement with the finer mesh but convergence is slow too. The results from Godunov upwind (bottom) are again a little better than the previous two schemes, but not by an appreciable margin.

In summary, the results from the three first-order methods are rather disappointing for this problem with smooth solution. Their main feature is *excessive numerical diffusion* that *clips extrema*. Mesh refinement helps but convergence is very slow. This means that getting accurate results from first-order methods may prove very expensive, as *extremely fine meshes* will be necessary to attain expected, prescribed small errors.

**Second-Order Methods for Test 1**

Results are shown in Fig. 9.8. Starting from the Lax-Wendroff method (top), we observe that the coarse mesh result on the left is very inaccurate ($M = 25$ cells).

## 9.3 Computational Results

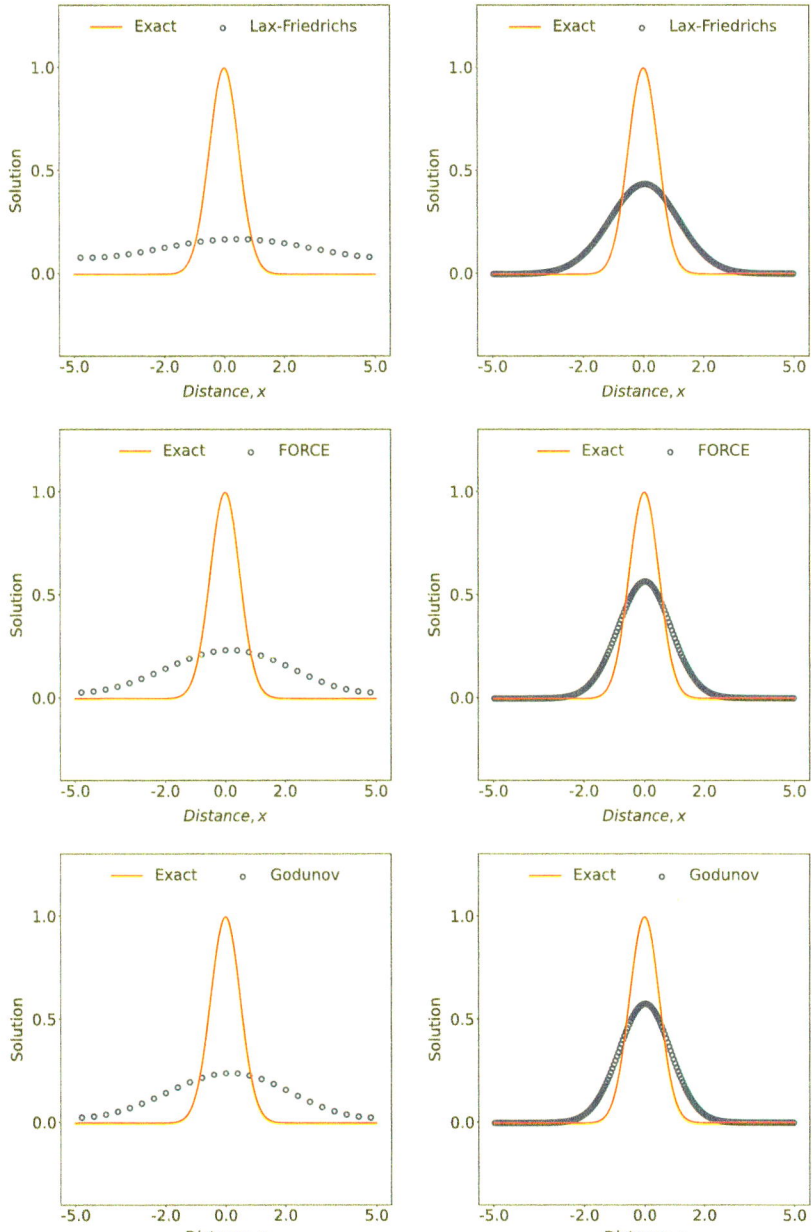

**Fig. 9.7 Test 1 solved with three first-order methods**: Lax-Friedrichs (top), FORCE (middle) and Godunov upwind (bottom). Results are shown at output time $T_{out} = 100$ and for two meshes: $M = 25$ cells (left column) and $M = 200$ cells (right column); the CFL coefficient is $C_{cfl} = 0.9$. The exact solution is shown by full line and the numerical solution is shown by symbols

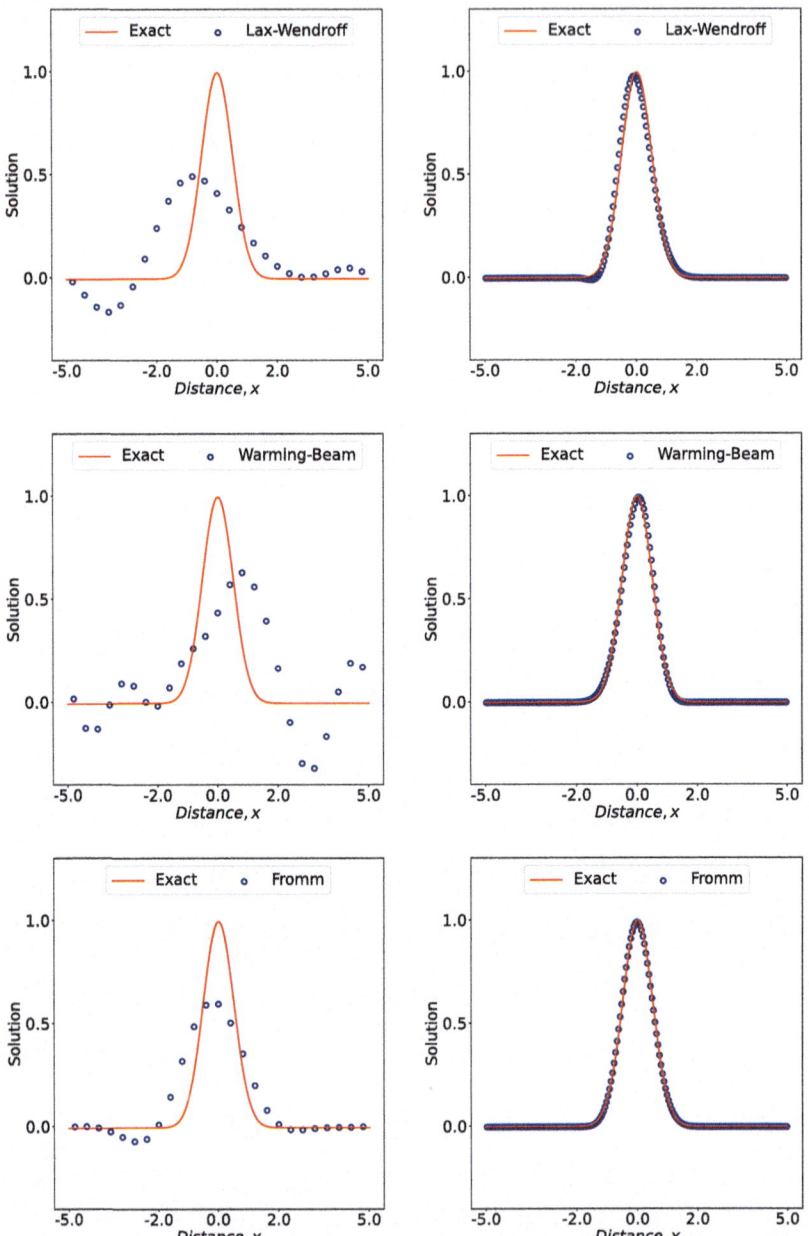

**Fig. 9.8 Test 1 solved with three second-order methods**: Lax-Wendroff (top), Warming-Beam (middle) and Fromm (bottom). Results are shown at output time $T_{out} = 100$ and for two meshes: $M = 25$ cells (left column) and $M = 200$ cells (right column); the CFL coefficient is $C_{cfl} = 0.9$. The exact solution is shown by full line and the numerical solution is shown by symbols

## 9.3 Computational Results

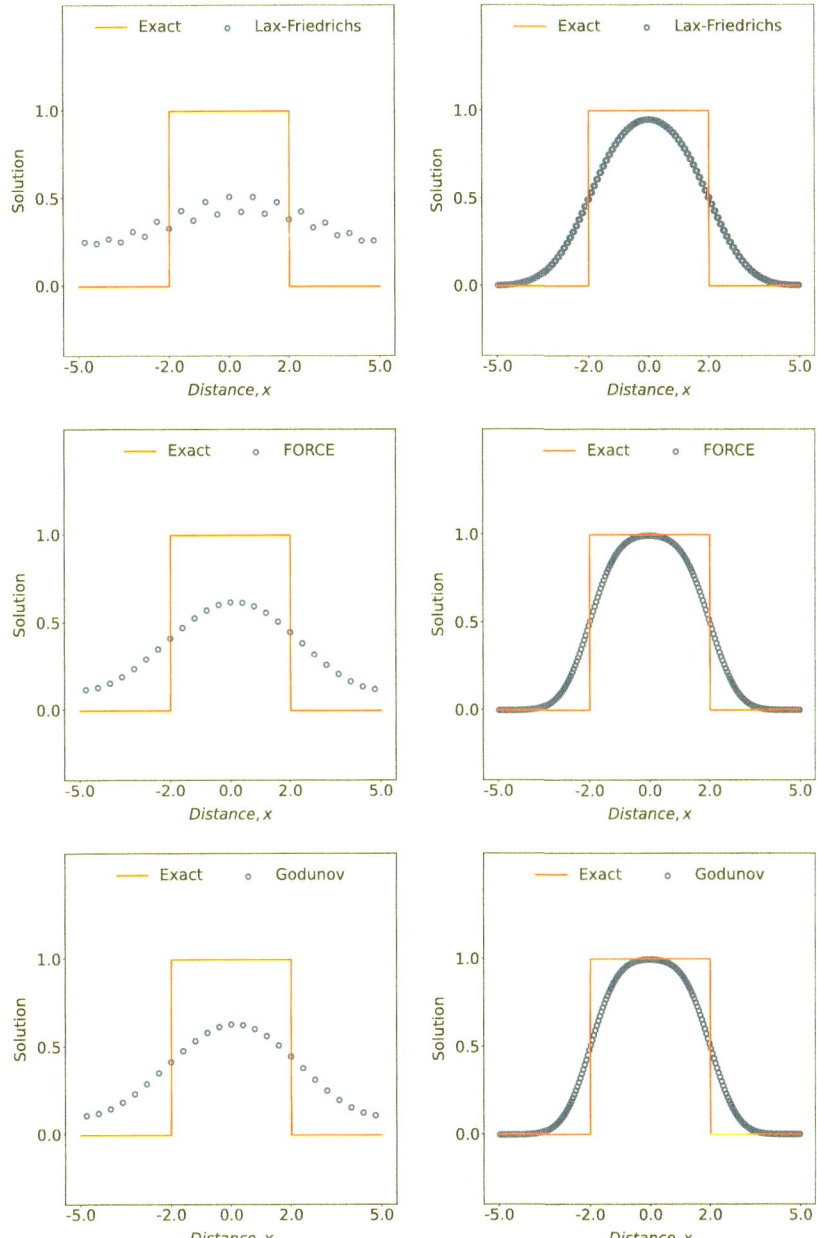

**Fig. 9.9 Test 2 solved with three first-order methods**: Lax-Friedrichs (top), FORCE (middle) and Godunov upwind (bottom). Results are shown at output time $T_{out} = 100$ and for two meshes: $M = 25$ cells (left column) and $M = 200$ cells (right column); the CFL coefficient is $C_{cfl} = 0.9$. The exact solution is shown by full line and the numerical solution is shown by symbols

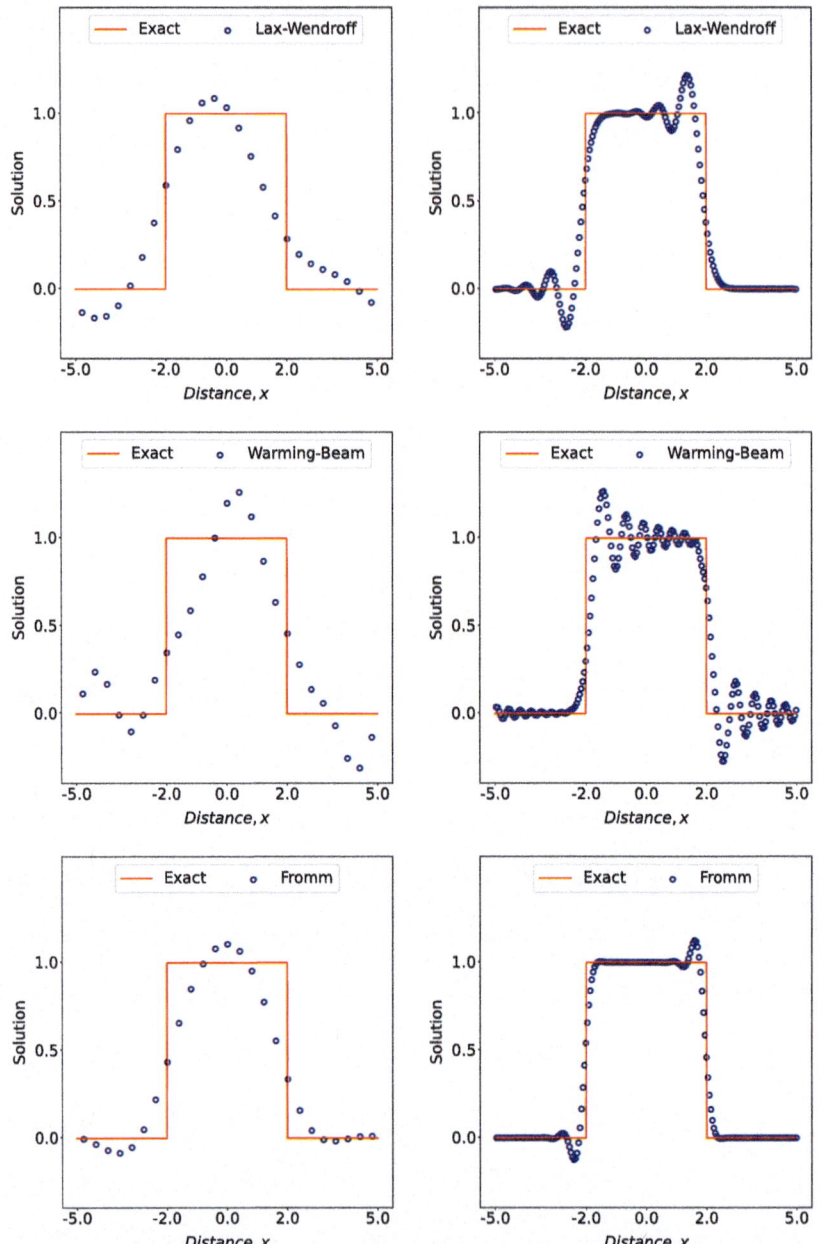

**Fig. 9.10 Test 2 solved with three second-order methods**: Lax-Wendroff (top), Warming-Beam (middle) and Fromm (bottom). Results are shown at output time $T_{out} = 100$ and for two meshes: $M = 25$ cells (left column) and $M = 200$ cells (right column); the CFL coefficient is $C_{cfl} = 0.9$. The exact solution is shown by full line and the numerical solution is shown by symbols

## 9.3 Computational Results

This is disappointing for a second-order method applied to a problem with smooth solution. Note the *undershoots* below zero and *overshoots* above zero. These spurious oscillations are typical of (linear) second-order methods and are present even for smooth problems that exhibit large gradients. This is contrary to popular belief that second-order methods (and higher order) are oscillation-free for smooth problems. We also observe a *phase error* in the form of erroneous position of the wave, which in this case trails behind the exact solution. The oscillations and the phase error are linked to the form of the *modified equation* for the Lax-Wendroff method, which is of *dispersive type*, leading to dispersive-type errors. As the mesh is refined by a factor of 8 the numerical solution improves dramatically (top right). But even so we still see a small undershoot at the tail of the wave.

The results from the Warming-Beam scheme (middle plots) are similar to those of the Lax-Wendroff method. A noticeable difference is that now the large overshoots for the coarse mesh case (middle left) are ahead of the wave. The finer mesh result (middle right) exhibits good agreement with the exact solution. To the eye, the dispersive error has decreased significantly, for the finer mesh computation.

Results from the Fromm scheme are shown in the bottom pictures of Fig. 9.8. This second-order method is distinctly better than the other two second-order methods. The coarser mesh case still shows visible undershoots ahead and behind the wave; the finer-mesh result looks very accurate. No appreciable dispersive errors are seen, which is typical of the Fromm scheme.

In summary, second-order methods for smooth problems are not as accurate as is sometimes believed. Spurious oscillations are present for smooth problems, even for reasonably fine meshes. These methods are indeed better than first-order methods, in the sense that the errors diminish faster than for first-order methods, as the mesh is refined.

### First-Order Methods for Test 2

Results are shown in Fig. 9.9. The exact solution is a square wave with two discontinuities. Subject to periodic boundary conditions, the initial profile has travelled towards the right, has left the domain through the right boundary and has re-entered the domain through the left boundary, completing a total of 10 cycles in 100 time units.

The two top profiles of Fig. 9.9 correspond to the Lax-Friedrichs method. Results are very inaccurate. The numerical method cannot represent the correct shape of the square wave with two discontinuities, even though the finer mesh results look better. A dominant feature of the numerical solution is *numerical diffusion*, linked to the diffusive character of the modified equation of the scheme. Each discontinuity is represented by virtually 100 cells (right frame), instead of none, in the exact solution. A very *puzzling* observation is the following: the solution from the Lax-Friedrichs method on the coarser mesh (left frame) exhibits visible *spurious oscillations*. This is in apparent contradiction with the fact that the method is monotone, strictly speaking.

Note however that the oscillations are *internal* to the interval containing the initial conditions. This is not a very well known feature of the Lax-Friedrichs method.

The FORCE method (middle) gives improved results relative to those of Lax-Friedrichs, but in general the quality of the numerical solution is still unsatisfactory, dominated by numerical diffusion. FORCE has eliminated the oscillations of the Lax-Friedrichs method. Refining the mesh by a factor of 8 has improved the result but this is still far from satisfactory. Again, convergence is very slow.

The Godunov upwind scheme (bottom pictures) shows again a slight improvement over the FORCE results, and a visibly improvement over the Lax-Friedrichs results. Again, numerical diffusion dominates the numerical solution. In summary, first order monotone methods for discontinuous problems have the unique feature of absence of spurious overshoots and undershoots, unlike second-order linear methods. The worst method is the Lax-Friedrichs method. Consistent with the theory, the best first-order method is the Godunov upwind method.

**Remark** (*First-order accuracy and monotonicity*) It is important to be aware that the first-order accuracy of a scheme is not a sufficient condition for that scheme to be monotone. The Godunov centred method is the typical example to show this. The scheme is first-order accurate but non-monotone.

**Second-Order Methods for Test 2**

Results from second-order methods for this problem with discontinuous solution are shown in Fig. 9.10. One statement that applies to all six results is that second-order (linear) schemes are wholly inadequate for problems with discontinuous solutions. While for the smooth solution case, mesh refinement made the difference, in the present case, mesh refinement does not help visibly. **The spurious oscillations will not vanish under mesh refinement**. Better methods for problems with discontinuities, or even large gradients, are needed, and these, according to Godunov's theorem, must be non-linear methods. In Chaps. 12 and 14 we construct non-linear high-order methods that to a large extent overcome these difficulties.

## 9.4 Conclusions and Further Reading

This chapter has presented some very basic notions on numerical methods for solving hyperbolic equations, utilising the simplest partial differential equation, namely the linear advection equation with constant wave propagation speed. The finite difference method has been introduced and examples of the most well-known finite difference methods have been given, including the Lax-Friedrichs method, the Lax-Wendroff method, the FORCE method, the Godunov centred method and the Godunov upwind method. Some of the main properties of these schemes have also been studied, including accuracy, monotonicity and linear stability. The finite difference schemes have been written in conservative form with a corresponding numerical flux. Some theoretical notions have also been included, notably the Godunov theorem, which is stated and proved. Godunov's theorem sets the theoretical bases for the construction of non-linear methods for hyperbolic equations. Some numerical results have been presented and discussed, noting the difference between problems with smooth solution and problems with discontinuous solution. The need for constructing non-linear methods has been highlighted. Non-linear methods of the TVD type (Total Variation Diminishing) are studied in Chap. 12 and higher-order, non-linear reconstruction-based fully discrete methods of the ADER type are studied in Chap. 14.

Usefull mathematical background is given in Chap. 2. As general references for further study I suggest the following: Smith [14], Anderson et al. [15], Mitchell and Griffiths [16], Roache [17], Richtmyer and Morton [18], Hoffmann [19] and Fletcher [20]. I particularly recommend the book by Morton and Mayers [21]. Very relevant textbooks to the main themes of this book are Sod [22], Holt [23], Hirsch [24, 25], LeVeque [1], Godlewski and Raviart [3], Kröner [26], Laney [27], Thomas [28] and Toro [2, 29, 30]. For those who are absolute beginners in the field I recommend the following self-study programme as a way of obtaining more benefit from this textbook: (a) read Chaps. 2–4 of the book by Hoffmann [19] and do the exercises therein, (b) read Chaps. 7–10 of the book by Hirsch [24] and do exercises therein. See also Chap. 5 of [2, 29] and various chapters in [31].

# References

1. R.J. LeVeque, *Finite Volume Methods for Hyperbolic Problems*, (Cambridge University Press, 2002)
2. E.F. Toro, *Riemann Solvers and Numerical Methods for Fluid Dynamics*, A Practical Introduction, 3rd edn. (Springer, 2009)
3. E. Godlewski, P.A. Raviart, *Numerical Approximation of Hyperbolic Systems of Conservation Laws*, 2nd edn. (Springer, 2021)
4. R. Courant, E. Isaacson, M. Rees, On the solution of nonlinear hyperbolic differential equations by finite differences. Comm. Pure. Appl. Math. **5**, 243–255 (1952)
5. S.K. Godunov, Finite difference methods for the computation of discontinuous solutions of the equations of fluid dynamics. Mat. Sb. **47**, 271–306 (1959)
6. P.D. Lax, B. Wendroff, Systems of Conservation Laws. Commun. Pure Appl. Math. **13**, 217–237 (1960)
7. S.J. Billett, E.F. Toro, On the accuracy and stability of explicit schemes for multidimensional linear homogeneous advection equations. J. Comput. Phys. **131**, 247–250 (1997)
8. S.K. Godunov, A.V. Zabrodin, G.P. Prokopov, A difference scheme for two-dimensional unsteady aerodynamics. J. Comput. Math. Math. Phys. USSR **2**(6), 1020–1050 (1961)
9. P.D. Lax, Weak solutions of nonlinear hyperbolic equations and their numerical computation. Commun. Pure. Appl. Math. **VII**, 159–193 (1954)
10. E.F. Toro, S.J. Billett, Centred TVD Schemes for Hyperbolic Conservation Laws. Technical Report MMU–9603, Department of Mathematics and Physics, Manchester Metropolitan University, UK (1996)
11. E.F. Toro, S.J. Billett, Centred TVD schemes for hyperbolic conservation laws. IMA J. Numer. Anal. **20**, 47–79 (2000)

12. R.F. Warming, R.W. Beam, Upwind second order difference schemes with applications in aerodynamic flows. AIAA J. **24**, 1241–1249 (1976)
13. J.E. Fromm, A method for reducing dispersion in convective difference schemes. J. Comput. Phys. **3**, 176–189 (1968)
14. G.D. Smith, *Numerical Solution of Partial Differential Equations* (Clarendon Press, Oxford, 1978)
15. D.A. Anderson, J.C. Tannehill, R.H. Pletcher, *Computational Fluid Mechanics and Heat Transfer* (Hemisphere Publishing Corporation, 1984)
16. A.R. Mitchell, D.F. Griffiths, The Finite Difference Method in Partial Differential Equations. (Wiley, 1980)
17. P.J. Roache, *Computational Fluid Dynamics*. (Hermosa Publishers, 1982)
18. R.D. Richtmyer, K.W. Morton, *Difference Methods for Initial Value Problems* (Interscience-Wiley, New York, 1967)
19. K.A. Hoffmann, *Computational Fluid Dynamics for Engineers* Engineering Education Systems. (Austin, Texas, USA, 1989)
20. C.A.J. Fletcher. *Computational Techniques for Fluid Dynamics, Vols. I and II.* (Springer, 1988)
21. K.W. Morton, D.F. Mayers. *Numerical Solution of Partial Differential Equations*. (Cambridge University Press, 1994)
22. G.A. Sod. *Numerical Methods in Fluid Dynamics*. (Cambridge University Press, 1985)
23. M. Holt, *Numerical Methods in Fluid Dynamics*. (Springer, 1984)
24. C. Hirsch, *Numerical Computation of Internal and External Flows, Vol. I: Fundamentals of Numerical Discretization*. (Wiley, 1988)
25. C. Hirsch, *Numerical Computation of Internal and External Flows, Computational Methods for Inviscid and Viscous Flows, Vol. II* (Wiley, 1990)
26. D. Kröner, *Numerical Schemes for Conservation Laws*. (Wiley Teubner, 1997)
27. C.B. Laney, *Computational Gasdynamics*. (Cambridge University Press, 1998)
28. J.W. Thomas, *Numerical Partial Differential Equations, Texts in Applied Mathematics 22.* (Springer, 1998)
29. E.F. Toro, *Riemann Solvers and Numerical Methods for Fluid Dynamics. A Practical Introduction.* (Springer, 1997)
30. E.F. Toro, *Riemann Solvers and Numerical Methods for Fluid Dynamics. A Practical Introduction*, 2nd edn. (Springer, 1999)
31. E.F. Toro, Müller LO. *Computational Bodily Fluid Dynamics. Models and Algorithms (to appear)* (Springer, 2024)

# Chapter 10
# First-Order Methods for Systems

**Abstract** First-order monotone numerical methods for non-linear systems of hyperbolic conservation laws are studied. Some of these form the bases for constructing higher-order schemes in successive chapters. Algorithms for one-dimensional PDEs are first presented; these include the Godunov upwind scheme [1] in conjunction with the exact Riemann solver; the Random Choice Method (RCM) of Glimm [2]; Flux-Vector Splitting (FVS) methods; and centred schemes, such as the Lax–Friedrichs [3], Lax–Wendroff [3] and FORCE methods [4, 5]. There follows the finite volume framework for PDEs in multiple space dimensions on unstructured meshes, including definitions for semi-discrete and fully discrete schemes; the rotated Riemann problem from the rotational invariance of the two-dimensional shallow water equations; the intercell numerical flux and determination of edge lengths, normals and cell areas in two space dimensions, in order to completely specify the numerical methods. Some examples are shown, in order to illustrate the performance of selected schemes studied, as compared against exact solutions for a suite of carefully selected test problems.

## 10.1 The Finite Volume Framework

In this section we set the finite volume framework for one-dimensional non-linear systems of hyperbolic balance laws, starting from the integral forms of the PDEs.

### 10.1.1 Balance Laws in Integral Form

Consider the general first order PDE written in differential form

$$\partial_t q(x,t) + \partial_x f(q(x,t)) = s(q(x,t)). \quad (10.1)$$

Here $q(x,t)$ is the unknown, $f(q(x,t))$ is the flux function and $s(q(x,t))$ is the source term. Equation (10.1) is called a *balance law* and when the source term

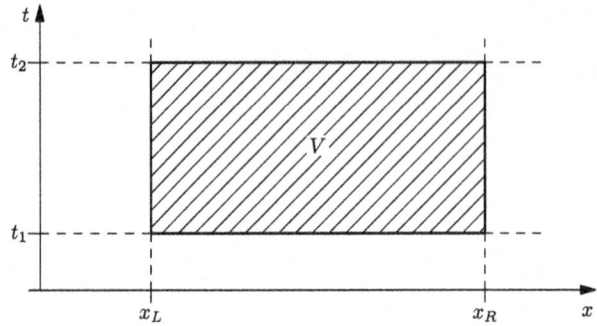

**Fig. 10.1** Control volume $V = [x_L, x_R] \times [t_1, t_2]$ in x-t space

$s(q(x, t)) = 0$ is called a *conservation law*. From the mathematical and numerical points of view it is convenient to express the PDE (10.1) in integral form, as done in Chap. 2 on *control volume V* in the x-t plane

$$V = [x_L, x_R] \times [t_1, t_2], \tag{10.2}$$

as depicted in Fig. 10.1. As seen in Chap. 2 it is also convenient to obtain an averaged version of the integral form of the PDE (10.1), namely

$$\left. \begin{array}{l} \dfrac{1}{\Delta x} \displaystyle\int_{x_L}^{x_R} q(x, t_2) dx = \dfrac{1}{\Delta x} \displaystyle\int_{x_L}^{x_R} q(x, t_1) dx \\ \qquad\qquad - \dfrac{\Delta t}{\Delta x} \left[ \dfrac{1}{\Delta t} \displaystyle\int_{t_1}^{t_2} f(q(x_R, t)) dt - \dfrac{1}{\Delta t} \displaystyle\int_{t_1}^{t_2} f(q(x_L, t)) dt \right] \\ \qquad\qquad + \dfrac{\Delta t}{\Delta x \Delta t} \displaystyle\int_{x_L}^{x_R} \displaystyle\int_{t_1}^{t_2} s(q(x, t)) dx dt \, . \end{array} \right\} \tag{10.3}$$

Then, the averaged integral expression (10.3) can be written as

$$q^{new} = q^{old} - \dfrac{\Delta t}{\Delta x} \left[ f_{right} - f_{left} \right] + \Delta t \, s_{vol} \, . \tag{10.4}$$

This expression is exact, with the following definitions

$$\left. \begin{array}{l} q^{new} = \dfrac{1}{\Delta x} \displaystyle\int_{x_L}^{x_R} q(x, t_2) dx \, , \quad q^{old} = \dfrac{1}{\Delta x} \displaystyle\int_{x_L}^{x_R} q(x, t_1) dx \, , \\ f_{right} = \dfrac{1}{\Delta t} \displaystyle\int_{t_1}^{t_2} f(q(x_R, t)) dt \, , \quad f_{left} = \dfrac{1}{\Delta t} \displaystyle\int_{t_1}^{t_2} f(q(x_L, t)) dt \, , \\ \qquad\qquad s_{vol} = \dfrac{1}{\Delta x \Delta t} \displaystyle\int_{x_L}^{x_R} \displaystyle\int_{t_1}^{t_2} s(q(x, t)) dt dx \, . \end{array} \right\} \tag{10.5}$$

## 10.1 The Finite Volume Framework

Numerical methods, called finite volume methods, use the *finite volume formula* (10.4) to compute approximate solutions to the PDE (10.1), in which $q^{old}$ is a known average of the solution at the old time level and the remaining terms on the right hand side of (10.4) are found by appropriate approximations to the integrals in (10.5). The computational parameters $\Delta t$ and $\Delta x$ must be prescribed in order to complete the scheme to compute $q^{new}$. See Chap. 2 for details.

### 10.1.2 The Finite Volume Formula

Consider the initial-boundary value problem (IBVP) for a generic system of $N$ non-linear hyperbolic balance laws

$$\left. \begin{array}{ll} \text{PDEs:} & \partial_t \mathbf{Q} + \partial_x \mathbf{F}(\mathbf{Q}) = \mathbf{S}(\mathbf{Q}), \quad x \in [a,b], \quad t > 0, \\ \text{ICs:} & \mathbf{Q}(x,0) = \mathbf{Q}^{(0)}(x), \quad x \in [a,b], \\ \text{BCs:} & \mathbf{Q}(a,t) = \mathbf{B}_L(t), \quad \mathbf{Q}(b,t) = \mathbf{B}_R(t), \quad t \geq 0. \end{array} \right\} \quad (10.6)$$

$\mathbf{Q}(x,t)$ is the vector of *conserved variables*; $\mathbf{F}(\mathbf{Q})$ is the flux vector, or *physical flux*; $\mathbf{S}(\mathbf{Q})$ is the *source term vector*; $\mathbf{Q}^{(0)}(x)$ is the initial condition; $\mathbf{B}_L(t)$ and $\mathbf{B}_R(t)$ are the boundary conditions on the left and right boundaries, respectively, two prescribed functions of time. The vectors $\mathbf{Q}$, $\mathbf{F}(\mathbf{Q})$ and $\mathbf{S}(\mathbf{Q})$ are as follows

$$\mathbf{Q} = [q_1, q_2, \ldots, q_N]^T, \quad \mathbf{F}(\mathbf{Q}) = [f_1, f_2, \ldots, f_N]^T, \quad \mathbf{S}(\mathbf{Q}) = [s_1, s_2, \ldots, s_N]^T. \quad (10.7)$$

The integral form of the conservation laws in (10.6) is written as

$$\left. \begin{array}{l} \int_{x_L}^{x_R} \mathbf{Q}(x,t_2) dx = \int_{x_L}^{x_R} \mathbf{Q}(x,t_1) dx \\ \qquad + \int_{t_1}^{t_2} \mathbf{F}(\mathbf{Q}(x_L,t)) dt - \int_{t_1}^{t_2} \mathbf{F}(\mathbf{Q}(x_R,t)) dt \\ \qquad + \int_{t_1}^{t_2} \int_{x_L}^{x_R} \mathbf{S}(\mathbf{Q}(x,t)) dx dt. \end{array} \right\} \quad (10.8)$$

Compare with (10.3) for the scalar case. In the finite volume approach the computational domain is discretised into finite volumes

$$V_i^n = [x_{i-\frac{1}{2}}, x_{i+\frac{1}{2}}] \times [t_n, t_{n+1}], \quad (10.9)$$

with $\Delta x = x_{i+\frac{1}{2}} - x_{i-\frac{1}{2}}$ and $\Delta t = t_{n+1} - t_n$. See finite volume mesh of Fig. 10.2. Direct application of the integral form (10.8) in averaged form in $V_i^n$ gives the formula

$$\mathbf{Q}_i^{n+1} = \mathbf{Q}_i^n - \frac{\Delta t}{\Delta x}(\mathbf{F}_{i+\frac{1}{2}} - \mathbf{F}_{i-\frac{1}{2}}) + \Delta t \mathbf{S}_i, \quad (10.10)$$

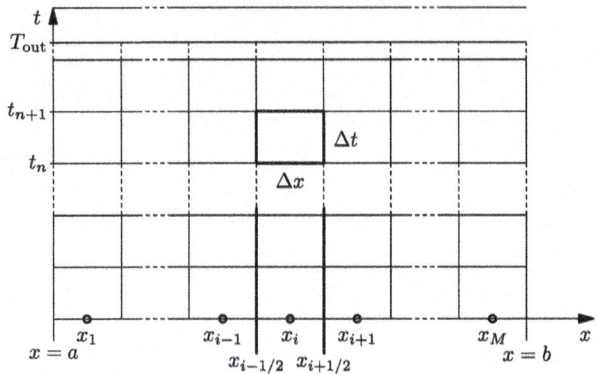

**Fig. 10.2** Control volume $V_i^n = [x_{i-\frac{1}{2}}, x_{i+\frac{1}{2}}] \times [t_n, t_{n+1}]$ in a finite volume mesh in $x$-$t$ space

with definitions

$$\left.\begin{aligned}
\mathbf{Q}_i^n &= \frac{1}{\Delta x} \int_{x_{i-\frac{1}{2}}}^{x_{i+\frac{1}{2}}} \mathbf{Q}(x, t_n) dx \; , \\
\mathbf{F}_{i+\frac{1}{2}} &= \frac{1}{\Delta t} \int_{t_n}^{t_{n+1}} \mathbf{F}(\mathbf{Q}(x_{i+\frac{1}{2}}, t)) dt \; , \\
\mathbf{S}_i &= \frac{1}{\Delta t \Delta x} \int_{t_n}^{t_{n+1}} \int_{x_{i-\frac{1}{2}}}^{x_{i+\frac{1}{2}}} \mathbf{S}(\mathbf{Q}(x, t)) dx dt \; .
\end{aligned}\right\} \quad (10.11)$$

Formula (10.10) with definitions (10.11) is *exact* and generalises formula (10.4). Finite volume numerical methods emerge from (10.10) if integrals in (10.11) are suitably approximated. In the *approximated* interpretation of (10.10) $\mathbf{F}_{i+\frac{1}{2}}$ is the *numerical flux* and $\mathbf{S}_i$ is the *numerical source*. These emerge from the second and third lines in (10.11) as suitable approximations to the respective integrals. The cell integral average $\mathbf{Q}_i^n$ is computed explicitly only at the initial time from the initial conditions; for later times this results automatically from formula (10.10), which becomes a one-step numerical method. Figure 10.3 depicts the finite volume method as applied to the system of balance laws in (10.6).

**Fig. 10.3** Finite volume scheme (10.10) to solve the system of hyperbolic equations in (10.6) with source terms. The scheme requires numerical fluxes at interfaces and the numerical source within the control volume

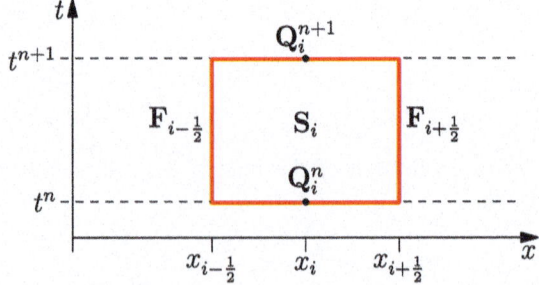

## 10.2 The Godunov Upwind Method

In this section we describe the Godunov upwind method [1] in conservative form (10.10) to solve the IBVP (10.6). We assume the source term is $\mathbf{S}(\mathbf{Q}) = \mathbf{0}$ and therefore the only item to determine is the numerical flux. Godunov's method [1] approximates the flux integral in (10.11) by means of the solution of a Riemann problem, which furnishes an approximation to the integrand.

### 10.2.1 The Numerical Flux from the Riemann Problem

Godunov's method [1] computes the *numerical flux* $\mathbf{F}_{i+\frac{1}{2}}$ from the solution of the local Riemann problem at the cell interface $x = x_{i+\frac{1}{2}}$ given as

$$\begin{aligned}
\text{PDEs:} & \ \partial_t \mathbf{Q} + \partial_x \mathbf{F}(\mathbf{Q}) = \mathbf{0} \ , \\
\text{ICs:} & \ \mathbf{Q}(x, 0) = \begin{cases} \mathbf{Q}_i^n & \text{if } x < x_{i+\frac{1}{2}} \ , \\ \mathbf{Q}_{i+1}^n & \text{if } x > x_{i+\frac{1}{2}} \ . \end{cases}
\end{aligned} \quad (10.12)$$

We remark that for convenience, the IVP (10.12) in global coordinates centred at $(x_{i+\frac{1}{2}}, t_n)$ is transformed to *local coordinates* $(\bar{x}, \bar{t})$ with origin at $(0, 0)$, by applying the coordinate transformation:

$$\begin{aligned}
\bar{x} = x - x_{i+\frac{1}{2}} & \ , \ \bar{t} = t - t_n \ , \\
x \in [x_i, x_{i+1}] & \ , \ t \in [t_n, t_{n+1}] \ , \\
\bar{x} \in [-\tfrac{1}{2}\Delta x, \tfrac{1}{2}\Delta x] & \ , \ \bar{t} \in [0, \Delta t] \ .
\end{aligned} \quad (10.13)$$

See Fig. 10.4. Then in practice we still keep the notation $(x, t)$ to mean the local coordinates $(\bar{x}, \bar{t})$.

The similarity solution of (10.12) is denoted as $\mathbf{Q}_{i+\frac{1}{2}}(x/t)$. To compute the numerical flux one first evaluates $\mathbf{Q}_{i+\frac{1}{2}}(x/t)$ at $x/t = 0$, that is along the interface (along

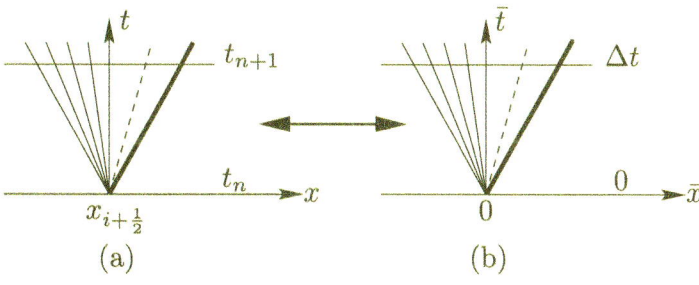

**Fig. 10.4** Correspondence between the global (**a**) and local (**b**) frames of reference for the solution of the Riemann problem

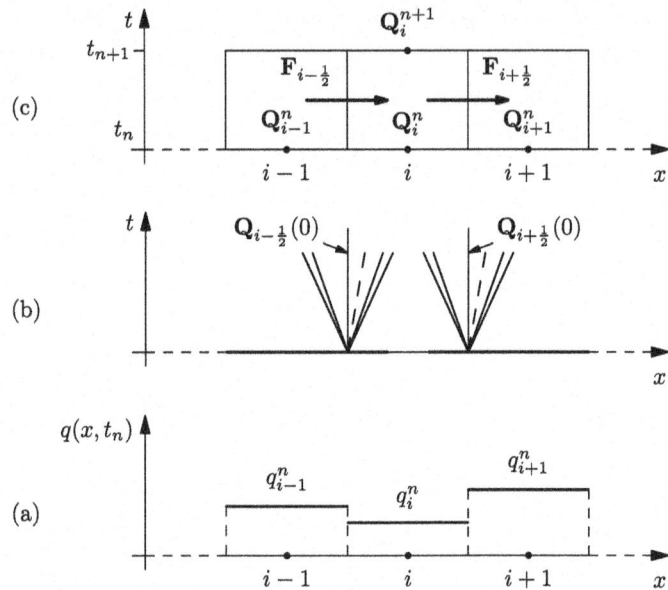

**Fig. 10.5** Godunov's method for a homogeneous hyperbolic system. Frame **a**: integral averages give piece-wise constant data at time $t_n$; plot represents a generic component $q$ of **Q**. Frame **b**: structure of solutions of Riemann problems at the intercell boundaries. Frame **c**: representation of finite volume formula (10.10) to update the cell average in cell $I_i$ using intercell numerical fluxes

the $t$-axis), giving $\mathbf{Q}_{i+\frac{1}{2}}(0)$, called the *Godunov state*. As seen in Chap. 6 $\mathbf{Q}_{i+\frac{1}{2}}(0)$ is constant along the interface. Therefore the flux integral in (10.11) is trivially computed to obtain the Godunov flux as

$$\mathbf{F}_{i+\frac{1}{2}} = \frac{1}{\Delta t} \int_0^{\Delta t} \mathbf{F}(\mathbf{Q}(x_{i+\frac{1}{2}}, t)) dt = \mathbf{F}(\mathbf{Q}_{i+\frac{1}{2}}(0)) \;. \tag{10.14}$$

Figure 10.5 represents the Godunov method as applied to a homogeneous hyperbolic system, $\mathbf{S} = \mathbf{0}$. Frame (a) represents integral averages in three successive cells at time $t_n$ for a generic component $q$ of vector $\mathbf{Q}$. Frame (b) shows the structure of solutions of two neighbouring Riemann problems at the intercell boundaries $x_{i-\frac{1}{2}}$ and $x_{i+\frac{1}{2}}$. Frame (c) represents the finite volume formula (10.10) to update cell averages in cell $I_i$ using intercell numerical fluxes $\mathbf{F}_{i-\frac{1}{2}}$ and $\mathbf{F}_{i+\frac{1}{2}}$.

### 10.2.2 Godunov's Method for the Linear Advection Equation

Godunov's method was introduced in Chap. 9 for the linear advection equation as a finite difference method. In this section we derive the method as a finite volume method via the solution of a local Riemann problem to compute the intercell numerical flux. We first consider the homogeneous case, $s = 0$, namely

## 10.2 The Godunov Upwind Method

$$\partial_t q(x,t) + \partial_x f(q(x,t)) = 0, \quad \text{flux function: } f(q) = \lambda q. \tag{10.15}$$

Here $\lambda$ is a constant wave propagation speed. The finite volume version of the Godunov method departs from (10.10), with definitions (10.11), which applied to (10.15) may be written as

$$q_i^{n+1} = q_i^n - \frac{\Delta t}{\Delta x}\left(f_{i+\frac{1}{2}} - f_{i-\frac{1}{2}}\right). \tag{10.16}$$

To compute the numerical flux $f_{i+\frac{1}{2}}$ in (10.16) from formula (10.14) one requires the Godunov state $q_{i+\frac{1}{2}}(0)$. This is obtained by solving the local Riemann problem

$$\left.\begin{array}{l} \text{PDE: } \partial_t q + \partial_x (\lambda q) = 0, \\ \text{ICs: } q(x,0) = \begin{cases} q_i^n & \text{if } x < x_{i+\frac{1}{2}}, \\ q_{i+1}^n & \text{if } x > x_{i+\frac{1}{2}}. \end{cases} \end{array}\right\} \tag{10.17}$$

The solution of this problem, in local coordinates, at the cell interface is

$$q(x/t) = \begin{cases} q_i^n & \text{if } \dfrac{x}{t} < \lambda, \\ q_{i+1}^n & \text{if } \dfrac{x}{t} > \lambda. \end{cases} \tag{10.18}$$

To determine the Godunov state $q_{i+\frac{1}{2}}(0)$ we need to perform a solution sampling procedure to correctly identify the solution value along the $t$-axis. See Chap. 6. From (10.18) we see that for the linear advection equation the sought *Godunov state* $q_{i+\frac{1}{2}}(0)$ is simply given as

$$q_{i+\frac{1}{2}}(0) = \begin{cases} q_i^n & \text{if } \lambda > 0, \\ q_{i+1}^n & \text{if } \lambda < 0. \end{cases} \tag{10.19}$$

For background see Chap. 2.

The *upwind* character of the Godunov method becomes evident; the sought Godunov state $q_{i+\frac{1}{2}}(0)$ depends on the *direction of the wind*. When $\lambda > 0$ the wind blows from the left and hence the Godunov state is given by the value $q_i^n$ that moved with the wind from the left. Otherwise the Godunov state is $q_{i+1}^n$. Therefore, from (10.14), the Godunov numerical flux $f_{i+\frac{1}{2}}$ is

$$f_{i+\frac{1}{2}} = \begin{cases} \lambda q_i^n & \text{if } \lambda > 0, \\ \lambda q_{i+1}^n & \text{if } \lambda < 0. \end{cases} \tag{10.20}$$

Similarly, for the left interface the corresponding numerical flux is

$$f_{i-\frac{1}{2}} = \begin{cases} \lambda q_{i-1}^n & \text{if } \lambda > 0, \\ \lambda q_i^n & \text{if } \lambda < 0. \end{cases} \tag{10.21}$$

After inserting the numerical fluxes (10.20)–(10.21) into the finite volume scheme (10.16) we obtain

$$q_i^{n+1} = q_i^n - \begin{cases} \dfrac{\Delta t}{\Delta x} \left( \lambda q_i^n - \lambda q_{i-1}^n \right) & \text{if } \lambda > 0, \\ \dfrac{\Delta t}{\Delta x} \left( \lambda q_{i+1}^n - \lambda q_i^n \right) & \text{if } \lambda < 0. \end{cases} \quad (10.22)$$

By introducing the Courant number $c = \frac{\lambda \Delta t}{\Delta x}$ in (10.22) we obtain

$$q_i^{n+1} = q_i^n - \begin{cases} c \left( \lambda q_i^n - \lambda q_{i-1}^n \right) & \text{if } \lambda > 0, \\ c \left( \lambda q_{i+1}^n - \lambda q_i^n \right) & \text{if } \lambda < 0, \end{cases} \quad (10.23)$$

which is identical in form to the finite difference form of the Godunov method presented in Chap. 9. The difference lies in the interpretation of the approximate values $q_i^n$. In the finite difference approach they are *point values* and in the finite volume approach they are *cell integral averages*.

The Godunov method for the scalar conservation law (no source term) has been described just requiring the numerical flux from the exact solution of a local Riemann problem, to completely determine the conservative scheme (10.16).

### 10.2.3 Godunov's Method and the Source Term

Here we explore potential extensions of Godunov's method to include the source term in the PDE (10.15). To this end we consider the linear advection equation with a linear source term

$$\partial_t q + \partial_x f(q) = s(q), \quad f(q) = \lambda q, \quad s(q) = \beta q, \quad (10.24)$$

with $\lambda$ a constant wave propagation speed and $\beta \leq 0$ a constant reaction rate. The corresponding finite volume formula (10.10) with the numerical source included becomes

$$q_i^{n+1} = q_i^n - \frac{\Delta t}{\Delta x} \left( f_{i+\frac{1}{2}} - f_{i-\frac{1}{2}} \right) + \Delta t s_i, \quad (10.25)$$

where $s_i$ is the *numerical source*, an approximation to the volume integral average in (10.11).

**The Simplest Numerical Source**

The simplest approximation assumes that the integrand is given in terms of the initial data at time $t^n$, that is $s(q(x,t)) = \beta q_i^n$, a constant. Then, from (10.11) the numerical source is simply

## 10.2 The Godunov Upwind Method

$$s_i = \beta q_i^n. \tag{10.26}$$

If we assume that the numerical flux for the inhomogeneous Eq. (10.24) is the same as that for the homogeneous Eq. (10.15), then the *extended* Godunov scheme is

$$q_i^{n+1} = q_i^n - \begin{cases} c\left(\lambda q_i^n - \lambda q_{i-1}^n\right) + \Delta t \beta q_i^n & \text{if } \lambda > 0, \\ c\left(\lambda q_{i+1}^n - \lambda q_i^n\right) + \Delta t \beta q_i^n & \text{if } \lambda < 0. \end{cases} \tag{10.27}$$

The simple numerical source $s_i$ in (10.27) is called *centred* in the literature, in the sense that it does not depend on the wave propagation speed $\lambda$.

We have used the term *extended* for Godunov's scheme (10.27), because source terms were not included in the original formulation of the Godunov method. The approximation to the numerical source proposed here is simply for illustrative purposes, it is not recommended for practical use. More sophisticated methods to determine the numerical source are required in practice.

### Upwinded Numerical Source

Recall that in order to evaluate the source term in (10.10), or specifically (10.25), we need to calculate an approximation to the volume integral in (10.11), namely

$$s_i = \frac{1}{\Delta x \Delta t} \int_{x_{i-\frac{1}{2}}}^{x_{i+\frac{1}{2}}} \int_0^{\Delta t} \beta q(x,t) dx dt. \tag{10.28}$$

Let us assume that the integrand is the solution of the two neighbouring Riemann problems in the control volume $V_i^n = [x_{i-\frac{1}{2}}, x_{i+\frac{1}{2}}] \times [t_n, t_{n+1}]$ introduced in (10.9). For $\lambda > 0$ only the left Riemann problem at $x_{i-\frac{1}{2}}$ propagates the solution inside the integration volume $V_i^n$; in the rest of $V_i^n$ the solution remains $q_i^n$. For $\lambda < 0$ only the right Riemann problem at $x_{i+\frac{1}{2}}$ propagates the value $q_{i+1}^n$ inside the integration volume $V_i^n$; in the rest of $V_i^n$ the solution remains $q_i^n$.

Let us consider the details for the case $\lambda > 0$. Then clearly, the Riemann problem at $x_{i-\frac{1}{2}}$ with data $q_{i-1}^n$ to the left of $x_{i-\frac{1}{2}}$ and $q_i^n$ to the right of $x_{i-\frac{1}{2}}$ propagates the value $q_{i-1}^n$ into a triangular portion of the control volume $V_i^n = \Delta t \times \Delta x$ equal to $V_1 = \frac{1}{2} c \Delta t \times \Delta x$. The remaining part of $V_i^n$ has area $V_2 = (1 - \frac{1}{2}c)\Delta t \times \Delta x$, in which the solution remains $q_i^n$. Then, evaluation of (10.28) yields

$$s_i = \frac{1}{2} c \beta q_{i-1}^n + (1 - \frac{1}{2}c)\beta q_i^n. \tag{10.29}$$

Compare with (10.26). Now a contribution to the numerical source $s_i$ comes from the upwinded value $q_{i-1}^n$. As will be seen in Chap. 13 the upwinded numerical source (10.29) gives a scheme (10.25) with a larger stability region than that with the centred numerical source (10.26).

**Exercise 10.1** Calculate the upwinded numerical source for $\lambda < 0$.

Next we describe Godunov's upwind method as applied to the one-dimensional shallow water equations.

## 10.2.4 Godunov's Method for Shallow Water

Let us consider the homogeneous $x$-split two-dimensional shallow water equations

$$\partial_t \mathbf{Q} + \partial_x \mathbf{F}(\mathbf{Q}) = \mathbf{0} \,, \tag{10.30}$$

where the vectors of conserved variables and fluxes are

$$\mathbf{Q} = \begin{bmatrix} h \\ hu \\ hv \end{bmatrix}, \quad \mathbf{F} = \begin{bmatrix} hu \\ hu^2 + \frac{1}{2}gh^2 \\ huv \end{bmatrix} . \tag{10.31}$$

Having already described Godunov's method for a generic hyperbolic one-dimensional system in Sect. 10.2.1, here we present the method for the shallow water equations in concise algorithmic form. The following steps are involved for computing the numerical flux $\mathbf{F}_{i+\frac{1}{2}}$:

1. *The Riemann problem.* Solve exactly the local Riemann problem at the cell interface $x = x_{i+\frac{1}{2}}$

$$\left. \begin{array}{l} \text{PDEs: } \partial_t \mathbf{Q} + \partial_x \mathbf{F}(\mathbf{Q}) = \mathbf{0} \,, \\ \text{ICs: } \quad \mathbf{Q}(x, 0) = \begin{cases} \mathbf{Q}_i^n & \text{if } x < x_{i+\frac{1}{2}} \,, \\ \mathbf{Q}_{i+1}^n & \text{if } x > x_{i+\frac{1}{2}} \,. \end{cases} \end{array} \right\} \tag{10.32}$$

Here $\mathbf{Q}$ and $\mathbf{F}(\mathbf{Q})$ are as for the shallow water Eqs. (10.30)–(10.31). The exact Riemann solver is given in Chap. 6 for the wet-bed case and in Chap. 7 for the dry-bed case. Approximate Riemann solvers will be presented in Chap. 11.

2. *Depth in the Star Region.* From the exact Riemann solver, first find the star value $h_*$ in the *Star Region* by solving iteratively the algebraic equation

$$f(h) = f_L(h) + f_R(h) + \Delta u = 0 \quad \text{with} \quad \Delta u = u_R - u_L \,. \tag{10.33}$$

3. *Velocity in the Star Region.* Having found $h_*$ compute $u_*$ as

$$u_* = \frac{1}{2}(u_L + u_R) - \frac{1}{2}(f_R(h_*) - f_L(h_*)) \,. \tag{10.34}$$

4. *Solution sampling.* Sample the solution as described in Chap. 6 to find the Godunov state $\mathbf{Q}_{i+\frac{1}{2}}(0)$, which is the solution right at the intercell boundary.

## 10.2 The Godunov Upwind Method

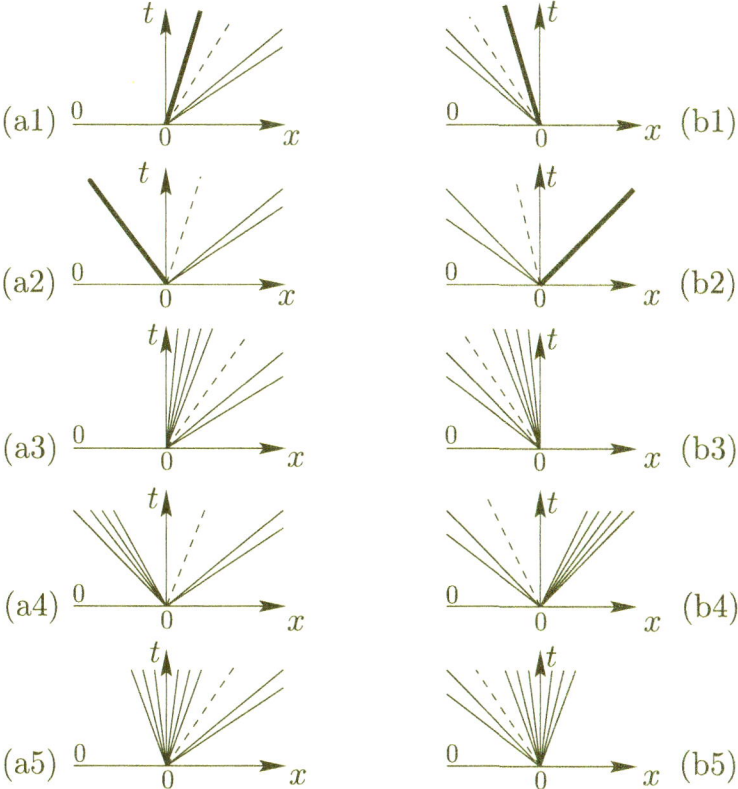

**Fig. 10.6** Ten possible wave patterns in evaluating $\mathbf{Q}_{i+\frac{1}{2}}(0)$ to compute the Godunov flux for the *augmented* one-dimensional case. Wave patterns shown are those for a wet bed

Recall that this process involves finding one out of ten possible wave patterns, as depicted in Fig. 10.6.

5. *Numerical flux.* Compute the Godunov numerical flux as

$$\mathbf{F}_{i+\frac{1}{2}} = \mathbf{F}(\mathbf{Q}_{i+\frac{1}{2}}(0)) \ . \tag{10.35}$$

6. *Conservative update formula.* Once all the intercell fluxes $\mathbf{F}_{i+\frac{1}{2}}$ have been computed, these are used in the conservative update formula

$$\mathbf{Q}_i^{n+1} = \mathbf{Q}_i^n - \frac{\Delta t}{\Delta x}(\mathbf{F}_{i+\frac{1}{2}} - \mathbf{F}_{i-\frac{1}{2}}) \tag{10.36}$$

to advance the solution to the next time level $n+1$.

7. *Source terms.* Methods to deal with the source terms $\mathbf{S}(\mathbf{Q})$ will be discussed in Chaps. 13 and 14.

Simplifications of the Godunov method result from using approximate Riemann solvers, which are discussed in Chap. 11. One can also use some suitable centred methods to determine the numerical flux.

## 10.3 Initial, Boundary and Stability Conditions

There are some practical issues to address, before the numerical method (10.36) can be implemented to solve the initial-boundary value problem (10.6). These are initial conditions at the initial time, boundary conditions at the domain ends and stability conditions to select the correct size of the time step $\Delta t$ in (10.36).

### 10.3.1 Initial Conditions

At the initial time $t = 0$, say, initial conditions are to be implemented. Strictly speaking, from the prescribed initial data $\mathbf{Q}(x, 0)$ one must provide discrete cell average values $\mathbf{Q}_k^0$ from integral averages of $\mathbf{Q}(x, 0)$ in each computing cell $k$, namely

$$\mathbf{Q}_k^0 = \frac{1}{\Delta x} \int_{x_{k-\frac{1}{2}}}^{x_{k+\frac{1}{2}}} \mathbf{Q}(x, 0) dx , \text{ for } k = 1, \ldots, M . \tag{10.37}$$

See finite volume mesh shown in Fig. 10.2. For low-order methods one can also use point values at the cell centre $x_k$ of each cell. But this is not correct for high-order methods of order greater than two, see Chap. 14.

### 10.3.2 Boundary Conditions

Boundary conditions are necessary to determine boundary fluxes in the finite volume mesh of Fig. 10.2. The update formula is well defined for updating all cells $k$, $k = 2, \ldots, M - 1$. The updating of cells 1 and $M$ requires the application of boundary conditions at the domain boundaries $x = a$ and $x = b$ in order to compute boundary numerical fluxes $\mathbf{F}_{\frac{1}{2}}$ at $x = a$ and $\mathbf{F}_{M+\frac{1}{2}}$ at $x = L$. It is worth remarking that the application of boundary conditions is fundamentally a physical problem. Consider first the left boundary at $x = a$. Numerically, we create a *fictitious state* $\mathbf{Q}_0^n$ in such a way that the solution of the Riemann problem with data $(\mathbf{Q}_0^n, \mathbf{Q}_1^n)$ gives a solution $\mathbf{W}_{\frac{1}{2}}(x/t)$ such that the Godunov state $\mathbf{W}_{\frac{1}{2}}(0)$ fulfils the physical requirements at the boundary $x = a$. Similar remarks apply to the right boundary $x = b$, where a fictitious state $\mathbf{Q}_{M+1}^n$ is required to define a boundary Riemann problem with data $(\mathbf{Q}_M^n, \mathbf{Q}_{M+1}^n)$, the solution of which satisfies the physically desirable boundary condition. Here we consider only three types of boundary conditions.

## 10.3 Initial, Boundary and Stability Conditions

**Transmissive Boundary**

To simulate *transmissive or transparent boundaries*, we set the following boundary conditions

$$\left. \begin{array}{l} \text{Left boundary } x = a: \ h_0^n = h_1^n, \ u_0^n = u_1^n, \\ \text{Left boundary } x = b: \ h_{M+1}^n = h_M^n, \ u_{M+1}^n = u_M^n. \end{array} \right\} \quad (10.38)$$

Note that the corresponding boundary Riemann problems generate trivial waves at the boundaries, which hopefully will allow the undisturbed passage of waves through the boundaries. This type of boundary conditions are usually applied when, for example, one wants to perform computations on a reduced section of a long channel. In one space dimension, these boundary conditions work reasonably well, but they are not perfect.

**Solid Reflective Boundary**

Solid, reflective, impermeable boundaries are handled by the boundary conditions

$$\left. \begin{array}{l} \text{Left boundary } x = a: \ h_0^n = h_1^n, \ u_0^n = -u_1^n, \\ \text{Left boundary } x = b: \ h_{M+1}^n = h_M^n, \ u_{M+1}^n = -u_M^n. \end{array} \right\} \quad (10.39)$$

The reader is encouraged to calculate the solution of the boundary Riemann problems, to note that the solution in the Godunov state, right at the boundary, has particle velocity $u_* = 0$, that is, there is not flow through the boundary. The resulting value for $h_*$, consistent with the *no flow through condition* $u_* = 0$, emerges automatically from the solution of the Riemann problem, as a bonus.

**Moving Solid Reflective Boundaries**

*Moving solid, impermeable reflective boundaries* are treated with the boundary conditions

$$\left. \begin{array}{l} \text{Left boundary } x = a: \ h_0^n = h_1^n, \ u_0^n = -u_1^n + 2u_{WL}, \\ \text{Left boundary } x = b: \ h_{M+1}^n = h_M^n, \ u_{M+1}^n = -u_M^n + 2u_{WR}. \end{array} \right\} \quad (10.40)$$

Here $u_{WL}$ and $u_{WR}$ are the speeds of the left and right boundaries, assumed to be known.

Again, the reader is encouraged to calculate the solution of the boundary Riemann problems, to note that the solution in the Godunov state, right at the boundary, has particle velocity $u_* = u_{WL}$ for the left boundary and $u_* = u_{WR}$ for the right boundary. Again, the value for $h_*$, consistent with the condition of *flow with the boundary*, emerges automatically from the solution of the Riemann problem.

So far we have described the Godunov method as used in conjunction with the exact Riemann solver. In Chap. 11 we also discuss the implementation of Godunov's method with *approximate Riemann solvers*.

### 10.3.3 Stability Condition

The practical problem is to calculate a time $\Delta t$ in formula (10.10), or (10.36), of a size that satisfies the constraint imposed by the Courant stability condition. See Chap. 9 for the linearised stability condition for various well-known numerical methods.

**The Linear Advection Equation**

We first consider the linear advection equation with constant wave speed $\lambda$, as solved by the Godunov upwind method; see Sect. 10.2.2. Recall from Chap. 9 that the stability condition for the Godunov upwind method applied to LAE is

$$|c| = \frac{|\lambda|\Delta t}{\Delta x} \leq C_{lim} = 1 \,. \tag{10.41}$$

Here $c$ is the Courant number and $C_{lim} = 1$ is the stability limit for the method. Note that for the Godunov upwind schemed applied to LAE in one space dimension $C_{lim} = 1$, but $C_{lim}$ may be different for other methods; see Chap. 9. Then from (10.41) the time step $\Delta t$ must obey

$$\Delta t \leq C_{lim} \frac{\Delta x}{|\lambda|} \,. \tag{10.42}$$

In practice, one selects the time step according to

$$\Delta t = S_c \times C_{lim} \frac{\Delta x}{|\lambda|}, \quad 0 < S_c \leq 1 \,. \tag{10.43}$$

The parameter $S_c$ is a *safety coefficient* that may take care of (i) round-off errors and (ii) uncertainties in determining the maximum wave speed emerging from the equations. For the linear advection equation, the speed $\lambda$ is well defined, but not necessarily so for hyperbolic systems. By defining

$$C_{cfl} = S_c \times C_{lim} \tag{10.44}$$

the time step is chosen according to

$$\Delta t = C_{cfl} \frac{\Delta x}{|\lambda|} \,. \tag{10.45}$$

## 10.3 Initial, Boundary and Stability Conditions

It is desirable to select $S_c$ as close to unity as possible, in order to aid the efficiency of the method and to avoid accumulation of truncation errors by performing an excessive number of time steps to reach the output time.

**Remark** (*Courant number of a computation*) In a practical setting one speaks of a computation with Courant number $C_{cfl} = S_c \times C_{lim}$, as defined in (10.44). Often in the literature, the meaning of $C_{cfl}$ is mistaken with that of $S_c$, and $C_{lim}$ is assumed to be unity, which is not always the case, even in one space dimension; see Chap. 9.

As already remarked, for the linear advection equation, the speed $\lambda$ is well defined. However, for a non-linear conservation law, such as that in (10.1), the characteristic speed $\hat{\lambda} = df(q)/dq$ is analogous to $\lambda$, but it is not identical to the sought wave speed, which needs to be estimated. See Chap. 2 for relevant definitions. Then, the time step is computed from

$$\Delta t = C_{\text{cfl}} \times \frac{\Delta x}{s_{\max}^n} , \qquad (10.46)$$

where $s_{\max}^n$ is an *estimate* for the maximum wave speed, in absolute value, of the true wave speeds in the full computational domain at the time level $n$. The estimation of $\Delta t$ from (10.46) is subject to the uncertainties derived from estimating $s_{\max}^n$. This is good reason for the role of the safety coefficient $S_c$.

The case of systems of equations will be dealt with next in the context of the one-dimensional shallow water equations.

### The Time Step for the Shallow Water Equations

For the one-dimensional shallow water equations we choose the time step $\Delta t$ according to formula

$$\Delta t = C_{\text{cfl}} \times \frac{\Delta x}{S_{max}^n} , \qquad (10.47)$$

where $S_{max}^n$ is to be estimated. If the estimate for $S_{max}^n$ in (10.47) is poor and subject to uncertainties, the scheme may become unstable, leading to crashing of the code, due to the wrong value for the time step $\Delta t$.

A popular way of estimating the maximum wave speed $S_{max}^n$ is by using the eigenvalues of the equations. But recall that for non-linear systems, wave speeds are not identical to eigenvalues. Nonetheless the following choice is often made

$$S_{max}^n = max_i \left\{ |u_i^n| + a_i^n \right\} , \quad C_{cfl} = S_c \times C_{lim} , \quad 0 < S_c \leq 1 . \qquad (10.48)$$

In practice one takes $S_c = 0.9$, $S_c = 0.5$ or even $S_c = 0.1$, in desperation!

It is easy to check that $S_{max}^n$ in (10.48) is not a bound for the maximum wave speed that would emerge at cell interfaces from the solution of local Riemann problems. For example, if the initial data includes $u = 0$ everywhere, then only the celerity $a_i^n$

would account for $S_{max}^n$. But a shock emerging from such boundaries would have speed larger than $S_{max}^n$ and therefore the application of (10.48) would fail.

As a final remark we note that the subject of estimating wave speeds for non-linear systems has not been exhausted. Theoretical bounds for interface wave speeds for the shallow water equations, and other hyperbolic systems, were proposed in the recent paper by Toro et al. [6]; see also the earlier work of Guermond and Popov [7]. The reader is encouraged to use these new results for choosing the time step $\Delta t$ reliably. Chapter 11 will also provide additional information on a reliable choices for $S_{max}^n$ in (10.47).

## 10.4 The Random Choice Method

The Random Choice Method (RCM), or Glimm's method, is based on a constructive proof of existence of solutions of hyperbolic conservation laws, due to Glimm [2]. Chorin [8] was the first to successfully implement the scheme as a practical computational tool. General background on the Random Choice Method is found in Chap. 7 of [9–11] and references therein. The method has been applied to a variety of problems in gas dynamics; see for instance the works of Chorin [12], Sod [13], Concus [14], Colella [15, 16], Gottlieb [17], Shi and Gottlieb [18], Toro [19], Saito and Glass [20], Takano [21], Singh and Clarke [22], Dawes [23], Olivier and Grönig [24], and Marshall and Plohr [25]. For the shallow water equations, Marshall and Méndez [26] appear to be the first authors to have applied the Random Choice Method. See also the work of Li and Holt [27].

Here we restrict the presentation to just one version of RCM, namely the *non-staggered grid version*. This is closely linked to Godunov's method, described previously in Sect. 10.2.4. The RCM updates $\mathbf{Q}_i^n$ to $\mathbf{Q}_i^{n+1}$, see Fig. 10.7, in the following two steps:

Step I: Solve the Riemann problems $RP(\mathbf{W}_{i-1}^n, \mathbf{W}_i^n)$ and $RP(\mathbf{W}_i^n, \mathbf{W}_{i+1}^n)$ to find their respective solutions $\mathbf{W}_{i-\frac{1}{2}}(x/t)$ and $\mathbf{W}_{i+\frac{1}{2}}(x/t)$, as depicted in Fig. 10.7. Figure 10.6 shows the 10 possible wave patterns emerging from the intercell boundaries $x_{i-\frac{1}{2}}$ and $x_{i+\frac{1}{2}}$.

Step II: Random sample these solutions at time $\Delta t$ (local) within cell $I_i$ to pick up a state and assign it to cell $I_i$. The random sampling range $[A, B]$ is shown in Fig. 10.7 by a thick horizontal line. The picked up state depends on a random, or quasi-random, number $\theta^n$ in the interval $[0, 1]$. The updated solution is then

$$\mathbf{W}_i^{n+1} = \begin{cases} \mathbf{W}_{i-\frac{1}{2}}(\theta^n \Delta x/\Delta t) & \text{if } 0 \leq \theta^n \leq \frac{1}{2}, \\ \mathbf{W}_{i+\frac{1}{2}}((\theta^n - 1)\Delta x/\Delta t) & \text{if } \frac{1}{2} < \theta^n \leq 1. \end{cases} \quad (10.49)$$

The case in which $0 \leq \theta^n \leq \frac{1}{2}$ is illustrated in Fig. 10.8. Here the updated solution depends on the random sampling procedure applied to the *right side of the left* Riemann problem solution $\mathbf{W}_{i-\frac{1}{2}}(x/t)$. The particular randomly chosen state is returned

## 10.4 The Random Choice Method

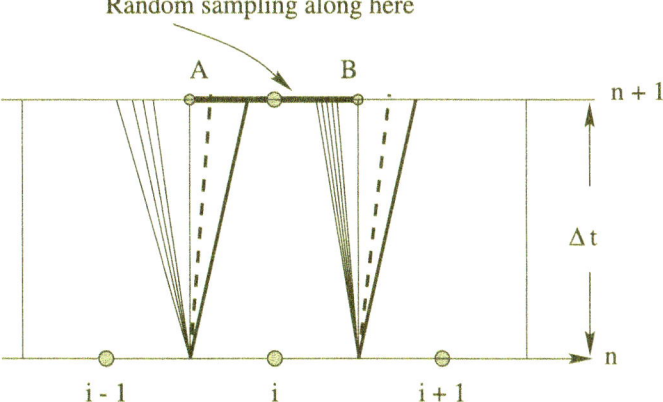

**Fig. 10.7** RCM on non-staggered grid. Solution is updated to time level $n+1$ by random sampling solutions of two Riemann problems $RP(\mathbf{W}_{i-1}^n, \mathbf{W}_i^n)$ and $RP(\mathbf{W}_i^n, \mathbf{W}_{i+1}^n)$ within cell $I_i$ at time $t_{n+1}$

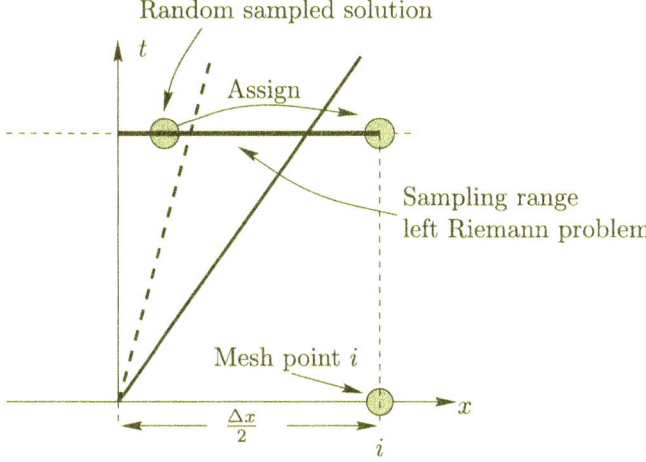

**Fig. 10.8** RCM sampling of the right-hand side of the left Riemann problem solution, when $0 \leq \theta^n \leq \frac{1}{2}$. The sampled state is assigned to the centre of cell $i$

by the *same* sampling routine used to evaluate the exact Riemann problem solution; see Chaps. 6 and 7. Here the sampling routine is called with argument

$$S \equiv x/t = \frac{\theta^n \Delta x}{\Delta t} \ . \tag{10.50}$$

The resulting state is then assigned to the grid point $i$ (centre of the cell) and is regarded as the solution in cell $I_i$ for the next time level; see Fig. 10.8. A similar procedure is applied if $\frac{1}{2} \leq \theta^n \leq 1$. In this case one samples the *left side of the right Riemann problem solution* $\mathbf{W}_{i+\frac{1}{2}}(x/t)$. When programing the non-staggered grid version of the Random Choice Method, there are many ways to organise the tasks: (I) solving the Riemann problems and (II) random sampling their solutions.

Concerning the use of random numbers in the scheme, Chorin [8] established that one only requires a single random number $\theta^n$ for a complete time level $n$. In Glimm's proof [2] one may take one random number per time step per cell. Full details are found in Chap. 7 of [9–11], in which Sect. 7.5 discusses generation of random numbers and their properties.

Finally, we note the relationship between the RCM scheme on a non-staggered grid to obtain the updated value $\mathbf{W}_i^{n+1}$ and the Godunov upwind method. Compare Fig. 10.7 for RCM with Fig. 10.5 for the Godunov method. Within cell $I_i$ at time $t_{n+1}$, on the equivalent to line $AB$ in Fig. 10.7, the Godunov scheme takes an integral average of the two relevant Riemann problem solutions; see Fig. 10.5c. Instead, RCM performs a random sampling procedure in $AB$ to pick up a single state.

Boundary conditions for RCM are the same as for Godunov's method described in Sect. 10.3. The linearised stability condition for RCM is

$$0 \leq |c| = \frac{|\lambda| \Delta t}{\Delta x} \leq C_{lim} = \frac{1}{2}. \tag{10.51}$$

Provided that a reliable choice for $S_{max}^n$ in (10.47) has been made, condition (10.51) ensures that waves from left and right Riemann problems in Fig. 10.7 do not interact within the chosen time step $\Delta t$. In practice one implements (10.48) with $S_c = 0.9$, which gives $C_{cfl} = S_c \times C_{lim} = 0.45$.

**RCM is not conservative.** It is important to point out that the Random Choice Method of Glimm is not conservative. One consequence is the wrong speed of shock waves, as computed by RCM. On average, however, shock positions tend to be quite accurate.

The previously described two methods, namely Godunov and Random Choice, utilise explicitly, the exact solution of the Riemann problem to update the solution and are thus referred to as *upwind methods*. Next we describe alternative numerical approaches.

## 10.5 Alternative Conservative Schemes

Here we briefly review two alternative numerical approaches to solve the shallow water equations, namely the Flux Vector Splitting approach and centred schemes.

### 10.5.1 The Flux Vector Splitting Approach

The flux vector splitting approach (FVS) for solving hyperbolic systems includes several schemes [28–32]. Most of these schemes have been applied preferentially to

## 10.5 Alternative Conservative Schemes

the compressible Euler equations of gas dynamics and are reviewed in Chap. 8 of [11]. Application of FVS schemes to the shallow water equations are more rare. Early works on shallow water applications of FVS include Bermúdez, Vázquez-Cendón and collaborators [33, 34], and Morel and Fey [35]. More recent works are based on the splitting method proposed by Toro and Vázquez [32] for the Euler equations, called the TV splitting. Briefly, in the TV approach as applied to a generic system

$$\partial_t \mathbf{Q} + \partial_x \mathbf{F}(\mathbf{Q}) = \mathbf{0} \tag{10.52}$$

the flux is *split* into an *advection flux* $\mathcal{A}(\mathbf{Q})$ and a *pressure flux* $\mathcal{P}(\mathbf{Q})$ as follows

$$\mathbf{F}(\mathbf{Q}) = \mathcal{A}(\mathbf{Q}) + \mathcal{P}(\mathbf{Q}) . \tag{10.53}$$

Corresponding numerical fluxes $\mathcal{A}_{i+\frac{1}{2}}$ and $\mathcal{P}_{i+\frac{1}{2}}$ are sought such that the numerical flux $\mathbf{F}_{i+\frac{1}{2}}$ for the full system (10.52) is split thus

$$\mathbf{F}_{i+\frac{1}{2}} = \mathcal{A}_{i+\frac{1}{2}} + \mathcal{P}_{i+\frac{1}{2}} . \tag{10.54}$$

The fluxes $\mathcal{A}_{i+\frac{1}{2}}$ and $\mathcal{P}_{i+\frac{1}{2}}$ are obtained from corresponding Riemann problems for the following two systems

$$\left. \begin{array}{l} \text{Advection system:} \quad \partial_t \mathbf{Q} + \partial_x \mathcal{A}(\mathbf{Q}) = \mathbf{0} \to \mathcal{A}_{i+\frac{1}{2}} \; . \\ \text{Pressure system:} \quad \partial_t \mathbf{Q} + \partial_x \mathcal{P}(\mathbf{Q}) = \mathbf{0} \to \mathcal{P}_{i+\frac{1}{2}} \; . \end{array} \right\} \tag{10.55}$$

For the one-dimensional shallow water equations the following splitting is recommended [36]

$$\mathcal{A}(\mathbf{Q}) = \begin{bmatrix} 0 \\ hu^2 \end{bmatrix}, \quad \mathcal{P}(\mathbf{Q}) = \begin{bmatrix} hu \\ \frac{1}{2}gh^2 \end{bmatrix} . \tag{10.56}$$

Splitting (10.56) differs slightly from the original TV splitting [32]. The evidence so far indicates that this variation performs well for general systems that do not include an equation for total energy, such as the shallow water equations and the blood flow equations [37, 38].

The TV splitting approach is also applicable to systems in non-conservative form, in which case it becomes an *advection-pressure splitting*. A distinguishing feature of this method is that it completely separates all pressure terms from the remaining terms of the flux. In this manner, fast pressure waves are separated from slow advection waves, a property that may be exploited numerically, see Busto et al. [39] and Boscheri et al. [40]. Other recent applications of the approach include [41–43]. Extension of the TV splitting to the shallow water equations including sediment transport is reported in [44].

## 10.5.2 Centred Methods

The schemes reviewed here do not require the (explicit) solution of the Riemann problem, as is the case for the Godunov upwind method of Sect. 10.2.4, or some other forms of upwinding, such as the Flux Vector Splitting Schemes of Sect. 10.5.1. Therefore, centred schemes are not biased by the wave propagation direction, which distinguishes *upwind* methods, and they are called *centred* or *symmetric* schemes. Centred schemes were introduced in Chap. 9 in the context of the linear advection equation. Here we review the most well-known centred schemes for general systems, including the shallow water equations.

**The Lax–Friedrichs Scheme**

The Lax–Friedrichs scheme [3], as applied to (10.52), has numerical flux

$$\mathbf{F}_{i+\frac{1}{2}} = \frac{1}{2}\left[\mathbf{F}(\mathbf{Q}_i^n) + \mathbf{F}(\mathbf{Q}_{i+1}^n)\right] - \frac{1}{2}\frac{\Delta x}{\Delta t}(\mathbf{Q}_{i+1}^n - \mathbf{Q}_i^n) \ . \tag{10.57}$$

As seen in Chap. 9, this method is first-order accurate, monotone and has linear stability condition $0 \leq |c| \leq C_{lim} = 1$.

**The Godunov Centred Scheme**

The Godunov centred scheme [45] has numerical flux (not the upwind Godunov flux) defined in a two-step procedure as follows

$$\left. \begin{array}{c} \mathbf{F}_{i+\frac{1}{2}}^{GodC} = \mathbf{F}(\mathbf{Q}_{i+\frac{1}{2}}^{GodC}) \ , \\ \mathbf{Q}_{i+\frac{1}{2}}^{GodC} = \frac{1}{2}\left(\mathbf{Q}_i^n + \mathbf{Q}_{i+1}^n\right) - \frac{\Delta t}{\Delta x}[\mathbf{F}(\mathbf{Q}_{i+1}^n) - \mathbf{F}(\mathbf{Q}_i^n)] \ . \end{array} \right\} \tag{10.58}$$

This method is first order accurate and has linear stability condition

$$0 \leq |c| \leq C_{lim} = \frac{1}{2}\sqrt{2} \ . \tag{10.59}$$

Regarding monotonicity, for LAE the scheme exhibits the following behaviour

$$\left. \begin{array}{ll} \text{Non-monotone in the range:} & 0 \leq |c| \leq \frac{1}{2} \ , \\ \text{Monotone in the range:} & \frac{1}{2} \leq |c| \leq \frac{1}{2}\sqrt{2} \ . \end{array} \right\} \tag{10.60}$$

This scheme furnishes an example of a first-order method that is not monotone.

## 10.5 Alternative Conservative Schemes

### The Lax–Wendroff Scheme

One version of the Lax–Wendroff flux is analogous to the Godunov centred flux, namely

$$\left.\begin{array}{l}\mathbf{F}^{LWa}_{i+\frac{1}{2}} = \mathbf{F}(\mathbf{Q}^{LWa}_{i+\frac{1}{2}})\,,\\[4pt] \mathbf{Q}^{LWa}_{i+\frac{1}{2}} = \tfrac{1}{2}\left(\mathbf{Q}^{n}_{i} + \mathbf{Q}^{n}_{i+1}\right) - \tfrac{1}{2}\dfrac{\Delta t}{\Delta x}[\mathbf{F}(\mathbf{Q}^{n}_{i+1}) - \mathbf{F}(\mathbf{Q}^{n}_{i})]\,.\end{array}\right\} \qquad (10.61)$$

Note that the only difference between (10.58) and (10.61) is the coefficient of $\frac{\Delta t}{\Delta x}$, which is 1 in the former and $\frac{1}{2}$ the latter. This is the most well-known version of the Lax–Wendroff flux, sometimes known as the two-step Lax–Wendroff flux. A second version of the Lax–Wendroff flux is

$$\mathbf{F}^{LWb}_{i+\frac{1}{2}} = \frac{1}{2}[\mathbf{F}(\mathbf{Q}^{n}_{i}) + \mathbf{F}(\mathbf{Q}^{n}_{i+1})] - \frac{1}{2}\frac{\Delta t}{\Delta x}\mathbf{A}_{i+\frac{1}{2}}[\mathbf{F}(\mathbf{Q}^{n}_{i+1}) - \mathbf{F}(\mathbf{Q}^{n}_{i})]\,. \qquad (10.62)$$

There are two choices to compute the interface matrix $\mathbf{A}_{i+\frac{1}{2}}$ in (10.62), namely

$$\mathbf{A}_{i+\frac{1}{2}} = \mathbf{A}\left(\frac{1}{2}\left(\mathbf{Q}^{n}_{i} + \mathbf{Q}^{n}_{i+1}\right)\right) \text{ and } \mathbf{A}_{i+\frac{1}{2}} = \frac{1}{2}[\mathbf{A}(\mathbf{Q}^{n}_{i}) + \mathbf{A}(\mathbf{Q}^{n}_{i+1})]\,. \qquad (10.63)$$

Here $\mathbf{A}(\mathbf{Q})$ is the Jacobian matrix of system (10.52). Note that flux (10.62) can also be written as

$$\mathbf{F}^{LWb}_{i+\frac{1}{2}} = \frac{1}{2}\left(\mathbf{I} + \frac{\Delta t}{\Delta x}\mathbf{A}_{i+\frac{1}{2}}\right)\mathbf{F}(\mathbf{Q}^{n}_{i}) + \frac{1}{2}\left(\mathbf{I} - \frac{\Delta t}{\Delta x}\mathbf{A}_{i+\frac{1}{2}}\right)\mathbf{F}(\mathbf{Q}^{n}_{i+1})\,, \qquad (10.64)$$

where $\mathbf{I}$ is the identity matrix. A third version of the Lax–Wendroff flux is

$$\mathbf{F}^{LWc}_{i+\frac{1}{2}} = \frac{1}{2}[\mathbf{F}(\mathbf{Q}^{n}_{i}) + \mathbf{F}(\mathbf{Q}^{n}_{i+1})] - \frac{1}{2}\frac{\Delta t}{\Delta x}\mathbf{A}^{2}_{i+\frac{1}{2}}\left(\mathbf{Q}^{n}_{i+1} - \mathbf{Q}^{n}_{i}\right)\,. \qquad (10.65)$$

As to the interface matrix $\mathbf{A}_{i+\frac{1}{2}}$, the two choices (10.63) are available for use in (10.65). The Lax–Wendroff scheme is second-order accurate in space and time. The scheme is non-monotone and has linear stability condition $0 \le |c| \le C_{lim} = 1$.

### The FORCE Scheme

The FORCE scheme (First-Order Centred) proposed by Toro [4, 5] has numerical flux

$$\mathbf{F}^{FO}_{i+\frac{1}{2}} = \frac{1}{2}(\mathbf{F}^{LF}_{i+\frac{1}{2}} + \mathbf{F}^{LWx}_{i+\frac{1}{2}})\,, \qquad (10.66)$$

where $\mathbf{F}^{LF}_{i+\frac{1}{2}}$ is the Lax–Friedrichs flux and $\mathbf{F}^{LWx}_{i+\frac{1}{2}}$ is any of three versions of the Lax–Wendroff method. The FORCE scheme is monotone and has linearised stability condition $0 \le |c| \le C_{lim} = 1$. See Chap. 9. Originally [4, 5], the FORCE flux was derived from replacing random values in the staggered-grid version of the Random Choice Method of Glimm by integral averages. Unexpectedly, it turned out to be a simple mean value of the Lax–Friedrichs and the two-step Lax–Wendroff fluxes. Convergence of the FORCE scheme for the one-dimensional shallow water equations and the equations of isentropic gas dynamics was proved by Chen and Toro [46, 47].

The FORCE scheme, as described, may be extended to two and three space dimensions on structured meshes through dimensional splitting, see Chap. 13. The scheme is unstable if directly applied through a *simultaneous update formula* in multiple space dimensions. A general, stable multidimensional extension of FORCE on structured and unstructured meshes was reported in [48, 49]. See [50] for an ambitious application of the mutidimensional FORCE method to geophysical flows. The recently communicated FORCE-$\alpha$ scheme [51] emerges as a simplification of the mutidimensional FORCE method [48, 49]. The scheme is stable in multiple space dimensions and is particularly accurate for slowly moving waves.

Other centred schemes, not studied here, are those advocated by Tadmor and collaborators [52–54] and the *composite schemes* advocated by Liska and Wendroff [55, 56].

In Chap. 9 we showed some numerical results from the centred methods studied in this chapter, as applied to the linear advection equation. The reader is encouraged to view these numerical results, as they graphically represent the schemes's properties, also valid for non-linear systems.

## 10.6 Finite Volume Schemes in Multidimensions

Here we study the finite volume approach to hyperbolic balance laws in multiple space dimensions on unstructured meshes and specialise the schemes to the two-dimensional shallow water equations. The Cartesian mesh case is obtained as a particular example of the general finite volume approach.

### 10.6.1 Unstructured Meshes

Let us consider a single 2D computational cell for a general unstructured mesh, as depicted in Fig. 10.9. Denoting a computational cell as $V_k$ we define its integral cell average as

$$\mathbf{Q}_k = \frac{1}{|V_k|} \int\int_{V_k} \mathbf{Q}\, dV \;, \tag{10.67}$$

## 10.6 Finite Volume Schemes in Multidimensions

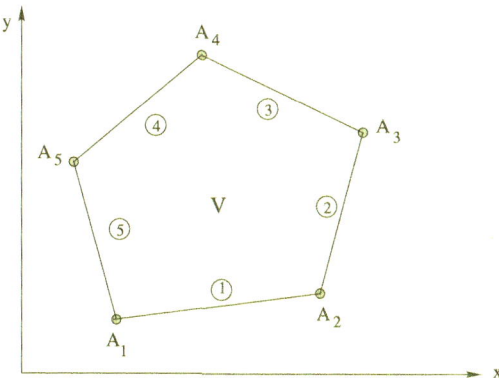

**Fig. 10.9** Example of a two-dimensional computational cell $V$ of an unstructured mesh in $x$-$y$ space. The cell is defined by the position of five vertices and corresponding five edges

where $|V_k|$ is the volume of $V_k$ (or area in the 2D case). Then, an integral form of the governing equations, see Chap. 4, leads to a *semidiscrete* scheme

$$\frac{d}{dt}\mathbf{Q}_k = -\frac{1}{|V_k|}\sum_{s=1}^{N_{fv}} \mathcal{F}_s \ , \qquad (10.68)$$

where $N_{fv}$ is the total number of sides, or edges, of cell $V_k$ and $\mathcal{F}_s$ are the corresponding intercell fluxes. Alternatively, by choosing the appropriate integral form one derives the *fully discrete* scheme

$$\mathbf{Q}_k^{n+1} = \mathbf{Q}_k^n - \frac{\Delta t}{|V_k|}\sum_{s=1}^{N_{fv}} \mathcal{F}_s \ , \qquad (10.69)$$

where $\Delta t$ is the time step. Compare (10.69) with the corresponding one-dimensional finite volume scheme (10.10). The fluxes $\mathcal{F}_s$ in (10.68) and (10.69) are

$$\mathcal{F}_s = \int_{A_s}^{A_{s+1}} \mathbf{T}_s^{-1} \mathbf{F}\left(\mathbf{T}_s(\mathbf{Q})\right) dA \ . \qquad (10.70)$$

This is an integral on the segment $A_s A_{s+1}$, denoted as $s$, of the computational cell; see Fig. 10.9. In the integrand in (10.70) $\mathbf{T}_s$ is the rotation matrix in the rotational invariance property of the equations, and $\mathbf{T}_s^{-1}$ is its inverse. See Chap. 4. $\mathbf{T}_s(\mathbf{Q})$ denotes the vector of *rotated* conserved variables by applying the rotation matrix $\mathbf{T}_s$ to the original vector of variables $\mathbf{Q}$. This is now aligned with the new rotated Cartesian frame $(\hat{x}, \hat{y})$ depicted in Fig. 10.10. Here $\hat{x}$ is the normal direction (normal to the intercell boundary), while $\hat{y}$ is the tangential direction (parallel to the intercell boundary). The rotation procedure rests on the choice of an underlying reference direction; here we have chosen the Cartesian $x$-direction as the reference direction, from which angles $\theta_s$ are measured, as shown in Fig. 10.10.

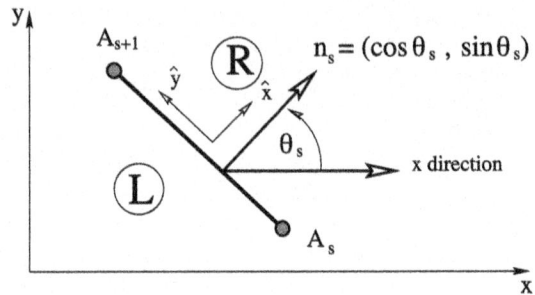

**Fig. 10.10** General intercell boundary (edge) $s$ between vertices $A_s$ and $A_{s+1}$ in two-dimensional $x$-$y$ space. The chosen reference direction here is the $x$-direction, but alternatively, one could have chosen $y$ as the reference direction

We consistently adhere to the convention of moving along the boundary of the computing cell $V_k$ so that the interior of the volume always lies on the left hand side of the boundary. Let us now consider any intercell boundary $s$ as shown in Fig. 10.10, where $L$ (left) denotes the interior side of the control volume $V$ and $R$ (right) denotes the state exterior to $V$ and adjacent to side $s$. The first problem is to determine the direction of the outward unit vector $\mathbf{n}_s$ normal to side $s$; this is needed to evaluate the integral terms in (10.70). Our chosen reference direction is the $x$-direction, $\mathbf{n}_s$ is the vector normal to the edge $s$ and $\theta_s$ is the angle formed by the $x$-direction and the normal vector $\mathbf{n}_s$. The components of $\mathbf{n}_s$ can be found in terms of the angle $\theta_s$ and it is easy to see that

$$\mathbf{n}_s = (cos\theta_s, sin\theta_s) \,. \tag{10.71}$$

Note that in schemes (10.68) and (10.69) the computational elements, or cells, are numbered by a single index $k$. This is what distinguishes structured from unstructured meshes. Further details are found in Chap. 16 of [9–11].

### 10.6.2 The Numerical Flux

To compute a numerical flux across the intercell boundary $s$ we use the one-dimensional equations in the rotated frame $(\hat{x}, \hat{y})$, see Fig. 10.10, namely

$$\partial_t \hat{\mathbf{Q}} + \partial_{\hat{x}} \hat{\mathbf{F}} = \mathbf{0} \,, \tag{10.72}$$

where $\hat{\mathbf{F}} = \mathbf{F}(\hat{\mathbf{Q}})$ is the flux for the augmented equations including the two components of velocity. Each term in (10.68) and (10.69) is approximated as

$$\mathcal{F}_s = \int_{A_s}^{A_{s+1}} \mathbf{T}_s^{-1} \mathbf{F}(\mathbf{T}_s \mathbf{Q}) \, dA \approx \mathcal{L}_s \mathbf{T}_s^{-1} \hat{\mathbf{F}}_s \,, \tag{10.73}$$

where $\mathcal{L}_s$ is the length of segment $A_s A_{s+1}$ and $\hat{\mathbf{F}}_s$ is the intercell flux corresponding to the augmented one-dimensional system (10.72).

10.6 Finite Volume Schemes in Multidimensions

To compute Godunov first-order upwind flux $\hat{\mathbf{F}}_s$ we solve exactly the rotated local Riemann problem in the *normal* direction to the boundary, at the cell interface $x = x_{i+\frac{1}{2}}$ (locally at 0)

$$\begin{aligned}\text{PDEs:} \quad & \partial_t \hat{\mathbf{Q}} + \partial_{\hat{x}} \hat{\mathbf{F}}(\hat{\mathbf{Q}}) = \mathbf{0}\,, \\ \text{ICs:} \quad & \hat{\mathbf{Q}}(x,0) = \begin{cases} \hat{\mathbf{Q}}_L = \mathbf{T}_s(\mathbf{Q}_L) \text{ if } \hat{x} < 0\,, \\ \hat{\mathbf{Q}}_R = \mathbf{T}_s(\mathbf{Q}_R) \text{ if } \hat{x} > 0\,. \end{cases}\end{aligned} \qquad (10.74)$$

Here $\mathbf{Q}_L$ is the state in the interior of the finite volume $V$ and $\mathbf{Q}_R$ is the state outside $V$ separated from $\mathbf{Q}_L$ by the intercell boundary $A_s A_{s+1}$. See Fig. 10.10. For the shallow water equations the exact Riemann solver is given in Chap. 6 for the wet bed case and in Chap. 7 for the dry-bed case. Approximate Riemann solvers for (10.74) will be presented in Chap. 11. Alternatively, one can also use centred fluxes for (10.74).

### Lengths, Normals and Areas

To complete the computation of $\mathcal{F}_s$ in (10.73) we give formulae for the lengths $\mathcal{L}_s$ in (10.73), for the components of the outward unit normals $\mathbf{n}_s$ and for the areas of co-planar quadrilateral finite volumes, with coordinates $(x_s, y_s)$ for the vertices $A_s$. We first define

$$\Delta x_s \equiv x_{s+1} - x_s\,, \quad \Delta y_s \equiv y_{s+1} - y_s\,. \qquad (10.75)$$

Then, the length $\mathcal{L}_s$ of side $s$ is given by

$$\mathcal{L}_s = \sqrt{\Delta x_s^2 + \Delta y_s^2}\,. \qquad (10.76)$$

The components $cos\theta_s$ and $sin\theta_s$ of the outward unit normal $\mathbf{n}_s$ in (10.71) are

$$cos\theta_s = \frac{\Delta y_s}{\Delta s}\,, \quad sin\theta_s = -\frac{\Delta x_s}{\Delta s}\,. \qquad (10.77)$$

Finally, the area of the co-planar quadrilateral $V$ is given by

$$|V| = \frac{1}{2} |(x_3 - x_1) \times (y_4 - y_2) - (y_3 - y_1) \times (x_4 - x_2)|\,. \qquad (10.78)$$

Given a mesh, structured or unstructured, all the necessary items have been specified and one can proceed to implement the fully discrete scheme (10.69). For the semi-discrete scheme (10.68) one requires the choice of a solver for the system of ordinary differential equations, see [38] and references therein. See also Chap. 13 on ODE solvers especially designed for semidiscrete methods.

### 10.6.3 The Cartesian Case

Assume Cartesian cells $I_{i,j}$ of area $\Delta x \times \Delta y$, as shown in Fig. 10.11. The finite volume formula (10.69) becomes

$$Q_{i,j}^{n+1} = Q_{i,j}^n - \frac{\Delta t}{\Delta x}[\mathbf{F}_{i+\frac{1}{2},j} - \mathbf{F}_{i-\frac{1}{2},j}] - \frac{\Delta t}{\Delta y}[\mathbf{G}_{i,j+\frac{1}{2}} - \mathbf{G}_{i,j-\frac{1}{2}}]. \tag{10.79}$$

To verify this *simultaneous update formula* note that the outward unit normals for sides 1 to 4 are

$$\mathbf{n}^1 = [0,-1], \quad \mathbf{n}^2 = [1,0], \quad \mathbf{n}^3 = [0,1], \quad \mathbf{n}^4 = [-1,0]. \tag{10.80}$$

See Fig. 10.11. The fluxes (multiplied by a length) corresponding to each cell side are

$$\left.\begin{aligned}
\mathcal{F}_1 &= \int_{A_1}^{A_2} (\mathbf{F},\mathbf{G}) \cdot \mathbf{n}^1 \, dA = -\Delta x \times \mathbf{G}_{i,j-\frac{1}{2}}, \\
\mathcal{F}_2 &= \int_{A_2}^{A_3} (\mathbf{F},\mathbf{G}) \cdot \mathbf{n}^2 \, dA = \Delta y \times \mathbf{F}_{i+\frac{1}{2},j}, \\
\mathcal{F}_3 &= \int_{A_3}^{A_4} (\mathbf{F},\mathbf{G}) \cdot \mathbf{n}^3 \, dA = \Delta x \times \mathbf{G}_{i,j+\frac{1}{2}}, \\
\mathcal{F}_4 &= \int_{A_4}^{A_1} (\mathbf{F},\mathbf{G}) \cdot \mathbf{n}^4 \, dA = -\Delta y \times \mathbf{F}_{i-\frac{1}{2},j}.
\end{aligned}\right\} \tag{10.81}$$

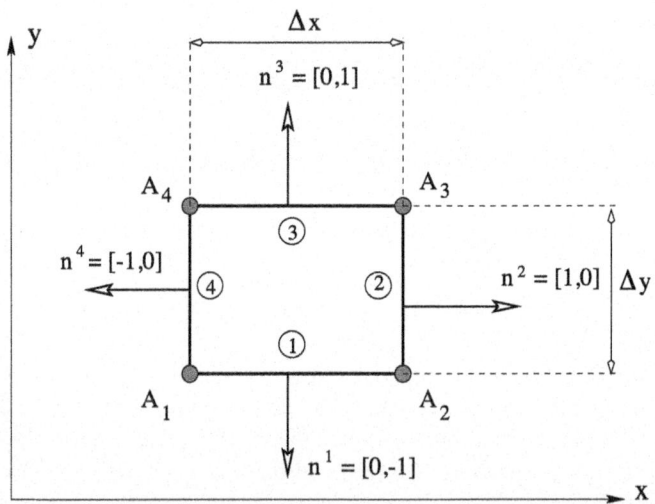

**Fig. 10.11** Cartesian control volume $V$ in $x$-$y$ space. The element has four edges, with corresponding four intercell numerical fluxes. Outward unit normal vectors for each side are shown

10.6 Finite Volume Schemes in Multidimensions

A conservative method such as (10.79) is completely determined once the fluxes $\mathbf{F}_{i+\frac{1}{2},j}$ and $\mathbf{G}_{i,j+\frac{1}{2}}$ are specified.

**Remark** (*stability*) In using the simultaneous update formula (10.79) it is tempting to compute the intercell fluxes by simply applying well-known one-dimensional fluxes, such as the ones studied here. Great care is required in doing so, as not all 1D fluxes give a stable scheme. The Godunov upwind method is successful, with stability restriction $0 \leq |c| \leq C_{lim} = \frac{1}{2}$ in two dimensions and $0 \leq |c| \leq C_{lim} = \frac{1}{3}$ in three dimensions. Direct use of the Lax–Friedrichs or FORCE fluxes, as presented, will lead to unstable schemes in multiple space dimensions. See [51] for simple, stable centred methods in multiple space dimensions and related issues.

## 10.6.4 The Telescopic Property

In most of this book we are concerned with conservative numerical schemes based on the *conservative formula*

$$\mathbf{Q}_i^{n+1} = \mathbf{Q}_i^n - \frac{\Delta t}{\Delta x}(\mathbf{F}_{i+\frac{1}{2}} - \mathbf{F}_{i-\frac{1}{2}}) \,. \tag{10.82}$$

Such schemes are completely determined once the numerical fluxes $\mathbf{F}_{i+\frac{1}{2}}$ have been specified. In general, the numerical flux has the form

$$\mathbf{F}_{i+\frac{1}{2}} = \mathbf{F}_{i+\frac{1}{2}}(\mathbf{Q}_{i-k_L}^n, \ldots, \mathbf{Q}_{i+k_R}^n) \,, \tag{10.83}$$

where the non-negative integers $k_L$ and $k_R$ depend on the particular choice of numerical flux. In *explicit methods* the arguments are evaluated at the *data* time level $n$. In *implicit methods* arguments are evaluated at the unknown values at the new time level $n + 1$, but may also include data values at time level $n$. We note that an important property to be satisfied by the intercell numerical flux is *consistency*. This means that if all the arguments of the numerical flux (10.83) are equal to a constant value $\hat{\mathbf{Q}}$, then the value of the numerical flux at $\hat{\mathbf{Q}}$ must coincide with the value of the *physical* (exact) flux $\mathbf{F}(\hat{\mathbf{Q}})$ of the PDEs evaluated at $\hat{\mathbf{Q}}$, namely

$$\mathbf{F}_{i+\frac{1}{2}}(\hat{\mathbf{Q}}, \ldots, \hat{\mathbf{Q}}) = \mathbf{F}(\hat{\mathbf{Q}}) \,. \tag{10.84}$$

Conservative methods (10.82) enjoy the so-called *telescopic property*. This property says that the intercell flux $\mathbf{F}_{i+\frac{1}{2}}$ used to update the cell average $\mathbf{Q}_i^n$ must be identical to the intercell flux used to update the cell average $\mathbf{Q}_{i+1}^n$. Hence, on summation of $\mathbf{Q}_i^n$ and $\mathbf{Q}_{i+1}^n$ the flux at the boundary between the cells $i$ and $i + 1$ cancels out. In other words, at every interface $x_{i+\frac{1}{2}}$ there can only be *one intercell flux* to update both the left cell and the right cell.

More generally, assume the discretised domain of interest has leftmost and rightmost cells $i_{Left}$ and $i_{Right}$ and leftmost and rightmost intercell boundary fluxes $\mathbf{F}_{Left}$ and $\mathbf{F}_{Right}$. Then, summation applied to formula (10.82), premultiplied by $\Delta x_i$, gives

$$\sum_{i=i_{Left}}^{i=i_{Right}} \Delta x_i \mathbf{Q}_i^{n+1} = \sum_{i=i_{Left}}^{i=i_{Right}} \Delta x_i \mathbf{Q}_i^{n} - \Delta t \left[ \mathbf{F}_{Right} - \mathbf{F}_{Left} \right] . \tag{10.85}$$

That is, the total amount of the conserved variable $\mathbf{Q}$ changes only because of the fluxes through the *end boundaries*. The telescopic property of conservative methods also holds in multiple space dimensions.

## 10.7 Numerical Results

Here we present numerical results from some of the methods studied, namely the Random Choice Method (RCM), the Godunov upwind method, and the FORCE method. In order to assess the performance of the methods we use the test problems with exact solution given in Table 7.1 of Chap. 8. For all cases we consider a channel of length $L = 50\ m$, discretised with a regular mesh of $M = 100$ cells, and safety coefficient $S_c = 0.9$. Recall that $C_{cfl} = S_c \times C_{lim}$; $C_{lim} = 1$ for Godunov and FORCE but $C_{lim} = \frac{1}{2}$ for the Random Choice Method. Results for Test 1 are shown at time $T_{out} = 7$ s in Figs. 10.15, 10.16 and 10.17. Results for Test 2 are shown at time $T_{out} = 2.5$ s in Figs. 10.12, 10.13 and 10.14. Tests 3, 4 and 5 involve dry-bed portions, for which we only show results from the Random Choice Method. See Figs. 10.18, 10.19 and 10.20.

For Test 1, all methods used run successfully, see Figs. 10.12, 10.13 and 10.14. The Godunov upwind method used in conjunction with exact Riemann solver, Fig. 10.12, performs as expected from a first-order, monotone upwind method. There are no visible oscillations in the vicinity of the shock wave on the right hand side; note

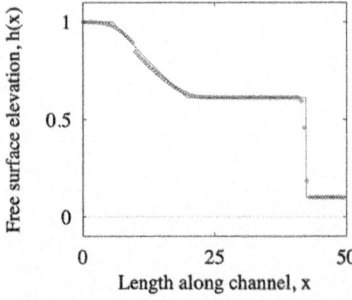 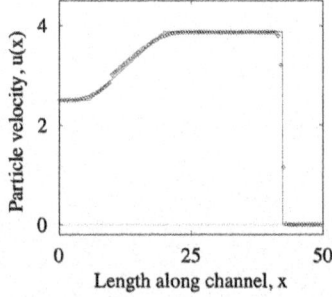

**Fig. 10.12** Test 1: Godunov's upwind method with exact Riemann solver (symbol) and exact solution (line). Mesh $M = 100$ and $T_{out} = 7$ s

## 10.7 Numerical Results

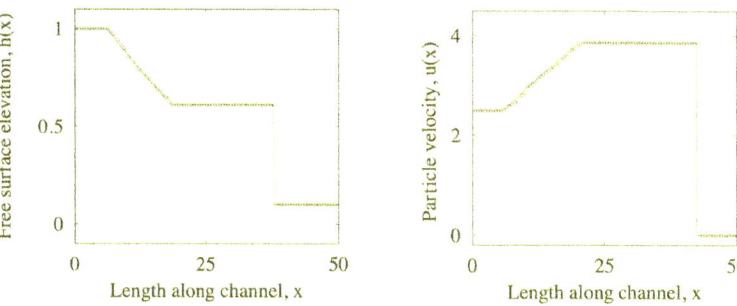

**Fig. 10.13** Test 1: RCM (symbol) and exact solution (line). Mesh $M = 100$ and $T_{out} = 7$ s

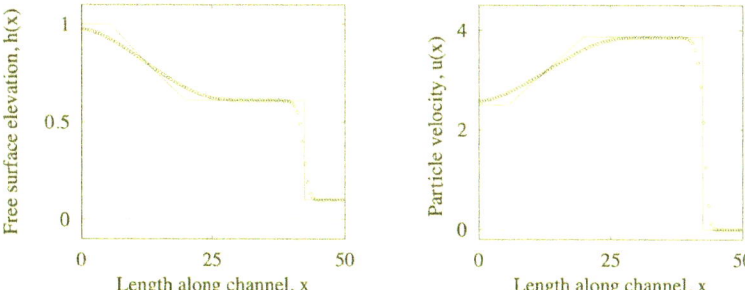

**Fig. 10.14** Test 1: FORCE method (symbol) and exact solution (line). Mesh $M = 100$ and $T_{out} = 7$ s

also that the shock wave is, on average, in the correct position, a consequence of the conservative character of the method. The discontinuity is resolved quite sharply, within two to three cells. The left rarefaction wave is reasonably well resolved, apart from the so called *entropy glitch*, which is a small jump whose position coincides with the occurrence of *critical flow* (or sonic flow), that is when $u = a$. It is known that the size of the jump in the entropy glitch in the Godunov method with the exact Riemann solver tends to zero, as the mesh is refined. Linearised Riemann solvers, in place of the exact Riemann solver used here, will give a very large jump in the presence of critical flow and will not vanish with mesh refinement. In this case the solution, locally, looks like a shock wave. But such shock wave is unphysical, as it does not satisfy the entropy condition, for which it is called *rarefaction shock* or *entropy-violating shock*.

The results from the Random Choice Method shown in Fig. 10.13 are very accurate. The discontinuity is resolved with infinite resolution (true discontinuity), though the smooth part of the flow exhibits random noise, typical of RCM. Results from the FORCE method are shown in Fig. 10.14. These are typical of a *centred method* (not upwind). There are no signs of spurious oscillations in the vicinity of the

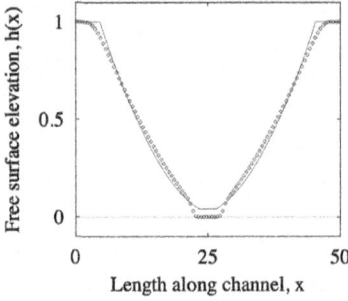 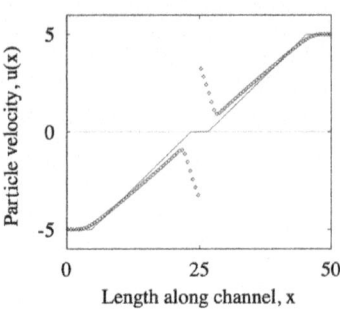

**Fig. 10.15** Test 2: Godunov's upwind method with exact Riemann solver (symbol) and exact solution (line). Mesh $M = 100$ and $T_{out} = 2.5$ s

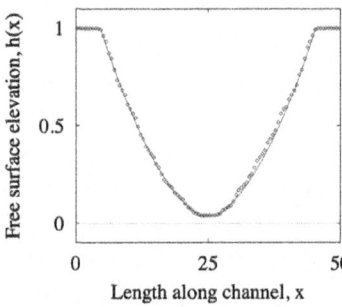 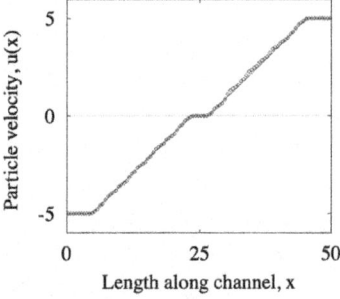

**Fig. 10.16** Test 2: RCM (symbol) and exact solution (line). Mesh $M = 100$ and $T_{out} = 2.5$ s

shock, as expected from a monotone method, though the resolution of the discontinuity is poor, within 8 to 9 cells. Compare with RCM and the Godunov upwind method. Although the resolution of the left rarefaction wave is inaccurate, there is no entropy glitch at the critical point. The FORCE method is entropy satisfying, as is the Godunov upwind method with a suitable Riemann solver. Again, compare results with Godunov upwind and RCM for the same problem. The performance of FORCE is inferior to that of the other two methods. The strength of FORCE is its simplicity and generality. It can be applied to any first-order system in one space dimension; its multidimensional extension is more involved [48, 49]. See also [51].

The results for Test 2 shown in Figs. 10.15, 10.16 and 10.17 add new elements to the discussion. The exact solution is continuous and consists of two rarefaction waves. However, there is a region of nearly zero water depth in the middle, as the result of the two strong rarefaction waves propagating in opposite directions. Some of the existing methods in the literature actually crush for this test, as a consequence of producing negative depths, which in turn results in complex celerity $a = \sqrt{gh}$. All three methods tested run successfully for this test problem. We note the inaccuracy of the Godunov upwind method for the velocity distribution, in the region of zero

## 10.7 Numerical Results

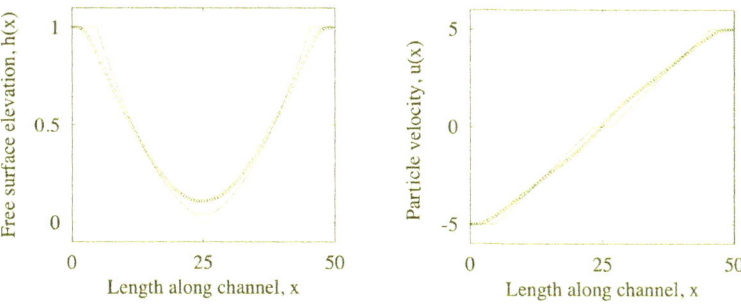

**Fig. 10.17** Test 2: FORCE method (symbol) and exact solution (line). Mesh $M = 100$ and $T_{out} = 2.5$ s

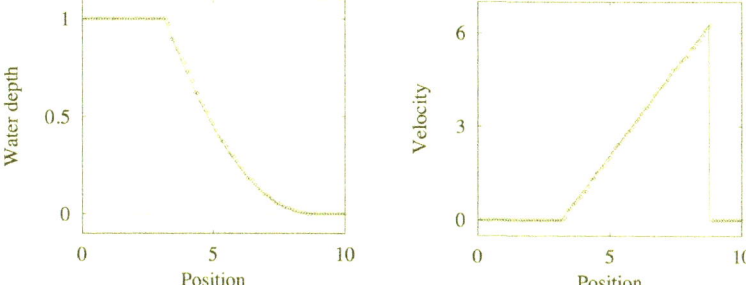

**Fig. 10.18** Test 3: RCM (symbol) and exact solution (line). Mesh $M = 100$ and $T_{out} = 4$ s

velocity in the exact solution. The best result is that of RCM, see Fig. 10.16. The FORCE method of Fig. 10.17, due to its large numerical diffusion, successfully avoids the problem of *near vacuum*. However, the velocity profile is very inaccurate in the region of nearly zero velocity.

Results for Tests 3, 4 and 5 are shown in Figs. 10.18, 10.19 and 10.20. The exact solution for these test problems involve dry-bed regions. In Test 3, the dry-bed region is the initial state on the right hand side. In Test 4 the dry-bed region is the initial state on the left hand side. In Test 5 with non-vacuum initial conditions, vacuum appears as the result of the interaction of the two initial states, in the middle, creating two wet/dry fronts. Numerical results are shown only for RCM, as all other methods are not suitable to run on this kind of test problems.

Currently there is considerable interest in adapting/improving main-stream methods to deal with these special but practical situations. It is worth remarking that for one-dimensional problems, such as Tests 3 to 5, RCM gives very good results, particularly for the wet/dry front; see Figs. 10.18, 10.19 and 10.20. However, RCM suffers from a major limitation, it does not work for multidimensional non-linear systems under dimensional splitting [16]. However, Toro and Roe [57, 58] noted that RCM could potentially work for linearly degenerate fields in multiple space

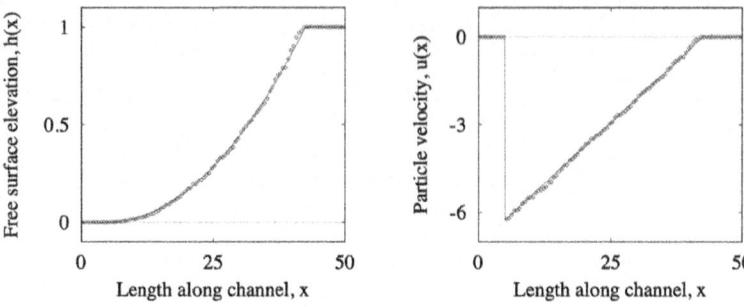

**Fig. 10.19** Test 4: RCM (symbol) and exact solution (line). Mesh $M = 100$ and $T_{out} = 4$ s

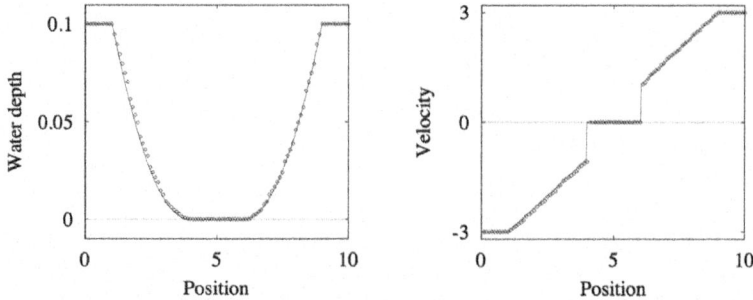

**Fig. 10.20** Test 5: RCM (symbol) and exact solution (line). Mesh $M = 100$ and $T_{out} = 5$ s

dimensions. As demonstrated by Ivings et al. [59], the resolution of wet/dry fronts in the two-dimensional shallow water equations may be improved significantly via hybrid RCM-based schemes, relative to other existing methods.

## 10.8 Conclusions

First-order methods for non-linear systems have been presented. Some of such methods will form the bases for the construction of non-linear higher-order schemes in successive chapters. Schemes for one-dimensional hyperbolic balance laws were first presented; these included the Godunov upwind method [1] in conjunction with exact Riemann solver of Chaps. 6 and 7; the Random Choice Method (RCM) of Glimm [2]; Flux-Vector Splitting (FVS) methods; and centred methods, such as the Lax–Friedrichs [3], Lax–Wendroff [3] and FORCE methods [5]. The finite volume framework for PDEs in multiple space dimensions on unstructured meshes is studied in some detail, including the definition of semi-discrete and fully discrete schemes; the rotated Riemann problem resulting from the rotational invariance property of the two-dimensional shallow water equations; the definition of the intercell numerical flux and determination of edge lengths, normals and computing cells areas in two space dimensions. Some numerical examples are shown, in order to illustrate the performance of selected schemes studied, as compared with exact solutions for a suite of test problems.

# References

1. S.K. Godunov, Finite difference methods for the computation of discontinuous solutions of the equations of fluid dynamics. Mat. Sb. **47**, 271–306 (1959)
2. J. Glimm, Solution in the large for nonlinear hyperbolic systems of equations. Comm. Pure. Appl. Math. **18**, 697–715 (1965)
3. P.D. Lax, B. Wendroff, Systems of conservation laws. Comm. Pure Appl. Math. **13**, 217–237 (1960)
4. E.F. Toro, On Glimm-related schemes for conservation laws. Technical Report MMU–9602, Department of Mathematics and Physics, Manchester Metropolitan University, UK (1996)
5. E.F. Toro, S.J. Billett, Centred TVD schemes for hyperbolic conservation laws. IMA J. Numer. Anal. **20**, 47–79 (2000)
6. E.F. Toro, L.O. Müller, A. Siviglia, Bounds for wave speeds in the Riemann problem: direct theoretical estimates. Comput. Fluids **209**(104640) (2020)
7. J.L. Guermond, B. Popov, Fast estimation of the maximum wave speed in the Riemann problem for the Euler equations. J. Comp. Phys. **321**, 908–926 (2016)
8. A.J. Chorin, Random choice solutions of hyperbolic systems. J. Comput. Phys. **22**, 517–533 (1976)
9. E.F. Toro, *Riemann Solvers and Numerical Methods for Fluid Dynamics. A Practical Introduction* (Springer, 1997)
10. E.F. Toro, *Riemann Solvers and Numerical Methods for Fluid Dynamics. A Practical Introduction*, 2nd edn. (Springer, 1999)
11. E.F. Toro, *Riemann Solvers and Numerical Methods for Fluid Dynamics. A Practical Introduction*, 3rd edn. (Springer, 2009)
12. A.J. Chorin, Random choice methods with applications to reacting gas flow. J. Comput. Phys. **25**, 253–272 (1977)
13. G.A. Sod, A numerical study of a converging cylindrical shock. J. Fluid Mech. **83**, 785–794 (1977)
14. P. Concus, W. Proskurowski, Numerical solution of a nonlinear hyperbolic equation by the random choice method. J. Comput. Phys. **30**, 153–166 (1979)
15. P. Colella, An analysis of the effect of operator splitting and of the sampling procedure on the accuracy of Glimm's method. Ph.D. thesis, Department of Mathematics, University of California, USA (1978)
16. P. Colella, Glimm's method for gas dynamics. SIAM J. Sci. Stat. Comput. **3**(1), 76–110 (1982)
17. J.J. Gottlieb, Staggered and non-staggered grids with variable node spacing and local timestepping for random choice method. J. Comput. Phys. **78**, 160–177 (1988)
18. Z.C. Shi, J.J. Gottlieb, Random choice method for two dimensional planar and axisymmetric steady supersonic flows. Technical Report 297, UTIAS, University of Toronto, Canada (1985)
19. E.F. Toro, A fast Riemann solver with constant covolume applied to the random choice method. Int. J. Numer. Meth. Fluids **9**, 1145–1164 (1989)
20. T. Saito, I.I. Glass, Application of random choice to problems in shock and detonation wave dynamics. Technical Report UTIAS 240, Institute for Aerospace Studies, University of Toronto (1979)
21. Y. Takano, An application of the random choice method to a reactive gas with many chemical species. J. Comput. Phys. **67**(1), 173–187 (1986)
22. G. Singh, J.F. Clarke, Transient phenomena in the initiation of a mechanically driven plane detonation. Proc. Roy. Soc. Lond. A **438**, 23–46 (1992)

23. A.S. Dawes, Natural Coordinates and High Speeed Flows. A Numerical Method for Reactive Gases. Ph.D. thesis, College of Aeronautics, Cranfield Institute of Technology, UK (1992)
24. H. Olivier, H. Grönig, The random choice method applied to two-dimensional shock focusing and diffraction. J. Comput. Phys. **63**, 85–106 (1986)
25. G. Marshall, B. Plohr, A random choice method for two-dimensional steady supersonic shock wave diffraction problems. J. Comput. Phys. **56**, 410–427 (1984)
26. E. Marshall, R. Méndez, Computational aspects of the random choice method for shallow water equations. J. Comput. Phys. **39**, 1–21 (1981)
27. K.M. Li, M. Holt, Numerical solutions to water waves generated by shallow underwater explosions. Phys. Fluids **24**, 816–824 (1981)
28. J.L. Steger, R.F. Warming, Flux vector splitting of the inviscid gasdynamic equations with applications to finite difference methods. J. Comput. Phys. **40**, 263–293 (1981)
29. B. van Leer, Flux–vector splitting for the Euler equations. Technical Report ICASE 82-30, NASA Langley Research Center, USA (1982)
30. M.S. Liou, C.J. Steffen, A new flux splitting scheme. J. Comput. Phys. **107**, 23–39 (1993)
31. G.-C. Zha, E. Bilgen, Numerical solution of Euler equations by a new flux vector splitting scheme. Int. J. Numer. Meth. Fluids **17**, 115–144 (1993)
32. E.F. Toro, M.E. Vázquez-Cendón, Flux splitting schemes for the Euler equations. Comput. Fluids **70**, 1–12 (2012)
33. L. Bermúdez, M.E. Vázquez, Upwind methods for hyperbolic conservation laws with source terms. Comput. Fluids **23**, 1049–1071 (1994)
34. L. Bermúdez, A. Dervieux, J.A. Desideri, M.E. Vázquez, Upwind schemes for the two-dimensional shallow water equations with variable depth using unstructured meshes. Comput. Methods Appl. Mech. **155**, 49–72 (1998)
35. A.T. Morel, M. Fey, Multi-dimensional method of transport for the shallow water equations. Priv. Commun. (1994)
36. E.F. Toro, C.E. Castro, D. Vanzo, A. Siviglia, A flux-vector splitting scheme for the shallow water equations extended to high-order on unstructured meshes. Int. J. Numer. Methods Fluids (2022) https://doi.org/10.1002/fld.5099
37. E.F. Toro, A. Siviglia, A. Spilimbergo, L.O. Müller, Advection-pressure splitting schemes for the equations of blood flow in arteries. Conservative and non-conservative forms. East Asian J. Appl. Math. (2024)
38. E.F. Toro, L.O. Müller, *Computational Bodily Fluid Dynamics. Models and Algorithms (to appear)* (Springer, 2024)
39. S. Busto, M. Dumbser, A staggered semi-implicit hybrid finite volume/finite element scheme for the shallow water equations at all Froude numbers. Appl. Numer. Math. **175**, 108–132 (2022)
40. W. Boscheri, M. Tavelli, C.E. Castro, An all Froude high order IMEX scheme for the shallow water equations on unstructured Voronoi meshes. Appl. Numer. Math. **185**, 311–335 (2023)
41. E.F. Toro, A. Hidalgo, S.A. Tokareva, On HLL-type schemes for hyperbolic equations: wave-speed estimates. Monotonicity and stability. Comput. Fluids (2023)
42. A. Spilimbergo, E.F. Toro, A. Siviglia, L.O. Müller, Flux vector splitting schemes applied to a conservative 1D blood flow model with transport for arteries and veins. Comput. Fluids **271**, 106165 (2024)
43. A. Lucca, S. Busto, L.O. Müller, E.F. Toro, M. Dumbser, A semi-implicit finite volume scheme for blood flow in elastic and viscoelastic vessels. J. Comput. Phys. **495**, 112530 (2023)
44. A. Siviglia, D. Vanzo, E.F. Toro, A splitting scheme for the coupled Saint Venant-Exner model. Adv. Water Res. **159**, 104062 (2022)
45. S.K. Godunov, A.V. Zabrodin, G.P. Prokopov, A difference scheme for two-dimensional unsteady aerodynamics. J. Comp. Math. Math. Phys. USSR **2**(6), 1020–1050 (1961)
46. G.Q. Chen, E.F. Toro, Centred schemes for non-linear hyperbolic equations. Technical report, Isaac Newton Institute for Mathematical Sciences, University of Cambridge, UK (2003)
47. G.Q. Chen, E.F. Toro, Centred schemes for non-linear hyperbolic equations. J. Hyperb. Diff. Equ. **1**(1), 531–566 (2004)

48. E.F. Toro, A. Hidalgo, M. Dumbser, FORCE Schemes on unstructured meshes I: conservative hyperbolic systems. Isaac Newton Institute for Mathematical Sciences, University of Cambridge, UK. Preprint NI09005-NPA, September 2008. Submitted to J. Comput. Phys. (2008)
49. M. Dumbser, M.J. Castro, C. Parés, E.F. Toro, A. Hidalgo, FORCE schemes on unstructured meshes II: non-conservative hyperbolic systems. Comput. Methods Appl. Mech. Eng. **199**(9–12), 625–647 (2010)
50. M. Dumbser, M.J. Castro, C. Parés, E.F. Toro, ADER schemes on unstructured meshes for nonconservative hyperbolic systems: applications to geophysical flows. Comput. Fluids **38**(9), 731–1748 (2009)
51. E.F. Toro, B. Saggiorato, S. Tokareva, A. Hidalgo, Low-dissipation centred schemes for hyperbolic equations in conservative and non-conservative form. J. Comput. Phys. **416**(109545) (2020)
52. H. Nessyahu, E. Tadmor, Non-oscillatory central differencing for hyperbolic conservation laws. J. Comput. Phys. **87**, 408–463 (1990)
53. G.S. Jiang, E. Tadmor, Non-oscillatory central schemes for multi-dimensional hyperbolic conservation laws. SIAM J. Sci. Comput. **19**(6), 1892–1917 (1998)
54. F. Bianco, G. Puppo, G. Russo, High order central schemes for hyperbolic systems of conservation laws. SIAM J. Sci. Comput. **21**, 294–322 (1999)
55. R. Liska, B. Wendroff, Analysis and computation with stratified fluid models. J. Comput. Phys. **137**, 212–244 (1997)
56. R. Liska, B. Wendroff, Two-dimensional shallow water equations by composite schemes. Int. J. Numer. Methods Fluids **30**, 461–479 (1999)
57. E.F. Toro, P.L. Roe, A hybrid scheme for the Euler equations using the random choice and Roe's methods, in *Numerical Methods for Fluid Dynamics III. The Institute of Mathematics and its Applications Conference Series*, New Series No. 17, ed. by Morton, Baines (Oxford University Press, New York, 1988), pp. 391–402
58. E.F. Toro, Random choice based hybrid schemes for one and two-dimensional gas dynamics, in *Proceedings of the Second International Conference on Hyperbolic Problems, Aachen, Germany, March 1988. Non–linear Hyperbolic Equations-Theory, Computation Methods and Applications*. Notes on Numerical Fluid Mechanics, vol. 24, ed. by R. Jeltsch, J. Ballmann, pp. 630–639 (Vieweg, Braunschweig, 1989)
59. M.J. Ivings, E.F. Toro, D.M. Webber, Numerical schemes for 2D shallow water equations including dry fronts. J. Comput. Fluid Dyn. **12**(1), 41–52 (2003)

# Chapter 11
# Approximate Riemann Solvers

**Abstract** A full range of 10 approximate Riemann solvers for the shallow water equations written in conservative and non-conservative forms have been presented, along with a discussion on criteria to judge their quality in comparison with the exact solver. A first criterion is whether a solver is linear or non-linear and whether a linear solver is accompanied with an effective entropy fix to exclude unphysical entropy-violating shocks. A second criterion is whether a solver is complete, that is whether or not the number of waves $W$ in its wave model matches that of the exact solver $N$. For incomplete Riemann solvers $W < N$, which is usually associated with the exclusion of intermediate characteristic fields in the wave model, thus resulting in excessive numerical dissipation for contact waves and vortical flows. Also included in this chapter are FORCE-type centred methods especially adapted to systems in non-conservative form. These methods do not require the explicit solution of the Riemann problem and may be placed at the bottom of the hierarchy of Riemann solvers. All the methods studied give directly a first-order accurate method of the Godunov type and can be extended to construct non-linear high order schemes in the frameworks of finite volumes and discontinuous Galerkin finite elements.

## 11.1 Recalling the Godunov Upwind Method

To compute numerical solutions by Godunov-type methods, one can use the exact Riemann solvers of Chaps. 6 or 7, or approximate Riemann solvers, which is the subject of this chapter. Approximate Riemann solvers, if used judiciously, can provide effective computational tools at a competitive cost. The theoretical bases for each of the approximate Riemann solvers given here are found in the textbooks [1–3]. Choosing between the exact and approximate Riemann solvers is motivated by (i) computational cost, (ii) simplicity and ease of implementation, and (iii) correctness. Correctness should be the overriding criterion. For the shallow water equations, the argument of computational cost is not as strong as for other fields of application, compressible fluids with general equations of state, for example.

We are concerned with solving numerically the general initial-boundary value problem (IBVP) for the *augmented* one-dimensional shallow water equations

PDEs: $\partial_t \mathbf{Q} + \partial_x \mathbf{F}(\mathbf{Q}) = \mathbf{S}(\mathbf{Q})$, $x \in [a, b]$, $t > 0$,
ICs: $\mathbf{Q}(x, 0) = \mathbf{Q}^{(0)}(x)$, $x \in [a, b]$, (11.1)
BCs: $\mathbf{Q}(a, t) = \mathbf{B}_L(t)$, $\mathbf{Q}(b, t) = \mathbf{B}_R(t)$, $t \geq 0$.

$\mathbf{Q}(x, t)$ is the vector of *conserved variables*; $\mathbf{F}(\mathbf{Q})$ is the flux vector, or *physical flux*; $\mathbf{S}(\mathbf{Q})$ is the *source term vector*; $\mathbf{Q}^{(0)}(x)$ is the initial condition; $\mathbf{B}_L(t)$ and $\mathbf{B}_R(t)$ are the *boundary conditions* on the left ($x = a$) and right ($x = b$) boundaries, respectively, two prescribed functions of time. The aim is to apply Godunov-type methods based on the explicit conservative formula for the homogenous version of the system in (11.1), namely

$$\mathbf{Q}_i^{n+1} = \mathbf{Q}_i^n - \frac{\Delta t}{\Delta x}[\mathbf{F}_{i+\frac{1}{2}} - \mathbf{F}_{i-\frac{1}{2}}]. \tag{11.2}$$

In Chap. 10 we defined the Godunov upwind [4] intercell numerical flux as

$$\mathbf{F}_{i+\frac{1}{2}} = \mathbf{F}(\mathbf{Q}_{i+\frac{1}{2}}(0)), \tag{11.3}$$

in which $\mathbf{Q}_{i+\frac{1}{2}}(0)$ is the exact similarity solution $\mathbf{Q}_{i+\frac{1}{2}}(x/t)$ of the Riemann problem

$$\partial_t \mathbf{Q} + \partial_x \mathbf{F}(\mathbf{Q}) = 0,$$
$$\mathbf{Q}(x, 0) = \begin{cases} \mathbf{Q}_L = \mathbf{Q}_i^n & \text{if } x < x_{i+\frac{1}{2}}, \\ \mathbf{Q}_R = \mathbf{Q}_{i+1}^n & \text{if } x > x_{i+\frac{1}{2}}. \end{cases} \tag{11.4}$$

As noted in Chap. 10, the Riemann problem (11.4) is solved in local coordinates, still denoted by $x$ and $t$, and the Godunov state results from evaluating the solution $\mathbf{Q}_{i+\frac{1}{2}}(x/t)$ at $x/t = 0$.

For the $x$-split two-dimensional (2D) shallow water equations or for the one-dimensional (1D) equations with an extra equation for a passive scalar $\psi(x, t)$ the vectors of conserved variables and fluxes are

$$\mathbf{Q} = \begin{bmatrix} h \\ hu \\ h\psi \end{bmatrix}, \quad \mathbf{F} = \begin{bmatrix} hu \\ hu^2 + \frac{1}{2}gh^2 \\ hu\psi \end{bmatrix}. \tag{11.5}$$

It is worth remarking that for the $x$-split 2D shallow water equations $\psi = v$; otherwise $\psi$ has the meaning of a passive scalar to represent the concentration variable for a chemical species. In both cases $\psi$ has the same mathematical behaviour in the equations, as seen in Chap. 4. For this reason we shall tend to use $\psi$, as in (11.5), to represent both situations.

In terms of the vector primitive variables $\mathbf{W} = [h, u, \psi]^T$, the piece-wise constant initial data are

$$\mathbf{W}_L = \begin{bmatrix} h_L \\ u_L \\ \psi_L \end{bmatrix}, \quad \mathbf{W}_R = \begin{bmatrix} h_R \\ u_R \\ \psi_R \end{bmatrix}. \tag{11.6}$$

## 11.1 Recalling the Godunov Upwind Method

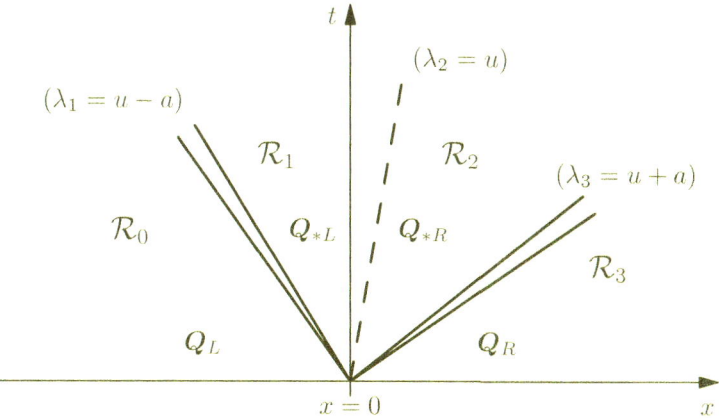

**Fig. 11.1** Structure of the exact solution of the Riemann problem for the $x$-split two-dimensional shallow water equations. There are three wave families *associated* with the eigenvalues $\lambda_1 = u - a$, $\lambda_2 = u$ and $\lambda_3 = u + a$. There are four constant regions separated by three wave families

The quantity $\psi$ gives rise to the middle eigenvalue $\lambda_2 = u$ and a corresponding middle, discontinuous wave. Figure 11.1 displays the structure of the solution of the Riemann problem (11.4) for the $x$-split 2D shallow water equations. There are four constant regions $\mathcal{R}_k$ separated by three wave families associated with three eigenvalues $\lambda_1 = u - a$, $\lambda_2 = u$ and $\lambda_3 = u + a$. The value of the solution along $x/t = 0$ corresponds to the $t$-axis and is the value required for the computation of the Godunov flux. See Chaps. 6 and 7 for details on the exact solution of the Riemann problem.

There are essentially two ways of extracting approximate information from the Riemann problem to be used in the computation of the intercell numerical flux in (11.2). One way is to look for an approximation to the Godunov state to then use in the flux evaluation (11.3). Approximations of this type are called *approximate-state* Riemann solvers. Another way, perhaps more popular, is to directly look for an approximation to the flux.

### Criteria for Judging Approximate Riemann Solvers

Before entering the subject of approximate Riemann solvers it is relevant to point out two important criteria to consider when approximating the solution. A first criterion is whether a solver is linear or non-linear and whether a linear solver is accompanied with an effective entropy fix to exclude the computation of unphysical entropy-violating shocks. A second criterion is whether a solver is complete, that is whether or not the number of waves $W$ in its wave model matches that of the exact solver $N$. For incomplete Riemann solvers $W < N$, which results in neglecting intermediate characteristic fields, such as those associated with contact waves in the case of species transport equations added to the one-dimensional shallow water equations and in the case of vortical flows in the two-dimensional equations. Incomplete Riemann solvers result in excessive numerical dissipation for the neglected characteristic fields.

In the next section we study *approximate-state* Riemann solvers.

## 11.2 Approximate-State Riemann Solvers

This section is concerned with finding an approximate solution for a state $\mathbf{W}_{i+\frac{1}{2}}(x/t)$ to the Riemann problem (11.4). We fist describe the general framework for doing so.

### 11.2.1 The Framework

To implement an approximate-state Riemann solver one first computes an approximate solution $\mathbf{W}_{i+\frac{1}{2}}(x/t)$ to the Riemann problem (11.4). Then an approximate numerical flux is obtained by evaluating the physical flux

$$\mathbf{F}_{i+\frac{1}{2}} = \mathbf{F}(\mathbf{W}_{i+\frac{1}{2}}(0)) \,, \tag{11.7}$$

where $\mathbf{W}_{i+\frac{1}{2}}(0)$ is the appropriate value along the $t$-axis. The process of finding the approximate solution $\mathbf{W}_{i+\frac{1}{2}}(x/t)$ has two steps. In the first step one computes an approximate solution for the variables $h, u, \psi$ in the Star Region, see Fig. 11.1. In the second step one samples the solution to obtain the correct values along the $t$-axis and evaluate the Godunov flux as in (11.7).

Recall that in the exact Riemann solver, see Chap. 6, the solution for $h_*$ in the Star Region is iterative; the solution for $u_*$ follows directly from the solution $h_*$ and so does the solution for $\psi$. As a matter of fact, the complete solution for $\psi(x, t)$ is simply

$$\psi(x, t) = \begin{cases} \psi_L & \text{if } \dfrac{x}{t} \leq u_* \,, \\ \psi_R & \text{if } \dfrac{x}{t} > u_* \,. \end{cases} \tag{11.8}$$

In fact the problem of finding approximate solutions for $h_*$ and $u_*$ in the Star Region is decoupled from that of finding the solution to $\psi$. The resulting approximate Riemann solvers will thus provide a solution for $\psi$ that is in a sense *exact*; the only approximation lies in the approximation to the normal velocity component $u_*$ in the Star Region. The fact that the solution for $\psi$ is virtually exact has important consequences when it comes to computing vortical flows in two dimensions, $\psi = v$, and contact discontinuities in one-dimensional pollutant transport models. The approximate-state Riemann solvers provide good resolution of these waves associated with *linearly-degenerate fields*. Some of the well known approximate Riemann solvers give the incorrect solution for these waves, leading to incorrect or inaccurate numerical results.

Having found an approximate solution for the Star Region, one then proceeds to sample the complete solution $\mathbf{W}_{i+\frac{1}{2}}(x/t)$, as done for the exact Riemann problem solution in Chap. 6. For the wet bed case this requires the identification of one out of ten wave patterns, see Chap. 10, to find the sought value $\mathbf{W}_{i+\frac{1}{2}}(0)$ and thus to evaluate the Godunov flux as in (11.7). Note that in the presence of a *sonic* or *critical* rarefaction the approximate Godunov method will pick up the solution inside

## 11.2 Approximate-State Riemann Solvers

the rarefaction, along the $t$-axis, in which case the solution is exact and is given in closed form. From results of Chap. 6, for left and right *critical* or *sonic* rarefactions we obtain

$$\left.\begin{array}{l}\text{Left critical rarefaction:} \quad a = \tfrac{1}{3}(u_L + 2a_L) \;,\; u = \tfrac{1}{3}(u_L + 2a_L) \;, \\ \text{Right critical rarefaction:} \quad a = \tfrac{1}{3}(-u_R + 2a_R) \;,\; u = \tfrac{1}{3}(u_R - 2a_R) \;.\end{array}\right\} \quad (11.9)$$

Next we study four ways of finding approximate solutions for $h_*$ and $u_*$, and hence for $\psi_*$, in the Star Region.

### 11.2.2 A Primitive Variable Riemann Solver

We utilise the primitive variable formulation of the equations introduced in Chap. 4

$$\mathbf{W}_t + \mathbf{A}(\mathbf{W})\mathbf{W}_x = \mathbf{0} \;, \tag{11.10}$$

where

$$\mathbf{W} = \begin{bmatrix} h \\ u \\ \psi \end{bmatrix}, \quad \mathbf{A}(\mathbf{W}) = \begin{bmatrix} u & h & 0 \\ g & u & 0 \\ 0 & 0 & u \end{bmatrix} . \tag{11.11}$$

By assuming that the data states

$$\mathbf{W}_L = \begin{bmatrix} h_L \\ u_L \\ \psi_L \end{bmatrix}, \quad \mathbf{W}_R = \begin{bmatrix} h_R \\ u_R \\ \psi_R \end{bmatrix} \tag{11.12}$$

are sufficiently close to their mean value

$$\hat{\mathbf{W}} = \tfrac{1}{2}(\mathbf{W}_L + \mathbf{W}_R) \;, \tag{11.13}$$

we approximate the non-linear problem (11.10)–(11.11) by the linear constant coefficient system

$$\mathbf{W}_t + \hat{\mathbf{A}}\mathbf{W}_x = \mathbf{0} \;, \quad \hat{\mathbf{A}} = \mathbf{A}(\hat{\mathbf{W}}) \;. \tag{11.14}$$

We now solve this approximate system *exactly* by using any of the techniques studied in Chaps. 3, 4 and 5. The solution in the Star Region is

$$\left.\begin{array}{l} h_* = \tfrac{1}{2}(h_L + h_R) + \tfrac{1}{2}(u_L - u_R)/\overline{C} \;, \\ u_* = \tfrac{1}{2}(u_L + u_R) + \tfrac{1}{2}(h_L - h_R)\overline{C} \;, \\ \overline{C} = \sqrt{2g/(h_L + h_R)} \;. \end{array}\right\} \tag{11.15}$$

This Riemann solver is very simple indeed but not sufficiently robust to be used in all circumstances. This is particularly so for cases involving strong rarefactions leading to *very shallow water*; this primitive variable Riemann solver will give negative depths well before the depth positivity condition in the exact Riemann solver of Chap. 6 has been reached. A simple correction to the Riemann solver (11.15) produces a surprisingly robust scheme; this is the subject of the next section.

### 11.2.3 Riemann Solver Based on Exact Depth Positivity

Motivated by the simplicity of the approximate Riemann solver (11.15), we now present a new approximate Riemann solver that preserves the simplicity of (11.15), while adding two important new properties. The first property is that it can deal very well with situations involving very shallow water, produced perhaps, by strong rarefactions. As a matter of fact, the construction of the scheme is based on allowing the limiting case contained in the exact solution of the Riemann problem. The second property is that unlike the Riemann solver (11.15), the new scheme is found to be very robust in dealing with shock waves. From Chap. 6 we recall the *depth positivity condition*

$$u_R - u_L \leq 2(a_L + a_R) \,. \tag{11.16}$$

Allowing the constant $\overline{C}$ in (11.15) to be a parameter, yet to be found, and imposing the condition $h_* \geq 0$ in (11.15) leads to

$$u_R - u_L \leq \overline{C}(h_L + h_R) \,. \tag{11.17}$$

By comparing the right-hand sides of expressions (11.16) and (11.17), we obtain

$$\overline{C} = \frac{2(a_L + a_R)}{h_L + h_R} \,,$$

which leads to solutions $h_*$ and $u_*$ for depth and particle velocity in the Star Region, namely

$$\left. \begin{array}{l} h_* = \frac{1}{2}(h_L + h_R) - \frac{1}{4}(u_R - u_L)(h_L + h_R)/(a_L + a_R) \,, \\ u_* = \frac{1}{2}(u_L + u_R) - (h_R - h_L)(a_L + a_R)/(h_L + h_R) \,. \end{array} \right\} \tag{11.18}$$

This solver has, by construction, the same *depth-positivity condition* as the exact Riemann solver. Our limited experience suggests that this solver is also very robust for shock waves. The solution for the passive scalar $\psi(x, t)$ is given by (11.8) and in the presence of critical flow one uses the exact solution given by (11.9).

## 11.2.4 A Two-Rarefaction Riemann Solver

This approximate Riemann solver is based on the exact Riemann solver presented in Chap. 6. If in the depth function $f(h)$ one assumes the rarefaction branches, then a closed-form solution $h_*$ for $f(h) = 0$ is obtained. Under the same assumption, a closed-form solution is obtained for $u_*$. The result is

$$\left. \begin{array}{l} h_* = \dfrac{1}{g}\left[\dfrac{1}{2}(a_L + a_R) + \dfrac{1}{4}(u_L - u_R)\right]^2 , \\ u_* = \dfrac{1}{2}(u_L + u_R) + a_L - a_R . \end{array} \right\} \tag{11.19}$$

The solution for $\psi(x, t)$ is given by (11.8), and in the presence of critical flow one uses the exact solution given by (11.9).

## 11.2.5 A Two-Shock Riemann Solver

This approximate Riemann solver results from evaluating the depth function $f(h)$ in the exact Riemann solver of Chap. 6 under the assumption that both non-linear waves are shocks. We obtain

$$\left. \begin{array}{l} h_* = \dfrac{g_L(h_0)h_L + g_R(h_0)h_R + u_L - u_R}{g_L(h_0) + g_R(h_0)} , \\ u_* = \tfrac{1}{2}(u_L + u_R) + \tfrac{1}{2}[(h_* - h_R)g_R(h_0) - (h_* - h_L)g_L(h_0)] . \end{array} \right\} \tag{11.20}$$

Here

$$g_K(h_0) = \sqrt{\dfrac{1}{2}g\dfrac{h_0 + h_K}{h_0 h_K}} , \tag{11.21}$$

with $K = L$ or $K = R$ and $g_K$ evaluated at an estimate $h_0$ of $h_*$, which may be obtained from any of the previous, simpler approximations. I particularly recommend the choice $h_0 = h_*$ from (11.18) or (11.19). The solution for $\psi(x, t)$ is given by (11.8). In the presence of critical flow one uses the exact solution given by (11.9).

**Solution Sampling**

For the approximate-state Riemann solvers studied, once the star values are obtained approximately, one must then perform a solution sampling procedure to find the correct approximation to the Godunov state for flux evaluation (11.7). Such sampling involves the use of the solution for $\psi(x, t)$ as given by (11.8) and the detection of critical flow, for which one uses the exact solution given by (11.9). To implement the sampling procedure I suggest to follow Chap. 6.

Next we study approximate Riemann solvers in which an approximation to the numerical flux is obtained *directly*.

## 11.3 HLL Riemann Solvers

Harten, Lax and van Leer [5] suggested a way of solving the Riemann problem approximately by finding directly an approximation to the numerical flux $\mathbf{F}_{i+\frac{1}{2}}$. The mathematical bases of the approach are given in Chap. 10 of [1–3], where Riemann solvers for the three-dimensional Euler equations are derived.

The HLL (Harten, Lax and van Leer) approach assumes estimates $S_L$ and $S_R$ for the smallest and largest signal velocities in the solution of the Riemann problem with data $\mathbf{U}_L \equiv \mathbf{U}_i^n$, $\mathbf{U}_R \equiv \mathbf{U}_{i+1}^n$ and corresponding fluxes $\mathbf{F}_L \equiv \mathbf{F}(\mathbf{U}_L)$, $\mathbf{F}_R \equiv \mathbf{F}(\mathbf{U}_R)$; see Fig. 11.2. Intermediate waves, such as shear waves and contact discontinuities, arising when species equations are added to the basic shallow water equations, are ignored in this approach. By applying the integral form of the conservation laws in appropriate control volumes, one derives the flux $\mathbf{F}^{hll}$ for the intermediate region of Fig. 11.2. Therefore the HLL flux, after sampling, becomes

$$\mathbf{F}_{i+\frac{1}{2}} = \begin{cases} \mathbf{F}_L & \text{if } S_L \geq 0, \\ \mathbf{F}^{hll} \equiv \dfrac{S_R \mathbf{F}_L - S_L \mathbf{F}_R + S_R S_L (\mathbf{U}_R - \mathbf{U}_L)}{S_R - S_L} & \text{if } S_L \leq 0 \leq S_R, \\ \mathbf{F}_R & \text{if } S_R \leq 0. \end{cases} \quad (11.22)$$

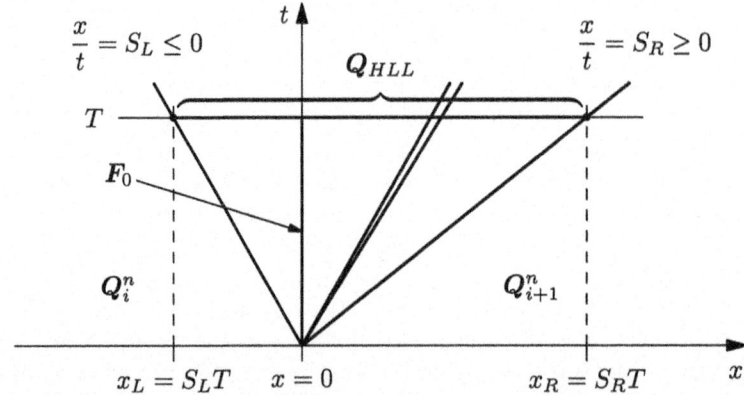

**Fig. 11.2** HLL two-wave model for the Riemann problem solution. Estimates $S_L$ and $S_R$ for the left and right waves need to be prescribed

## 11.4 HLLC Riemann Solvers

As to the wave speed estimates $S_L$ and $S_R$, there are several possible choices available. The positive experience gained from gas dynamics [6, 7] suggests the following choice

$$S_L = u_L - a_L q_L , \quad S_R = u_R + a_R q_R , \quad (11.23)$$

where $q_K$ ($K = L, R$) is given by

$$q_K = \begin{cases} \sqrt{\dfrac{1}{2}\left[\dfrac{(\hat{h}+h_K)\hat{h}}{h_K^2}\right]} & \text{if } \hat{h} > h_K , \\ 1 & \text{if } \hat{h} \leq h_K . \end{cases} \quad (11.24)$$

Here $\hat{h}$ is an estimate for the exact solution $h_*$ in the Star Region. Various choices for $\hat{h}$ suggest themselves from the previously given approximate-state Riemann solvers, but I strongly recommend (11.19). This is theoretically justified in the recent work of Toro et al. [8], where this choice is proved theoretically to bound the maximal and minimal wave speeds in the Riemann problem for the shallow water equations. See also the important, earlier works of Guermond and Popov [9].

The two-wave model ($W = 2$) of the HLL Riemann solver is *complete* for the purely one-dimensional problem without extra species-like equations. For the *augmented system* allowing for shear waves or species equations ($N = 3$) the HLL solver is *incomplete* and is thus inadequate; it ignores the middle wave. This results in excessive smearing of vortices and contact discontinuities; see Chap. 10 of [1–3]. This unsatisfactory situation motivates the introduction of the HLLC modification, which is discussed next.

## 11.4 HLLC Riemann Solvers

The HLLC approximate Riemann solver [6, 7] is a modification of the basic HLL scheme to account for the presence of intermediate waves. Full details on the HLLC approach are given in Chap. 10 of [1–3]. A review of HLLC is presented in [10], where extension to the shallow water equations is included. Figure 11.3 illustrates the assumed three-wave model of the HLLC Riemann solver.

There are now two distinct fluxes for the Star Region; compare this with the wave structure of the HLL solver of Fig. 11.2. In addition to the wave speed estimates $S_L$ and $S_R$ in the HLL solver, we now need an estimate $S_*$ for the speed of the middle wave. In the exact Riemann solver $S_* = u_*$. Assuming all wave speed estimates are available, we can write the HLLC numerical flux in sampled form as

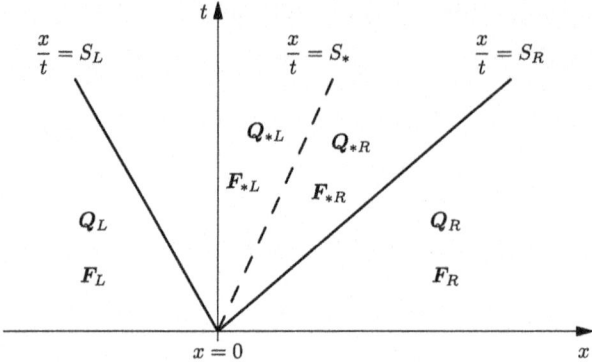

**Fig. 11.3** HLLC Riemann solver for $x$-split 2D shallow water equations

$$\mathbf{F}_{i+\frac{1}{2}}^{hllc} = \begin{cases} \mathbf{F}_L & \text{if} \quad 0 \leq S_L, \\ \mathbf{F}_{*L} & \text{if} \quad S_L \leq 0 \leq S_*, \\ \mathbf{F}_{*R} & \text{if} \quad S_* \leq 0 \leq S_R, \\ \mathbf{F}_R & \text{if} \quad 0 \geq S_R. \end{cases} \tag{11.25}$$

Here

$$\left. \begin{aligned} \mathbf{F}_{*L} &= \mathbf{F}_L + S_L(\mathbf{Q}_{*L} - \mathbf{Q}_L), \\ \mathbf{F}_{*R} &= \mathbf{F}_R + S_R(\mathbf{Q}_{*R} - \mathbf{Q}_R), \end{aligned} \right\} \tag{11.26}$$

in which the states $\mathbf{Q}_{*L}, \mathbf{Q}_{*R}$ are given by

$$\mathbf{Q}_{*K} = h_K \left( \frac{S_K - u_K}{S_K - S_*} \right) \begin{bmatrix} 1 \\ S_* \\ \psi_K \end{bmatrix}. \tag{11.27}$$

These are obtained from manipulating (11.26) and making assumptions consistent with the exact solution. See [10] for details.

The wave speed estimates for $S_L$ and $S_R$ are as in (11.23)–(11.24) for the HLL Riemann solver. Estimates for the middle wave speed $S_*$ can be provided in a variety of ways. For instance one can employ the approximate solution values for the particle speed $u_*$ given by (11.18), (11.19) or (11.20). An attractive choice is furnished by

$$S_* = \frac{S_L h_R(u_R - S_R) - S_R h_L(u_L - S_L)}{h_R(u_R - S_R) - h_L(u_L - S_L)}. \tag{11.28}$$

This is obtained from manipulations of (11.26). The first component of the first vector equation gives an expression for $h_{*L}$; similarly, the first component of the second vector equation gives an expression for $h_{*R}$. Assuming $h_{*L} = h_{*R}$, which is actually correct in the exact Riemann solver, expression (11.28) for $S_*$ follows. This wave speed estimate for $S_*$ has the remarkable property of being exact when one of the data states is a dry bed state, in which case $S_*$ is the speed of the wet/dry front; see Chap. 7. In Sect. 11.10 of this chapter we shall discuss the issue of approximate Riemann solvers for the dry-bed case in more detail.

## 11.5 The Rusanov and Lax-Friedrichs Schemes

The HLL approximate Riemann solver studied in Sect. 11.3 provides a useful framework to discuss the relationship between upwind and centred methods. Suppose a positive wave speed estimate $S^+$ is available. Then we set

$$S_L = -S^+, \quad S_R = S^+ . \tag{11.29}$$

Wave speed estimates for $S_L$ and $S_R$ were given in Sect. 11.3, from which a very simple estimate for $S^+$ is

$$S^+ = \max\{|S_L|, |S_R|\} . \tag{11.30}$$

By substituting the speeds $S_L$ and $S_R$ defined by (11.29)–(11.30) into the expression for $\mathbf{F}^{hll}$ in (11.22) we obtain the Rusanov flux

$$\mathbf{F}^{Rus}_{i+\frac{1}{2}} = \frac{1}{2}(\mathbf{F}_L + \mathbf{F}_R) - \frac{1}{2}S^+(\mathbf{Q}_R - \mathbf{Q}_L) . \tag{11.31}$$

This upwind scheme has the lowest explicit level of upwinding. The approximate structure for the corresponding Riemann problem contains a single wave. Rusanov's method is based on a one-wave model and is therefore an incomplete Riemann solver for any system of PDEs.

The choice of $S^+$ for (11.31) is restricted; it must bound all signals present at the time level $n$. For stability one requires

$$S^+ \leq \frac{\Delta x}{\Delta t} . \tag{11.32}$$

Here $\frac{\Delta x}{\Delta t}$ is the *mesh speed* and must bound all other wave speeds. Choosing the largest possible speed consistent with stability, namely

$$S^+ = \frac{\Delta x}{\Delta t} , \tag{11.33}$$

results in the Lax-Friedrichs flux

$$\mathbf{F}^{LF}_{i+\frac{1}{2}} = \frac{1}{2}(\mathbf{F}_L + \mathbf{F}_R) - \frac{1}{2}\frac{\Delta x}{\Delta t}(\mathbf{Q}_R - \mathbf{Q}_L) . \tag{11.34}$$

The Lax-Friedrichs method, a centred method, can then be seen as the extreme case of a Godunov method with the lowest possible *level of upwinding*. See Chap. 10.

## 11.6 Roe's Approximate Riemann Solver

Roe's approximate Riemann solver was first constructed for the Euler equations [11]. Details are given in Chap. 11 of [1–3]. Glaister [12] appears to be the first to have applied the Roe approach to the shallow water equations. He followed the Roe-Pike approach [13] to derive the approximate Riemann solver. See also [14].

### 11.6.1 The Basic Scheme

The Roe Riemann solver begins by approximating the non-linear system in (11.4),

$$\partial_t \mathbf{Q} + \partial_x \mathbf{F}(\mathbf{Q}) \equiv \partial_t \mathbf{Q} + \mathbf{A}(\mathbf{Q}) \partial_x \mathbf{Q} = \mathbf{0} \;, \tag{11.35}$$

where $\mathbf{A}(\mathbf{Q})$ is the Jacobian matrix, by a linear system with constant coefficients, namely

$$\partial_t \mathbf{Q} + \tilde{\mathbf{A}} \partial_x \mathbf{Q} = \mathbf{0} \;. \tag{11.36}$$

The *constant* coefficient matrix $\tilde{\mathbf{A}}$, the Roe matrix, is an approximation to the Jacobian matrix $\mathbf{A}$, and is found in terms of the data states $\mathbf{Q}_L$ and $\mathbf{Q}_R$ of the Riemann problem. In practice it is more convenient to work with the primitive variable vectors $\mathbf{W}_L$ and $\mathbf{W}_R$, with $\mathbf{W} = [h, u, \psi]^T$, where $\psi$ is a passive scalar.

The basic step in the Roe approximate Riemann solver is to find average values $\tilde{h}$, $\tilde{a}$, $\tilde{u}$ and $\tilde{\psi}$ for the depth $h$, the celerity $a$, the velocity component $u$ and the passive scalar $\psi$. The Roe-Pike approach [13] yields the following averages, called the Roe averages:

$$\left. \begin{array}{l} \tilde{u} = \dfrac{u_L \sqrt{h_L} + u_R \sqrt{h_R}}{\sqrt{h_L} + \sqrt{h_L}} \;, \\[6pt] \tilde{\psi} = \dfrac{\psi_L \sqrt{h_L} + \psi_R \sqrt{h_R}}{\sqrt{h_L} + \sqrt{h_L}} \;, \\[6pt] \tilde{h} = \sqrt{h_L h_R} \;, \\[6pt] \tilde{a} = \sqrt{\frac{1}{2}(a_L^2 + a_R^2)} \;. \end{array} \right\} \tag{11.37}$$

The average eigenvalues are then

$$\tilde{\lambda}_1 = \tilde{u} - \tilde{a} \;, \quad \tilde{\lambda}_2 = \tilde{u} \;, \quad \tilde{\lambda}_3 = \tilde{u} + \tilde{a} \;, \tag{11.38}$$

## 11.6 Roe's Approximate Riemann Solver

with corresponding right eigenvectors

$$\tilde{\mathbf{K}}^{(1)} = \begin{bmatrix} 1 \\ \tilde{u} - \tilde{a} \\ \tilde{\psi} \end{bmatrix}, \quad \tilde{\mathbf{K}}^{(2)} = \begin{bmatrix} 0 \\ 0 \\ 1 \end{bmatrix}, \quad \tilde{\mathbf{K}}^{(3)} = \begin{bmatrix} 1 \\ \tilde{u} + \tilde{a} \\ \tilde{\psi} \end{bmatrix}. \quad (11.39)$$

The *wave strengths* $\tilde{\alpha}_j$, in terms of the Roe averages, are

$$\left. \begin{aligned} \tilde{\alpha}_1 &= \tfrac{1}{2}[\Delta h - \frac{\tilde{h}}{\tilde{a}}\Delta u], \\ \tilde{\alpha}_2 &= \tilde{h}\Delta\psi, \\ \tilde{\alpha}_3 &= \tfrac{1}{2}[\Delta h + \frac{\tilde{h}}{\tilde{a}}\Delta u]. \end{aligned} \right\} \quad (11.40)$$

The computation of the corresponding Godunov intercell numerical flux with the Roe solver is accomplished via the direct application of the theory of linear systems with constant coefficients; see Chap. 3. The relevant state along the $t$-axis is found to be

$$\mathbf{Q}_{i+\frac{1}{2}}(0) = \mathbf{Q}_i^n + \sum_{\tilde{\lambda}_j \leq 0} \tilde{\alpha}_j \tilde{\mathbf{R}}^{(j)}, \quad (11.41)$$

or

$$\mathbf{Q}_{i+\frac{1}{2}}(0) = \mathbf{Q}_{i+1}^n - \sum_{\tilde{\lambda}_j \geq 0} \tilde{\alpha}_j \tilde{\mathbf{R}}^{(j)}. \quad (11.42)$$

It may also ba taken as a mean value of the above one-sided expressions

$$\mathbf{Q}_{i+\frac{1}{2}}(0) = \frac{1}{2}(\mathbf{Q}_i^n + \mathbf{Q}_{i+1}^n) - \frac{1}{2}\sum_{j=1}^{3} \text{sign}(\tilde{\lambda}_j)\tilde{\alpha}_j \tilde{\mathbf{R}}^{(j)}. \quad (11.43)$$

The corresponding sought *numerical flux* is found to be

$$\mathbf{F}_{i+\frac{1}{2}} = \mathbf{F}(\mathbf{Q}_i^n) + \sum_{\tilde{\lambda}_j \leq 0} \tilde{\alpha}_j \tilde{\lambda}_j \tilde{\mathbf{R}}^{(j)}, \quad (11.44)$$

or

$$\mathbf{F}_{i+\frac{1}{2}} = \mathbf{F}(\mathbf{Q}_{i+1}^n) - \sum_{\tilde{\lambda}_j \geq 0} \tilde{\alpha}_j \tilde{\lambda}_j \tilde{\mathbf{R}}^{(j)}, \quad (11.45)$$

or

$$\mathbf{F}_{i+\frac{1}{2}} = \frac{1}{2}\left[\mathbf{F}(\mathbf{Q}_i^n) + \mathbf{F}(\mathbf{Q}_{i+1}^n)\right] - \frac{1}{2}\sum_{j=1}^{3} \tilde{\alpha}_j |\tilde{\lambda}_j| \tilde{\mathbf{R}}^{(j)}. \quad (11.46)$$

This numerical flux used in conjunction with the conservative formula (11.2), will give a Godunov-type method of first-order accuracy. Being a linearised solver, the Roe solver requires an *entropy fix* to be used in practice; this is the subject of the next section.

## 11.6.2 Entropy Fix for the Roe Solver

Here we present the Harten and Hyman entropy fix [15], details of which as applied to the Euler equations are given in Chap. 11 of [1–3]. See also the book by LeVeque [16].

Suppose the Roe approximate eigenvalue $\tilde{\lambda}_1 = \tilde{u} - \tilde{a}$ corresponding to the left non-linear wave is close to zero ($|\tilde{\lambda}_1| \approx 0$). Then the wave can be a slowly moving shock wave or a *sonic* or *critical* rarefaction. It is the rarefactions that require the entropy fix. An obvious way of discriminating between the two cases is to compare the characteristic speeds on the left and right of the wave. We compute

$$\lambda_{1L} = u_L - a_L , \quad \lambda_{1R} = u_* - a_* . \tag{11.47}$$

We note that finding $\lambda_{1R}$ requires an estimate for the state $\mathbf{Q}_*$ in the Star Region; details to be given later.

The entropy fix is needed only if the left wave is a rarefaction, that is $\lambda_{1L} \leq \lambda_{1R}$, and is *sonic* or *critical*, namely

$$\lambda_{1L} = u_L - a_L < 0 , \quad \lambda_{1R} = u_* - a_* > 0 . \tag{11.48}$$

In order to implement the entropy fix, one splits the total jump across the left wave propagating at speed $\tilde{\lambda}_1$ into two smaller jumps propagating with speeds $\lambda_{1L}$ and $\lambda_{1R}$. As a result, the Roe average eigenvalue $\tilde{\lambda}_1$ appearing in the expression (11.46) for the intercell flux is replaced by the modified, averaged eigenvalue

$$\bar{\lambda}_1 = \tilde{\lambda}_1 \left[ \frac{\lambda_{1R} - \tilde{\lambda}_1}{\lambda_{1R} - \lambda_{1L}} \right] . \tag{11.49}$$

For a right sonic rarefaction of the family $\lambda_3 = u + a$, the procedure is entirely analogous. If $|\lambda_3|$ is close to zero, we consider the speeds

$$\lambda_{3L} = u_* + a_* , \quad \lambda_{3R} = u_R + a_R . \tag{11.50}$$

If in addition $\lambda_{3L} < 0$ and $\lambda_{3R} > 0$ then a right *sonic* or *critical* rarefaction is present and the entropy fix must be enforced by replacing the Roe average eigenvalue $\tilde{\lambda}_3$ appearing in the expression (11.46) for the intercell flux by the *modified* eigenvalue

$$\bar{\lambda}_3 = \tilde{\lambda}_3 \left[ \frac{\lambda_{3R} - \tilde{\lambda}_3}{\lambda_{3R} - \lambda_{3L}} \right] . \qquad (11.51)$$

The implementation of the entropy fix requires estimates for the state $\mathbf{Q}_*$ in the Star Region. In particular, we require $u_*$ and $a_*$ in (11.48) and (11.50). There are several ways of obtaining these values. One obvious possibility is to compute the state $\mathbf{Q}_*$ arising from the Roe linearisation by using the methods of Chap. 3 for linear systems with constant coefficients. Alternatively, we may use approximate-state Riemann solvers from Sect. 11.2.

A final point related to the entropy fix concerns a criterion for deciding when $|\tilde{\lambda}_1|$ and $|\tilde{\lambda}_3|$ are small. The magnitude of a wave speed is best assessed with reference to the distances and times involved. An obvious possibility is to relate $\tilde{\lambda}_k$ to the mesh sizes $\Delta x$, $\Delta t$ by defining a Courant number

$$\tilde{c}_k = \frac{|\tilde{\lambda}_k| \Delta t}{\Delta x} . \qquad (11.52)$$

We regard $|\tilde{\lambda}_k|$ as being small if

$$\tilde{c}_k \leq TOL , \qquad (11.53)$$

where $TOL$ is some small positive number. Recall that the Courant number is effectively the proportion of the cell width $\Delta x$ covered in time $\Delta t$. Therefore, as a rule of thumb I would suggest using $TOL = 0.1$.

A strong point of the Roe solver is that it is complete, that is the number $W$ of waves in its wave model matches $N$, the number of waves in the exact Riemann solver. As it stands, however, the Roe solver cannot strictly handle situations involving dry bed states. The trick of *wetting the bed* a little might seem to work but it is incorrect. This point will be discussed in more detail in Sect. 11.10. See the work of Dodd [17].

## 11.7 The Riemann Solver of Osher and Solomon

The Osher-Solomon approximate Riemann solver [18, 19] was presented for a general system of hyperbolic conservation laws in Chap. 12 of [1–3]. Details of the solver for the Burgers equation, for the isothermal equations and for the time-dependent Euler equations are also given there. The scheme for the shallow water equations is given here in a very succinct form.

Let us assume a hyperbolic system of $N$ equations in (11.1). The Osher-Solomon Riemann solver begins by splitting the Jacobian matrix $\mathbf{A}(\mathbf{Q})$ as

$$\mathbf{A}(\mathbf{Q}) = \mathbf{A}^+(\mathbf{Q}) + \mathbf{A}^-(\mathbf{Q}) , \qquad (11.54)$$

where $\mathbf{A}^+(\mathbf{Q})$ has *non-negative* eigenvalues and $\mathbf{A}^-(\mathbf{Q})$ has *non-positive* eigenvalues. The scheme then assumes that there exist vector-valued functions $\mathbf{F}^+(\mathbf{Q})$ and $\mathbf{F}^-(\mathbf{Q})$ that satisfy

$$\mathbf{F}(\mathbf{Q}) = \mathbf{F}^+(\mathbf{Q}) + \mathbf{F}^-(\mathbf{Q}) \tag{11.55}$$

and

$$\frac{\partial \mathbf{F}^+}{\partial \mathbf{Q}} = \mathbf{A}^+(\mathbf{Q}) \;, \quad \frac{\partial \mathbf{F}^-}{\partial \mathbf{Q}} = \mathbf{A}^-(\mathbf{Q}) \;. \tag{11.56}$$

For given initial data $\mathbf{Q}_i^n$ and $\mathbf{Q}_{i+1}^n$, it is conventional in the Osher-Solomon scheme to define

$$\mathbf{Q}_0 \equiv \mathbf{Q}_i^n \;, \quad \mathbf{Q}_1 \equiv \mathbf{Q}_{i+1}^n \;. \tag{11.57}$$

The numerical flux $\mathbf{F}_{i+\frac{1}{2}}$ is then taken as

$$\mathbf{F}_{i+\frac{1}{2}} = \mathbf{F}^+(\mathbf{Q}_0) + \mathbf{F}^-(\mathbf{Q}_1) \;. \tag{11.58}$$

Using the relations in phase space

$$\int_{\mathbf{Q}_0}^{\mathbf{Q}_1} \mathbf{A}^-(\mathbf{Q}) \, d\mathbf{Q} = \mathbf{F}^-(\mathbf{Q}_1) - \mathbf{F}^-(\mathbf{Q}_0)$$

and

$$\int_{\mathbf{Q}_0}^{\mathbf{Q}_1} \mathbf{A}^+(\mathbf{Q}) \, d\mathbf{Q} = \mathbf{F}^+(\mathbf{Q}_1) - \mathbf{F}^+(\mathbf{Q}_0) \;,$$

one can express (11.58) in three different forms, namely

$$\mathbf{F}_{i+\frac{1}{2}} = \mathbf{F}(\mathbf{Q}_0) + \int_{\mathbf{Q}_0}^{\mathbf{Q}_1} \mathbf{A}^-(\mathbf{Q}) \, d\mathbf{Q} \;, \tag{11.59}$$

or

$$\mathbf{F}_{i+\frac{1}{2}} = \mathbf{F}(\mathbf{Q}_1) - \int_{\mathbf{Q}_0}^{\mathbf{Q}_1} \mathbf{A}^+(\mathbf{Q}) \, d\mathbf{Q} \;, \tag{11.60}$$

or the average

$$\mathbf{F}_{i+\frac{1}{2}} = \frac{1}{2}[\mathbf{F}(\mathbf{Q}_0) + \mathbf{F}(\mathbf{Q}_1)] - \frac{1}{2}\int_{\mathbf{Q}_0}^{\mathbf{Q}_1} |\mathbf{A}(\mathbf{Q})| \, d\mathbf{Q} \;. \tag{11.61}$$

Here we shall adopt (11.59) for the Osher-Solomon flux. The integral requires an integration path in phase space. For the Osher-Solomon scheme one selects *partial* integration paths $I_1(\mathbf{Q})$, $I_2(\mathbf{Q})$ and $I_3(\mathbf{Q})$ such that they intersect at a single intersection point

$$\mathbf{Q}_{i/N} = I_i(\mathbf{Q}) \cap I_{i+1}(\mathbf{Q}) \tag{11.62}$$

## 11.7 The Riemann Solver of Osher and Solomon

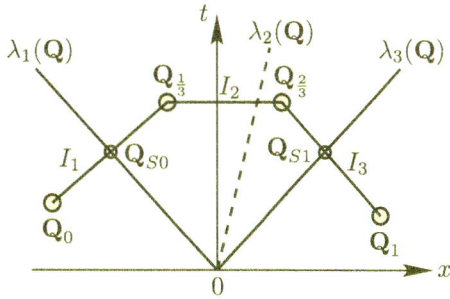

**Fig. 11.4** Integration paths, intersection points and sonic points for the Osher-Solomon Riemann solver applied to the augmented shallow water equations (N = 3)

and the total integration path is

$$I(\mathbf{Q}) = I_1(\mathbf{Q}) \cup I_2(\mathbf{Q}) \cup I_3(\mathbf{Q}) \,. \tag{11.63}$$

Figure 11.4 illustrates the choice of integration paths for the case of the *augmented* shallow water equations. The vectors $\mathbf{Q}_0$, $\mathbf{Q}_{1/3}$, $\mathbf{Q}_{2/3}$ and $\mathbf{Q}_1$ are vectors in phase space and can be thought of as being the four constant states arising in the exact solution to the corresponding Riemann problem represented in $x$-$t$ space, as in Fig. 11.4. The points $\mathbf{Q}_{S0}$ and $\mathbf{Q}_{S1}$ are representations of potential *sonic points* that may arise from the non-linear fields associated with the eigenvalues $\lambda_1(\mathbf{Q})$ and $\lambda_3(\mathbf{Q})$.

The determination of the *intersection points* $\mathbf{Q}_{i/N}$ and the *sonic points* $\mathbf{Q}_{S0}$ and $\mathbf{Q}_{S1}$ requires, to some extent, information about the solution of the Riemann problem. We find these points using Riemann invariants; see Chap. 4. In order to compute the numerical flux $\mathbf{F}_{i+\frac{1}{2}}$ using the Osher-Solomon Riemann solver, there are a total of 16 cases to take care of. These are given in Table 11.1. We omit the details of the derivation. The intersection points $\mathbf{Q}_{\frac{1}{3}}$ and $\mathbf{Q}_{\frac{2}{3}}$ are found from applying Riemann invariants, see Chap. 4. These give the solution for the unknown vectors

$$\mathbf{W}_{\frac{1}{3}} = \begin{bmatrix} h_{\frac{1}{3}} \\ u_{\frac{1}{3}} \\ \psi_{\frac{1}{3}} \end{bmatrix}, \quad \mathbf{W}_{\frac{2}{3}} = \begin{bmatrix} h_{\frac{2}{3}} \\ u_{\frac{2}{3}} \\ \psi_{\frac{2}{3}} \end{bmatrix}. \tag{11.64}$$

as

$$\left.\begin{array}{l} h_{\frac{1}{3}} = h_{\frac{2}{3}} = [\frac{1}{2}(a_L + a_R) - \frac{1}{4}(u_R - u_L)]^2/g \,, \\ u_{\frac{1}{3}} = u_{\frac{2}{3}} = \frac{1}{2}(u_L + u_R) + a_L - a_R \,, \\ \psi_{\frac{1}{3}} = \psi_0 \,, \\ \psi_{\frac{2}{3}} = \psi_1 \,. \end{array}\right\} \tag{11.65}$$

**Table 11.1** The Osher-Solomon flux formulae for the *augmented* shallow water equations using P-ordering of integration paths, $\mathbf{F}_k \equiv \mathbf{F}(\mathbf{Q}_k)$

|  | $u_0 - a_0 \geq 0$<br>$u_1 + a_1 \geq 0$ | $u_0 - a_0 \geq 0$<br>$u_1 + a_1 \leq 0$ | $u_0 - a_0 \leq 0$<br>$u_1 + a_1 \geq 0$ | $u_0 - a_0 \leq 0$<br>$u_1 + a_1 \leq 0$ |
|---|---|---|---|---|
| $u^* \geq 0$<br>$u^* - a_{\frac{1}{3}} \geq 0$ | $\mathbf{F}_0$ | $\mathbf{F}_0 + \mathbf{F}_1$<br>$-\mathbf{F}_{S1}$ | $\mathbf{F}_{S0}$ | $\mathbf{F}_{S0} - \mathbf{F}_{S1}$<br>$+\mathbf{F}_1$ |
| $u^* \geq 0$<br>$u^* - a_{\frac{1}{3}} \leq 0$ | $\mathbf{F}_0 - \mathbf{F}_{S0}$<br>$+\mathbf{F}_{\frac{1}{3}}$ | $\mathbf{F}_0 - \mathbf{F}_{S0}$<br>$+\mathbf{F}_{\frac{1}{3}} - \mathbf{F}_{S1}$<br>$+\mathbf{F}_1$ | $\mathbf{F}_{\frac{1}{3}}$ | $\mathbf{F}_1 + \mathbf{F}_{\frac{1}{3}}$<br>$-\mathbf{F}_{S1}$ |
| $u^* \leq 0$<br>$u^* + a_{\frac{2}{3}} \geq 0$ | $\mathbf{F}_0 - \mathbf{F}_{S0}$<br>$+\mathbf{F}_{\frac{2}{3}}$ | $\mathbf{F}_0 - \mathbf{F}_{S0}$<br>$+\mathbf{F}_{\frac{2}{3}} - \mathbf{F}_{S1}$<br>$+\mathbf{F}_1$ | $\mathbf{F}_{\frac{2}{3}}$ | $\mathbf{F}_{\frac{2}{3}} - \mathbf{F}_{S1}$<br>$+\mathbf{F}_1$ |
| $u^* \leq 0$<br>$u^* + a_{\frac{2}{3}} \leq 0$ | $\mathbf{F}_0 - \mathbf{F}_{S0}$<br>$+\mathbf{F}_{S1}$ | $\mathbf{F}_0 - \mathbf{F}_{S0}$<br>$+\mathbf{F}_1$ | $\mathbf{F}_{S1}$ | $\mathbf{F}_1$ |

The sonic points are found to be

$$\left.\begin{array}{l} u_{S0} = \frac{1}{3}(u_0 + 2a_0)\,, \\ a_{S0} = u_{S0}\,, \\ h_{S0} = a_{S0}^2/g\,, \\ \psi_{S0} = \psi_0\,, \end{array}\right\} \quad (11.66)$$

and

$$\left.\begin{array}{l} u_{S1} = \frac{1}{3}(u_1 - 2a_1)\,, \\ a_{S1} = -u_{S1}\,, \\ h_{S1} = a_{S1}^2/g\,, \\ \psi_{S1} = \psi_1\,. \end{array}\right\} \quad (11.67)$$

In order to compute the Godunov flux (11.3) using the Osher-Solomon Riemann solver, one must form the flux as the summation of all the *partial* fluxes in the appropriate entry of Table 11.1. Clearly this is a rather expensive process, as first one needs to identify the correct case out of 16 possible cases, then compute all the appropriate partial fluxes and then form the sum. The worst cases have 5 partial fluxes.

We remark that the Osher-Solomon scheme was originally presented in terms of a different ordering of the integration paths, the so called O-ordering, or Original ordering. Computational practice in the field of compressible gas dynamics suggests that the P-ordering, or Physical ordering, presented here leads to better schemes but there appears to be little experience in using the Osher-Solomon scheme for the shallow water equations for any of the two possible ordering of the integration paths.

Full details of the original Osher-Solomon Riemann solver are found in Chap. 12 of Toro [1–3]. Applications of the Osher-Solomon approach to the shallow water equations have been reported by Zhao et al. [20, 21].

We conclude by noting that the Osher-Solomon approximate Riemann solvers is non-linear and complete. In the next section we present a simplification of the original Osher-Solomon Riemann solver that is more generally applicable, including systems of non-linear equations in non-conservative form.

## 11.8 The Dumbser-Osher-Toro Riemann Solver: DOT

Here we present a modification of the Osher-Solomon Riemann solver [19] due to Dumbser and Toro [22, 23]. Such modification makes the approach much more practical and applicable to any hyperbolic system for which the complete eigenstructure is known, either analytically of numerically. The resulting scheme is non-linear and complete.

### 11.8.1 Notation

Consider the Riemann problem for the PDE system in (11.1) of $N$ equations with initial condition

$$\mathbf{Q}(x,0) = \begin{cases} \mathbf{Q}_0 & \text{if } x < x_{i+\frac{1}{2}}, \\ \mathbf{Q}_1 & \text{if } x > x_{i+\frac{1}{2}}. \end{cases} \quad (11.68)$$

Recall that hyperbolicity of the system in (11.1) is equivalent to saying that the Jacobian matrix $\mathbf{A}(\mathbf{Q})$ of the flux $\mathbf{F}(\mathbf{Q})$ is diagonalizable, that is

$$\mathbf{A}(\mathbf{Q}) = \mathbf{R}(\mathbf{Q})\Lambda(\mathbf{Q})\mathbf{R}^{-1}(\mathbf{Q}), \quad (11.69)$$

where $\mathbf{R}(\mathbf{Q})$ is the matrix formed by the right eigenvectors $\mathbf{R}_i(\mathbf{Q})$, $\mathbf{R}^{-1}(\mathbf{Q})$ is its inverse and $\Lambda(\mathbf{Q})$ is the diagonal matrix whose diagonal entries are the eigenvalues $\lambda_i(\mathbf{Q})$. We introduce the definitions

$$\lambda_i^+(\mathbf{Q}) = \max(\lambda_i(\mathbf{Q}), 0), \quad \lambda_i^-(\mathbf{Q}) = \min(\lambda_i(\mathbf{Q}), 0) \quad (11.70)$$

and consider the associated diagonal matrices $\Lambda^+(\mathbf{Q})$, $\Lambda^-(\mathbf{Q})$ and $|\Lambda^-(\mathbf{Q})|$, whose diagonal entries are $\lambda_i^+(\mathbf{Q})$, $\lambda_i^-(\mathbf{Q})$ and $|\lambda_i(\mathbf{Q})|$ respectively. Note that

$$|\lambda_i(\mathbf{Q})| = \lambda_i^+(\mathbf{Q}) - \lambda_i^-(\mathbf{Q}) \quad (11.71)$$

and hence
$$|A(\mathbf{Q})| = \Lambda^+(\mathbf{Q}) - \Lambda^-(\mathbf{Q}) . \tag{11.72}$$

Then we introduce
$$|\mathbf{A}(\mathbf{Q})| = \mathbf{R}(\mathbf{Q})|\Lambda(\mathbf{Q})|\mathbf{R}^{-1}(\mathbf{Q}) . \tag{11.73}$$

As seen in (11.61) one choice for the Osher-Solomon [19] numerical flux is the average
$$\mathbf{F}_{i+\frac{1}{2}} = \frac{1}{2}[\mathbf{F}(\mathbf{Q}_0) + \mathbf{F}(\mathbf{Q}_1)] - \frac{1}{2}\int_{\mathbf{Q}_0}^{\mathbf{Q}_1} |\mathbf{A}(\mathbf{Q})|d\mathbf{Q} . \tag{11.74}$$

The flux can also be defined by one-sided expressions, such as (11.59) and (11.60). As a matter of fact, one-sided implementations are bound to be more efficient.

The flux requires the evaluation of an integral in phase space, which depends on the chosen integration path joining $\mathbf{Q}_0$ to $\mathbf{Q}_1$. Originally, Osher and Solomon proposed two ways of choosing integration paths *so as to make the actual integration tractable*, (a) the P-ordering and (b) the O-ordering. However, the analytical calculations to be performed are still too involved for general hyperbolic systems. See Sect. 11.7. Full details of the original Osher-Solomon Riemann solver are found in Chap. 12 of Toro [1–3].

### 11.8.2 The DOT Numerical Flux

Dumbser and Toro [22, 23] made two simple but effective suggestions: (i) choose any path, without considerations regarding computational tractability of the scheme; (ii) evaluate matrices by numerical integration in phase space. The simplest path to evaluate the integral in (11.74) is the *canonical path*

$$\phi(s; \mathbf{Q}_0, \mathbf{Q}_1) = \mathbf{Q}_0 + s(\mathbf{Q}_1 - \mathbf{Q}_0) , \quad s \in [0, 1] . \tag{11.75}$$

Obviously, other choices are available. Then, under a change of variables we obtain

$$\mathbf{F}_{i+\frac{1}{2}} = \frac{1}{2}(\mathbf{F}(\mathbf{Q}_0) + \mathbf{F}(\mathbf{Q}_1)) - \frac{1}{2}\left(\int_0^1 |\mathbf{A}(\phi(s; \mathbf{Q}_0, \mathbf{Q}_1))|\,ds\right)(\mathbf{Q}_1 - \mathbf{Q}_0) . \tag{11.76}$$

Finally, the integral in (11.76) is computed *numerically* along the path $\phi$ using a Gauss-type quadrature rule with $G$ points $s_j$ and associated weights $\omega_j$ in the unit interval $I = [0, 1]$. We obtain

$$\mathbf{F}_{i+\frac{1}{2}} = \frac{1}{2}(\mathbf{F}(\mathbf{Q}_0) + \mathbf{F}(\mathbf{Q}_1)) - \frac{1}{2}\left(\sum_{j=1}^{G}\omega_j\left|\mathbf{A}(\phi(s_j;\mathbf{Q}_0,\mathbf{Q}_1))\right|\right)(\mathbf{Q}_1 - \mathbf{Q}_0) \,.$$
(11.77)

Note that $\left|\mathbf{A}(\phi(s_j;\mathbf{Q}_0,\mathbf{Q}_1))\right|$ must be decomposed as in (11.73) for each $s_j$.

For details on the definition of the *absolute value matrix* are given in the original references [22, 23]. Recommended choices for the computational parameters in (11.77) are $G = 3$ and

$$s_1 = \frac{1}{2} - \frac{\sqrt{15}}{10} \,,\, s_2 = \frac{1}{2} \,,\, s_3 = \frac{1}{2} + \frac{\sqrt{15}}{10} \,,\, \omega_1 = \frac{5}{18} \,,\, \omega_2 = \frac{8}{18} \,,\, \omega_3 = \frac{5}{18} \,.$$
(11.78)

The upwind DOT method has been applied in several areas of study, including the global, closed loop model for the human circulation in [24–27].

**Remarks on the DOT Scheme:**

1. The complete eigenstructure of the system is needed and is used at each integration point in (11.77).
2. The scheme is *non-linear and complete*, as it contains all characteristic fields of the exact problem.
3. The scheme is very general. The original version of Osher and Solomon was restricted to very simple hyperbolic systems.
4. The new DOT scheme also applies to non-conservative hyperbolic systems [22, 23].

We have completed the presentation of approximate Riemann solvers for conservative systems. In the next section we deal with the related subject of schemes for systems written in non-conservative form.

## 11.9 Path-Conservative Methods

Path-conservative methods [28] have arisen as an effective approach to solve systems written in non-conservative form. In this section we give a brief introduction to this methodology, still in development.

### 11.9.1 Non-conservative Methods

We are concerned with systems of PDEs written in the non-conservative form

$$\partial_t \mathbf{Q} + \mathbf{A}(\mathbf{Q})\partial_x \mathbf{Q} = \mathbf{S}(\mathbf{Q}) \,.$$
(11.79)

Here **Q** could be the vector of physically conserved variables but the system is still written in non-conservative form. As a matter of fact, conservative systems via the chain rule can be expressed as (11.79), where **A(Q)** is the Jacobian matrix of a physical flux function. However there are systems that lack a physical flux function and these cannot be written in conservation-law form, even if the vector of unknowns **Q** is the vector of physically conserved variables and the equations have been derived from physical conservation principles. The vector of unknowns **Q** in system (11.79) could be a vector of primitive or physical variables.

It is well known that *primitive variable* schemes, even in conservative form, and *non-conservative* schemes will compute shock waves with the wrong strength and thus the wrong speed of propagation. See Chap. 5 for a proof of this for the case of shocks in the shallow water equations. For smooth flows, these schemes may be adequate, depending on the details of the particular scheme. There is a large body of work concerned with primitive-variable schemes and their combination with shock-fitting techniques; see for instance the works of Moretti and co-workers [29–32]. Upwind TVD primitive variable schemes have been constructed by Karni [33], Toro [34–37] and by Abgrall [38]. See also the PRICE scheme [39].

In what follows we briefly introduce a new class of schemes for solving PDEs in non-conservative form (11.79). These schemes depend on the choice of a path $\phi$ in phase space and have been termed *path-conservative schemes*. For details see [28] and the many references therein.

### 11.9.2  The Framework

The update formula for a *high-order* path-conservative scheme to solve (11.79) is

$$\mathbf{Q}_i^{n+1} = \mathbf{Q}_i^n - \frac{\Delta t}{\Delta x}[\mathbf{D}_{i-\frac{1}{2}}^+ + \mathbf{D}_{i+\frac{1}{2}}^-] + \Delta t \mathbf{S}_i - \Delta t \mathbf{H}_i^n \ . \tag{11.80}$$

Here $\mathbf{D}_{i+\frac{1}{2}}^-$ and $\mathbf{D}_{i+\frac{1}{2}}^+$ are called *fluctuations, or increments* [40], and are related to the numerical flux in the case of a conservative system. $\mathbf{S}_i$ is the usual numerical source. The term $\mathbf{H}_i^n$ is present in schemes (11.80) only if the accuracy is greater than one, and is defined as

$$\mathbf{H}_i^n = \frac{1}{\Delta t \Delta x} \int_{t_n}^{t_{n+1}} \int_{x_{i-\frac{1}{2}}}^{x_{i+\frac{1}{2}}} \mathbf{A}\left(\mathbf{P}_i(x,t)\right) \partial_x \mathbf{P}_i(x,t) \, dx dt \ . \tag{11.81}$$

Here $\mathbf{P}_i(x,t)$ is a polynomial representation of the solution in cell $I_i$, which results from a *spatial reconstruction procedure*, see Chap. 14. In the first-order case the reconstruction reduces to piece-wise constant cell averages; hence $\partial_x \mathbf{P}_i(x,t) = \mathbf{0}$ and therefore $\mathbf{H}_i^n = \mathbf{0}$ in (11.80). For convenience we also neglect the source term.

## 11.9 Path-Conservative Methods

The fluctuations $\mathbf{D}^-_{i+\frac{1}{2}}$ and $\mathbf{D}^+_{i+\frac{1}{2}}$, accounting for the absence of a flux function, are expected to satisfy the *consistency condition*

$$\mathbf{D}^-_{i+\frac{1}{2}}(\mathbf{Q},\dots,\mathbf{Q}) = \mathbf{0}\,,\quad \mathbf{D}^+_{i+\frac{1}{2}}(\mathbf{Q},\dots,\mathbf{Q}) = \mathbf{0} \qquad (11.82)$$

and the *compatibility condition*

$$\mathbf{D}^-_{i+\frac{1}{2}} + \mathbf{D}^+_{i+\frac{1}{2}} = \int_0^1 \mathbf{A}(\phi(s;\mathbf{Q}^n_i,\mathbf{Q}^n_{i+1}))\frac{\partial}{\partial s}\phi(s;\mathbf{Q}^n_i,\mathbf{Q}^n_{i+1})ds\,. \qquad (11.83)$$

The path function $\phi(s;\mathbf{Q}^n_i,\mathbf{Q}^n_{i+1})$, with $s \in [0,1]$, joins $\mathbf{Q}^n_i$ to $\mathbf{Q}^n_{i+1}$ satisfying

$$\phi(0;\mathbf{Q}^n_i,\mathbf{Q}^n_{i+1}) = \mathbf{Q}^n_i\,,\quad \phi(1;\mathbf{Q}^n_i,\mathbf{Q}^n_{i+1}) = \mathbf{Q}^n_{i+1}\,. \qquad (11.84)$$

Many choices for the path are available. The simplest case is furnished by the *canonical path*

$$\phi(s;\mathbf{Q}^n_i,\mathbf{Q}^n_{i+1}) = \mathbf{Q}^n_i + s(\mathbf{Q}^n_{i+1} - \mathbf{Q}^n_i)\,. \qquad (11.85)$$

In analogy to conservative methods, which require an intercell numerical flux $\mathbf{F}_{i+\frac{1}{2}}$ to be determined, path-conservative methods (11.80) are defined once the path $\phi(s;\mathbf{Q}^n_i,\mathbf{Q}^n_{i+1})$ and the fluctuations $\mathbf{D}^-_{i+\frac{1}{2}}$, $\mathbf{D}^+_{i+\frac{1}{2}}$ are specified. A Riemann solver may enter the discussion when defining the fluctuations.

### 11.9.3 DOT Path-Conservative Scheme

Here we present a specific path-conservative scheme (11.80) to solve (11.79), which is a variant of the conservative Dumbser-Osher-Toro (DOT) scheme presented in Sect. 11.8. See [22, 23] for details. In the DOT path-conservative scheme the fluctuations are defined as follows

$$\mathbf{D}^\pm_{i+\frac{1}{2}} = \frac{1}{2}\int_0^1 [\mathbf{A}(\phi(s;\mathbf{Q}_0,\mathbf{Q}_1)) \pm |\mathbf{A}(\phi(s;\mathbf{Q}_0,\mathbf{Q}_1))|]\frac{\partial \phi}{\partial s}ds\,. \qquad (11.86)$$

For relevant definitions see Sect. 11.8. The resulting path-conservative method makes use of the local Riemann problem, the data of which is denoted as $\mathbf{Q}_0$ and $\mathbf{Q}_1$. The path function $\phi$ in (11.86) is completely general. However, if the canonical path (11.85) is chosen, then (11.86) becomes

$$\mathbf{D}^\pm_{i+\frac{1}{2}} = \frac{1}{2}\left(\int_0^1 [\mathbf{A}(\phi(s;\mathbf{Q}_0,\mathbf{Q}_1)) \pm |\mathbf{A}(\phi(s;\mathbf{Q}_0,\mathbf{Q}_1))|]ds\right)(\mathbf{Q}_1 - \mathbf{Q}_0)\,. \qquad (11.87)$$

Now the integrals in (11.87) are computed *numerically* along the path $\phi$ using a Gauss quadrature rule with $G$ points $s_j$ and associated weights $\omega_j$ in the unit interval $[0, 1]$. We obtain

$$\mathbf{D}^\pm_{i+\frac{1}{2}} = \frac{1}{2}\left(\sum_{j=1}^{G} \omega_j \left[\mathbf{A}(\phi(s_j; \mathbf{Q}_0, \mathbf{Q}_1)) \pm |\mathbf{A}(\phi(s_j; \mathbf{Q}_0, \mathbf{Q}_1))|\right]\right)(\mathbf{Q}_1 - \mathbf{Q}_0). \tag{11.88}$$

More details are given in the original references [22, 23]. Recommended choices for the computational parameters in (11.88) are given in (11.78). The upwind, path-conservative method (11.80) with (11.88 has been applied in several areas of study. Here we emphasise its utilization in what could be regarded as an unusual application, namely the global, closed loop model for the entire human circulation in [24, 25, 27].

### 11.9.4 FORCE-$\alpha$ Path-Conservative Scheme

Here we present a *centred* (non-upwind) path-conservative scheme, in which the fluctuations are provided by the centred FORCE approach, presented for conservative systems in Chap. 10. For details on the original FORCE for conservative systems see [41, 42]; see also [43–45]. For general non-conservative systems (11.79), a FORCE-type scheme was presented by Canestrelli et al. [46]. See also [47].

Here we present a generalization of the FORCE approach to solve PDEs (11.79) in non-conservative form following the recently proposed FORCE-$\alpha$ variant [45]. The fluctuations are given as

$$\mathbf{D}^-_{i+\frac{1}{2}} = \mathbf{A}^-_{i+\frac{1}{2}}(\mathbf{Q}^n_{i+1} - \mathbf{Q}^n_i), \quad \mathbf{D}^+_{i+\frac{1}{2}} = \mathbf{A}^+_{i+\frac{1}{2}}(\mathbf{Q}^n_{i+1} - \mathbf{Q}^n_i), \tag{11.89}$$

with

$$\left.\begin{aligned}\mathbf{A}^-_{i+\frac{1}{2}} &= \frac{1}{2}\hat{\mathbf{A}}_{i+\frac{1}{2}} - \frac{1}{4}\frac{\alpha \Delta t}{\Delta x}\left[\hat{\mathbf{A}}^2_{i+\frac{1}{2}} + \left(\frac{\Delta x}{\alpha \Delta t}\right)^2 \mathbf{I}\right], \\ \mathbf{A}^+_{i+\frac{1}{2}} &= \frac{1}{2}\hat{\mathbf{A}}_{i+\frac{1}{2}} + \frac{1}{4}\frac{\alpha \Delta t}{\Delta x}\left[\hat{\mathbf{A}}^2_{i+\frac{1}{2}} + \left(\frac{\Delta x}{\alpha \Delta t}\right)^2 \mathbf{I}\right]. \end{aligned}\right\} \tag{11.90}$$

In (11.90) $\mathbf{I}$ is the identity matrix and $\hat{\mathbf{A}}_{i+\frac{1}{2}}$ is a path integral of the coefficient matrix $\mathbf{A}$ in the governing equations, which is approximated numerically as

$$\hat{\mathbf{A}}_{i+\frac{1}{2}} = \sum_{j=1}^{G} \omega_j \mathbf{A}(\phi(s_j; \mathbf{Q}^n_i, \mathbf{Q}^n_{i+1})) \approx \int_0^1 \mathbf{A}(\phi(s; \mathbf{Q}^n_i, \mathbf{Q}^n_{i+1}))ds. \tag{11.91}$$

## 11.9.5 Choosing α: Accuracy Versus Size of Time Step

The parameter $\alpha$ in (11.90) is chosen in the range $\alpha \in [1, \infty)$. For $\alpha = 1$ one obtains the scheme of Canestrelli et al. [46]. More sophisticated options admit $\alpha > 1$ [45]. Each value of $\alpha$ gives a scheme with different properties regarding stability and numerical viscosity. These schemes enhance accuracy, which is particularly evident for slowly-moving intermediate waves. We recall that these are the most challenging features for centred methods and for incomplete Riemann solvers. The penalty of the enhanced accuracy of the FORCE-$\alpha$ schemes is the reduction of the stability range, as $\alpha$ increases. The stability and monotonicity condition is

$$0 \leq |c| \leq C_{lim} = \frac{1}{\alpha}\sqrt{2\alpha - 1}. \qquad (11.92)$$

Table 11.2 shows for some selected values of $\alpha$, the numerical values of the Courant number $C_{lim}$ and the corresponding numerical viscosity function $d_0$ for stationary waves (the worse case scenario). The value $\alpha = 1$ simply reproduces the classical FORCE scheme in 1D and the scheme of Canestrelli et al. [46]. Note the remarkable decrease of $d_0$ from $d_0 = 1$ for the Lax-Friedrichs scheme to $d_0 = 0.1667$ for $\alpha = 3$.

**Table 11.2** Variation of the Courant number $C_{lim}$ and numerical viscosity function $d_0$ for stationary waves as functions of selected, increasing values of the parameter $\alpha$

| $\alpha$ | Courant number $C_{lim}$ | Viscosity for stationary waves $d_0$ |
|---|---|---|
| 1 | 1.0000 | 0.5000 |
| 2 | 0.8660 | 0.2500 |
| 3 | 0.7454 | 0.1667 |
| 4 | 0.6614 | 0.1250 |
| 5 | 0.6000 | 0.1000 |
| 10 | 0.4359 | 0.0500 |
| 20 | 0.3122 | 0.0250 |
| 25 | 0.2800 | 0.0200 |
| 30 | 0.2560 | 0.0167 |
| 40 | 0.2222 | 0.0125 |
| 50 | 0.1990 | 0.0100 |
| 60 | 0.1818 | 0.0083 |
| 199.4987 | 0.1000 | 0.0025 |

For $\alpha = 20$ a reduction of $d_0$ relative to the Lax-Friedrichs scheme is by a factor of 40, while the corresponding reduction of $C_{lim}$ is by a factor of 3.2. For $\alpha = 25$ these numbers are respectively 50 and 3.57. For large $\alpha$, for example for $\alpha = 199.4987$, the reduction of $C_{lim}$ is by a factor of 10, to $C_{lim} = 0.1$, but the corresponding reduction of $d_0$ is by a factor of 400. For practical applications, values in the range $\alpha \in [3, 5]$ are a good compromise between accuracy, small $d_0$, and efficiency, that is large $C_{lim}$.

The discussion has so far been restricted to schemes in the finite volume framework. Next we consider the above issues in the context of discontinuous Galerkin (DG) finite element methods.

### 11.9.6 FORCE-$\alpha$ in DG Finite Element Methods

The construction of higher-order schemes based on the Riemann solvers of this chapter will be the subject for Chaps. 12 and 14. However, it is useful at this stage to note the relationship between increased gains of accuracy but at the cost of reduced time-step size in the FORCE-$\alpha$ schemes and discontinuous Galerkin (DG) finite element methods [48–54].

We note that explicit DG finite element methods have a very strong stability restriction on the Courant number $c$, which depends on the chosen order of accuracy of the method. For accuracy $m + 1$ with polynomial of degree $m$ one has the stability restriction

$$0 < |c| \le C_{lim} = \frac{1}{2m + 1} \,. \tag{11.93}$$

Table 11.3 shows the decreasing behaviour of $C_{lim}$ as a function of the order of accuracy, or the degree of the associated polynomial $m$. Hence for DG methods the upper limit of the stability range of most schemes discussed in this chapter will not be fully exploited. For example, a third-order DG method will only require a scheme with maximum stable Courant number of $C_{lim} = 0.2$, rather than $C_{lim} = 1$ for most finite volume methods in one space dimension. Therefore the FORCE-$\alpha$ schemes presented here are fully satisfactory for DG methods. In fact for any given polynomial degree $m$ we can always find a value $\alpha$ and the associated FORCE-$\alpha$ flux (or fluctuations) that exactly matches the stability requirement of the DG method of order of accuracy $m + 1$; see fourth column of Table 11.3. As a matter of fact the stability limit $C_{lim}$ for FORCE-$\alpha$ in the finite volume framework amply exceeds that of the DG method. Again for the third-order DG method with $C_{lim} = 0.2$, Table 11.2 shows that any $\alpha > 40$ will give a sufficient generous stability range for the DG scheme. As a final remark we note that the stability conditions of the FORCE-$\alpha$ schemes are unexpectedly generous in multiple space dimensions. See [45] for full details.

11.10 Computation of Wet/Dry Fronts

**Table 11.3** Variation of Courant number stability limit $C_{lim}$ with order of accuracy $m+1$ for polynomials of degree $m$ in the discontinuous Galerkin finite element method. The fourth column shows the corresponding value for the parameter $\alpha$ that would match the DG method stability requirement. The last column shows $d_0$ for the first-order finite volume scheme corresponding to the tabulated values of $\alpha$ in the third column

| Degree $m$ | Order $m+1$ | $C_{lim} = 1/(2m+1)$ | $\alpha$ | $d_0$ |
|---|---|---|---|---|
| 0 | 1 | 1 | 1.0 | 0.5 |
| 1 | 2 | $1/3 \approx 0.3333$ | 17.4852 | 0.02859 |
| 2 | 3 | $1/5 \approx 0.2000$ | 49.4949 | 0.01010 |
| 3 | 4 | $1/7 \approx 0.1429$ | 97.4974 | 0.00513 |
| 4 | 5 | $1/9 \approx 0.1111$ | 161.4984 | 0.00309 |
| 5 | 6 | $1/11 \approx 0.0909$ | 241.4989 | 0.00207 |
| 6 | 7 | $1/13 \approx 0.0769$ | 366.4992 | 0.001485 |
| 7 | 8 | $1/15 \approx 0.0667$ | 449.4994 | 0.00111 |
| 8 | 9 | $1/17 \approx 0.0588$ | 577.4995 | 0.00086 |
| 9 | 10 | $1/19 \approx 0.0028$ | 721.4996 | 0.00069 |

## 11.10 Computation of Wet/Dry Fronts

It is generally accepted that the numerical computation of wet/dry fronts separating zones of water from dry zones, remains a very challenging task. In other areas of computational fluid mechanics this is known as the vacuum problem. The purpose of this section is to discus some of the underlying issues regarding wet/dry fronts and potential ways of designing effective numerical schemes. Here we first discuss some sources of errors arising from the numerical computation of wet/dry fronts. There appear to be two main sources: the *artificial bed wetting* practice and the *conservative updating* formula. We shall deal with them separately.

### 11.10.1 Artificial Bed Wetting

To illustrate the problem, we consider a Riemann problem for the shallow water equations with initial left and right data as follows

$$\text{Wet state: } \mathbf{W}_L = \begin{bmatrix} h_L \\ u_L \end{bmatrix}, \quad \text{Dry state: } \mathbf{W}_R = \begin{bmatrix} h_R \\ u_R \end{bmatrix} = \begin{bmatrix} 0 \\ 0 \end{bmatrix}. \quad (11.94)$$

In Chap. 7 we studied the exact solution of this type of Riemann problems. Recalll that the solution in this case consists of a single left rarefaction associated with the *left* eigenvalue $\lambda_1 = u - a$. The *expected* right shock associated with the eigenvalue

$\lambda_2 = u + a$ in the case of a wet right state is *absent*. There is instead a wet/dry front corresponding to the tail of the left rarefaction and has exact propagation speed

$$S_{*L} = u_L + 2a_L \ . \tag{11.95}$$

Note that this front is very fast. Its speed is larger than the speed associated with any of the eigenvalues $\lambda_1 = u - a$ and $\lambda_2 = u + a$ for the case of a wet bed. This fact should be taken into account when enforcing the Courant stability condition for the particular numerical method being used [55, 56].

A popular way of dealing with this kind of problems is by *artificially wetting the dry bed*, that is by replacing the zero water depth on the right-hand side in our problem by some small positive tolerance $\epsilon$, namely by setting $h_R = \epsilon > 0$. Having done this, the solution of the corresponding Riemann problem has a different structure to that of the exact problem for dry-bed conditions, see Chaps. 6 and 7. Now the solution contains the expected left rarefaction, if $h_L > \epsilon$, and a relatively weak, right-facing shock wave of speed $S_R$, which is meant to represent the wet/dry front speed. The speed $S_R$ of this shock is considerably slower than that of the wet/dry front $S_{*L}$. Of course, in the limit as $\epsilon$ tends to zero, the two speeds coincide. However, for practical values of the *artificial bed-wetting parameter* $\epsilon$, the errors can be large, significantly slowing down the wet/dry front, as we now illustrate for a sequence of cases.

Table 11.4 shows computed values for the solution of the *artificially wetted* Riemann problem, in which we take $h_L = 1$ m, $u_L = 0$ m/s for the initial wet left state. Column 1 shows the range of chosen values for the artificial bed-wetting param-

**Table 11.4** Variation of wet/dry front speed (column 4) $S_R$ with artificial bed-wetting parameter $\epsilon$ in m (column 1). The exact front speed is $S_{*L} = 6.260999$. Also shown are corresponding variations of depth and particle velocity in the Star Region

| Wetting parameter $\epsilon$ | $h_*$ | $u_*$ | $S_R$ |
|---|---|---|---|
| 0.2 | 0.5078714 | 1.799089 | 2.967817 |
| 0.15 | 0.4575845 | 2.025744 | 3.013640 |
| 0.1 | 0.3961748 | 2.320172 | 3.103551 |
| 0.075 | 0.3579379 | 2.515171 | 3.181882 |
| 0.05 | 0.3100852 | 2.774539 | 3.307930 |
| 0.04 | 0.2863394 | 2.910691 | 3.383322 |
| 0.03 | 0.2581136 | 3.080102 | 3.485176 |
| 0.02 | 0.2224406 | 3.308082 | 3.634902 |
| 0.01 | 0.1711789 | 3.670582 | 3.898316 |
| 0.005 | 0.1303973 | 4.000112 | 4.159608 |
| 0.001 | 0.0668297 | 4.642433 | 4.712955 |
| 0.0001 | 0.0239567 | 5.291917 | 5.314099 |
| 0.00001 | 0.0081422 | 5.696035 | 5.703039 |
| 0.000001 | 0.0026824 | 5.936720 | 5.938933 |

## 11.10 Computation of Wet/Dry Fronts

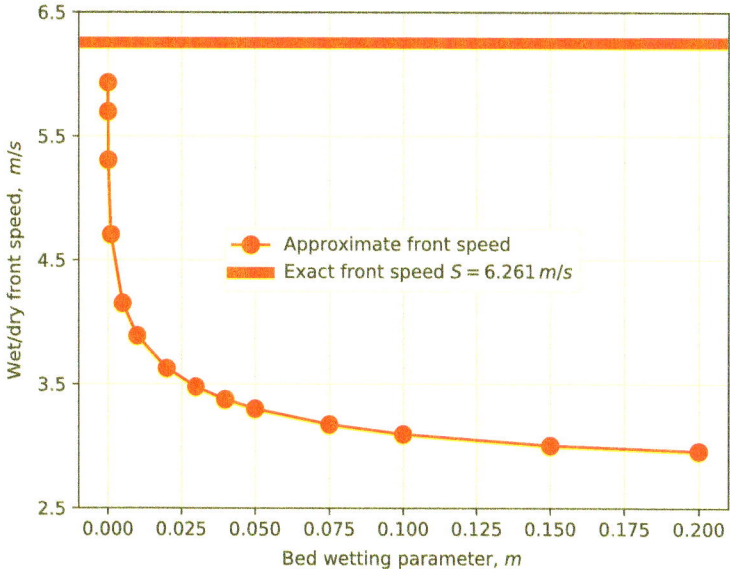

**Fig. 11.5** Variation of wet/dry front speed with artificial bed-wetting parameter $\epsilon$ shown by symbols and full line. Exact value $S = 6.261$ is shown by full horizontal line. See Table 11.4

eter $\epsilon = h_R$ (in m) and column 4 shows the corresponding shock speeds, meant to approximate the wet/dry front speed, which in this case is $S_{*L} = 6.260999$. Also shown are the values for depth (column 2) and particle velocity (column 3) in the *Star Region*, which in the exact problem does not exist. Figure 11.5 depicts the *approximated front speed*, shown by symbols and line, versus the *wetting parameter* $\epsilon$. The thick line at the top level $S_{*L} = 6.260999$ denotes the exact speed of the wet/dry front. Note that the variations are very rapid for small $\epsilon$. In the larger range of $\epsilon$ values, very small variations of the front speed are seen to take place. This is the region of large relative errors in the *predicted* speed of the front, of the order of 50%. For example, the value $\epsilon = 0.15$ m in water depth gives a front speed of 3.014 m/s, with a relative error of 51%. For $\epsilon = 0.01$ m in water depth the front speed is 3.898 m/s, which is in error by 38%. The choice of $\epsilon = 0.001$ m in water depth (one millimetre of water) gives an error of 25% in propagation speed. And even an artificial water depth of ten microns, still gives an error close to 10%.

In conclusion, the bed-wetting practice is computationally very convenient but hides potentially large errors in wet/dry front propagation speed. This has a practical value when simulating, for example, the propagation of an inundation wave down a valley of tens of kilometres, resulting from the collapse of a dam. The accurate prediction of the arrival time of the front has human population safety implications.

## 11.10.2 Conservative-Form Induced Errors

A separate problem concerns errors induced by conservative methods, see Chaps. 9 and 10. When using conservative methods one first computes values for the conserved variables $h_i^{n+1}$ (mass) and $(hu)_i^{n+1}$ (momentum). Then the particle velocity $u_i^{n+1}$ is computed as

$$u_i^{n+1} = \frac{(hu)_i^{n+1}}{h_i^{n+1}}. \tag{11.96}$$

Near the wet/dry front, both the numerator and the denominator in (11.96) are small. Moreover, they might also be in error due to the usual discretisation errors in the numerical method but they might also be in error due to the *artificial bed-wetting* referred to above. Therefore the operation (11.96) will give an erroneous result for the particle velocity at the front. In addition, division by zero in (11.96) must of course be avoided. The procedure by which this is done can again introduce more errors. Errors due to the *artificial bed-wetting* practice can combine with those induced by the conservative updating. The errors in front propagation speed can thus be very large and are bound to grow with time. Therefore to predict the evolution of wet/dry bed fronts for long evolution times, the errors will most certainly be large. Of course in practice there will be more *physics* in the problem, such as bed friction and complicated geometries. In this case the exact solution is unknown and we have no reliable ways of assessing the errors in computed front propagation speeds.

Finally we remark that the results of Table 11.4 and Fig. 11.5, and associated discussion, are for a single Riemann problem. In numerical computations the problem is present in a set of Riemann problems solved along the wet/dry front over many time steps.

## 11.10.3 Dry-Bed Approximate Riemann Solvers

It is not easy to fix the problem discussed above, see Dodd [17]. The HLL and HLLC approaches presented in Sects. 11.3 and 11.4 offer simple ways of dealing with wet/dry fronts [56]. Restricting the discussion to the purely one-dimensional case without additional equations for passive scalars, the HLL Riemann solver has numerical flux as given by (11.22), which in turn requires wave speed estimates $S_L$ and $S_R$. In the presence of a wet/dry front, we propose to set the wave speed estimates to the exact dry front speed, namely

$$S_L = \begin{cases} u_R - 2a_R & \text{if } h_L = 0, \\ \text{usual estimate} & \text{if } h_L > 0, \end{cases} \tag{11.97}$$

or

$$S_R = \begin{cases} u_L + 2a_L & \text{if } h_R = 0, \\ \text{usual estimate} & \text{if } h_R > 0. \end{cases} \quad (11.98)$$

In the case of additional equations, for the HLLC Riemann solver the procedure is analogous. If the wave speed estimate $S_*$ is as given in Eq. (11.28), then for a wet/dry front the estimate $S_*$ coincides with the front speed, which is correct. This approach was first reported by Fraccarollo and Toro [56] and appeared to work well.

Front tracking is probably the best way of dealing with wet/dry fronts. This approach is, however quite complicated for multidimensional problems. Of the methods studied in this book the Random Choice Method of Glimm, see Chap. 10, appears to be the best for dealing with wet/dry fronts. See numerical results in Chap. 10. To be noted however, RCM does not extend to multiple space dimensions under dimensional splitting or other known forms. On the other hand, a hybrid approach as suggested by Toro and Roe [57, 58] for gas dynamics, might be productive. RCM under operator splitting can successfully handle linearly degenerate fields. Preliminary results from an application of RCM to the two-dimensional shallow water equations with wet/dry fronts were reported by Ivings, Toro and Weber [59].

## 11.11 Concluding Remarks

A full range of 10 approximate Riemann solvers for the shallow water equations written in conservative and non-conservative forms have been presented. A major feature that groups Riemann solvers into two categories is whether the number of waves $W$ included in their wave model matches the number $N$ of the exact Riemann solver. *Complete Riemann solvers* are those in which $W = N$ and their wave models account for all characteristic fields in the the equations. Examples include the exact solver of course, the Roe solver, the Osher-Solomn solver, the Dumbser-Osher-Toro solver and some approximate-state solvers. Approximate Riemann solvers in which $W < N$ are called *incomplete Riemann solvers*. Their wave models exclude some of the wave fields from their structure, thus sacrificing the accuracy of those fields in their numerical implementations. Linearly degenerate fields are particularly sensitive to exclusion from the wave models. Examples of incomplete Riemann solvers includes the Rusanov flux for $N > 1$, the bottom of the hierarchy, the HLL solver for $N > 2$ and the HLLC solver for $N > 3$. We note however that HLLC does capture multiple linearly degenerate fields with coincident eigenvalues, such as in multiple space dimensions and systems that include transport of chemical species, or both. For example, for the two-dimensional shallow water equations with $K$ species transport equations added, $N = 3 + K$. The three waves of HLLC capture all wave fields, as the intermediate field with eigenvalue $\lambda = u$ has multiplicity $K + 1$. Therefore HLLC is complete for the two-dimensional shallow water equations with $K$ species equations added. Another criterion is whether the approximate Riemann solver is linear or non-linear. Recall that linear solvers require an additional entropy fix in order to avoid the computation of unphysical entropy-violating shocks. Linearised solvers will also fail in the presence of near vacuum states (wet/dry fronts); they have a depth-positivity condition that is much more restrictive than that of the exact solver. We have also included in this chapter FORCE-like centred methods. These do not require the explicit solution of the Riemann problem, a reason for which they are attractive. But they are less accurate than complete non-linear Riemann solvers, especially for linearly degenerate fields. The reader is warned that an incomplete approximate Riemann solver may be less accurate than a good centred method [45]. The reason for this is the incomplete character of the Riemann solver.

Not studied in this book are refinements to the schemes that make them *well-balanced*. This issue arises in the presence of source terms, particularly of the *geometric type*, such as in variable bathymetry in water flows. In steady state or nearly steady state regimes, the equations express *balance* between the flux gradient and the source term. Numerical methods experience difficulties in respecting this balance. Pioneering work in this area goes back to Bermúdez and Vázquez [60]. This topic has become an important area of research in many applications. More recent works relevant to the shallow water equations include [61–68], and many more.

All the methods studied here give directly a first-order accurate method of the Godunov type. All the monotone methods can be extended to non-linear high order schemes in the frameworks of finite volumes and discontinuous Galerkin finite element methods. Chap. 12 deals with second-order non-linear schemes via the Total Variation Diminishing (TVD) and ENO criteria. Chapter 14 deals with higher-order, non-linear fully discrete schemes of theoretically unlimited accuracy, based on non-linear spatial reconstructions and the solution of a generalised Riemann problem.

## References

1. E.F. Toro, *Riemann Solvers and Numerical Methods for Fluid Dynamics. A Practical Introduction*. (Springer, 1997)
2. E.F. Toro. *Riemann Solvers and Numerical Methods for Fluid Dynamics. A Practical Introduction*, 2nd edn. (Springer, 1999)
3. E.F. Toro, *Riemann Solvers and Numerical Methods for Fluid Dynamics. A Practical Introduction*, 3rd edn. (Springer, 2009)
4. S.K. Godunov, Finite difference methods for the computation of discontinuous solutions of the equations of fluid dynamics. Mat. Sb. **47**, 271–306 (1959)
5. A. Harten, P.D. Lax, B. van Leer, On upstream differencing and Godunov-type schemes for hyperbolic conservation laws. SIAM Rev. **25**(1), 35–61 (1983)
6. E.F. Toro, A. Chakraborty, Development of an approximate Riemann solver for the steady supersonic Euler equations. Aeronaut. J. **98**, 325–339 (1994)
7. E.F. Toro, M. Spruce, W. Speares, Restoration of the contact surface in the HLL-Riemann solver. Shock Waves **4**, 25–34 (1994)
8. E.F. Toro, L.O. Müller, A. Siviglia, Bounds for wave speeds in the Riemann problem: direct theoretical estimates. Comput. Fluids **209**(104640) (2020)
9. J.L. Guermond, B. Popov, Fast estimation of the maximum wave speed in the Riemann problem for the Euler equations. J. Comput. Phys. **321**, 908–926 (2016)
10. E.F. Toro, The HLLC Riemann solver. Shock Waves **29**, 1065–1082 (2019)

11. P.L. Roe, Approximate Riemann solvers, parameter vectors, and difference schemes. J. Comput. Phys. **43**, 357–372 (1981)
12. P. Glaister, Difference Schemes for the Shallow Water Equations. Technical Report 9/87, Department of Mathematics, University of Reading, England (1987)
13. P.L. Roe, J. Pike, Efficient construction and utilisation of approximate Riemann solutions, in *Computing Methods in Applied Science and Engineering*. North–Holland (1984)
14. E.F. Toro, Riemann problems and the WAF method for solving two-dimensional shallow water equations. Philos. Trans. Roy. Soc. Lond. **A338**, 43–68 (1992)
15. A. Harten, J.M. Hyman, Self adjusting grid methods for one-dimensional hyperbolic conservation laws. J. Comput. Phys. **50**, 235–269 (1983)
16. R.J. LeVeque. *Numerical Methods for Conservation Laws* (Birkhäuser Verlag, 1992)
17. N. Dodd, A numerical model of wave runn–up, overtopping and regeneration. J. Water Port, Coast and Ocean Eng., ASCE **124**(2), 73–81 (1998)
18. B. Engquist, S. Osher, One sided difference approximations for nonlinear conservation laws. Math. Comput. **36**(154), 321–351 (1981)
19. S. Osher, F. Solomon, Upwind difference schemes for hyperbolic conservation laws. Math. Comput. **38**(158), 339–374 (1982)
20. D.H. Zhao, H.W. Shen, G.Q. Tabios III, Finite–volume two–dimensional unsteady–flow model for rive basins. J. Hydraulic Eng. ASCE **120**(7), 863–883 (1994)
21. D.H. Zhao, H.W. Shen, J.S. Lai, G.Q. Tabios III, Approximate Riemann solvers in FVM for 2D hydraulic shock wave modelling. J. Hydraulic Eng. ASCE **122**(12), 692–702 (1996)
22. M. Dumbser, E.F. Toro, A simple extension of the Osher Riemann solver to general non-conservative hyperbolic systems. J. Sci. Comput. **48**, 70–88 (2011)
23. M. Dumbser, E.F. Toro, On universal Osher-type schemes for general nonlinear hyperbolic conservation laws. Commun. Comput. Phys. **10**, 635–671 (2011)
24. L.O. Müller, E.F. Toro, A global multi-scale model for the human circulation with emphasis on the venous system. Int. J. Numer. Methods Biomed. Eng. **30**(7), 681–725 (2014)
25. L.O. Müller, Mathematical modelling and simulation of the human circulation with emphasis on the venous system: application to the CCSVI condition. Ph.D. thesis, Doctoral School in Environmental Engineering, University of Trento, Italy (2014)
26. L.O. Müller, E.F. Toro, Enhanced global mathematical model for studying cerebral venous blood flow. J. Biomech. **47**(13), 3361–3372 (2014)
27. E.F. Toro, M. Celant, Q Zhang, C. Contarino, N. Agarwal, A.A. Linninger, L.O. Müller, Cerebrospinal fluid dynamics coupled to the global circulation in holistic setting: Mathematical models, numerical methods and applications. Int. J. Numer. Methods Biomed. Eng. **26**, e3532 (2021). https://doi.org/10.1002/cnm.3532
28. C. Parés, Numerical methods for non-conservative hyperbolic systems. A theoretical framework. SIAM J. Num. Anal. **44**(1), 300–321 (2006)
29. G. Moretti, G. Bleich, A time-dependent computational method for blunt-body flows. AIAA J. **4**, 2136–2141 (1966)
30. G. Moretti, The $\lambda$-scheme. Comput. Fluids **7**, 191–205 (1979)
31. G. Moretti, Computation of flows with strong shocks. Ann. Rev. Fluid Mech. **19**, 313–337 (1987)
32. T. DeNeff, G. Moretti, Shock fitting for everyone. Comput. Fluids **8**, 327–334 (1980)
33. S. Karni, Multicomponent flow calculations using a consistent primitive algorithm. J. Comput. Phys. **112**(1), 31–43 (1994)
34. E.F. Toro, Defects of Conservative Approaches and Adaptive Primitive–Conservative Schemes for Computing Solutions to Hyperbolic Conservation Laws. Technical Report MMU 9401, Department of Mathematics and Physics, Manchester Metropolitan University, UK (1994)
35. E.F. Toro, On Adaptive Primitive–Conservative Schemes for Conservation Laws, in *Sixth International Symposium on Computational Fluid Dynamics: A Collection of Technical Papers*, ed. by M.M. Hafez, vol. 3, pp. 1288–1293, Lake Tahoe, Nevada, USA, September 4–8 (1995)
36. E.F. Toro, MUSCL–Type Primitive Variable Schemes. Technical Report MMU–9501, Department of Mathematics and Physics, Manchester Metropolitan University, UK (1995)

37. E.F. Toro, Primitive, conservative and adaptive schemes for hyperbolic conservation laws, in *Numerical Methods for Wave Propagation*, ed. by. E.F. Toro, J.F. Clarke (Kluwer Academic Publishers, 1998), pp. 323–385
38. R. Abgrall, How to prevent pressure oscillations in multicomponent flow calculations: a quasiconservative approach. J. Comput. Phys. **125**, 150–160 (1996)
39. E.F. Toro, A. Siviglia, PRICE: primitive centred schemes for hyperbolic systems. Int. J. Numer. Meth. in Fluids **42**, 1263–1291 (2003)
40. P.L. Roe, Fluctuation and signals-a framework for numerical evolution problems, in *Numerical Methods for Fluid Dynamics*, pp. 219–257 (1982)
41. E.F. Toro, On Glimm–Related Schemes for Conservation Laws. Technical Report MMU–9602, Department of Mathematics and Physics, Manchester Metropolitan University, UK (1996)
42. E.F. Toro, S.J. Billett, Centred TVD schemes for hyperbolic conservation laws. IMA J. Numer. Anal. **20**, 47–79 (2000)
43. G.Q. Chen, E.F. Toro, Centred schemes for non-linear hyperbolic equations. J. Hyperbolic Differ. Equ. **1**(1), 531–566 (2004)
44. E.F. Toro, A. Hidalgo, ADER finite volume schemes for diffusion-reaction equations. Appl. Numer. Math. **59**, 73–100 (2009)
45. E.F. Toro, B. Saggiorato, S. Tokareva, A. Hidalgo, Low-dissipation centred schemes for hyperbolic equations in conservative and non-conservative form. J. Comput. Phys. **416**(109545) (2020)
46. A. Canestrelli, A. Siviglia, M. Dumbser, E.F. Toro, Well-balanced high-order centred schemes for non-conservative hyperbolic systems. Applications to shallow water equations with fixed and mobile bed. Adv. Water Res. **32**, 834–844 (2009)
47. M. Dumbser, M.J. Castro, C. Parés, E.F. Toro, A. Hidalgo, FORCE schemes on unstructured meshes II: non-conservative hyperbolic systems. Comput. Methods Appl. Mech. Eng. **199**(9–12), 625–647 (2010)
48. B. Cockburn, C.W. Shu, TVB Runge–Kutta local projection discontinuous Galerkin method for conservation laws II: general framework. Math. Comput. **52**(–), 411– (1989)
49. B. Cockburn, S. Hou, C.W. Shu, The Runge-Kutta local projection discontinuous Galerkin finite element method for conservation laws IV: the multidimensional case. Math. Comput. **54**, 545–581 (1990)
50. B. Cockburn, C.W. Shu, The Runge-Kutta local projection P1-discontinuous Galerkin finite element method for scalar conservation laws. Math. Model. Numer. Anal. **25**, 337–361 (1991)
51. B. Cockburn, C.W. Shu, The Runge–Kutta discontinuous Galerkin method for conservation laws. J. Comput. Phys. **141**(–), 199– (1998)
52. B. Cockburn, G.E. Karniadakis, C.W. Shu, *Discontinuous Galerkin Methods*. Lecture Notes in Computational Science and Engineering (Springer, 2000)
53. M. Dumbser, O. Zanotti, R. Loubère, S. Diot, A posteriori subcell limiting of the discontinuous Galerkin finite element method for hyperbolic conservation laws. J. Comput. Phys. **278**, 47–75 (2014)
54. S. Busto, M. Dumbser, C. Escalante, S. Gavrilyuk, N. Favrie, On high order ADER discontinuous Galerkin schemes for first order hyperbolic reformulations of nonlinear dispersive systems. J. Sci. Comput. **87**, 48 (2021)
55. L. Fraccarollo, E.F. Toro, A shock–capturing method for two dimensional dam–break problems, in *Proceedings of the Fifth International Symposium in Computational Fluid Dynamics, Sendai, Japan* (1993)
56. L. Fraccarollo, E.F. Toro, Experimental and numerical assessment of the shallow water model for two-dimensional dam-break type problems. J. Hydraul. Res. **33**, 843–864 (1995)
57. E.F. Toro, P.L. Roe, A hybrid scheme for the Euler equations using the random choice and Roe's methods, in *Numerical Methods for Fluid Dynamics III. The Institute of Mathematics and its Applications Conference Series, New Series No. 17*, Morton and Baines (Editors) (Oxford University Press, New York, 1988), pp. 391–402
58. E.F Toro, Random choice based hybrid schemes for one and two–dimensional gas dynamics, in *Proceedings of the Second International Conference on Hyperbolic Problems, Aachen,*

*Germany, March 1988. Non–linear Hyperbolic Equations–Theory, Computation Methods and Applications. Notes on Numerical Fluid Mechanics*, ed. by R. Jeltsch, J. Ballmann, vol. 24. (Vieweg, Braunschweig, 1989), pp. 630–639
59. M.J. Ivings, E.F. Toro, D.M. Webber, Numerical schemes for 2D shallow water equations including dry fronts. J. Comput. Fluid Dyn. **12**(1), 41–52 (2003)
60. L. Bermúdez, M.E. Vázquez, Upwind methods for hyperbolic conservation laws with source terms. Comput. Fluids **23**, 1049–1071 (1994)
61. R.J. LeVeque, Balancing source terms and flux gradients in high-resolution Godunov methods. J. Comput. Phys. **146**, 346–365 (1998)
62. M.E. Vázquez-Cendón, Improved treatment of source terms in upwind schemes for the shallow water equations in channels with irregular geometry. J. Comput. Phys. **148**, 497–526 (1999)
63. E.D. Fernández-Nieto, D. Bresch, J. Monnier, A consistent intermediate wave speed for a well-balanced HLLC solver. C. R. Acad. Sci. Paris **346**, 795–800 (2008)
64. L.O. Müller, C. Parés, E.F. Toro, Well-balanced high-order numerical schemes for one-dimensional blood flow in vessels with varying mechanical properties. J. Comput. Phys. **242**(7), 53–85 (2013)
65. A. Navas-Montilla, J. Murillo, 2D well-balanced augmented ADER schemes for the shallow water equations with bed elevation and extension to the rotating frame. J. Comput. Phys. **372**, 316–348 (2018)
66. E. Guerrero-Fernández, M.J. Castro-Díaz, M. Dumbser, T. Morales de Luna, An arbitrary high order well-balanced ADER-DG numerical scheme for the multilayer shallow-water model with variable density. J. Sci. Comput. **9**, 52 (2022). https://doi.org/10.1007/s10915-021-01734-2
67. L. Martaud, C. Berthon, Fully well-balanced entropy stable Godunov numerical schemes for the shallow water equations with the topography source term. Technical Report HAL Id: hal-04394378, HAL open science (2024)
68. V. González-Tabernero, M.J. Castro, J.A. García-Rodríguez, High-order well-balanced numerical schemes for one-dimensional shallow-water systems with Coriolis terms. Appl. Math. Comput. **469**(128528) (2024)

# Chapter 12
# Second-Order Non-linear Methods

**Abstract** This chapter is devoted to the construction of second-order accurate numerical methods as applied to the shallow water equations. The schemes are non-linear, in order to circumvent Godunov's theorem, which is concerned with the phenomenon of spurious oscillations in the vicinity of large spatial gradients, shocks in particular. For the class of flux limiter methods considered, the non-linear character of the schemes results from enforcing total variation diminishing (TVD) criteria. A second class of non-linear second-order methods is based either on TVD criteria or on non-linear spatial reconstruction of the Essentially Non-Oscillatory (ENO) type. Regarding the underlying first-order methods, two classes of methods are constructed, namely upwind methods based on the Riemann solvers studied in previous chapters and centred methods based on the FORCE framework. Numerical results from some of the methods are shown, followed by analysis and discussion on the performance of the methods tested.

## 12.1 Introduction

In this chapter we present second-order *non-linear methods* for solving the time-dependent non-linear shallow water equations numerically. The schemes are given for the split two-dimensional equations, so that extension to the two-dimensional case, as described in Chap. 13, can be easily implemented. For one class of non-linear methods considered here, called *flux limiter methods*, the non-linear character of the schemes results from enforcing total variation diminishing (TVD) criteria. For background information on TVD methods and their rationale, the reader is referred to Chap. 13 of the textbooks [1–3] and the many references therein; see also [4]. For a review on Godunov-type methods for shallow water flows see Toro and García-Navarro [5]. For a review on numerical fluxes for hyperbolic equations and the shallow water equations in particular, see Chaps. 6, 10 and 11 of this book.

The Initial-Boundary Value Problem (IBVP) to solve reads

$$\left.\begin{array}{ll} \text{PDEs:} & \partial_t \mathbf{Q} + \partial_x \mathbf{F}(\mathbf{Q}) = \mathbf{S}(\mathbf{Q}) \,, \quad x \in [a,b] \,, \quad t > 0 \,, \\ \text{ICs:} & \mathbf{Q}(x,0) = \mathbf{Q}^{(0)}(x) \,, \quad x \in [a,b] \,, \\ \text{BCs:} & \mathbf{Q}(a,t) = \mathbf{B}_L(t) \,, \quad \mathbf{Q}(b,t) = \mathbf{B}_R(t) \,, \quad t \geq 0 \,. \end{array}\right\} \quad (12.1)$$

$\mathbf{Q}(x,t)$ is a vector of $N$ *conserved variables*; $\mathbf{F}(\mathbf{Q})$ is the flux vector, or *physical flux*; $\mathbf{S}(\mathbf{Q})$ is the *source term vector*; $\mathbf{Q}^{(0)}(x)$ is the initial condition; $\mathbf{B}_L(t)$ and $\mathbf{B}_R(t)$ are the *boundary conditions* on the left ($x = a$) and right ($x = b$) boundaries, respectively, two prescribed functions of time. The computational domain is discretizesd via a finite volume mesh, in which the spatial domain interval $[a, b]$ is partitioned into $M$ regular computational cells of size $\Delta x$, namely

$$\left.\begin{array}{ll} \text{Computational cells:} & I_i = [x_{i-\frac{1}{2}}, x_{i+\frac{1}{2}}] \,, \quad i = 1, \ldots, M \,, \\ \text{Cell boundaries:} & x_{\frac{1}{2}} = a \,, \ldots x_{i+\frac{1}{2}} = a + i \Delta x \,, \ldots, x_{M+\frac{1}{2}} = b \,, \\ \text{Mesh size:} & \Delta x = (b-a)/M = x_{i+\frac{1}{2}} - x_{i-\frac{1}{2}} \,, \\ \text{Cell centre:} & x_i = \frac{1}{2}(x_{i-\frac{1}{2}} + x_{i+\frac{1}{2}}) \,. \end{array}\right\} \quad (12.2)$$

Boundary conditions are applied at $x = x_{\frac{1}{2}} = a$ and $x = x_{M+\frac{1}{2}} = b$.

Most of the schemes of this chapter are formulated for solving the augmented one-dimensional (or split two-dimensional) shallow equations

$$\left.\begin{array}{l} \partial_t \mathbf{Q} + \partial_x \mathbf{F}(\mathbf{Q}) = \mathbf{0} \,, \\ \mathbf{Q} = \begin{bmatrix} h \\ hu \\ hv \end{bmatrix} \,, \quad \mathbf{F}(\mathbf{Q}) = \begin{bmatrix} hu \\ hu^2 + \frac{1}{2}gh^2 \\ huv \end{bmatrix} \,. \end{array}\right\} \quad (12.3)$$

Strictly speaking, this is a one-dimensional problem but the third equation is for the tangential component of velocity $v$ written as conservation law for $hv$; see Chap. 1. The $x$-direction is the direction *normal* to the surface of the relevant control volume; see Chap. 4. The role of the tangential velocity component is mathematically and numerically identical to that of a *passive scalar* $\psi$ advected with velocity $u$. In pollutant transport models, for instance, the passive scalar $\psi$ is the concentration of a chemical species. In any case, the scalar $\psi$, or $v$, gives rise to an extra eigenvalue $\lambda_2 = u$, associated with a *linearly degenerate field*, and an extra *middle* wave in the Riemann problem solution structure; see Fig. 12.1.

The schemes to solve (12.3) studied here are based on the explicit conservative formula

$$\mathbf{Q}_i^{n+1} = \mathbf{Q}_i^n - \frac{\Delta t}{\Delta x} \left[ \mathbf{F}_{i+\frac{1}{2}} - \mathbf{F}_{i-\frac{1}{2}} \right] \,. \quad (12.4)$$

12.2 The Weighted Average Flux (WAF) Method

$\mathbf{F}_{i+\frac{1}{2}}$ is the intercell numerical flux corresponding to the intercell boundary at $x = x_{i+\frac{1}{2}}$ between cells $i$ and $i+1$. Next we study a particular flux limiter second-order method by defining the numerical flux.

## 12.2 The Weighted Average Flux (WAF) Method

The weighted average flux (WAF) method was first put forward for the Euler equations in [6, 7]. A detailed derivation of the scheme as applied to the three-dimensional Euler equations is given in Chaps. 14 and 16 of the textbooks [1–3]. The method has been applied to the two-dimensional shallow water equations by Toro [8], Watson, Peregrine and Toro [9], Fraccarollo and Toro [10], Siviglia and Toro [11], Ata et al. [12], and many more.

### 12.2.1 The Basic WAF Scheme

Here we first present the basic WAF scheme without the non-linear TVD modification. As such, the scheme is oscillatory and should not be used in practice. In the following section we introduce the non-linear TVD modification to deal with shock waves and other situations involving steep gradients.

For the $x$-split shallow water system (12.3) the WAF method defines a numerical flux via the *integral average*

$$\mathbf{F}^{waf}_{i+\frac{1}{2}} = \frac{1}{t_2 - t_1} \frac{1}{x_2 - x_1} \int_{t_1}^{t_2} \int_{x_1}^{x_2} \mathbf{F}\left(\mathbf{Q}^*(x,t)\right) \, dx \, dt \;, \qquad (12.5)$$

where choices for the integration volume $[x_1, x_2] \times [t_1, t_2]$ in the $x$-$t$ plane and the integrand must be made. This general formulation of the WAF flux is presented in detail in [13, 14]. For the *integration volume* the choice

$$x_1 = -\frac{\Delta x}{2} \;, \quad x_2 = \frac{\Delta x}{2} \;, \quad t_1 = 0 \;, \quad t_2 = \Delta t \qquad (12.6)$$

has led to good numerical methods and is the one we adopt here. Further, we assume that the vector $\mathbf{Q}^*(x, t)$ in the integrand is the solution $\mathbf{Q}_{i+\frac{1}{2}}(x/t)$ of the Riemann problem with *piece-wise constant* data $(\mathbf{Q}_i^n, \mathbf{Q}_{i+1}^n)$. If in addition we approximate the time integration in (12.5) by the mid-point rule in time we obtain

$$\mathbf{F}^{waf}_{i+\frac{1}{2}} = \frac{1}{\Delta x} \int_{-\frac{1}{2}\Delta x}^{\frac{1}{2}\Delta x} \mathbf{F}\left(\mathbf{Q}_{i+\frac{1}{2}}(x, \frac{1}{2}\Delta t)\right) dx \;. \qquad (12.7)$$

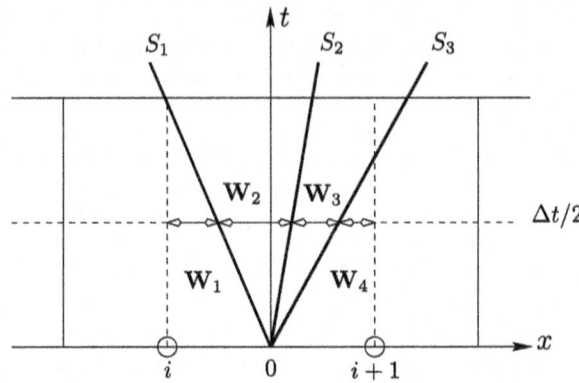

**Fig. 12.1** WAF flux is a weighted average across the wave structure of the solution of the Riemann problem at local time $t = \frac{1}{2}\Delta t$. Here $N = 3$ is the number of conservation laws in system (12.3)

For a wave structure as depicted in Fig. 12.1, the integral (12.7) becomes the summation

$$\mathbf{F}^{waf}_{i+\frac{1}{2}} = \sum_{k=1}^{N+1} w_k \mathbf{F}^{(k)}_{i+\frac{1}{2}} , \qquad (12.8)$$

where the *weights* $w_k$ are given by

$$w_k = \tfrac{1}{2}(c_k - c_{k-1}) , \quad c_0 = -1 , \quad c_{N+1} = 1 , \qquad (12.9)$$

with

$$c_k = \frac{\Delta t\, S_k}{\Delta x} \; : \; \text{Courant number for wave } k . \qquad (12.10)$$

Here $S_k$ is the speed of wave $k$ and $N$ is the number of conservation laws in system (12.1) or the number of waves in the solution of the Riemann problem, if the Riemann solver is complete; $\mathbf{F}^{(k)}_{i+\frac{1}{2}}$ is the value of the flux vector in the interval $k$ of length $w_k$. The numerical flux becomes a *weighted average* of fluxes in the intervals $k$. By inserting the weights (12.9) into (12.8) we obtain the alternative expression

$$\mathbf{F}^{waf}_{i+\frac{1}{2}} = \frac{1}{2}(\mathbf{F}_i + \mathbf{F}_{i+1}) - \frac{1}{2}\sum_{k=1}^{N} c_k \Delta \mathbf{F}^{(k)}_{i+\frac{1}{2}} , \qquad (12.11)$$

where

$$\Delta \mathbf{F}^{(k)}_{i+\frac{1}{2}} = \mathbf{F}^{(k+1)}_{i+\frac{1}{2}} - \mathbf{F}^{(k)}_{i+\frac{1}{2}} \qquad (12.12)$$

is the flux jump across wave $k$.

The whole derivation of the flux can also be made in terms of a *weighted average state* $\mathbf{Q}^{waf}_{i+\frac{1}{2}}$ given by

## 12.2 The Weighted Average Flux (WAF) Method

$$\mathbf{Q}^{waf}_{i+\frac{1}{2}} = \frac{1}{2}(\mathbf{Q}_i + \mathbf{Q}_{i+1}) - \frac{1}{2}\sum_{k=1}^{N} c_k \Delta \mathbf{Q}^{(k)}_{i+\frac{1}{2}}, \quad (12.13)$$

leading to the numerical flux defined as

$$\mathbf{F}^{waf}_{i+\frac{1}{2}} = \mathbf{F}(\mathbf{Q}^{waf}_{i+\frac{1}{2}}). \quad (12.14)$$

This expression for the flux resembles Godunov's method, which evaluates the flux function at the *Godunov state*. See Chap. 10. Here we use the WAF state instead of the Godunov state. See Chap. 14 of the textbooks [1–3] for details on the *weighted average state* version of the approach. For application to the two-dimensional shallow water equations see [8].

### 12.2.2 TVD Version of the WAF Scheme

For the model linear advection equation, see Chap. 2, the scheme just described reproduces identically the method of Lax–Wendroff, which is second-order accurate in space and time. See Chap. 9. According to Godunov's theorem studied in Chap. 9, *spurious oscillations* in the vicinity of large gradients are expected. We therefore require a *non-linear* modification of the WAF flux to avoid such unphysical, or spurious, oscillations. This is carried out by enforcing a total variation diminishing (TVD) constraint on the scheme. Details are found in Chap. 13 of [1–3]. The resulting TVD WAF flux is

$$\mathbf{F}^{tvd}_{i+\frac{1}{2}} = \frac{1}{2}(\mathbf{F}_i + \mathbf{F}_{i+1}) - \frac{1}{2}\sum_{k=1}^{N} \text{sign}(c_k) A_k \Delta \mathbf{F}^{(k)}_{i+\frac{1}{2}}. \quad (12.15)$$

This expression looks virtually identical to the non-TVD version (12.11) of the scheme. In the TVD WAF flux (12.15) $A_k$ is a WAF limiter function. There are various choices for the limiter function. Here we give the following possibilities:

$$A_{sa}(r, |c|) = \begin{cases} 1 & \text{if } r \leq 0, \\ 1 - 2(1 - |c|)r & \text{if } 0 \leq r \leq \frac{1}{2}, \\ |c| & \text{if } \frac{1}{2} \leq r \leq 1, \\ 1 - (1 - |c|)r & \text{if } 1 \leq r \leq 2, \\ 2|c| - 1 & \text{if } r \geq 2. \end{cases} \quad (12.16)$$

$$A_{vl}(r, |c|) = \begin{cases} 1 & \text{if } r \leq 0, \\ 1 - \dfrac{(1 - |c|)2r}{1 + r} & \text{if } r \geq 0. \end{cases} \quad (12.17)$$

$$A_{va}(r, |c|) = \begin{cases} 1 & \text{if } r \leq 0, \\ 1 - \dfrac{(1-|c|)r(1+r)}{1+r^2} & \text{if } r \geq 0. \end{cases} \quad (12.18)$$

$$A_{ma}(r, |c|) = \begin{cases} 1 & \text{if } r \leq 0, \\ 1 - (1-|c|)r & \text{if } 0 \leq r \leq 1, \\ |c| & \text{if } r \geq 1. \end{cases} \quad (12.19)$$

For convenience, subscripts and superscripts have been omitted. The WAF limiter functions are entirely equivalent to conventional flux limiters $B(r)$, and they are related as follows

$$A(r) = 1 - (1-|c|)B(r). \quad (12.20)$$

The limiter $A_{sa}$ is related to SUPERBEE, $A_{vl}$ is related to van Leer's limiter, $A_{va}$ is related to van Albada's limiter and $A_{ma}$ is related to MINBEE. See Sect. 13.7.1 in [1], [2, 3] for details on the derivation of limiter functions for the WAF method. The WAF limiter functions depend on the argument

$$r^{(k)} = \begin{cases} \dfrac{\Delta q_{i-\frac{1}{2}}^{(k)}}{\Delta q_{i+\frac{1}{2}}^{(k)}} \equiv \dfrac{q_i^{(k)} - q_{i-1}^{(k)}}{q_{i+1}^{(k)} - q_i^{(k)}} & \text{if } c_k > 0, \\[1em] \dfrac{\Delta q_{i+\frac{3}{2}}^{(k)}}{\Delta q_{i+\frac{1}{2}}^{(k)}} \equiv \dfrac{q_{i+2}^{(k)} - q_{i+1}^{(k)}}{q_{i+1}^{(k)} - q_i^{(k)}} & \text{if } c_k < 0. \end{cases} \quad (12.21)$$

This is the ratio of the *upwind change* $\Delta q_{i-\frac{1}{2}}^{(k)}$ if $c_k > 0$ or $\Delta q_{i+\frac{3}{2}}^{(k)}$ if $c_k < 0$, to the *local change* $\Delta q_{i+\frac{1}{2}}^{(k)}$ in a scalar quantity $q$. For background on the theory of flux limiter methods see Sweby [15, 16].

In the limiting procedure for the $x$-split two-dimensional shallow water equations we choose $q = h$ for the non-linear waves and $q = v$, the tangential velocity component, for the shear wave. For other passive scalars $\psi$ we choose $q = \psi$ for contacts.

### 12.2.3 Handling Critical Flow

The summation (12.11) or (12.15) is exact for a wave configuration such as that of Fig. 12.1. In the presence of a rarefaction wave the summation is an approximation; this approximation works well when the rarefaction is *non-critical* and one chooses the speed of the *head* of the wave as the wave speed for that wave family. One

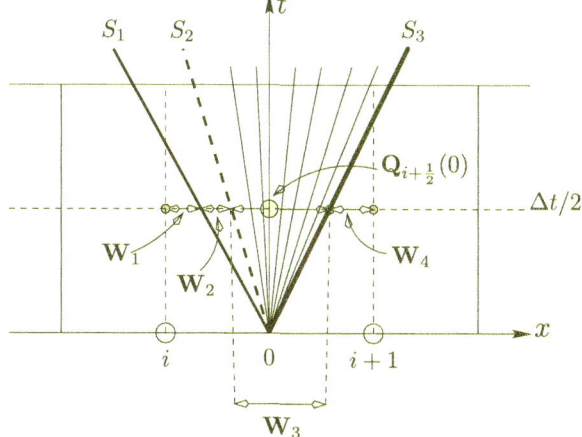

**Fig. 12.2** Evaluating the WAF flux in the presence of a right critical rarefaction, when using exact or approximate state Riemann solvers

effectively *lumps* the rarefaction wave together with the closest of the star states $\mathbf{Q}_{*K}$ into a single state; see Chap. 6. The flux for that state in the summation is then

$$\mathbf{F}^{(k)}_{i+\frac{1}{2}} = \mathbf{F}(\mathbf{Q}_{*K}) \,. \tag{12.22}$$

In the presence of a *critical rarefaction*, care is required in choosing the state for evaluating $\mathbf{F}^{(k)}_{i+\frac{1}{2}}$. Figure 12.2 illustrates the situation in which the right wave is a critical rarefaction. We merge the rarefaction zone with the closest star state, but the state used for evaluating the flux in that enlarged region is $\mathbf{Q}_{i+\frac{1}{2}}(0)$; this is the solution of the Riemann problem along the $t$-axis; it is in fact the Godunov state used for evaluating the Godunov flux; see Chap. 10. Actually, the Godunov flux is always one term in summations (12.8), (12.11), (12.13) and (12.15) for the WAF flux. Strictly speaking, in the presence of a critical rarefaction one should evaluate the integral (12.7) across the rarefaction wave, thus introducing an extra weight in the WAF flux (12.8). Watson et al. [9] implemented this approach and reported similar results to those from the simpler version presented here.

The special treatment of critical rarefaction waves is necessary only when using the exact Riemann solver, see Chaps. 6 and 7, or approximate-state Riemann solvers Chap. 11. For entropy-satisfying approximate Riemann solvers in which the flux is provided directly, such as the HLL and HLLC Riemann solvers in Chap. 11, this special treatment is not required. I particularly recommend the HLLC approximate Riemann solver; this performs very well at critical points and handles wet/dry fronts reasonably well if the wave speeds are chosen appropriately [10]. See discussion in Chap. 11.

## 12.3 The MUSCL-Hancock Scheme

The MUSCL-Hancock scheme is a typical member of another class of second-order extensions of the Godunov upwind method, attributed to S. Hancock in [17]. Full details of the scheme as applied to any system of hyperbolic conservation laws are given in Chaps. 13 and 14 of [1–3]. Unlike the WAF extension to second-order accuracy, in the MUSCL-Hancock scheme one achieves second-order accuracy in space and time by (i) a spatial reconstructing procedure via piece-wise first-degree polynomials, (ii) by evolving boundary extrapolated values in time and (iii) by solving a conventional piece-wise constant data Riemann problem with initial data consisting of evolved boundary extrapolated values. Supplementary material is found in Chap. 14, in which the basic elements of the approach are generalised to unlimited accuracy.

### 12.3.1 The Basic Linear Scheme

We first present the basic linear scheme that results in a second-order linear method, and thus oscillatory.

**Spatial Reconstruction Step**

As usual we assume a finite volume mesh as defined in (12.2). Then consider the set of cell integral averages $\{\mathbf{Q}_i^n\}$ at time $t_n$, for $i = 1, \ldots, M$. These cell averages provide a piece-wise constant distribution in the mesh at time $t_n$. For the MUSCL-Hancock method one then *reconstructs* a piece-wise linear polynomial representation of the data by defining the set of polynomials

$$\mathbf{P}_i(x) = \mathbf{Q}_i^n + \frac{(x - x_i)}{\Delta x} \Delta_i , \quad x \in I_i , \quad \text{for } i = 1, \ldots, M . \tag{12.23}$$

Here $\Delta_i / \Delta x$ is the slope of the first-degree polynomial, yet to be specified. Figure 12.3 depicts the component $p_i(x)$ of the local polynomial vector $\mathbf{P}_i(x)$ corresponding to cell $I_i$.

**Generalised Riemann Problem $GRP_1$**

The polynomial distribution (12.23) defines at each intercell boundary $x = x_{i+\frac{1}{2}}$ a **generalised Riemann problem** with piece-wise linear initial conditions, namely

## 12.3 The MUSCL-Hancock Scheme

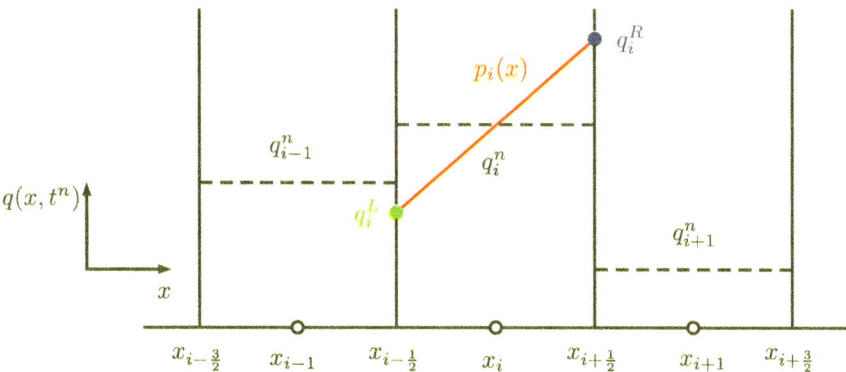

**Fig. 12.3** Illustration of the component $p_i(x)$ of the reconstructed first-degree polynomial $\mathbf{P}_i(x)$ in cell $I_i = [x_{i-\frac{1}{2}}, x_{i+\frac{1}{2}}]$ centred at $x_i$. Note the corresponding boundary values $q_i^L$ at $x_{i-\frac{1}{2}}$ and $q_i^R$ at $x_{i+\frac{1}{2}}$

$$\left. \begin{array}{l} \partial_t \mathbf{Q} + \partial_x \mathbf{F}(\mathbf{Q}) = \mathbf{0} \, , \\[4pt] \mathbf{Q}(x, 0) = \begin{cases} \mathbf{P}_i(x) & \text{if } x < x_{i+\frac{1}{2}} \, , \\ \mathbf{P}_{i+1}(x) & \text{if } x > x_{i+\frac{1}{2}} \, . \end{cases} \end{array} \right\} \quad (12.24)$$

In theory, the sought intercell numerical flux should emerge from the solution of this generalized Riemann problem denoted as $GRP_1$, in which the initial data are polynomials of degree $m = 1$. See Chap. 14 in which the generalised Riemann problem $GRP_m$ is studied, where $m$ is the degree of the reconstruction polynomials that define the initial conditions. In the MUSCL-Hancock method, this challenging problem is simplified through the following steps, described below in the conventional way.

### Time-Evolution of Extrapolated Values

The limiting values of the polynomial $\mathbf{P}_i(x)$ within cell $I_i$ at the cell boundaries are

$$\mathbf{Q}_i^L = \mathbf{P}_i(x_{i-\frac{1}{2}}) = \mathbf{Q}_i^n - \frac{1}{2}\Delta_i \, , \quad \mathbf{Q}_i^R = \mathbf{P}_i(x_{i+\frac{1}{2}}) = \mathbf{Q}_i^n + \frac{1}{2}\Delta_i \, . \quad (12.25)$$

These are often called *boundary extrapolated values* in the literature, but in fact there is no extrapolation. See Fig. 12.3. Then, in each cell $I_i$ one evolves in time the boundary-extrapolated values $\mathbf{Q}_i^L$ and $\mathbf{Q}_i^R$ by a time $t = \frac{1}{2}\Delta t$ according to the following formulae:

**Fig. 12.4** Evolved boundary extrapolated values form the initial conditions for a conventional piece-wise constant data Riemann problem at $x = x_{i+\frac{1}{2}}$, as in the Godunov method

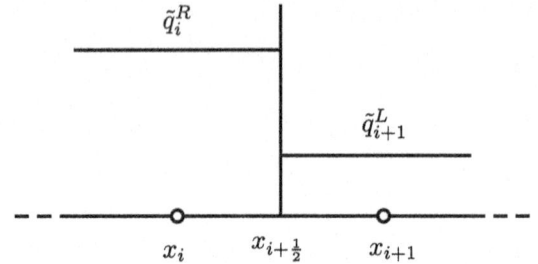

$$\left.\begin{aligned}\tilde{\mathbf{Q}}_i^L &= \mathbf{Q}_i^L - \frac{1}{2}\frac{\Delta t}{\Delta x}\left[\mathbf{F}(\mathbf{Q}_i^R) - \mathbf{F}(\mathbf{Q}_i^L)\right], \\ \tilde{\mathbf{Q}}_i^R &= \mathbf{Q}_i^R - \frac{1}{2}\frac{\Delta t}{\Delta x}\left[\mathbf{F}(\mathbf{Q}_i^R) - \mathbf{F}(\mathbf{Q}_i^L)\right].\end{aligned}\right\} \quad (12.26)$$

The flux vector is evaluated at the two boundary extrapolated vector values (12.25) to provide a flux difference. The above step looks like the updating via a conservative method but in fact it is not. In Chap. 14 we discuss the meaning of this step, and that of the entire MUSCL-Hancock scheme.

**Conventional Riemann Problem $GRP_0$**

Now at each intercell position $x = x_{i+\frac{1}{2}}$ the evolved boundary extrapolated values define a pair of constant states, as illustrated in Fig. 12.4 for a single component of the vector $\mathbf{Q}$. One therefore defines a conventional Riemann problem $GRP_0$ as follows:

$$\left.\begin{aligned}\partial_t \mathbf{Q} + \partial_x \mathbf{F}(\mathbf{Q}) &= \mathbf{0}, \\ \mathbf{Q}(x, 0) &= \begin{cases} \tilde{\mathbf{Q}}_i^R & \text{if } x < x_{i+\frac{1}{2}}, \\ \tilde{\mathbf{Q}}_{i+1}^L & \text{if } x > x_{i+\frac{1}{2}}.\end{cases}\end{aligned}\right\} \quad (12.27)$$

This Riemann problem is called *conventional* because there are no source terms and the initial conditions are piece-wise constant ($m = 0$), as for the Godunov upwind method, see Chap. 10.

**Numerical Flux**

As in the Godunov method, see Chap. 10, one solves the Riemann problem (12.27) to find the similarity solution $\tilde{\mathbf{Q}}_{i+\frac{1}{2}}(x/t)$. After the usual solution sampling procedure the Godunov state along the $t$-axis $\tilde{\mathbf{Q}}_{i+\frac{1}{2}}(0)$, in locall coordinates, is identified.

## 12.3 The MUSCL-Hancock Scheme

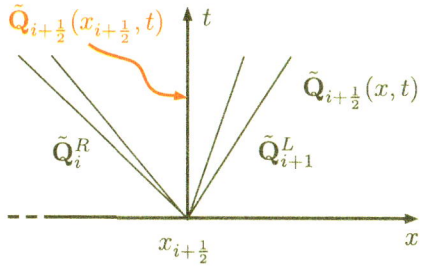

**Fig. 12.5** The Godunov state along the $t$-axis identified after solution sampling, to evaluate the Godunov flux

Figure 12.5 depicts the similarity solution $\tilde{\mathbf{Q}}_{i+\frac{1}{2}}(x/t)$ in global spatial coordinates, positioned at the interface $x_{i+\frac{1}{2}}$. Then, the intercell numerical flux is

$$\mathbf{F}_{i+\frac{1}{2}} = \mathbf{F}\left(\tilde{\mathbf{Q}}_{i+\frac{1}{2}}(0)\right) . \tag{12.28}$$

The complete, linear scheme resulting from applying all the previous steps is second-order accurate in space and time. Obviously, the Godunov flux (12.28) can be computed by using the exact Riemann solver for (12.27), or any of the approximate Riemann solvers studied in Chap. 11. Exact Riemann solvers for the shallow water equations are found in Chaps. 6 and 7, for the wet and dry bed cases respectively.

**Remarks on the MUSCL-Hancock scheme.** Before proceeding to the construction of the non-linear version of the MUSCL-Hancock scheme, a couple of remarks are in order.

1. **Rationale behind the construction of the method.** The very construction of the MUSCL-Hancock method appears somehow obscure, especially the time evolution of the boundary extrapolated values. In Chap. 14 we provide a rational justification and also provide a clear way to incorporate source terms.
2. **Kolgan's scheme.** In 1972, Kolgan [18] proposed a similar method to the one just described, with the difference that the time-evolution of the boundary extrapolated values was excluded. The linear version of the scheme turned out to be unconditionally unstable. Kolgan stabilised the scheme by effectively using a non-linear choice of the slopes. In this manner Kolgan appears to have been the first to propose non-linear schemes to circumvent Godunov's theorem.

### 12.3.2 TVD Version of the MUSCL-Hancock Scheme

Here we construct a non-linear version of the MUSCL-Hancock scheme, to circumvent Godunov's theorem. The practical aim is to avoid, or much reduce, potential spurious oscillations in the vicinity of large gradients, typically shock waves. We do so here by enforcing Total Variation Diminishing (TVD) criteria. This is done in the data reconstruction procedure by *limiting* the differences $\Delta_i$ in the reconstruction polynomials (12.23), which effectively defines the slopes of the polynomials (12.23), as $\Delta x$ is constant. In what follows we shall use *slopes* to mean *differences*, such as the following

$$\Delta_{i-\frac{1}{2}} = \mathbf{Q}_i^n - \mathbf{Q}_{i-1}^n , \quad \Delta_{i+\frac{1}{2}} = \mathbf{Q}_{i+1}^n - \mathbf{Q}_i^n . \qquad (12.29)$$

For some of the subsequent analysis we shall use the weighted average $\Delta_i$ defined as

$$\Delta_i = \frac{1}{2}(1+\omega)\Delta_{i-\frac{1}{2}} + \frac{1}{2}(1-\omega)\Delta_{i+\frac{1}{2}} . \qquad (12.30)$$

The parameter $\omega$ is allowed to lie in the interval $[-1, 1]$. For most applications we take $\omega = 0$. The application proceeds in a component-by-component fashion. In what follows we assume $q$ to be a componet of $\mathbf{Q}$ and $\Delta_i$ to be the corresponding component of $\Delta_i$.

**Limited Slopes: Method 1**

One choice for a *limited slope* is the following

$$\overline{\Delta}_i = \begin{cases} \max[0, \min(\beta\Delta_{i-\frac{1}{2}}, \Delta_{i+\frac{1}{2}}), \min(\Delta_{i-\frac{1}{2}}, \beta\Delta_{i+\frac{1}{2}})] , & \Delta_{i+\frac{1}{2}} > 0 , \\ \min[0, \max(\beta\Delta_{i-\frac{1}{2}}, \Delta_{i+\frac{1}{2}}), \max(\Delta_{i-\frac{1}{2}}, \beta\Delta_{i+\frac{1}{2}})] , & \Delta_{i+\frac{1}{2}} < 0 , \end{cases} \qquad (12.31)$$

where, in agreement with (12.29)

$$\Delta_{i-\frac{1}{2}} = q_i^n - q_{i-1}^n , \quad \Delta_{i+\frac{1}{2}} = q_{i+1}^n - q_i^n . \qquad (12.32)$$

The meaning of the functions $min(a, b)$ and $max(a, b)$ is the obvious one. The value $\beta = 1$ reproduces the MINBEE (or MINMOD) flux limiter and $\beta = 2$ reproduces the SUPERBEE flux limiter. See Sweby [15, 16].

**Limited Slopes: Method 2**

Another approach is to find a *slope limiter* $\xi_i$ such that

$$\overline{\Delta}_i = \xi_i \Delta_i , \qquad (12.33)$$

with $\Delta_i$ as given by (12.30), for example. This approach leads to a TVD region for $\xi(r)$ defined as follows:

$$\xi(r) = 0 \quad \text{for } r \leq 0 , \quad 0 \leq \xi(r) \leq \min\{\xi_L(r), \xi_R(r)\} \quad \text{for } r > 0 , \qquad (12.34)$$

## 12.3 The MUSCL-Hancock Scheme

where

$$\left.\begin{array}{l}\xi_L(r) = \dfrac{2\beta_{i-\frac{1}{2}} r}{1-\omega+(1+\omega)r} \ , \\[2mm] \xi_R(r) = \dfrac{2\beta_{i+\frac{1}{2}}}{1-\omega+(1+\omega)r} \ , \\[2mm] r = \dfrac{\Delta_{i-\frac{1}{2}}}{\Delta_{i+\frac{1}{2}}} \ , \end{array}\right\} \quad (12.35)$$

and

$$\beta_{i-\frac{1}{2}} = \frac{2}{1+c} \ , \quad \beta_{i+\frac{1}{2}} = \frac{2}{1-c} \ . \quad (12.36)$$

The coefficients $\beta_{i-\frac{1}{2}}$ and $\beta_{i+\frac{1}{2}}$ are derived for the scalar case in Section 13.8.3 of [1–3] and $c$ is the Courant number for the single wave present. By considering the limiting values of $\beta_{i-\frac{1}{2}}$ and $\beta_{i+\frac{1}{2}}$ one may eliminate the dependence on $c$ and simply set $\beta_{i-\frac{1}{2}} = \beta_{i+\frac{1}{2}} = 1$ in (12.35).

A more refined approach would be to adopt the *characteristic limiting* approach described in Chap. 14 of [1], [2, 3]. This requires solving extra Riemann problems to find jumps across each wave and their respective Courant numbers, so that one may define coefficients $\beta_{i+\frac{1}{2}}$ as functions of Courant numbers $c_{i+\frac{1}{2}}^{(k)}$.

**Slope Limiters**

Slope limiters $\xi_i$ for use in (12.33) must now be defined. In Chap. 13 of [1–3] slope limiters were constructed that are analogous to conventional flux limiters, such as SUPERBEE and MINBEE. But they are only analogous, not equivalent. In what follows we list a few possible choices.

A slope limiter that is analogous to the SUPERBEE flux limiter is

$$\xi_{sb}(r) = \begin{cases} 0 & \text{if } r \leq 0 \ , \\ 2r & \text{if } 0 \leq r \leq \frac{1}{2} \ , \\ 1 & \text{if } \frac{1}{2} \leq r \leq 1 \ , \\ \min\{r, \xi_R(r), 2\} & \text{if } r \geq 1 \ . \end{cases} \quad (12.37)$$

A van Leer type slope limiter is

$$\xi_{vl}(r) = \begin{cases} 0 & \text{if } r \leq 0 \ , \\ \min\left\{\dfrac{2r}{1+r}, \xi_R(r)\right\} & \text{if } r \geq 0 \ . \end{cases} \quad (12.38)$$

A van Albada type slope limiter is

$$\xi_{va}(r) = \begin{cases} 0 & \text{if } r \leq 0, \\ \min\left\{\dfrac{r(1+r)}{1+r^2}, \xi_R(r)\right\} & \text{if } r \geq 0. \end{cases} \quad (12.39)$$

A MINBEE type slope limiter is

$$\xi_{mb}(r) = \begin{cases} 0 & \text{if } r \leq 0, \\ r & \text{if } 0 \leq r \leq 1, \\ \min\{1, \xi_R(r)\} & \text{if } r \geq 1. \end{cases} \quad (12.40)$$

The above limiters are distinguished by their inherent numerical diffusion. We have listed these slope limiters (12.37) to (12.40) in order of increasing numerical diffusion.

### 12.3.3 ENO Version of the MUSCL-Hancock Scheme

An alternative, simpler approach to construct non-linear versions of the MUSCL-Hancock scheme results from the application of the non-linear reconstruction method called Essentially Non-Oscillatory (ENO) [19, 20]. See Chap. 14 and [21], and the many references therein. For a scalar mesh function $\{q_k\}$ the relevant differences are

$$\Delta_{i-\frac{1}{2}} = q_i^n - q_{i-1}^n \quad \text{and} \quad \Delta_{i+\frac{1}{2}} = q_{i+1}^n - q_i^n \quad (12.41)$$

and we choose the ENO slope as

$$\Delta_i^{ENO} = \begin{cases} \Delta_{i-\frac{1}{2}} & \text{if } |\Delta_{i-\frac{1}{2}}| < |\Delta_{i+\frac{1}{2}}|, \\ \Delta_{i+\frac{1}{2}} & \text{if } |\Delta_{i-\frac{1}{2}}| \geq |\Delta_{i+\frac{1}{2}}|. \end{cases} \quad (12.42)$$

The ENO method selects the polynomial with the smallest slope in absolute value. Some final remarks are in order.

1. **Higher-order ENO methods.** First we note that higher-order ENO methods exist [21]. In order to ensure conservation, these higher-order ENO methods require special procedures, omitted here.
2. **The WENO variant.** A variant of ENO, called WENO (for Weighted ENO) [22, 23] utilises linear combinations of potential ENO polynomials.
3. **Reconstruction in characteristic variables.** The reconstruction is usually performed in terms of *characteristic variables*, rather than conserved or physical variables. It is important to note that this is an essential ingredient of reconstruction-based higher-order methods, not studied here. For further details see [21] and the many references therein.

## 12.4 FORCE-Based TVD Schemes: The SLIC Method

In this section we present a second-order TVD scheme that is an extension of the first-order centred FORCE scheme [24, 25]. See Chaps. 10 and 11. The scheme is of the slope-limiter type and results from replacing the Godunov flux (12.28) from solving the Riemann problem (12.27) by the FORCE flux. First, from Chap. 11 we recall the Lax–Friedrichs, the Lax–Wendroff and the FORCE fluxes for the Riemann problem for a system of conservation laws

$$\left. \begin{array}{l} \partial_t \mathbf{Q} + \partial_x \mathbf{F}(\mathbf{Q}) = \mathbf{0} \;, \\ \mathbf{Q}(x,0) = \begin{cases} \mathbf{Q}_L & \text{if } x < x_{i+\frac{1}{2}} \;, \\ \mathbf{Q}_R & \text{if } x > x_{i+\frac{1}{2}} \;. \end{cases} \end{array} \right\} \quad (12.43)$$

The Lax–Friedrichs flux is

$$\mathbf{F}_{i+\frac{1}{2}}^{LF} = \frac{1}{2}[\mathbf{F}(\mathbf{Q}_L) + \mathbf{F}(\mathbf{Q}_R)] - \frac{1}{2}\frac{\Delta x}{\Delta t}(\mathbf{Q}_R - \mathbf{Q}_L) \;. \quad (12.44)$$

The Lax–Wendroff flux is

$$\left. \begin{array}{l} \mathbf{F}_{i+\frac{1}{2}}^{LW} = \mathbf{F}(\mathbf{Q}_{i+\frac{1}{2}}^{LW}) \;, \\ \mathbf{Q}_{i+\frac{1}{2}}^{LW} = \frac{1}{2}(\mathbf{Q}_L + \mathbf{Q}_R) - \frac{1}{2}\frac{\Delta t}{\Delta x}[\mathbf{F}(\mathbf{Q}_R) - \mathbf{F}(\mathbf{Q}_L)] \;. \end{array} \right\} \quad (12.45)$$

The FORCE flux [24, 25] is

$$\mathbf{F}_{i+\frac{1}{2}}^{force} = \mathbf{F}_{i+\frac{1}{2}}^{force}(\mathbf{Q}_L, \mathbf{Q}_R) = \frac{1}{2}[\mathbf{F}_{i+\frac{1}{2}}^{LF}(\mathbf{Q}_L, \mathbf{Q}_R) + \mathbf{F}_{i+\frac{1}{2}}^{LW}(\mathbf{Q}_L, \mathbf{Q}_R)] \;. \quad (12.46)$$

Then a second-order TVD centred scheme based on the FORCE flux (12.46) emerges from the scheme of the slope-limiter type and results from replacing the Godunov flux (12.28) from solving the Riemann problem (12.27) by the FORCE flux, namely

$$\mathbf{F}_{i+\frac{1}{2}}^{force} = \mathbf{F}_{i+\frac{1}{2}}^{force}(\tilde{\mathbf{Q}}_i^R, \tilde{\mathbf{Q}}_{i+1}^L) \;. \quad (12.47)$$

The two arguments of the FORCE flux function are the evolved boundary extrapolated values (12.26) in the MUSCL-Hancock method, namely

$$\mathbf{Q}_L \equiv \tilde{\mathbf{Q}}_i^R \;, \quad \mathbf{Q}_R \equiv \tilde{\mathbf{Q}}_{i+1}^L \;. \quad (12.48)$$

**SLIC and FLIC schemes.** The scheme presented was called in SLIC in [1]; it is exceeding simple, partly due to the fact that there is no need to explicitly solve the Riemann problem. This, apart from the ease of implementation, broadens the scope of scheme, as it is applicable to any hyperbolic system. An analogous scheme called FLIC was also proposed in [1]. The theoretical bases of the TVD versions of the schemes are given in [25]. In principle, one can also use SLIC or FLIC in combination with the ENO reconstruction procedure (12.42), rather than with TVD slope limiters.

## 12.5 Numerical Results

We show numerical results obtained from some of the numerical methods studied in this chapter; we compare the results with exact solutions. Two test problems are selected; these are Tests 1 and 2 introduced in Chap. 8, modified here to include a passive scalar here denoted as $c(x, t)$, as for a one-dimensional model to include pollutant transport. Thus, in addition to the initial conditions for the variables water depth $h(x, t)$ and particle velocity $u(x, t)$, we include initial conditions for the concentration variable $c(x, t)$. The initial conditions are given in Table 12.1. In addition, for each test, the position $x_0$ of the initial discontinuity at time $t = 0$ and the output time $T_{out}$ are specified. In all computations shown, the domain length is $L = 50$ m and is discretised by $M = 100$ cells; the safety CFL coefficient is $S_c = 0.9$. For the WAF results the flux limiter SUPERBEE (12.16) is used and for the MUSCL-Hancock and SLIC results the slope limiter (12.37) is used.

Figures 12.6, 12.7, 12.8 show numerical results compared with the exact solution. Figure 12.6 shows results for Test 1. The three challenges of this test for the numerical methods are (i) the right-propagating shock, (ii) the left *critical* rarefaction with a critical point at $x_0 = 10$, and (iii) the middle contact discontinuity. The left column of the three profiles at $t_{out} = 7.0$ correspond to the first-order Godunov method in conjunction with the HLLC approximate Riemann solver. These profiles are compared with those from the WAF TVD method, also in conjunction with the HLLC solver, on the right-hand side column, at the same output time. We note that the first-order HLLC scheme (left) gives reasonably accurate results, coping very well with all three challenges. Nonetheless, the improvements provided by the TVD extension (right) are visible. These are most evident in the resolution of the two discontinuities, particularly the contact discontinuity.

Figure 12.7 shows results for Test 1 from the FORCE scheme (left column) and the FORCE-based TVD SLIC scheme (right column). We note that FORCE (left) copes well with the shock wave and the critical point, but the resolution of the contact

**Table 12.1** Data for two test problems with exact solution. The length of the channel is 50 m, $x_0$ is position of initial discontinuity; $T_{out}$ is the output time in seconds

| Test | $h_L$ | $u_L$ | $c_L$ | $h_R$ | $u_R$ | $c_R$ | $x_0$ | $T_{out}$ |
|---|---|---|---|---|---|---|---|---|
| 1 | 1.0 | 2.5 | 1.0 | 0.1 | 0.0 | 0.0 | 10.0 | 7.0 |
| 2 | 1.0 | −5.0 | 1.0 | 1.0 | 5.0 | 0.0 | 25.0 | 2.5 |

## 12.5 Numerical Results

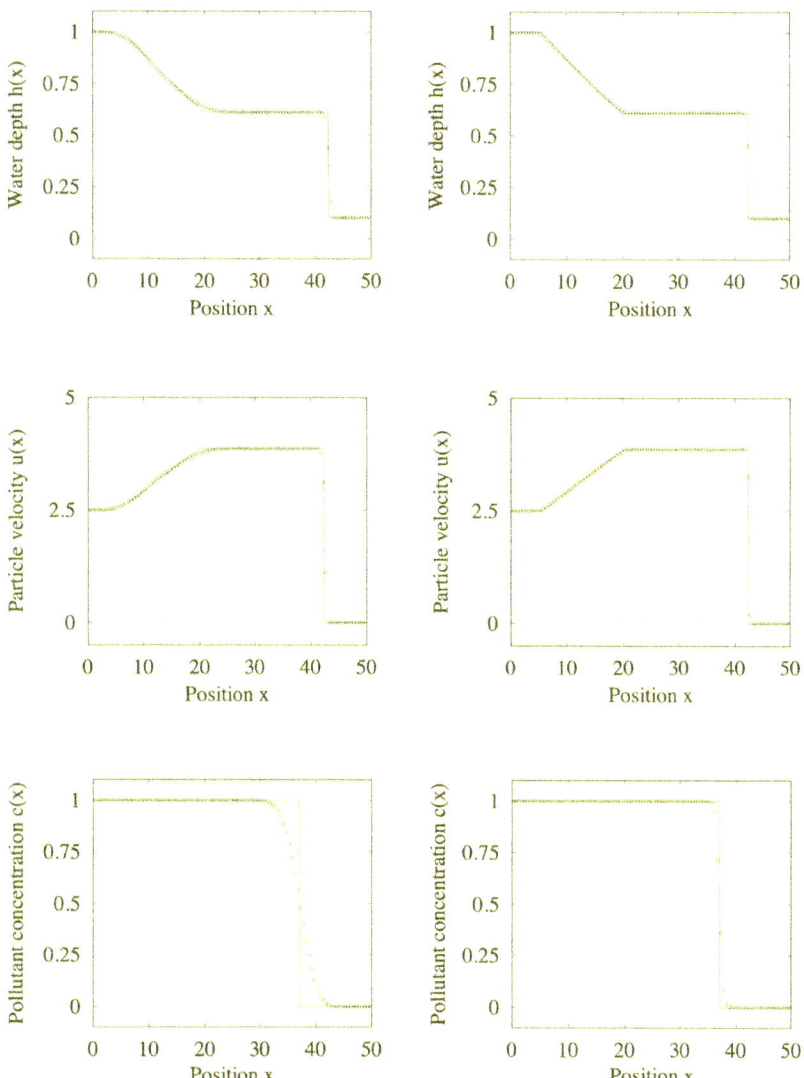

**Fig. 12.6** Test 1: Numerical results (symbol) and exact solution (line) at time $T_{out} = 7$ s. Left column: first-order HLLC approximate Riemann solver. Right column: second-order TVD WAF method in conjunction with HLLC

discontinuity is poor; this is expected from *centred methods*. The SLIC results on the right-hand side column show a very clear improvement relative to the first-order FORCE results on the left column. The improvements are most evident for the shock wave and for the rarefaction. The contact discontinuity is still rather *smeared*. Compare results between those of Figs. 12.6 and 12.7. In general, the upwind-based

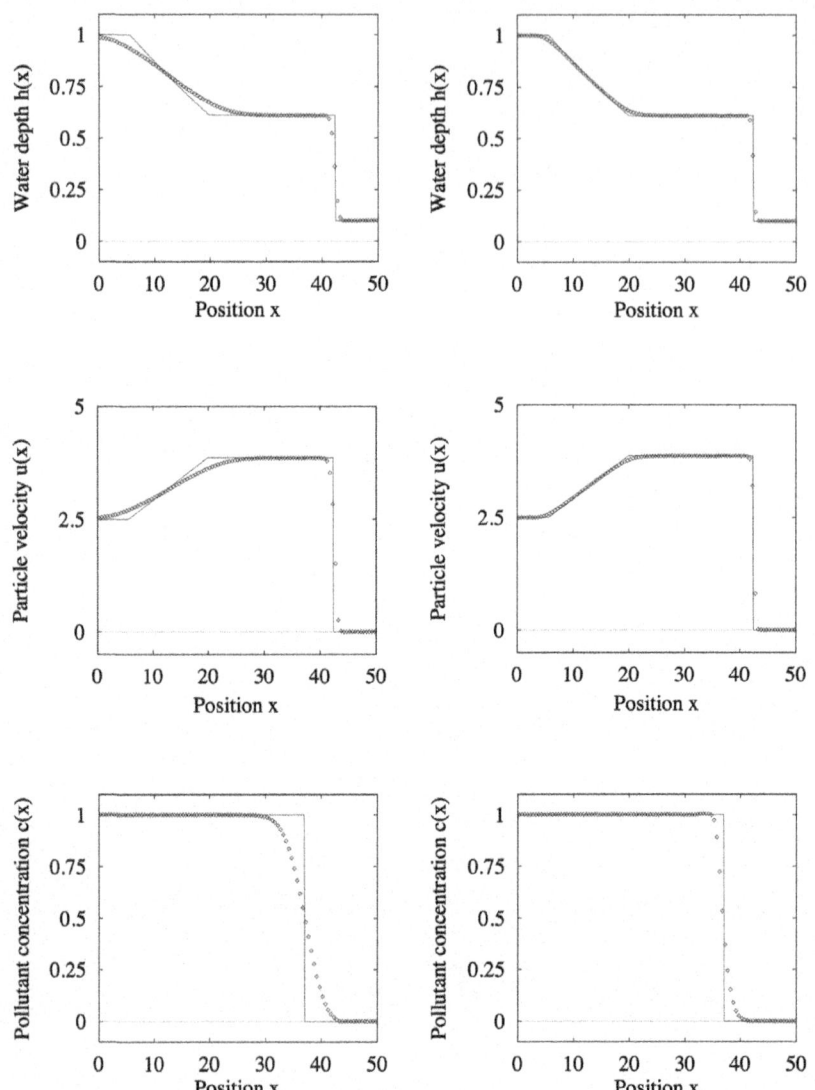

**Fig. 12.7** Test 1: Numerical results (symbol) and exact solution (line) at time $T_{out} = 7$ s. Left column: first-order FORCE method. Right column: second-order TVD SLIC method in conjunction with FORCE

method gives more accurate results than the centred-based method. The advantages of the FORCE-based method are simplicity and generality.

Results from Test 2 are shown in Fig. 12.8. This test problem contains two challenges: (i) the near-vacuum zone in the middle of the domain generated by the two strong rarefaction waves travelling in opposite directions, and (ii) the stationary con-

## 12.5 Numerical Results

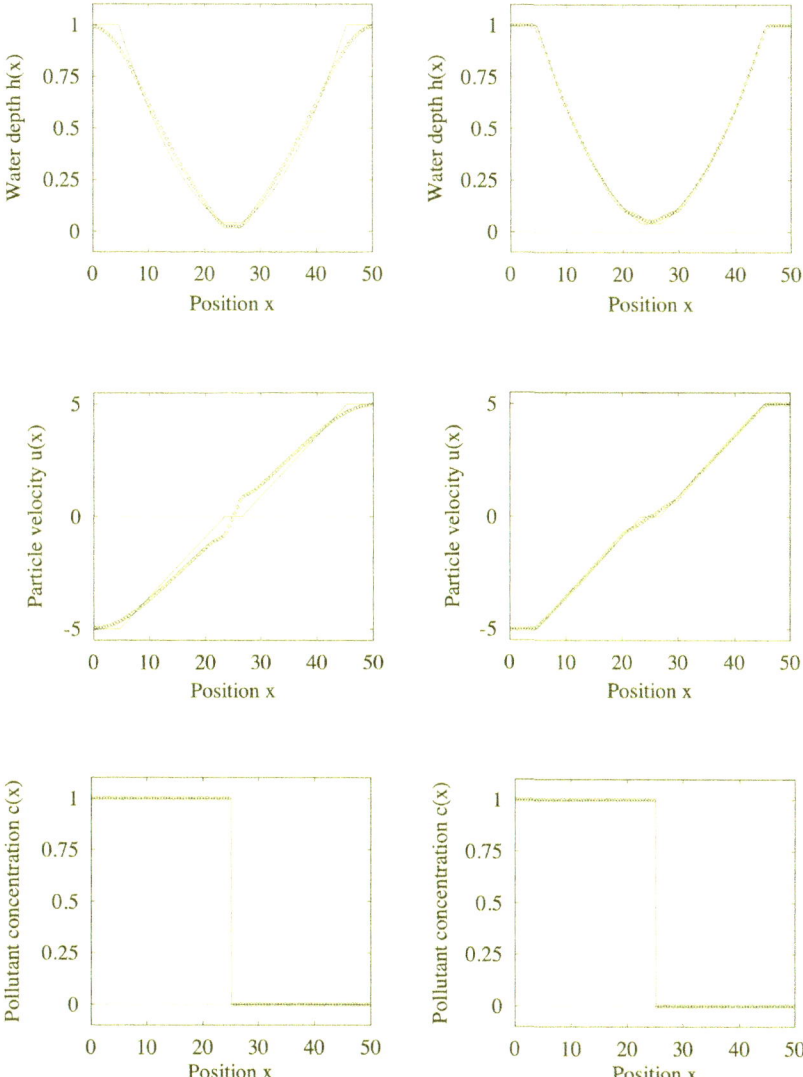

**Fig. 12.8** Test 2: Numerical results (symbol) and exact solution (line) at time $T_{out} = 2.5\,\text{s}$. Left column: first-order HLLC approximate Riemann solver. Right column: second-order TVD WAF method in conjunction with HLLC

tact discontinuity, the middle wave. We only show results for the upwind-based WAF method with the HLLC Riemann solver. The left column shows first-order results from the Godunov method with HLLC. The results are satisfactory, except for the resolution of the velocity profile in the vicinity of the near-vacuum region. The resolution of the stationary contact discontinuity is perfect, it is exact. The second-order

WAF method on the right-hand side shows some improvements, relative to the first-order results on the left. These are evident for the rarefactions and for the velocity in the near-vacuum region. For the contact discontinuity there is nothing to be improved, relative to the first-order HLLC based method on the left.

## 12.6 Conclusions

Second-order non-linear methods have been presented; these are extensions of first-order monotone schemes studied in Chaps. 10 and 11. The schemes have been tested for two problems that include a species equation for the concentration of a pollutant; this species equation adds a third wave to the solution, a contact discontinuity in the Riemann problem structure. It is seen that resolving contact discontinuities accurately is more difficult than resolving shock waves. It is also seen that upwind methods give generally more accurate results than centred methods; such advantage of upwind methods over centred methods is most evident for contact discontinuities. As the speed of the contact wave tends to zero, centred methods get worse and upwind methods improve their accuracy further, to the point that a stationary contact is resolved exactly. Note, however, that if an *incomplete* Riemann solver is used in the Godunov-type upwind methods, then this property of exact recognition of stationary contacts is lost and results from upwind methods are similar to those of centred methods.

Second-order non-linear methods, both upwind and centred based, show significant improvements relative to their first-order counterparts, and are thus recommended for practical applications. More sophisticated methods are available today. As an example, in Chap. 13 we introduce the family of ADER methods; its general framework allows for the construction of non-linear schemes of arbitrary order of accuracy in both space and time, on structured and unstructured meshes for multi-dimensional problems. Moreover, ADER operates in the frameworks of both finite volumes and discontinuous Galerkin finite elements.

## References

1. E.F. Toro, *Riemann Solvers and Numerical Methods for Fluid Dynamics. A Practical Introduction*. (Springer, 1997)
2. E.F. Toro, *Riemann Solvers and Numerical Methods for Fluid Dynamics. A Practical Introduction*, 2nd edn. (Springer, 1999)
3. E.F. Toro, *Riemann Solvers and Numerical Methods for Fluid Dynamics. A Practical Introduction*, 3rd edn. (Springer, 2009)
4. E.F. Toro, The Riemann problem: solvers and numerical fluxes, in *Elsevier Handbook of Numerical Methods for Hyperbolic Problems (Chap. 2)*, ed. by R. Abgrall, C.W. Shu (2016), pp. 17:19–54
5. E.F. Toro, P. García-Navarro, Godunov-type methods for free-surface shallow flows: a review. J. Hydraulic Res. **45**(6), 736–751 (2007)

# References

6. E.F. Toro, A weighted average flux method for hyperbolic conservation laws. Proc. Roy. Soc. Lond. **A423**, 401–418 (1989)
7. E.F. Toro, The weighted average flux method applied to the time-dependent Euler equations. Phil. Trans. Roy. Soc. Lond. **A341**, 499–530 (1992)
8. E.F. Toro, Riemann problems and the WAF method for solving two-dimensional shallow water equations. Phil. Trans. Roy. Soc. Lond. **A338**, 43–68 (1992)
9. G. Watson, D.H. Peregrine, E.F. Toro, Numerical solution of the shallow water equations on a beach using the weighted average flux method, in *Computational Fluid Dynamics 1992*, vol. 1 (Elsevier, 1992), pp. 495–502
10. L. Fraccarollo, E.F. Toro, Experimental and numerical assessment of the shallow water model for two-dimensional dam-break type problems. J. Hydraul. Res. **33**, 843–864 (1995)
11. A. Siviglia, E.F. Toro, The WAF method and splitting procedures for simulating hydro and thermal peaking waves in open channel flows. J. Hydraul. Eng. **135**(8), 651–662 (2009)
12. Sofiane Khelladi, Riadh Ata, Sara Pavan, Eleuterio F. Toro, A Weighted Average Flux (WAF) scheme applied to shallow water equations for real-life applications. Adv. Water Res. **62**, 155–172 (2013)
13. S.J. Billett, E.F. Toro, Implementing a three-dimensional finite volume WAF-type scheme for the Euler equations (1996)
14. S.J. Billett, E.F. Toro, WAF-type schemes for multidimensional hyperbolic conservation laws. J. Comp. Phys. **130**, 1–24 (1997)
15. P.K. Sweby, High resolution schemes using flux limiters for hyperbolic conservation laws. SIAM J. Numer. Anal. **21**, 995–1011 (1984)
16. P.K. Sweby, TVD schemes for inhomogeneous conservation laws, in *Notes on Numerical Fluid Mechanics*, vol. 24. Non–Linear Hyperbolic Equations-Theory, Computation Methods and Applications (Vieweg, 1989), pp. 599–607
17. B. van Leer, On the relation between the upwind-differencing schemes of Godunov, Enguist-Osher and Roe. SIAM J. Sci. Stat. Comput. **5**(1), 1–20 (1984)
18. V.P. Kolgan, Application of the principle of minimum derivatives to the construction of difference schemes for computing discontinuous solutions of gas dynamics (in Russian). Uch. Zap. TsaGI, Russia **3**(6), 68–77 (1972)
19. A. Harten, S. Osher, Uniformly high-order accurate nonoscillatory schemes I. SIAM J. Numer. Anal. **24**(2), 279–309 (1987)
20. A. Harten, B. Engquist, S. Osher, S.R. Chakravarthy, Uniformly high order accuracy essentially non-oscillatory schemes III. J. Comput. Phys. **71**, 231–303 (1987)
21. E.F. Toro, L.O. Müller, *Computational Bodily Fluid Dynamics. Models and Algorithms (to appear)* (Springer, 2024)
22. X.D. Liu, S. Osher, T. Chan, Weighted Essentially Non-oscillatory Schemes. J. Comput. Phys. **115**, 200–212 (1994)
23. G.S. Jiang, C.W. Shu, Efficient implementation of weigthed ENO schemes. Technical Report ICASE 95–73, NASA Langley Research Center, Hampton, USA (1995)
24. E.F. Toro, On Glimm-related schemes for conservation laws. Technical Report MMU–9602, Department of Mathematics and Physics, Manchester Metropolitan University, UK (1996)
25. E.F. Toro, S.J. Billett, Centred TVD schemes for hyperbolic conservation laws. IMA J. Numer. Anal. **20**, 47–79 (2000)

# Chapter 13
# Sources and Multidimensions

**Abstract** This chapter addresses two major issues when solving advection-reaction partial differential equations (PDEs) in general domains, namely the presence of source terms in the PDEs and the multidimensional character of the computational domain. We present simple but effective methods to deal with both situations. First, 1D advection-reaction systems are split into an advection system of PDEs and a reaction part consisting of a system of Ordinary Differential Equations (ODEs). We call the procedure *advection-reaction splitting*. For the advection part, there are several schemes presented in previous chapters that are directly applicable. For the reaction part, a section on numerical methods for ODEs is included. First and second-order advection-reaction splitting schemes are formulated. For multidimensional advection systems, we present two approaches. The first one relies on *dimensional splitting*, whereby 1D schemes applied to augmented 1D advection systems are solved in succession. First and second-order dimensional splitting schemes are presented. The second approach rests on *unsplit, simultaneous-update* finite volume formulations, whereby the updating includes contributions from all element-edge fluxes. This approach is suitable for unstructured meshes. A major issue enters the scene in this second class of methods, namely stability. A section on stability analysis for multidimensional schemes is included, and examples of stable schemes are given. Finally, we present a particular class of unsplit second-order finite volume methods of the WAF type, which exhibit very generous stability regions in 2D and 3D. Moreover, the WAF framework may be extended to other classes of schemes, including centred schemes.

## 13.1 Introductory Remarks

In this chapter we briefly discuss the potential extension of the numerical methods presented in Chaps. 10, 11 and 12 to deal with two mayor issues for computing solutions to advection-reaction PDEs in multiple space dimensions, namely source terms and multiple spatial dimensions. For one-dimensional non-linear advection-reaction systems we present an **advection-reaction splitting** approach that *splits* locally the processes of advection and reaction. For the advection part all the methods studied in previous chapters are applicable. For the reaction part we include a section on numerical methods for solving systems of Ordinary Differential Equations (ODEs), along with references for further study. Here we note the concept of **stiffness**, usually associated with chemical reactions, and its challenge to numerical methods. For multidimensional systems of PDEs we extend the splitting approach, in which spatial dimensions are split locally in time. This approach is also known as **dimensional splitting** in the literature. Again, all methods for (augmented) one-dimensional systems studied in previous chapters are directly applicable, maintaining the stability limit of the one-dimensional components of the overall scheme. The splitting methods presented, either advection-reaction or dimensional splitting, have an accuracy barrier; they are at most second-order accurate. More general schemes are introduced in Chap. 14.

An alternative approach to construct *unsplit schemes* based on *simultaneous update formulae* is presented. The presentation is largely based on structured meshes, Cartesian meshes in particular, even though some of the schemes are applicable to unstructured meshes. Schemes in this class exclude the direct application of several of the one-dimensional methods presented in previous chapters. The overriding issue here is **stability**. A section on von Neumann linear stability analysis for multidimensional simultaneous-update finite volume schemes is included, along with examples on stable schemes and their stability limits.

The schemes described in this chapter are applicable to three-dimensional non-linear systems including source terms. In most situations the schemes are likely to be successful. There are however situations is which the methods might need modifications or cautious use. The potential challenges may come from the source terms, apparently the simplest part of the whole advection-reaction systems. Source terms are generally algebraic (non differential) functions of the unknowns, the independent variables, and parameters of the physical problem. The source terms may give rise to **stiffness** of the system; this issue is normally resolved by the adoption of implicit methods. Also challenging are source terms of the *geometric type*, when the source term depends on parameters of the problem (not the unknowns of the PDEs) and more specifically, on gradients of the parameters. In the shallow water equations this kind of source terms is present via the derivatives of the bathymetry function $b(x, y)$ for rives, lakes or oceans. A class of successful schemes to deal with this kind of source terms are called **well-balanced schemes**. In the steady or nearly steady state, the PDEs become balance laws, which perfectly balance the flux gradient with the source terms. We do not deal with this class of schemes in this book. The reader is advised to revise the extensive and dynamic literature on the subject.

In Sect. 13.2 we deal with advection-reaction splitting to treat source terms in the PDEs; in Sect. 13.3 we study numerical methods for solving ODEs; in Sect. 13.4 we deal with dimensional splitting as applied to multidimensional systems; in Sect. 13.5 we study the von Neumann method for analysing the linear stability of multidimensional schemes; in Sect. 13.6 we introduce unsplit, simultaneous-update schemes. Finally, concluding remarks are found in Sect. 13.7.

General background on the subjects of this chapter is found in Chaps. 15 and 16 of the textbooks [1–3]. See also LeVeque [4] and Godlewski and Raviart [5].

## 13.2 Treatment of Source Terms by Splitting

Here we deal with *splitting methods* for solving non-linear systems of hyperbolic equations with source terms. The term **splitting** in the field of numerical methods for PDEs has different meanings in the literature and care is required in interpreting the term correctly. In this section we split the advection terms in the PDEs from the source terms (reaction). We call the procedure **advection-reaction splitting**.

### 13.2.1 Preliminary Notions

The objective is to solve numerically the following general initial-boundary value problem for an *inhomogeneous* non-linear system

$$\begin{aligned}
\text{PDEs:} \quad & \partial_t \mathbf{Q} + \partial_x \mathbf{F}(\mathbf{Q}) = \mathbf{S}(\mathbf{Q}), \quad x \in [a, b], \quad t > 0, \\
\text{ICs:} \quad & \mathbf{Q}(x, 0) = \mathbf{Q}^{(0)}(x), \quad x \in [a, b], \\
\text{BCs:} \quad & \mathbf{Q}(a, t) = \mathbf{B}_L(t), \quad \mathbf{Q}(b, t) = \mathbf{B}_R(t), \quad t \geq 0.
\end{aligned} \quad (13.1)$$

$\mathbf{Q}(x, t)$ is the vector of *conserved variables*; $\mathbf{F}(\mathbf{Q})$ is the flux vector, or *physical flux*; $\mathbf{S}(\mathbf{Q})$ is the *source term vector*; $\mathbf{Q}^{(0)}(x)$ is the initial condition; $\mathbf{B}_L(t)$ and $\mathbf{B}_R(t)$ are the *boundary conditions* on the left ($x = a$) and right ($x = b$) boundaries, respectively.

One of the *ideal* methods to solve (13.1) is the finite volume method based on the explicit conservative formula

$$\mathbf{Q}_i^{n+1} = \mathbf{Q}_i^n - \frac{\Delta t}{\Delta x}(\mathbf{F}_{i+\frac{1}{2}} - \mathbf{F}_{i-\frac{1}{2}}) + \Delta t \mathbf{S}_i. \quad (13.2)$$

This formula is derived from the integral form of the conservation laws and accounts simultaneously for the flux gradient via the numerical flux $\mathbf{F}_{i+\frac{1}{2}}$ and the source terms via the numerical source $\mathbf{S}_i$. The numerical flux is

$$\mathbf{F}_{i+\frac{1}{2}} = \frac{1}{\Delta t} \int_{t_n}^{t_{n+1}} \mathbf{F}(\mathbf{Q}(x_{i+\frac{1}{2}}, t)) dt \quad (13.3)$$

and the numerical source is

$$\mathbf{S}_i = \frac{1}{\Delta t \Delta x} \int_{t_n}^{t_{n+1}} \int_{x_{i-\frac{1}{2}}}^{x_{i+\frac{1}{2}}} \mathbf{S}(\mathbf{Q}(x, t)) dx dt. \quad (13.4)$$

Formula (13.2), with definitions (13.3)–(13.4), constitute a reference approach to solve the full problem (13.1).

Next we consider the simpler computational problem via simpler methods in which the determination of the numerical flux is separated from that of the numerical source. To this end we adopt the framework of the initial-value problem for an advection-reaction system, namely

$$\begin{aligned} \text{PDEs:} \quad & \partial_t \mathbf{Q} + \partial_x \mathbf{F}(\mathbf{Q}) = \mathbf{S}(\mathbf{Q}), \quad x \in [a,b], \quad t > 0, \\ \text{ICs:} \quad & \mathbf{Q}(x, t_n) = \mathbf{Q}^n = \{\mathbf{Q}_i^n\}, \quad x \in [a,b]. \end{aligned} \qquad (13.5)$$

The vector of sources $\mathbf{S}(\mathbf{Q})$ is in general an algebraic function of $\mathbf{Q}(x,t)$ and possibly other physical parameters of the problem at hand. In some situations the source term may also be an explicit function of the independent variables $x$ and $t$.

A simplified version of (13.5) results from neglecting the source term $\mathbf{S}(\mathbf{Q})$ to obtain a *homogeneous*, pure advection system of PDEs

$$\text{Advection system:} \quad \partial_t \mathbf{Q} + \partial_x \mathbf{F}(\mathbf{Q}) = \mathbf{0}. \qquad (13.6)$$

In Chaps. 9, 10, 11 and 12 we studied methods that are directly applicable to solve (13.6). Another simplification of (13.5) results from the assumption of no spatial variations, that is $\partial_x \mathbf{F}(\mathbf{Q}) = \mathbf{0}$. In this case one obtains a system of Ordinary Differential Equations (ODEs) in time $t$, for fixed $\hat{x}$, which we term reaction system

$$\text{Reaction system:} \quad \frac{d}{dt} \mathbf{Q}(\hat{x}, t) = \mathbf{S}(\mathbf{Q}(\hat{x}, t)). \qquad (13.7)$$

Section 13.3 gives a brief introduction to numerical methods for solving the reaction problem (13.7). Having methodologies available for each separate problem, the PDEs in (13.6) and the ODEs in (13.7), we now combine such methods to solve the full inhomogeneous problem (13.5), the details of which are explained in Sect. 13.2.2.

But before doing so we consider some very simple examples of hyperbolic conservation laws with source terms; these are useful in understanding the basic issues involved in solving numerically the more general inhomogeneous problem (13.5). To this end we assume model problems for a quantity $q(x,t)$ with $-\infty < x < \infty$ and $t > 0$.

**Model problem I.** The problem is

$$\begin{aligned} \text{PDE:} \quad & \partial_t q + \lambda \partial_x q = s(q) := \alpha q, \\ \text{ICs:} \quad & q(x, t_n) = q^{(n)}(x). \end{aligned} \qquad (13.8)$$

Here $\lambda$ is a constant wave propagation speed and $\alpha$ is a constant reaction-rate parameter of dimensions reciprocal of time. The exact solution of this problem is

$$q(x, t) = q^{(n)}(x - \lambda t) e^{\alpha t}. \qquad (13.9)$$

Note in particular that for $\alpha = 0$ we recover the exact solution to the homogeneous equation $\partial_t q + \lambda \partial_x q = 0$, namely $q^{(n)}(x - \lambda t)$. The solution (13.9) represents a

13.2 Treatment of Source Terms by Splitting

wave profile $q^{(n)}(x)$ travelling at constant speed $\lambda$ and being attenuated or amplified in time, depending on whether $\alpha < 0$ or $\alpha > 0$. For stability of the PDE one chooses $\alpha \leq 0$ in (13.8); see Chap. 2. From the numerical point of view, this very simple problem can be rather challenging, particularly in resolving the time evolution accurately; see Chap. 14 and [6], for example.

**Model problem II.** Another model problem is

$$\left.\begin{array}{l} \text{PDE:} \quad \partial_t q + \lambda \partial_x q = s(q) := b(x), \\ \text{IC:} \quad q(x, t_n) = q^{(n)}(x). \end{array}\right\} \quad (13.10)$$

Note that the source term here is $s(q) = b(x)$, assumed to be a prescribed function of $x$. This is analogous to problems involving bed or breadth variation in shallow water models, in which the bathymetry is a function of $x$ alone; see Chap. 1. The exact solution of (13.10) is

$$q(x, t) = q^{(n)}(x - \lambda t) + \frac{1}{\lambda} \int_{x-\lambda t}^{x} b(\xi) \, d\xi. \quad (13.11)$$

These simple problems with exact solution may be helpful in assessing and interpreting the performance of numerical methods intended for solving more general inhomogeneous problems.

### 13.2.2 Splitting for Systems with Source Terms

The inhomogeneous problem to solve is

$$\left.\begin{array}{l} \text{PDEs:} \quad \partial_t \mathbf{Q} + \partial_x \mathbf{F}(\mathbf{Q}) = \mathbf{S}(\mathbf{Q}), \quad x \in [a, b], \quad t > 0, \\ \text{ICs:} \quad \mathbf{Q}(x, t_n) = \mathbf{Q}^n = \{\mathbf{Q}_i^n\}, \quad x \in [a, b]. \end{array}\right\} \quad (13.12)$$

We want to evolve $\mathbf{Q}^n$ from time $t = t_n$ to the new value $\mathbf{Q}^{n+1}$ at $t = t_{n+1}$ by a time step $\Delta t = t_{n+1} - t_n$. A basic splitting scheme relies on solving in succession the following two IVPs:

$$\text{Advection IVP:} \begin{cases} \text{PDEs:} \quad \partial_t \mathbf{Q} + \partial_x \mathbf{F}(\mathbf{Q}) = \mathbf{0} \\ \text{ICs:} \quad \mathbf{Q}(x, t_n) = \mathbf{Q}^n \end{cases} \stackrel{\Delta t}{\Longrightarrow} \mathbf{Q}^{(adv)} \quad (13.13)$$

and

$$\text{Reaction IVP:} \begin{cases} \text{ODEs:} \quad \dfrac{d}{dt}\mathbf{Q} = \mathbf{S}(\mathbf{Q}) \\ \text{ICs:} \quad \mathbf{Q}(x, t_n) = \mathbf{Q}^{(adv)} \end{cases} \stackrel{\Delta t}{\Longrightarrow} \mathbf{Q}^{n+1}. \quad (13.14)$$

In the advection IVP (13.13) the initial condition of the homogeneous pure advection problem is the initial condition $\mathbf{Q}^n$ of the full problem (13.12) and the problem is solved for a time $\Delta t$, resulting in solution $\mathbf{Q}^{(adv)}$, which may be regarded as a *predicted* solution to the full problem (13.12).

In the reaction IVP (13.14) the initial condition is given by the solution $\mathbf{Q}^{(adv)}$ of the advection IVP, and the problem is solved also for a time $\Delta t$. The solution of the reaction IVP (13.14) is then regarded as the sought approximate solution to the full problem (13.12).

We call this splitting procedure **advection-reaction splitting**. To represent the advection IVP (13.13) and the reaction IVP (13.14), let us assume the *advection operator* $\mathcal{A}^{(t)}(\mathbf{Q})$ and the *source/reaction operator* $\mathcal{S}^{(t)}(\mathbf{Q})$, as follows:

$$\mathcal{A}^{(\Delta t)}: \quad \mathbf{Q}_i^{(adv)} = \mathbf{Q}_i^n - \frac{\Delta t}{\Delta x}\left[\mathbf{F}_{i+\frac{1}{2}} - \mathbf{F}_{i+\frac{1}{2}}\right] \tag{13.15}$$

and

$$\mathcal{S}^{(\Delta t)}: \quad \mathbf{Q}_i^{n+1} = \mathbf{Q}_i^{(adv)} + \Delta t\, \mathbf{S}(\mathbf{Q}_i^{(s)}). \tag{13.16}$$

$\mathbf{Q}_i^{(s)}$ is a value at which the source term vector $\mathbf{S}$ is evaluated, for which several possibilities are available. Then the splitting scheme (13.13)–(13.14) may be conveniently written thus

$$\mathbf{Q}^{n+1} = \mathcal{S}^{(\Delta t)} \mathcal{A}^{(\Delta t)}(\mathbf{Q}^n). \tag{13.17}$$

The above notation means that one first applies the operator $\mathcal{A}^{(\Delta t)}$ on the data $\mathbf{Q}^n$ for a time $\Delta t$, followed by the application of $\mathcal{S}^{(\Delta t)}$ to the result, also for a time $\Delta t$. An analogous scheme is

$$\mathbf{Q}^{n+1} = \mathcal{A}^{(\Delta t)} \mathcal{S}^{(\Delta t)}(\mathbf{Q}^n). \tag{13.18}$$

In this case one begins with the reaction operator, then followed by the advection operator. Both splitting schemes (13.17) and (13.18) are first-order accurate in time, if both $\mathcal{S}$ and $\mathcal{A}$ are at least first-order accurate in time.

A second-order accurate splitting scheme is

$$\mathbf{Q}^{n+1} = \mathcal{S}^{(\frac{1}{2}\Delta t)} \mathcal{A}^{(\Delta t)} \mathcal{S}^{(\frac{1}{2}\Delta t)}(\mathbf{Q}^n), \tag{13.19}$$

provided that $\mathcal{S}$ and $\mathcal{A}$ are at least second-order accurate in time. An analogous splitting scheme is

$$\mathbf{Q}^{n+1} = \mathcal{A}^{(\frac{1}{2}\Delta t)} \mathcal{S}^{(\Delta t)} \mathcal{A}^{(\frac{1}{2}\Delta t)}(\mathbf{Q}^n). \tag{13.20}$$

Clearly both second-order schemes (13.19) and (13.20) are 50% more costly than the first-order schemes (13.17) and (13.18), but the potential gains in accuracy are well worth it.

Note that the full scheme (13.15)–(13.16) may be written thus

## 13.2 Treatment of Source Terms by Splitting

$$\mathbf{Q}_i^{n+1} = \mathbf{Q}_i^n - \frac{\Delta t}{\Delta x}\left[\mathbf{F}_{i+\frac{1}{2}} - \mathbf{F}_{i-\frac{1}{2}}\right] + \Delta t\ \mathbf{S}(\mathbf{Q}_i^{(s)}). \tag{13.21}$$

Compare (13.21) with the *model* scheme (13.2). As already pointed out, there are potentially several choices for $\mathbf{Q}_i^{(s)}$. One obvious possibility is $\mathbf{Q}_i^{(s)} = \mathbf{Q}_i^n$, leading to the scheme

$$\mathbf{Q}_i^{n+1} = \mathbf{Q}_i^n - \frac{\Delta t}{\Delta x}\left[\mathbf{F}_{i+\frac{1}{2}} - \mathbf{F}_{i-\frac{1}{2}}\right] + \Delta t\ \mathbf{S}(\mathbf{Q}_i^n). \tag{13.22}$$

In the literature, scheme (13.22) is often called *centred*, in the sense that the numerical source does not have any influence from wave propagation, or upwinding. Scheme (13.22) may be interpreted as one version of the numerical source $\mathbf{S}_i$ in (13.2), in which the integrand in (13.4) is assumed to be constant and evaluated on the data $\mathbf{Q}_i^n$.

Practitioners, often claim the simple scheme (13.22) to be *unsplit*, merely because the scheme advances the solution in a single step. This is not so because (i) the computation of the numerical fluxes may not have included the influence of the source terms and (ii) the computation of the source term, which depends on $\mathbf{Q}$, may not have information from the advection part via the fluxes. Therefore scheme (13.23) is a splitting scheme, and as a matter of fact, it is the worst version, as we shall demonstrate via a model equation in Sect. 13.2.3.

Another possibility is to choose $\mathbf{Q}_i^{(s)} = \mathbf{Q}_i^{(adv)}$, or more generally, the linear combination

$$\mathbf{Q}_i^{(s)} = \beta \mathbf{Q}_i^{(adv)} + (1-\beta)\mathbf{Q}_i^n, \quad 0 \le \beta \le 1. \tag{13.23}$$

The value $\beta = 0$ leads to scheme (13.22) and the choice $\beta = 1$ leads to the scheme

$$\mathbf{Q}_i^{n+1} = \mathbf{Q}_i^n - \frac{\Delta t}{\Delta x}\left[\mathbf{F}_{i+\frac{1}{2}} - \mathbf{F}_{i-\frac{1}{2}}\right] + \Delta t\ \mathbf{S}(\mathbf{Q}_i^{(adv)}). \tag{13.24}$$

We shall show that scheme (13.24) is actually better than scheme (13.22) and constitutes what is regarded here as the *standard* advection-reaction splitting scheme.

**Exercise 13.1** (*Exact splitting*) It is worth remarking that the simple splitting scheme (13.17), when applied to the model problem (13.8), is actually exact when the operators $\mathcal{A}$ and $\mathcal{S}$ are exact. Details of the proof are given in Chap. 15 of [1–3]. The reader is invited to verify this.

Some closing remarks are in order.

1. **Simplicity and modularity**. The advection-reaction splitting schemes presented here are indeed very simple and benefit from the freedom available for choosing the numerical operators $\mathcal{A}$ for advection and $\mathcal{S}$ for reaction. In general, one may choose the best scheme for each type of sub-problem. For solving the advection (homogeneous) IVP (13.13) one can, for instance, apply any of the schemes studied in Chaps. 10–12, or some other method not studied here. For solving the ODEs (13.14) one may choose some appropriate ODE solver, for which well-documented libraries exist.

2. **ODE solvers**. Section 13.3 is devoted to a succinct review of numerical methods for solving systems of ordinary differential equations that can be used to tackle the reaction part of the advection-reaction splitting methodology presented here.
3. **Stiffness**. The advection-reaction splitting methodology is applicable to most practical problems but there are situations in which the methodology would need modifications. Stiff source terms are one issue. Choosing an appropriate ODE solver may help, but it may not be sufficient; see the early paper by LeVeque and Yee [7]. See also Dumbser et al. [8].
4. **Upwinding source terms**. As noted by Roe [9] and by Bermúdez and Vázquez [10], and many others since then, the concept of upwinding applied to the source term may be beneficial in dealing with complicated reaction terms.
5. **Advection-diffusion splitting**. To model water biochemistry in rivers, lakes and oceans, the shallow water equations are augmented with additional advection-diffusion reaction equations. The diffusion terms may be treated through an advection-diffusion splitting, analogous to the advection-reaction splitting just presented. See Brown and Toro [11, 12] and also the more recent paper by Siviglia and Toro [13].
6. **Well-balancing**. There are certain types of source terms for which standard methods, such as the ones presented here, do not work. Typical examples are geometric-type source terms, as for shallow water models with variable bathymetry and models for blood flow in vessels with variable mechanical properties. Such source terms require changes to the methodology. This is a relatively new subject of research and has become increasingly active. A few typical references on the subject include, amongst many others, Bermúdez and Vázquez [10], LeVeque [14], Vázquez [15], Castro et al. [16], Dumbser et al. [17], Canestrelli et al. [18], Castro et al. [19], Delestre and Lagrée [20], Müller, Parés and Toro [21], Müller and Toro [22], Ghitti et al. [23], Arpaia et al. [24], Pimentel et al. [25], Martaud and Berthon [26], González et al. [27], Navas-Montilla and Murillo [28], Barsukow and Berberich [29].

## 13.2.3 Upwinding and Advection-Reaction Splitting

Here we discuss some theoretical issues regarding the relationship between upwinding the source term and the advection-reaction splitting scheme just presented. We do so in terms of simple model problems.

The exact solutions to the model problems presented in Sect. 13.2.1 clearly showed the dependence of the solution on wave propagation direction. See also a preliminary discussion on this issue in Chap. 10. Consistent with this fact, Roe [9] put forward the idea of upwinding the source terms in inhomogeneous conservation laws, in a manner similar to that for constructing Godunov-type numerical fluxes for solving homogenous conservation laws. Further work in this direction was carried out by Bermúdez and Vázquez [10], who started by considering the model problem (13.8). For example, for $\lambda > 0$ one would expect the contribution $q_i^{(s)}$ in (13.16)

## 13.2 Treatment of Source Terms by Splitting

**Table 13.1** Relation between upwinding and splitting schemes for source terms and linear stability conditions

| $\omega$ (upwinding) | $\beta$ (splitting) | Linear stability range |
|---|---|---|
| 0 | 0 | $c + \frac{1}{2}r \leq 1$ |
| $\frac{1}{2}c$ | $\frac{1}{2}$ | $c \leq 1, \ r \leq 2$ |
| $c$ | 1 | $c \leq 1, \ r \leq 2$ |
| $\frac{1}{2}$ | $\frac{1}{2c}$ | $c \leq 1, \ r \leq 2$ |
| 1 | $\frac{1}{2c}$ | $c \leq 1, \ r \leq 2$ |

to the numerical source $s_i = s(q_i^{(s)})$ to depend on some combination of the *centred value* $q_i^n$ and the *upwind value* $q_{i-1}^n$. Bermúdez and Vázquez-Cendón expressed this dependence as

$$q_i^{(s)} = \omega q_{i-1}^n + (1-\omega) q_i^n, \quad 0 \leq \omega \leq 1. \tag{13.25}$$

They carried out a linear stability and positivity analysis of various schemes arising from various choices of the parameter $\omega$. These results are reproduced in Table 13.1; see the original Ref. [10] for more details.

As an illustrative example, we consider the Godunov first-order upwind scheme as the advection operator $\mathcal{A}$ applied for one time step $\Delta t$ to the homogeneous part in (13.8), analogous to (13.15). For $\lambda > 0$ this gives the predicted value as from the Godunov upwind scheme, namely

$$q_i^{(adv)} = q_i^n - c(q_i^n - q_{i-1}^n). \tag{13.26}$$

Here $c = \lambda \Delta t / \Delta x$ is the Courant number associated with the advection process governed by the homogeneous PDE; see Chap. 9. As the source operator $\mathcal{S}$ applied for one time step $\Delta t$ to the ODE in (13.8), analogous to (13.16), we may use the first-order Euler method, see Sect. 13.3. The result is

$$q_i^{n+1} = q_i^{(adv)} + r q_i^{(s)}, \tag{13.27}$$

where $r = \Delta t \, \alpha$ is a dimensionless number associated with the *reaction-like* process governed by the source term. As seen in (13.23) there are two obvious choices for $q_i^{(s)}$, namely $q_i^{(s)} = q_i^n$ and $q_i^{(s)} = q_i^{(adv)}$, leading respectively to the schemes

$$q_i^{n+1} = \left[q_i^n - c(q_i^n - q_{i-1}^n)\right] + r q_i^n \tag{13.28}$$

and

$$q_i^{n+1} = \left[q_i^n - c(q_i^n - q_{i-1}^n)\right] + r \left[q_i^n - c(q_i^n - q_{i-1}^n)\right]. \tag{13.29}$$

For the model problem considered and for the particular advection scheme applied, there is a clear relationship between the upwind schemes contained in (13.25) for various values of $\omega$ and the *generalised* splitting schemes for corresponding values of $\beta$ in (13.23). Table 13.1 shows the relationship between the various upwinding and splitting schemes and their corresponding linear stability region, derived in [10], as a function of the dimensionless numbers $c$, associated with pure *advection*, and $r$, associated with pure *reaction*. The so-called *centred scheme* for the source term is obtained when $\omega = 0$; this scheme has the smallest stability region of all the schemes considered in the analysis of [10]. This centred scheme corresponds to $\beta = 0$ in the splitting schemes, that is the source term is evaluated at the data time level $n$. The *standard* splitting scheme is obtained when $\beta = 1$, which corresponds to the *upwind* scheme with $\omega = c$. As seen in Table 13.1 this scheme has a very generous stability region.

We recall the preliminary analysis regarding the source in Chap. 9. There, we proposed to evaluate the numerical source $s_i$ from the volume integral (13.4), in which the integrand was calculated from the solution of the two neighbouring intercell boundary Riemann problems. The result was

$$s_i = \frac{1}{2} c (\alpha q_{i-1}^n) + \left(1 - \frac{1}{2} c\right) (\alpha q_i^n). \tag{13.30}$$

This turns out to be identical to the upwind scheme of Table 13.1 with $\omega = \frac{1}{2} c$ and to the splitting scheme with $\beta = \frac{1}{2}$. It is somehow reassuring that the simple analyses for the model problem (13.8) reveal that the first-order splitting approach reproduces the upwind approach, which is meant to be more fundamental.

It would be of interest to pursue this relationship between upwinding and splitting schemes for source terms in non-linear systems. This would probably involve a splitting approach with

1. transformation to characteristic variables for the unknowns and the source terms,
2. application of the source schemes to each characteristic variable and
3. transformation back to the original variables.

To end this section we mention early works on related topics to the treatment of source terms in the shallow water equations, such as those of LeVeque and Yee [7], Bermúdez and Vázquez-Cendón [10], Vázquez-Cendón [15, 30], Greenberg and LeRoux [31], LeVeque [14], Toro and Vázquez-Cendón [32], Garcia-Navarro et al. [33], Hubbard and García-Navarro [34] and Garcia-Navarro and Vázquez-Cendón [35] and Borthwick et al. [36]. In Chap. 14 we shall present the general high-order ADER methodology, in which fluxes and source terms are included in the scheme to unlimited accuracy, via the generalised Riemann problem $GRP_m$.

## 13.3 Solvers for Ordinary Differential Equations

Splitting schemes for solving inhomogeneous systems of PDEs require solvers for systems of Ordinary Differential Equations (ODEs). Systems of ODEs also appear when using semi-discrete methods, or methods of line, for solving PDEs, even for

## 13.3 Solvers for Ordinary Differential Equations

the homogenous case [37, 38]. Here, we study some numerical methods for solving systems of first-order ordinary differential equations, with the aim of treating source terms in advection-reaction systems, as described in Sect. 13.2.2.

### 13.3.1 First-Order Systems of ODEs

Here we recall some very basic facts about first-order systems of ODEs

$$\frac{d}{dt}\mathbf{Q}(t) = \mathbf{S}(t, \mathbf{Q}(t)) := \mathbf{Q}'. \tag{13.31}$$

$\mathbf{Q} = \mathbf{Q}(t)$ and $\mathbf{S}(t, \mathbf{Q}(t))$ are vector-valued functions of $N$ components

$$\mathbf{Q} = [q_1, q_2, \ldots, q_N]^T, \quad \mathbf{S} = [s_1, s_2, \ldots, s_N]^T \tag{13.32}$$

and the independent variable $t$ is a time-like variable. The Jacobian matrix $\mathbf{A}(\mathbf{Q})$ of the source function $\mathbf{S}(t, \mathbf{Q})$ is defined as

$$\mathbf{A}(\mathbf{Q}) = \partial \mathbf{S}/\partial \mathbf{Q} = \begin{bmatrix} \partial s_1/\partial q_1 & \cdots & \partial s_1/\partial q_N \\ \partial s_2/\partial q_1 & \cdots & \partial s_2/\partial q_N \\ \vdots & \vdots & \vdots \\ \partial s_N/\partial q_1 & \cdots & \partial s_N/\partial q_N \end{bmatrix}. \tag{13.33}$$

The entries $a_{ij} = \partial s_i/\partial q_j$ of $\mathbf{A}(\mathbf{Q})$ are partial derivatives of the components $s_i$ of the vector $\mathbf{S}$ with respect to the components $q_j$ of the vector of unknowns $\mathbf{Q}$. The eigenvalues $\alpha_i$ of $\mathbf{A}$ are the solutions of the *characteristic polynomial*

$$|\mathbf{A} - \alpha \mathbf{I}| = \det(\mathbf{A} - \alpha \mathbf{I}) = 0, \tag{13.34}$$

where $\mathbf{I}$ is the identity matrix. Generally, the eigenvalues are complex numbers. Trivially, the eigenvalue of the model ODE

$$q'(t) = s(t, q(t)) = \alpha q(t) \tag{13.35}$$

is $\alpha$. The behaviour of a system of ODEs (13.31) is, in the main, determined by the behaviour of its eigenvalues. For instance, the exact solution of (13.35) with initial condition $q(0) = 1$ is $q(t) = e^{\alpha t}$. For $t$ close to 0 the solution varies rapidly if the eigenvalue is negative and large in absolute value. For $t$ away from zero the solution is almost indistinguishable from 0.

**Stability of ODEs**

An important property of ODEs is **stability**. Generally speaking stable solutions are bounded. Note that the solution of the linear model ODE (13.35) is bounded only if $\alpha < 0$. Geometrically, a solution $\mathbf{Q}(t)$ is stable if any other solution of the ODE whose initial condition is sufficiently close to that of $\mathbf{Q}(t)$ remains in a *tube* enclosing $\mathbf{Q}(t)$. If the diameter of the tube tends to 0 as $t$ tends to $\infty$, the solution is said to be **asymptotically stable**. Stability of solutions $\mathbf{Q}(t)$ is characterised in terms of the eigenvalues $\alpha_j$ of the Jacobian matrix (13.33). In particular if the real part of every eigenvalue is negative the solution is asymptotically stable.

**Stiffness of ODEs**

Another feature of ODEs is **stiffness**, yet to be defined. Stiff ODEs are usually associated with processes operating on disparate time scales. Chemical kinetics is a classical source of stiff ODEs. The stiffness of a system is generally determined by the behaviour of the eigenvalues of the system. In addition, the time interval over which the solution is sought is also a consideration in determining the stiffness of the system. There will be intervals of *rapid variations* (transient) of the solution and intervals of *slow variation*. The single ODE (13.35) is stiff for $\alpha \ll 0$ and for time $t$ in the vicinity of 0.

Following Lambert [39], a nonlinear system of the form (13.31) is said to be stiff if

1. $Re(\alpha_j) < 0, \quad j = 1, 2, \ldots, N$ and
2. $\alpha_{max} \equiv \max_j |Re(\alpha_j)| \gg \alpha_{min} \equiv \min_j |Re(\alpha_j)|$.

Here $Re(\alpha_j)$ denotes the real part of the complex number $\alpha_j$. The stiffness ratio is defined as $R_{stif} = \alpha_{max}/\alpha_{min}$. Modest values of $R_{stif}$, e.g. 20, are sufficient to cause serious numerical difficulties to explicit methods. In real applications $R_{stif}$ may be as large as $10^6$.

Before thinking of numerical methods to solve ODEs, the very first task is to investigate whether the ODEs are stiff or not; this will determine the appropriate numerical methods to be used for solving the equations. See Kahaner, Moler and Nash [40], Gear [41] and Lambert [39].

### 13.3.2 Conventional Numerical Methods for ODEs

We are interested in solving the Initial-Value Problem for (13.31), that is

$$\left. \begin{array}{ll} \text{ODEs:} & \dfrac{d}{dt}\mathbf{Q}(t) = \mathbf{S}(t, \mathbf{Q}(t)) := \mathbf{Q}', \\ \text{ICs:} & \mathbf{Q}(t_0) = \mathbf{Q}_0. \end{array} \right\} \quad (13.36)$$

## 13.3 Solvers for Ordinary Differential Equations

One begins by discretising the domain of integration $[t_0, t_f]$ through the partition $t_0 < t_1 < t_2 \cdots < t_n < t_{n+1} \cdots < t_f$. One way of constructing numerical methods to solve the ODEs in (13.36) is by using Taylor series expansions, giving rise to the family of Taylor methods. Another way is to integrate the ODEs in (13.36) between $t_n$ and $t_{n+1}$ to obtain

$$\mathbf{Q}(t_{n+1}) = \mathbf{Q}(t_n) + \int_{t_n}^{t_{n+1}} \mathbf{S}(t, \mathbf{Q}(t)) dt. \tag{13.37}$$

Various numerical methods are obtained depending on the way the integral in (13.37) is evaluated. The *Euler method* results from evaluating the integral at the *old* time, obtaining

$$\mathbf{Q}^{n+1} = \mathbf{Q}^n + \Delta t \mathbf{S}(t_n, \mathbf{Q}^n). \tag{13.38}$$

Here $\Delta t = t_{n+1} - t_n$ is the time step and $\mathbf{Q}^n \approx \mathbf{Q}(t_n)$. The Euler method is *explicit* and first-order accurate. The *Backward Euler Method*, also first-order accurate but *implicit*, results from evaluating the integral in (13.37) at the *new* time $t_{n+1}$, namely

$$\mathbf{Q}^{n+1} = \mathbf{Q}^n + \Delta t \mathbf{S}(t_{n+1}, \mathbf{Q}^{n+1}). \tag{13.39}$$

A second-order implicit method results from a *trapezium rule* approximation to the integral in (13.37), yielding the *Trapezoidal Method*

$$\mathbf{Q}^{n+1} = \mathbf{Q}^n + \frac{1}{2} \Delta t [\mathbf{S}(t_n, \mathbf{Q}^n) + \mathbf{S}(t_{n+1}, \mathbf{Q}^{n+1})]. \tag{13.40}$$

*Runge–Kutta methods* constitute a classical family of numerical methods for solving ODEs. A second-order, two-stage (explicit) Runge–Kutta method is

$$\left. \begin{aligned} \mathbf{K}_1 &= \Delta t \mathbf{S}(t_n, \mathbf{Q}^n), \\ \mathbf{K}_2 &= \Delta t \mathbf{S}(t_n + \Delta t, \mathbf{Q}^n + \mathbf{K}_1), \\ \mathbf{Q}^{n+1} &= \mathbf{Q}^n + \tfrac{1}{2}[\mathbf{K}_1 + \mathbf{K}_2]. \end{aligned} \right\} \tag{13.41}$$

A fourth-order, four-stage explicit Runge–Kutta method is

$$\left. \begin{aligned} \mathbf{K}_1 &= \Delta t \mathbf{S}(t_n, \mathbf{Q}^n), \\ \mathbf{K}_2 &= \Delta t \mathbf{S}(t_n + \tfrac{1}{2}\Delta t, \mathbf{Q}^n + \tfrac{1}{2}\mathbf{K}_1), \\ \mathbf{K}_3 &= \Delta t \mathbf{S}(t_n + \tfrac{1}{2}\Delta t, \mathbf{Q}^n + \tfrac{1}{2}\mathbf{K}_2), \\ \mathbf{K}_4 &= \Delta t \mathbf{S}(t_n + \Delta t, \mathbf{Q}^n + \mathbf{K}_3), \\ \mathbf{Q}^{n+1} &= \mathbf{Q}^n + \tfrac{1}{6}[\mathbf{K}_1 + 2\mathbf{K}_2 + 2\mathbf{K}_3 + \mathbf{K}_4]. \end{aligned} \right\} \tag{13.42}$$

Most classical numerical analysis textbooks will contain chapters on conventional solvers for ODEs. Specialised books are, for example, those of Gear [41], Lambert [39] and Shampine [42].

## 13.3.3 TVD Runge–Kutta Schemes for ODEs

A new class of Runge–Kutta schemes for ODEs emerged from the numerical analysis of hyperbolic PDEs solved via *semidiscrete methods* [37]; see Chap. 10. These ODE solvers were originally called TVD (Total Variation Diminishing) ODE solvers and are currently called *strong stability preserving methods* [38]. Semi-discrete methods, as pointed out in Chap. 10, separate the discretization in time from that in space. In order to advance the solution from time $t_n$ to the next time $t_{n+1}$, semi-discrete methods rely on the solution of an initial-value problem for a system of ODEs. Classical methods for ODEs, including conventional Runge–Kutta methods, are potential candidates for this task. However, in view of the Godunov theorem, see Chap. 9, that task requires the construction of non-linear methods [43]; that is, special ODE solvers are needed. This need gave rise to TVD Runge–Kutta schemes for ODEs. In what follows we introduce two ODE solvers of the Runge–Kutta type that satisfy the TVD property. For details on their construction and properties see [37, 38].

### Second-Order Scheme (Heun's Method)

A second-order, two-stage TVD Runge–Kutta scheme to solve (13.36) is the following

$$\left. \begin{array}{ll} \text{Stage 1:} & \mathbf{Q}^{(1)} = \mathbf{Q}^n + \Delta t \mathbf{S}(\mathbf{Q}^n), \\ \text{Stage 2:} & \mathbf{Q}^{(2)} = \mathbf{Q}^{(1)} + \Delta t \mathbf{S}(\mathbf{Q}^{(1)}), \\ \text{Solution:} & \mathbf{Q}^{n+1} = \frac{1}{2}(\mathbf{Q}^n + \mathbf{Q}^{(2)}). \end{array} \right\} \quad (13.43)$$

Note that at each stage one applies the first-order explicit Euler scheme (13.38). For convenience, we have dropped the explicit dependence on time $t$ of $\mathbf{S}(t, \mathbf{Q}(t))$.

### Third-Order TVD Runge–Kutta Scheme

A third order, three-stage TVD Runge–Kutta method is the following

$$\left. \begin{array}{ll} \text{Stage 1:} & \mathbf{Q}^{(1)} = \mathbf{Q}^n + \Delta t \mathbf{S}(\mathbf{Q}^n), \\ \text{Stage 2:} & \hat{\mathbf{Q}}^{(2)} = \mathbf{Q}^{(1)} + \Delta t \mathbf{S}(\mathbf{Q}^{(1)}), \\ \text{Averaging:} & \mathbf{Q}^{(2)} = \frac{3}{4}\mathbf{Q}^n + \frac{1}{4}\hat{\mathbf{Q}}^{(2)}, \\ \text{Stage 3:} & \mathbf{Q}^{(3)} = \mathbf{Q}^{(2)} + \Delta t \mathbf{S}(\mathbf{Q}^{(2)}), \\ \text{Solution:} & \mathbf{Q}^{n+1} = \frac{1}{3}\mathbf{Q}^n + \frac{2}{3}\mathbf{Q}^{(3)}. \end{array} \right\} \quad (13.44)$$

Again, at each stage one applies the explicit first-order Euler operator (13.38). Further details on TVD Runge–Kutta schemes as applied in semi-discrete methods to solve PDEs are found in [44].

### 13.3.4 A Note on Stability

Stability of numerical methods is a fundamental issue. To illustrate this point we assume the model ODE (13.35) in IVP (13.36), as solved by the explicit Euler method (13.38). The scheme then reads

$$q^{n+1} = (1 + \Delta t \alpha) q^n. \tag{13.45}$$

Clearly, for stability one requires that the ODE itself be stable, $\alpha < 0$, and that the *amplification factor* satisfy

$$|1 + \alpha \Delta t| \leq 1. \tag{13.46}$$

Therefore, the time step $\Delta t$ must satisfy the stability restriction

$$\Delta t \leq \frac{2}{|\alpha|}. \tag{13.47}$$

For large $|\alpha|$ (stiff ODE) the time step $\Delta t$ can be extremely small, which means that the method becomes very inefficient or even useless in practice. On the other hand, the implicit trapezoidal method (13.40) gives

$$q^{n+1} = \frac{(1 + \frac{1}{2}\Delta t \alpha)}{(1 - \frac{1}{2}\Delta t \alpha)} q^n. \tag{13.48}$$

This scheme is stable for any $\Delta t$, provided $\alpha \leq 0$, that is whenever the ODE itself is stable.

Explicit methods are much simpler to use than implicit methods. The latter generally require the solution of non-linear algebraic equations at each time step and are therefore much more expensive. However, as illustrated, for stiff problems implicit methods are mandatory.

There is a vast literature on the subject of numerical methods for ODEs. For theoretical properties of ODEs see for example Brown [45], Ince and Sneddon [46], Sánchez [47] and Coddington and Levinson [48]. As to numerical methods, almost any textbook on numerical analysis will contain a chapter on schemes for ODEs. See for example Hildebrand [49], Mathews [50], Conte and de Boor [51], Maron and López [52], Johnson and Riess [53], Kahaner et al. [40]. Advanced textbooks are those of Gear [41], Lambert [39] and Shampine [42]. Advanced methods for ODEs in the context of PDEs are found in [37, 38]. As already noted, ODEs (13.7) may be **stiff**, in which case one must employ appropriate numerical methods; see Kahaner, Moler and Nash [40], Gear [41] and Lambert [39].

In the next section we study simple approaches for extending numerical methods developed for one-dimensional PDEs, with or without source terms, to problems involving multiple space dimensions.

## 13.4 Multidimensional Systems of PDEs

In this section we introduce finite volume schemes for solving numerically non-linear PDE systems in multiple space dimensions, the two-dimensional shallow water equations in particular. Two classes of schemes are considered, *dimesional split* schemes, analogous to those for source terms in Sect. 13.2.2, and unsplit or *simultaneous update* schemes, in which the solution at a cell is updated in a single step, accounting simultaneously for all intercell fluxes. Background on some of the schemes discussed is found in Chap. 16 of the textbooks [1–3]. See also the books by LeVeque [4], Vázquez [54] and Godlewski and Raviart [5].

Consider the two-dimensional initial-value problem for a system of PDEs

$$\left. \begin{array}{l} \text{PDEs:}\ \partial_t \mathbf{Q} + \partial_x \mathbf{F}(\mathbf{Q}) + \partial_y \mathbf{G}(\mathbf{Q}) = \mathbf{S}(\mathbf{Q}), \\ \text{ICs:}\ \ \ \mathbf{Q}(x, y, t_n) = \mathbf{Q}^n := \left\{ \mathbf{Q}_{i,j}^n \right\}. \end{array} \right\} \quad (13.49)$$

For the two-dimensional shallow water equations with variable bathymetry $\mathbf{Q}$, $\mathbf{F}(\mathbf{Q})$ and $\mathbf{G}(\mathbf{Q})$ are the vectors of conserved variables, fluxes in the $x$- and $y$-directions, and sources, given by

$$\left. \begin{array}{l} \mathbf{Q} = \begin{bmatrix} q_1 \\ q_2 \\ q_3 \end{bmatrix} = \begin{bmatrix} h \\ hu \\ hv \end{bmatrix},\ \mathbf{F}(\mathbf{Q}) = \begin{bmatrix} f_1 \\ f_2 \\ f_3 \end{bmatrix} = \begin{bmatrix} hu \\ hu^2 + \frac{1}{2}gh^2 \\ huv \end{bmatrix}, \\[2em] \mathbf{G}(\mathbf{Q}) = \begin{bmatrix} g_1 \\ g_2 \\ g_3 \end{bmatrix} = \begin{bmatrix} hv \\ hvu \\ hv^2 + \frac{1}{2}gh^2 \end{bmatrix},\ \mathbf{S}(\mathbf{Q}) = \begin{bmatrix} s_1 \\ s_2 \\ s_3 \end{bmatrix} = \begin{bmatrix} 0 \\ -gh\partial_x b \\ -gh\partial_y b \end{bmatrix}. \end{array} \right\}$$

(13.50)

Here $u(x, y, t)$ and $v(x, y, t)$ are respectively the $x$- and $y$-components of velocity, $h(x, y, t)$ is the *water depth*, $b(x, y)$ is the *bed elevation*, or bathymetry, with the *free surface* defined as $s(x, y, t) = b(x, y) + h(x, y, t)$; $g = 9.81\ m/s^2$ is the acceleration due to gravity, a constant here. See Chaps. 1 and 4 for further details.

### 13.4.1 Dimensional-Split Schemes

Dimensional-split schemes are presented as for the homogeneous version of (13.49), assuming $\mathbf{S}(\mathbf{Q}) = \mathbf{0}$. The initial data at time $t_n$ in (13.49) is given by the set $\mathbf{Q}^n$ of discrete cell average values $\mathbf{Q}_{i,j}^n$ on a structured mesh; index $i$ refers to the $x$-coordinate direction and index $j$ refers to the $y$-coordinate direction. Figure 13.1 depicts a typical Cartesian, regular mesh with computing cells labelled $I_{i,j}$, of dimensions $\Delta x \times \Delta y$.

The dimensional-splitting approach replaces the homogeneous version of IVP (13.49) by a pair of one-dimensional IVPs, which in the simplest version of the approach are:

## 13.4 Multidimensional Systems of PDEs

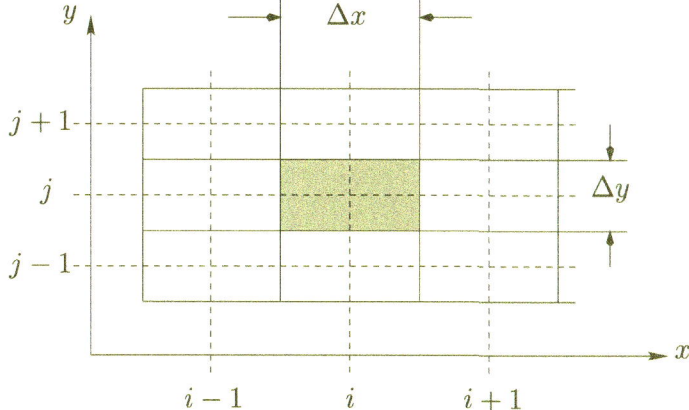

**Fig. 13.1** Cartesian mesh made up of regular cells $I_{i,j}$ of dimensions $\Delta x \times \Delta y$, with four intercell boundaries, all aligned with the Cartesian coordinate directions

$$x\text{-sweep IVP:} \begin{cases} \text{PDEs:} & \partial_t \mathbf{Q} + \partial_x \mathbf{F}(\mathbf{Q}) = \mathbf{0} \\ \text{ICs:} & \mathbf{Q}(x, t_n) = \mathbf{Q}^n \end{cases} \overset{\Delta t}{\Longrightarrow} \mathbf{Q}^{n+\frac{1}{2}} \quad (13.51)$$

and

$$y\text{-sweep IVP:} \begin{cases} \text{PDEs:} & \partial_t \mathbf{Q} + \partial_y \mathbf{G}(\mathbf{Q}) = \mathbf{0} \\ \text{ICs:} & \mathbf{Q}(x, t_n) = \mathbf{Q}^{n+\frac{1}{2}} \end{cases} \overset{\Delta t}{\Longrightarrow} \mathbf{Q}^{n+1}. \quad (13.52)$$

**The $x$-Seep IVP**

In the first IVP (13.51) one solves a one-dimensional problem in the $x$-direction for a time step $\Delta t$. The initial data is that of the full problem and its solution after a time $\Delta t$ is denoted by $\mathbf{Q}^{n+\frac{1}{2}}$. In this problem one solves the $x$-split one-dimensional problem (13.51) **for each strip of cells labelled** $j$; see Fig. 13.1. The 1D equations are **augmented** by the tangential velocity component $v$, a passive scalar; see Chap. 4. Any of the methods studied in the previous Chaps. 10, 11 and 12, is directly applicable to solve (13.51).

**The $y$-Seep IVP**

In the next IVP (13.52) one solves the augmented one-dimensional problem in the $y$-direction, also for a time step $\Delta t$. This is called the $y$-*sweep*; its initial data is the solution $\mathbf{Q}^{n+\frac{1}{2}}$ of the previous IVP (13.51) and its solution after a time $\Delta t$ is $\mathbf{Q}^{n+1}$. Here, **for each strip of cells labelled** $i$ one solves the augmented one-dimensional problem, augmented by the tangential velocity component $u$, a passive scalar. Again, any of the methods studied in previous chapters is directly applicable to solve (13.52). At the end of (13.52) one regards its solution $\mathbf{Q}^{n+1}$ as the solution to the full two-dimensional problem (13.49).

Let us now introduce the notation $\mathcal{X}^{(t)}$ and $\mathcal{Y}^{(t)}$ for the solution operators for IVPs (13.51) and (13.52), respectively. Then the splitting (13.51)–(13.52) of the original two-dimensional IVP (13.49) can be written thus

$$\mathbf{Q}^{n+1} = \mathcal{Y}^{(\Delta t)} \mathcal{X}^{(\Delta t)}(\mathbf{Q}^n). \tag{13.53}$$

There is generally no particular reason for applying the operators in the order just described. An analogous scheme is

$$\mathbf{Q}^{n+1} = \mathcal{X}^{(\Delta t)} \mathcal{Y}^{(\Delta t)}(\mathbf{Q}^n). \tag{13.54}$$

It can be shown, see Strang [55], that for general systems, both splitting schemes (13.53) and (13.54) are first-order accurate in time if the individual operators $\mathcal{X}^{(t)}$ and $\mathcal{Y}^{(t)}$ are at least first-order accurate in time.

There exist second-order dimensional splitting schemes in the literature. One example is

$$\mathbf{Q}^{n+1} = \frac{1}{2} \left[ \mathcal{X}^{(\Delta t)} \mathcal{Y}^{(\Delta t)} + \mathcal{Y}^{(\Delta t)} \mathcal{X}^{(\Delta t)} \right] (\mathbf{Q}^n). \tag{13.55}$$

This is a second-order accurate splitting scheme, provided that each of the solution operators is at least second-order accurate in time [55]. Note, however, that this scheme requires double the amount of work of the first-order schemes (13.53) and (13.54). More attractive second-order split schemes are

$$\mathbf{Q}^{n+1} = \mathcal{X}^{(\frac{1}{2}\Delta t)} \mathcal{Y}^{(\Delta t)} \mathcal{X}^{(\frac{1}{2}\Delta t)}(\mathbf{Q}^n) \tag{13.56}$$

and

$$\mathbf{Q}^{n+1} = \mathcal{Y}^{(\frac{1}{2}\Delta t)} \mathcal{X}^{(\Delta t)} \mathcal{Y}^{(\frac{1}{2}\Delta t)}(\mathbf{Q}^n). \tag{13.57}$$

These require about 50% more work than the first-order schemes (13.53) and (13.54). Two more second-order accurate schemes are

$$\mathbf{Q}^{n+2} = \mathcal{X}^{(\Delta t)} \mathcal{Y}^{(\Delta t)} \mathcal{Y}^{(\Delta t)} \mathcal{X}^{(\Delta t)}(\mathbf{Q}^n) \tag{13.58}$$

and

$$\mathbf{Q}^{n+2} = \mathcal{Y}^{(\Delta t)} \mathcal{X}^{(\Delta t)} \mathcal{X}^{(\Delta t)} \mathcal{Y}^{(\Delta t)}(\mathbf{Q}^n). \tag{13.59}$$

These can be shown to be second-order accurate every other time step [56], and they are as efficient as the first-order schemes (13.53) and (13.54).

### 13.4 Multidimensional Systems of PDEs

The dimensional splitting schemes presented for solving multi-dimensional PDEs have been described as for Cartesian meshes, in which the solution vector $\mathbf{Q}_{i,j}^n$ in cell $I_{i,j}$ is identified by two indexes $i$ and $j$ associated respectively with the Cartesian $x$- and $y$-directions. See Fig. 13.1. The methods described for Cartesian meshes are also applicable to *structured meshes*, in which computing cells are not necessarily Cartesian, but the $i$-$j$ indexing can still be kept.

In the next section we study numerical schemes that advance the solution by simultaneously accounting for all fluxes in a single time step. Such methods are potentially applicable to *unstructured meshes*. For background see Chap. 10.

#### 13.4.2 Unsplit Finite Volume Schemes

Consider again the time-dependent two-dimensional system of homogeneous conservation laws

$$\partial_t \mathbf{Q} + \partial_x \mathbf{F}(\mathbf{Q}) + \partial_y \mathbf{G}(\mathbf{Q}) = \mathbf{0} \tag{13.60}$$

to be solved on a Cartesian mesh, as depicted in Fig. 13.1. Consider a typical finite volume or computational cell $I_{i,j}$ of dimensions $\Delta x \times \Delta y$, as depicted in Fig. 13.2. The cell average $\mathbf{Q}_{i,j}^n$ is assigned to *the centre of the cell* $I_{i,j}$. To each intercell boundary there corresponds a numerical flux.

A **simultaneous-update** formula for an explicit finite volume scheme to solve (13.60) reads

$$\mathbf{Q}_{i,j}^{n+1} = \mathbf{Q}_{i,j}^n - \frac{\Delta t}{\Delta x}[\mathbf{F}_{i+\frac{1}{2},j} - \mathbf{F}_{i-\frac{1}{2},j}] - \frac{\Delta t}{\Delta y}[\mathbf{G}_{i,j+\frac{1}{2}} - \mathbf{G}_{i,j-\frac{1}{2}}]. \tag{13.61}$$

The cell average $\mathbf{Q}_{i,j}^n$ in cell $I_{i,j}$ at time level $n$ is updated to time level $n+1$ via (13.61) in a **single step**, involving simultaneous flux contributions *from all four*

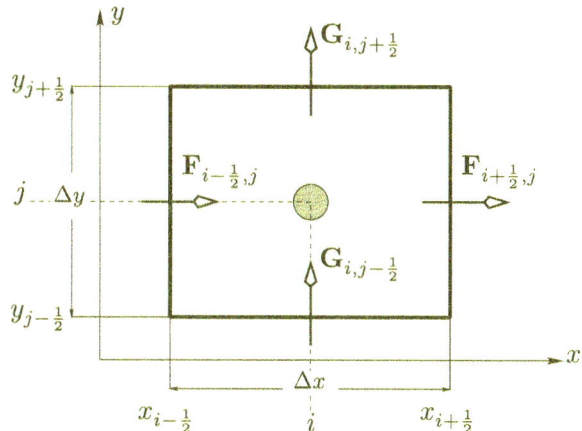

**Fig. 13.2** Finite volume discretisation of computational domain into Cartesian cells $I_{i,j}$ of dimensions $\Delta x \times \Delta y$. Typical computing cell $I_{i,j}$ in 2D has four intercell boundaries with corresponding intercell fluxes to be used simultaneously in formula (13.61)

*intercell boundaries*; see Fig. 13.2. The conservative formula (13.61) is the natural extension of the one-dimensional conservative formula (13.15), and is completely determined once the intercell numerical fluxes $\mathbf{F}_{i+\frac{1}{2},j}$ and $\mathbf{G}_{i,j+\frac{1}{2}}$ are specified.

When it comes to the design of numerical fluxes, there is a profound difference between the dimensional-split schemes of Sect. 13.4.1 and the simultaneous update schemes (13.61). A major issue is *stability*, a subject we address next.

## 13.5 Stability of Multi-dimensional Schemes

As already pointed out, for the dimensional-split schemes of Sect. 13.4.1 one can use any of the one-dimensional schemes presented in previous Chaps. 9, 10 and 11. The stability constraints of the overall 2D/3D split schemes are the same as for the one-dimensional scheme used in each sweep. This is an advantage of split schemes.

For the *simultaneous update* scheme (13.61) the situation is rather different. The reader is warned that the temptation to directly apply one-dimensional fluxes to each interface, as presented in previous chapters, could prove frustrating. The issue is **stability**. One-dimensional upwind schemes, such as the first-order Godunov scheme of Chaps. 10 and 11, can be used straightforwardly to specify each of the four intercell fluxes in (13.61), as depicted in Fig. 13.2. However, the resulting stability constraint is more severe than for the 1D schemes. For example, for the one-dimensional problem the linear stability constraint for the Godunov scheme is $0 \le |c| \le C_{lim} = 1$, where $c$ is the Courant number. For the 2D case $C_{lim} = 1/2$ and for 3D $C_{lim} = 1/3$. Conventional centred schemes, such as Lax-Friedrichs, Lax-Wendroff and FORCE presented in Chap. 10 are not directly usable. The resulting schemes are **unconditionally unstable**. The centred FORCE-$\alpha$ scheme [57], see Chap. 10, is one of the exceptions, as we shall see in Sect. 13.5.2.

### 13.5.1 Von Neumann Linear Stability Analysis

We discuss linear stability for 2D numerical schemes on regular Cartesian meshes. To this end we consider the linear advection equation in two space dimensions

$$\left.\begin{array}{l}\partial_t q(x,y,t) + \partial_x f(q(x,y,t)) + \partial_y g(q(x,y,t)) = 0, \\ f(q) = \lambda_x q, \quad g(q) = \lambda_y q.\end{array}\right\} \quad (13.62)$$

Here $\lambda_x$ and $\lambda_y$ are the $x$- and $y$-components of the constant wave propagation speed $V = (\lambda_x, \lambda_y)$; see Chap. 2. The simultaneous update numerical formula for (13.62) corresponding to (13.61) is

## 13.5 Stability of Multi-dimensional Schemes

$$q_{i,j}^{n+1} = q_{i,j}^n - \frac{\Delta t}{\Delta x}[f_{i+\frac{1}{2},j} - f_{i-\frac{1}{2},j}] - \frac{\Delta t}{\Delta y}[g_{i,j+\frac{1}{2}} - g_{i,j-\frac{1}{2}}]. \tag{13.63}$$

The numerical fluxes in the $x$- and $y$-directions are $f_{i+\frac{1}{2},j}$ and $g_{i,j+\frac{1}{2}}$. Specification of the fluxes into the scheme (13.63) gives $q_{i,j}^{n+1}$ as a linear combination of old values $q_{k,l}^n$ with constant coefficients $b_{k,l}$, which are functions of Courant numbers

$$c_x = \frac{\lambda_x \Delta t}{\Delta x}, \quad c_y = \frac{\lambda_y \Delta t}{\Delta y}. \tag{13.64}$$

See Chap. 9 for relevant background for the 1D case. There are various methods for analysing the stability of schemes for the linear advection equation. See Hirsch [58], Chaps. 7–10, for a comprehensive presentation of techniques. See also [4, 5, 59, 60]. Here we assume the reader to be familiar with the rudiments of the popular von Neumann linear stability analysis method.

### Von Neumann Analysis for the 1D Case

The von Neumann method begins by introducing a **trial function**, or **test function**

$$q_i^n = A^n e^{I\theta i}, \quad \text{with } -\pi \leq \theta \leq \pi, \tag{13.65}$$

where the amplitude $A$ is a complex number and $A^n$ means $A$ **raised to the power** $n$. For convenience, and contrary to convention, we use here $I := \sqrt{-1}$ to denote the complex unity, and keep $i$ to identify a mesh spatial position in the numerical method; $\theta = P\Delta x$ is an angle and $P$ is the **wave number** in the $x$-direction. For stability one imposes the **stability condition**

$$||A|| \leq 1. \tag{13.66}$$

In this manner, the amplitude raised to the power $n$ in (13.65), where $n$ is the time level, will not increase unboundedly, as time marching proceeds into the future.

As an example, let us consider the linear stability for the Godunov upwind method for $\lambda > 0$, that is for the method

$$q_i^{n+1} = q_i^n - c\left(q_i^n - q_{i-1}^n\right), \tag{13.67}$$

where $c$ is the Courant number. Substitution of $q_{i-1}^n$, $q_i^n$ and $q_i^{n+1}$ by $A^n e^{I\theta(i-1)}$, $A^n e^{I\theta i}$ and $A^{n+1} e^{I\theta i}$, according to (13.65), into the scheme (13.67) yields

$$A^{n+1} e^{I\theta i} = A^n e^{I\theta i} - c\left[A^n e^{I\theta i} - A^n e^{I\theta(i-1)}\right]. \tag{13.68}$$

Cancelling $A^n e^{I\theta i}$ we obtain the complex number

$$A = [1 + c(\cos\theta - 1)] + I[-c\sin\theta]. \tag{13.69}$$

Then, the squared modulus of $A$ is

$$||A||^2 = [1 + c(\cos\theta - 1)]^2 + c^2\sin^2\theta, \tag{13.70}$$

which after manipulations becomes

$$||A||^2 = 1 - 2c(1-c)(1-\cos\theta). \tag{13.71}$$

We impose the stability condition (13.66) in terms of $||A||^2 \leq 1$, which results in the following condition

$$1 - 2c(1-c)(1-\cos\theta) \leq 1. \tag{13.72}$$

Manipulation of (13.72) leads to the stability condition

$$0 \leq c \leq C_{lim} = 1. \tag{13.73}$$

It is easy to perform the analysis for the negative wave speed, $\lambda < 0$. The general stability condition that includes both positive and negatives wave speeds is

$$0 \leq |c| \leq C_{lim} = 1. \tag{13.74}$$

In general, the analysis may become algebraically intractable, for example for higher-order methods. In this case one resorts to a numerical evaluation of the amplification factor $A$ in order to get an *indication* of the stability region of the scheme of interest. For the one-dimensional case in which the coefficients of the scheme depend only on the Courant number $c_x$ the von Neumann method derives an algebraic expression for the amplification factor as

$$A = A(c_x, \theta_x). \tag{13.75}$$

where $c_x$ is the Courant number and $\theta_x$ is the *phase angle* in the range $0 \leq \theta_x \leq \pi$. For sufficiently simple schemes the *stability condition* (13.66) can be used to derive, algebraically, conditions on the Courant number for the scheme to be linearly stable, as just done for the Godunov upwind scheme, see condition (13.74). Otherwise one performs a numerical evaluation of $||A||$, the modulus of the complex number $A$, for a *large number* of values $(c_x, \theta_x)$ in a rectangle $[0, C_{x,upper}] \times [0, \pi]$; see [61]. Even though this numerical implementation of the von Neumann method does not constitute a mathematical proof for stability, the results are reliable, if done properly. In the numerical test one guesses $C_{x,upper}$ to be larger than the *expected* stability limit $C_{x,lim}$. One expects an interval $0 \leq |c_x| \leq C_{x,lim}$, for which $||A|| \leq 1, \forall \theta_x \in [0, \pi]$.

## 13.5 Stability of Multi-dimensional Schemes

**Von Neumann Analysis for the 2D Case**

For two-dimensional schemes such as (13.63), the amplification factor reads

$$A = A(c_x, c_y, \theta_x, \theta_y). \tag{13.76}$$

Here $c_x$ and $c_y$ are the Courant numbers in the $x$- and $y$- directions, as defined in (13.64), and $\theta_x$, $\theta_y$ are corresponding phase angles. Deriving algebraic conditions for the scheme to be stable, such as in (13.74), may now be much harder than in the one-dimensional case. The numerical implementation of the von Neumann method is a practical alternative [61] to provide an indication of stability. For a given pair $(c_x, c_y)$ of Courant numbers in the rectangle $[0, C_{x,upper}] \times [0, C_{y,upper}]$ compute $||A||$ for a *large number* $M_{ang}$ of values $(\theta_x, \theta_y)$ of phase angles in the rectangle $[0, \pi] \times [0, \pi]$. Then, record the number $M_{cfl}$ for which $||A|| \leq 1$ is satisfied and define the ratio

$$S = S(c_x, c_y) = \frac{M_{cfl}}{M_{ang}}. \tag{13.77}$$

If $S = 1$ the scheme is regarded as stable for the pair $(c_x, c_y)$. A contour plot of $S = S(c_x, c_y)$ for a large number of pairs $(c_x, c_y)$ in the range $[0, C_{x,upper}] \times [0, C_{y,upper}]$ will give an *indication* of the linear stability of the two-dimensional scheme. In three space dimensions, an analogous procedure is followed. The methodology is fully described in Appendix B of [62] and in Chap. 16 of [2, 3, 63]; see also [57].

**Remark 13.2** The numerical von Neumann stability analysis just described, useful in practice as it may be, does not constitute proof of linear stability, and caution is needed when quoting the conclusions from the study. The numerical implementation calls for *large numbers* of Courant numbers and phase angles, to increase the reliability of the conclusions.

In the next section we give examples of stable schemes in multiple space dimensions.

### 13.5.2 Examples of Stable Schemes in 2D/3D

Here we apply the von Neumann stability analysis method just described to establish the linear stability of the simultaneous update 2D scheme (13.63), in which the intercell fluxes are computed using one-dimensional schemes. Though not necessary, for convenience we assume $\lambda_x = \lambda_y$.

**Fig. 13.3 Stability regions for the 2D linear advection equation.** The stability region is shown in yellow for the simultaneous-update 2D scheme (13.61), in which the 1D Godunov upwind method is used to determine all four fluxes

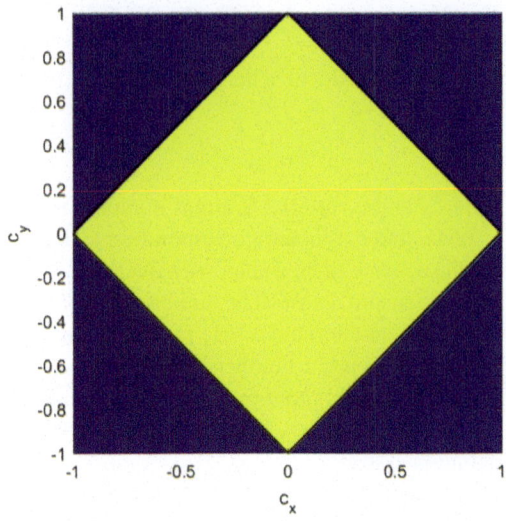

### The Godunov Upwind Scheme

Figure 13.3 shows the results for the case in which the 1D Godunov upwind method is used to determine all four fluxes in the simultaneous-update 2D scheme (13.63). The respective stability region in the $c_x$-$c_y$ plane are shown in yellow and can be described as follows

$$|c_x| + |c_y| \leq C_{lim} = 1, \quad C_{lim} = min\left\{C_{x,lim}, C_{y,lim}\right\}. \tag{13.78}$$

Here $C_{x,lim}$ is the stability limit for the one-dimensional case in the $x$-direction, with analogous meaning for $C_{y,lim}$. When $c_x = c_y \geq 0$ condition (13.78) implies that the Godunov upwind scheme has stability limit $C_{lim} = \frac{1}{2}$. Analogously, in three space dimensions $C_{lim} = \frac{1}{3}$. This represents a drastic reduction of the stability limit of the scheme, compared with the stability limit $C_{lim} = 1$ of the respective one-dimensional scheme and the associated dimensional-split schemes in Sect. 13.4.1.

### The FORCE-$\alpha$ Scheme

A centred, simultaneous-update 2D scheme (13.61) emerges from using one-dimensional FORCE-$\alpha$ fluxes for each edge. Figure 13.4 shows in yellow the circular stability regions of radius $R_{2D}$ in the $c_x$-$c_y$ plane for two FORCE-$\alpha$ schemes, that is for two values of the parameter $\alpha$. For the three-dimensional case one obtains a sphere of radius $R_{3D}$. For both 2D and 3D, under the assumption of equality of dimensional Courant numbers, the stability regions may be described as follows

**Fig. 13.4 Stability regions for the 2D linear advection equation.** Circular stability regions shown in yellow for the simultaneous-update 2D scheme (13.61), in which the 1D FORCE-$\alpha$ scheme is used to determine all four fluxes. Top frame $\alpha = 2$; bottom frame $\alpha = 3$

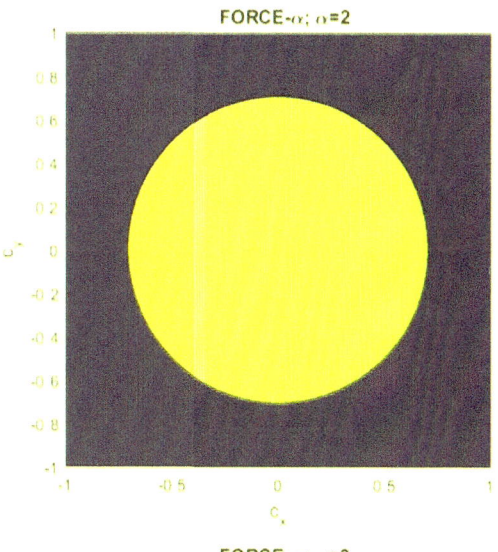

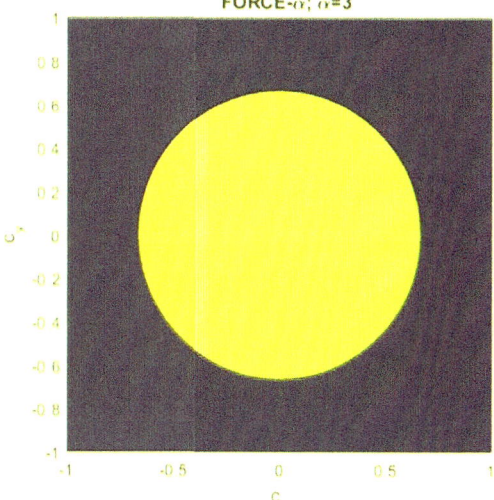

$$\left.\begin{array}{ll} \text{2D:} & c_x^2 + c_y^2 \leq R_{2D}^2 \rightarrow C_{lim} = R_{2D}/\sqrt{2}, \\ \text{3D:} & c_x^2 + c_y^2 + c_z^2 \leq R_{3D}^2 \rightarrow C_{lim} = R_{3D}/\sqrt{3}. \end{array}\right\} \quad (13.79)$$

The parameter $\alpha$ determines the radii of the stability circle in 2D and the stability sphere in 3D. These radii decrease as $\alpha$ increases. Table 13.2 shows the variation of the stability regions with $\alpha$, in 1D, 2D and 3D. Results for the 2D and 3D cases derive from the numerical von Neumann method, in which the radii are estimated from the graphs of the counterplots. For $\alpha = 1$ only the 1D scheme is stable, in which case the theoretical stability limit is $C_{lim} = \frac{1}{\alpha}\sqrt{2\alpha - 1}$. See [57] for details.

**Table 13.2** Radii of stability regions and corresponding Courant stability limit $C_{lim}$ as functions of the parameter $\alpha$ for $1D$, $2D$ and $3D$ simultaneous-update schemes. Results for the 2D and 3D cases derive from the numerical von Neuman method. For $\alpha = 1$ only the 1D scheme is stable, with theoretical stability limit $C_{lim} = \frac{1}{\alpha}\sqrt{2\alpha - 1}$

| $\alpha$ | $C_{lim}$ in $1D$ | $R_{2D}$ | $C_{lim}$ in $2D$ | $R_{3D}$ | $C_{lim}$ in $3D$ |
|---|---|---|---|---|---|
| 1.0 | 1.000 | ... | ... | ... | ... |
| 2.0 | 0.866 | 0.704 | 0.498 | 0.495 | 0.286 |
| 3.0 | 0.745 | 0.664 | 0.470 | 0.576 | 0.333 |
| 4.0 | 0.662 | 0.612 | 0.433 | 0.556 | 0.321 |
| 5.0 | 0.600 | 0.564 | 0.399 | 0.528 | 0.309 |
| 6.0 | 0.554 | 0.524 | 0.371 | 0.496 | 0.286 |
| 7.0 | 0.516 | 0.492 | 0.348 | 0.472 | 0.273 |
| 8.0 | 0.484 | 0.464 | 0.328 | 0.448 | 0.259 |
| 9.0 | 0.457 | 0.444 | 0.314 | 0.428 | 0.247 |
| 10.0 | 0.435 | 0.423 | 0.299 | 0.410 | 0.237 |

From the results of Table 13.2, the following comments are in order:

1. In 2D, for $\alpha = 2$, the stability limit is $C_{lim} = 0.498 \approx 1/2$. That is, the stability limit of FORCE-$\alpha$ is equivalent to that of the Godunov upwind scheme.
2. In 3D, for $\alpha = 3$, the stability limit is $C_{lim} = 0.333 \approx 1/3$. That is, the stability limit of FORCE-$\alpha$ is equivalent to that of the Godunov upwind scheme.
3. Obviously, as $\alpha$ increases, $C_{lim}$ decreases in both 2D and 3D, but not very drastically. This is attractive, as increasing $\alpha$ results in decreased numerical dissipation, which is especially beneficial for resolving slowly-moving intermediate waves, such as contact discontinuities and vortical flows. See [57] for detailed discussion.
4. The above observations have implications for discontinuous Galerkin (DG) methods, which have a very stringent stability constraint. This means that the FORCE-$\alpha$ schemes are well suited for use in conjunction with DG schemes.

## 13.6 Unsplit 2D/3D Second-Order WAF Schemes

Here we study a class of second-order, simultaneous-update schemes (13.61) based on the Weighted Average Flux (WAF) methodology, originally developed for one-dimensional problems [64], see Chap. 12. For multidimensional problems, dimensional-split WAF schemes were constructed for the Euler equations in [65] and for the 2D shallow water equations in [66]. See also Sect. 13.4.1.

In this section we study the simultaneous-update WAF schemes developed by Billett and Toro [67–69]. Details are found in Chap. 16 of the textbooks [1–3]. We note that, surprisingly, the direct extension of the one-dimensional WAF fluxes into the simultaneous update formula (13.61) gives an *unstable* scheme, and therefore alternative methods are needed. This is discussed next.

## 13.6 Unsplit 2D/3D Second-Order WAF Schemes

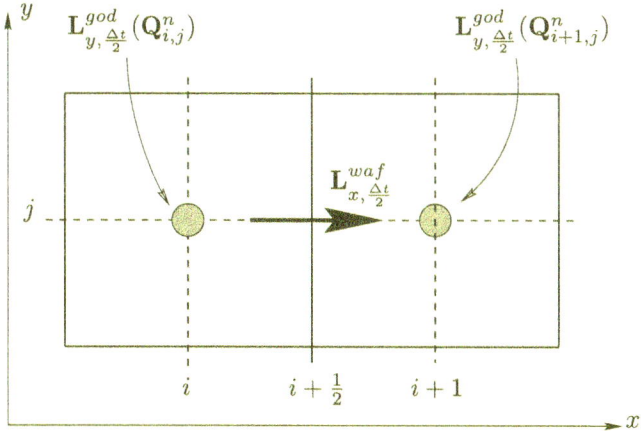

**Fig. 13.5** The intercell flux $\mathbf{F}_{i+\frac{1}{2},j}$ is obtained by applying the 1D WAF operator $\mathbf{L}^{waf}_{x,\frac{1}{2}\Delta t}$ in the $x$-direction by a time $\frac{1}{2}\Delta t$ to modified data in cells $I_{i,j}$ and $I_{i+1,j}$. The modification of the data is attained by applying the Godunov operator $\mathbf{L}^{god}_{y,\frac{1}{2}\Delta t}$ in the $y$-direction to the initial states $\mathbf{Q}^n_{i,j}$ and $\mathbf{Q}^n_{i+1,j}$ by a time $\frac{1}{2}\Delta t$

### 13.6.1 Construction of Schemes in 2D/3D

The extension of the WAF approach on structured meshes is presented in the general framework discussed below. For convenience we assume the simpler case of a 2D Cartesian mesh, as depicted in Fig. 13.2. The task is to determine the numerical fluxes $\mathbf{F}_{i+\frac{1}{2},j}$ in the $x$-direction and $\mathbf{G}_{i+\frac{1}{2},j}$ in the $y$-direction for use in the simultaneous update formula (13.61).

The computation of the $\mathbf{F}_{i+\frac{1}{2},j}$ flux is accomplished through the following two steps:

$$\left. \begin{array}{ll} \text{Step 1:} & \mathbf{Q}^{god}_{i,j} = \mathbf{L}^{god}_{y,\frac{1}{2}\Delta t}(\mathbf{Q}^n_{i,j}), \quad \mathbf{Q}^{god}_{i+1,j} = \mathbf{L}^{god}_{y,\frac{1}{2}\Delta t}(\mathbf{Q}^n_{i+1,j}), \\ \text{Step 2:} & \mathbf{F}_{i+\frac{1}{2},j} = \mathbf{L}^{waf}_{x,\frac{1}{2}\Delta t}(\mathbf{Q}^{god}_{i,j}, \mathbf{Q}^{god}_{i+1,j}). \end{array} \right\} \quad (13.80)$$

Operations (13.80), illustrated in Fig. 13.5, are executed as follows:

1. **Step 1:** the data vectors $\mathbf{Q}^n_{i,j}$ and $\mathbf{Q}^n_{i+1,j}$ on the left and right of the interface $(i+\frac{1}{2},j)$ are modified by applying a first-order Godunov step in the transverse $y$-direction by a time $\frac{1}{2}\Delta t$. This is represented by the operator $\mathbf{L}^{god}_{y,\frac{1}{2}\Delta t}$.
2. **Step 2:** The one-dimensional second-order WAF operator is applied on the modified data in the $x$-direction normal to the interface $(i+\frac{1}{2},j)$ by a time $\frac{1}{2}\Delta t$. This is just the simple second-order 1D WAF scheme described in Chap. 14 and represented by the operator $\mathbf{L}^{waf}_{x,\frac{1}{2}\Delta t}$.

Figure 13.5 illustrates the two operations (13.80) to compute $\mathbf{F}_{i+\frac{1}{2},j}$ in the $x$-direction.

The computation of the numerical flux $\mathbf{G}_{i,j+\frac{1}{2}}$ in the $y$-direction is performed analogously, that is

$$\begin{aligned}\text{Step 1:}\quad & \mathbf{Q}_{i,j}^{god}=\mathbf{L}_{x,\frac{1}{2}\Delta t}^{god}(\mathbf{Q}_{i,j}^{n}),\quad \mathbf{Q}_{i,j+1}^{god}=\mathbf{L}_{x,\frac{1}{2}\Delta t}^{god}(\mathbf{Q}_{i,j+1}^{n}),\\ \text{Step 2:}\quad & \mathbf{G}_{i,j+\frac{1}{2}}=\mathbf{L}_{y,\frac{1}{2}\Delta t}^{waf}(\mathbf{Q}_{i,j}^{god},\mathbf{Q}_{i,j+1}^{god}).\end{aligned} \quad (13.81)$$

Operations (13.81) are executed as follows:

1. **Step 1:** the data vectors $\mathbf{Q}_{i,j}^{n}$ and $\mathbf{Q}_{i,j+1}^{n}$ on the bottom and top of the interface $(i, j + \frac{1}{2})$ are modified by applying a first-order Godunov step in the transverse $x$-direction by a time $\frac{1}{2}\Delta t$. This is represented by the operator $\mathbf{L}_{x,\frac{1}{2}\Delta t}^{god}$.
2. **Step 2:** The one-dimensional second-order WAF operator is applied on the modified data in the $y$-direction, normal to the interface $(i, j + \frac{1}{2})$ by a time $\frac{1}{2}\Delta t$. This is represented by the operator $\mathbf{L}_{y,\frac{1}{2}\Delta t}^{waf}$.

The extension of the simultaneous-update second-order WAF scheme to three space dimensions is entirely analogous to the 2D case just described. Further details are found in Billett and Toro [67–69].

### 13.6.2 Stability of the Schemes in 2D/3D

The resulting second-order unsplit WAF scheme in two space dimensions has linearised stability condition [68]

$$\max\{|c_x|,|c_y|\}\leq C_{lim}=1. \quad (13.82)$$

This is double the stability limit $C_{lim}$ for the unsplit first-order scheme obtained by the direct application of the Godunov upwind scheme, and shown in Fig. 13.3. In three space dimensions the linear stability condition is

$$\max\{|c_x|,|c_y|,|c_z|\}\leq C_{lim}=2/3. \quad (13.83)$$

Again, this is double the stability limit $C_{lim}$ for the unsplit 3D scheme obtained by the direct application of the first-order Godunov upwind scheme. Further details on these WAF-type schemes are found in the original Refs. [67–70], and also in Chap. 16 of the textbooks [1–3]. For related multidimensional schemes see Colella [71], Fey et al. [72, 73], and LeVeque [74, 75].

### 13.6.3 Extensions of WAF Framework

The multidimensional WAF framework just described in Sect. 13.6.1 can be extended to construct simultaneous-update schemes in two and three space dimensions. The framework work turns out to be suitable for designing first and second-order methods by combining various 1D first and second order schemes. For example, Toro and Hu [76] proposed centred-type schemes based on the FORCE method ($\alpha = 1$) [62, 77] presented in Chap. 10 and its second-order extension the SLIC scheme [62], presented in Chap. 12.

The computation of the $\mathbf{F}_{i+\frac{1}{2},j}$ flux is accomplished through the following operations:

$$\left.\begin{array}{ll} \textbf{Step 1:} & \mathbf{Q}_{i,j}^{force} = \mathbf{L}_{y,\frac{1}{2}\Delta t}^{FO1d}(\mathbf{Q}_{i,j}^n), \quad \mathbf{Q}_{i+1,j}^{force} = \mathbf{L}_{y,\frac{1}{2}\Delta t}^{FO1d}(\mathbf{Q}_{i+1,j}^n), \\ \textbf{Step 2:} & \mathbf{F}_{i+\frac{1}{2},j} = \mathbf{L}_{x,\frac{1}{2}\Delta t}^{slic}(\mathbf{Q}_{i,j}^{force}, \mathbf{Q}_{i+1,j}^{force}). \end{array}\right\} \quad (13.84)$$

Operations (13.84) are executed as follows:

1. **Step 1:** the data vectors $\mathbf{Q}_{i,j}^n$ and $\mathbf{Q}_{i+1,j}^n$ on the left and right of the interface $(i + \frac{1}{2}, j)$ are modified by applying a first-order FORCE scheme in the transverse $y$-direction by a time $\frac{1}{2}\Delta t$. This is represented by the operator $\mathbf{L}_{y,\frac{1}{2}\Delta t}^{FO1d}$.
2. **Step 2:** The one-dimensional second-order SLIC operator is applied on the modified data in the $x$-direction normal to the interface $(i + \frac{1}{2}, j)$ by a time $\frac{1}{2}\Delta t$. This is just the simple second-order 1D SLIC scheme [62] described in Chap. 12 and represented here by the operator $\mathbf{L}_{x,\frac{1}{2}\Delta t}^{slic}$.

The computation of the numerical flux $\mathbf{G}_{i,j+\frac{1}{2}}$ in the $y$-direction is performed analogously, that is

$$\left.\begin{array}{ll} \textbf{Step 1:} & \mathbf{Q}_{i,j}^{force} = \mathbf{L}_{x,\frac{1}{2}\Delta t}^{FO1d}(\mathbf{Q}_{i,j}^n), \quad \mathbf{Q}_{i,j+1}^{force} = \mathbf{L}_{x,\frac{1}{2}\Delta t}^{FO1d}(\mathbf{Q}_{i,j+1}^n), \\ \textbf{Step 2:} & \mathbf{G}_{i,j+\frac{1}{2}} = \mathbf{L}_{y,\frac{1}{2}\Delta t}^{slic}(\mathbf{Q}_{i,j}^{force}, \mathbf{Q}_{i,j+1}^{force}). \end{array}\right\} \quad (13.85)$$

The procedure (13.84)–(13.85) to compute the fluxes $\mathbf{F}_{i+\frac{1}{2},j}$ and $\mathbf{G}_{i,j+\frac{1}{2}}$ for use in the simultaneous-update formula (13.61) leads to second-order schemes of the centred-type with good stability properties [76]. The approach extends to three space dimensions on structured meshes. Furthermore, there is room for combining various one-dimensional operators to optimise the resulting unsplit schemes.

An introduction to more advanced computational schemes for PDEs will be presented in Chap. 14.

## 13.7 Concluding Remarks

We have addressed two major themes in the subject of numerical methods for advection-reaction partial differential equations in general domains, namely the presence of source terms in the PDEs and the multidimensional definition of the computational domain. Simple but effective methods to deal with both situations have been presented.

One-dimensional advection-reaction systems are split into an advection system of PDEs and a reaction system of ODEs. For the advection part, several schemes presented in previous chapters are directly applicable. For the reaction part, a section on numerical methods for ODEs is included. The complete procedure is termed *advection-reaction splitting*. Schemes of first and second-order of accuracy for advection-reaction systems are formulated.

For multidimensional advection systems, two approaches have been presented. The first relies on *dimensional splitting*, whereby separate 1D systems in each spatial direction are solved in succession at each time level. Suitably extended 1D schemes presented in previous chapters are applicable to each augmented 1D system. First and second-order accurate dimensional-split schemes have been presented, which preserve the generous stability constraints of the 1D schemes. These methods are suitable for structured meshes, Cartesian meshes in particular.

A second approach to solve multidimensional advection systems rests on *unsplit, simultaneous-update* finite volume formulations, whereby the updating of the solution includes contributions from all element-edge numerical fluxes. This approach is suitable for unstructured meshes. However, in this framework, not all 1D fluxes are applicable. A major issue enters the scene, namely stability. A section on stability analysis for multidimensional schemes has been included, and examples of stable schemes have been given.

Finally, a particular class of second-order finite volume, upwind unsplit methods of the WAF type, have been presented. These schemes exhibit very generous stability regions in 2D and 3D. Moreover the WAF framework may be generalised to include other classes of stable schemes, such as centred schemes. The methods presented in this chapter are second-order accurate at most. More advanced, unsplit upwind and centred schemes in the fully-discrete ADER framework are introduced in Chap. 14.

## References

1. E.F. Toro, *Riemann Solvers and Numerical Methods for Fluid Dynamics. A Practical Introduction* (Springer, 1997)
2. E.F. Toro, *Riemann Solvers and Numerical Methods for Fluid Dynamics. A Practical Introduction*, 2nd edn. (Springer, 1999)
3. E.F. Toro, *Riemann Solvers and Numerical Methods for Fluid Dynamics. A Practical Introduction*, 3rd edn. (Springer, 2009)
4. R.J. LeVeque, *Finite Volume Methods for Hyperbolic Problems* (Cambridge University Press, 2002)

# References

5. E. Godlewski, P.A. Raviart, *Numerical Approximation of Hyperbolic Systems of Conservation Laws*, 2nd edn. (Springer, 2021)
6. E.F. Toro, V.A. Titarev, ADER schemes for scalar hyperbolic conservation laws with source terms in three space dimensions. J. Comput. Phys. **202**(1), 196–215 (2005)
7. R.J. LeVeque, H.C. Yee, A study of numerical methods for hyperbolic conservation laws with stiff source terms. J. Comput. Phys. **86**, 187–210 (1990)
8. M. Dumbser, C. Enaux, E.F. Toro, Finite volume schemes of very high order of accuracy for stiff hyperbolic balance laws. J. Comput. Phys. **227**(8), 3971–4001 (2008)
9. P.L. Roe, Upwind differencing schemes for hyperbolic conservation laws with source terms, in *Proceedings of the First International Conference on Hyperbolic Problems*, ed. by Carasso, Raviart and Serre. (Springer, 1986), pp. 41–51
10. L. Bermúdez, M.E. Vázquez, Upwind methods for hyperbolic conservation laws with source terms. Comput. Fluids **23**, 1049–1071 (1994)
11. R.E. Brown, Numerical solution of the 2D unsteady Navier–Stokes equations using viscous–convective operator splitting. MSc. Thesis, Department of Aerospace Science, Cranfield University, UK (1990)
12. E.F. Toro, R.E. Brown, The WAF method and splitting procedures for viscous, shocked flows, in *Proceedings of the 18th International Symposium on Shock Waves, Tohoku University, Sendai, Japan* ed. by K. Takayama (Springer, 1992), pp. 1119–1126
13. A. Siviglia, E.F. Toro, The WAF method and splitting procedures for simulating hydro and thermal peaking waves in open channel flows. J. Hydraul. Eng. **135**(8), 651–662 (2009)
14. R.J. LeVeque, Balancing source terms and flux gradients in high-resolution Godunov methods. J. Comput. Phys. **146**, 346–365 (1998)
15. M.E. Vázquez-Cendón, Improved treatment of source terms in upwind schemes for the shallow water equations in channels with irregular geometry. J. Comput. Phys. **148**, 497–526 (1999)
16. M.J. Castro, J.M. Gallardo, J.A. López, C. Parés, Well-balanced high order extensions of Godunov's method for semi-linear balance laws. SIAM J. Numer. Anal. **46**, 1012–1039 (2008)
17. M. Dumbser, M.J. Castro, C. Parés, E.F. Toro, ADER schemes on unstructured meshes for nonconservative hyperbolic systems: applications to geophysical flows. Comput. Fluids **38**(9), 731–1748 (2009)
18. A. Canestrelli, A. Siviglia, M. Dumbser, E.F. Toro, Well-balanced high-order centred schemes for non-conservative hyperbolic systems. Applications to shallow water equations with fixed and mobile bed. Adv. Water Res. **32**, 834–844 (2009)
19. M.J. Castro, A. Pardo, C. Parés, E.F. Toro, On some fast well-balanced first order solvers for Nonconservative systems. Math. Comput. **79**(271), 1427–1472 (2010)
20. O. Delestre, P.-Y. Lagrée, A well-balanced finite volume scheme for blood flow simulation. Int. J. Numer. Methods Fluids **72**, 177–205 (2013)
21. L.O. Müller, E.F. Toro, Well-balanced high-order solver for blood flow in networks of vessels with variable properties. Int. J. Numer. Methods Fluids **29**(12), 1388–1411 (2013)
22. L.O. Müller, C. Parés, E.F. Toro, Well-balanced high-order numerical schemes for one-dimensional blood flow in vessels with varying mechanical properties. J. Comput. Phys. **242**(7), 53–85 (2013)
23. B. Ghitti, C. Berthon, M.H. Le, E.F. Toro, A fully well-balanced scheme for the 1D blood flow equations with friction source term. J. Comput. Phys. **421**, 109750 (2020)
24. L. Arpaia, M. Ricchiuto, A.G. Filippini, R. Pedreros, An efficient covariant frame for the spherical shallow water equations: Well balanced DG approximation and application to tsunami and storm surge. Ocean Modell. **169**, 101915 (2022)
25. E. Pimentel-García, L.O. Müller, E.F. Toro, C. Parés, High-order fully well-balanced numerical methods for one-dimensional blood flow with discontinuous properties. J. Comput. Phys. **475**, 111869 (2023)
26. L. Martaud, C. Berthon, Fully well-balanced entropy stable Godunov numerical schemes for the shallow water equations with the topography source term. Technical Report HAL Id: hal-04394378, HAL open science (2024)

27. V. González-Tabernero, M.J. Castro, J.A. García-Rodríguez, High-order well-balanced numerical schemes for one-dimensional shallow-water systems with Coriolis terms. Appl. Math. Comput. **469**(128528) (2024)
28. A. Navas-Montilla, J. Murillo, 2D well-balanced augmented ADER schemes for the shallow water equations with bed elevation and extension to the rotating frame. J. Comput. Phys. **372**, 316–348 (2018)
29. W. Barsukow, J.P. Berberich, A well-balanced active flux method for the shallow water equations with wetting and drying. Commun. Appl. Math. Comput. (2022). https://doi.org/10.1007/s42967-022-00241-x
30. M.E. Vázquez-Cendón, Estudio de Esquemas Descentrados para su Aplicación a las Leyes de Conservación Hiperbólicas con Términos de Fuente. Ph.D. thesis, Departamento de Matemáticas Aplicadas, Universidad de Santiago de Compostela, España (1994)
31. J.M. Greenberg, A. LeRoux, A well-balanced scheme for the numerical processing of source terms in hyperbolic equations. SIAM J. Numer. Anal. **33**, 1–16 (1996)
32. E.F. Toro, M.E. Vázquez-Cendón, Model hyperbolic systems with source terms: exact and numerical solutions, in *Godunov Methods: Theory and Applications* ed. by E.F. Toro (Kluwer Academic/Plenum Publishers, 2001)
33. P. García-Navarro, M.E. Hubbard, P. Brufau, Multi-dimensional upwind schemes: application to hydraulics, in *Godunov Methods: Theory and Applications*, ed. by E.F. Toro (Kluwer Academic/Plenum Publishers, 2001)
34. M.E. Hubbard, P. García-Navarro, Balancing source terms and flux gradients in finite volume schemes, in *Godunov Methods: Theory and Applications* ed. by E.F. Toro (Kluwer Academic/Plenum Publishers, 2001)
35. P. García-Navarro, M.E. Vázquez-Cendón, On numerical treatment of source terms in the shallow water equations. Comput. Fluids **29**(8), 951–979 (2000)
36. A.G.L. Borthwick, M. Fujihara, B.D. Rogers, Godunov solution of shallow water equations on curvilinear and quadtree grids, in *Godunov Methods: Theory and Applications* ed. by E.F. Toro. (Kluwer Academic/Plenum Publishers, 2001)
37. C.W. Shu, S. Osher, Efficient implementation of essentially non-oscillatory shock-capturing schemes. J. Comput. Phys. **77**, 439–471 (1988)
38. S. Gottlieb, C.W. Shu, Total variation diminishing Runge–Kutta schemes. Technical Report ICASE 96–50, NASA Langley Research Center, Hampton, USA (1996)
39. J.D. Lambert, *Computational Methods in Ordinary Differential Equations* (Wiley, 1973)
40. D. Kahaner, C. Moler, S. Nash, *Numerical Methods and Software* (Prentice Hall, Englewood Cliffs, New Jersey, 1989)
41. C.W. Gear, *Numerical Initial-Value Problems in Ordinary Differential Equations* (Prentice-Hall, Englewood Cliffs, NJ, 1971)
42. L.F. Shampine, *Numerical Solution of Ordinary Differential Equations* (Chapman and Hall, London and New York, 1994)
43. S.K. Godunov, Finite difference methods for the computation of discontinuous solutions of the equations of fluid dynamics. Mat. Sb. **47**, 271–306 (1959)
44. E.F. Toro, L.O. Müller, *Computational Bodily Fluid Dynamics. Models and Algorithms (to appear)* (Springer, 2024)
45. M. Brown, *Differential Equations and their Applications* (Springer, 1975)
46. E.L. Ince, I.N. Sneddon, *The Solution of Ordinary Differential Equations* (Longman Scientific and Technical, 1987)
47. D.A. Sánchez, *Ordinary Differential Equations and Stability Theory* (Dover Publications, Inc., 1968)
48. E. Coddington, N. Levinson, *Theory of Ordinary Differential Equations* (McGraw-Hill, New York, 1955)
49. F.B. Hildebrand, *Introduction to Numerical Analysis* (Tata McGraw-Hill Publishing Co., Limited, New Delhi, 1974)
50. J.H. Mathews, *Numerical Methods* (Prentice–Hall International, Inc., 1987)
51. S.D. Conte, C. de Boor. *Elementary Numerical Analyis* (McGraw–Hill Kogakusha Ltd., 1980)

References 315

52. M.J. Maron, R.J. Lopez, *Numerical Analysis* (Wadsworth, 1991)
53. L.W. Johnson, R.D. Riess, *Numerical Analysis* (Addison–Wesley Publishing Company, 1982)
54. M.E. Vázquez-Cendón, *Solving Hyperbolic Equations with Finite Volume Methods* (Springer, 2015)
55. G. Strang, On the construction and comparison of difference schemes. SIAM J. Numer. Anal. **5**(3), 506–517 (1968)
56. R.F. Warming, R.W. Beam, Upwind second order difference schemes with applications in aerodynamic flows. AIAA J. **24**, 1241–1249 (1976)
57. E.F. Toro, B. Saggiorato, S. Tokareva, A. Hidalgo, Low-dissipation centred schemes for hyperbolic equations in conservative and non-conservative form. J. Comput. Phys. **416**(109545) (2020)
58. C. Hirsch, *Numerical Computation of Internal and External Flows, Vol. 1: Fundamentals of Numerical Discretization* (Wiley, 1988)
59. C.B. Laney, *Computational Gasdynamics* (Cambridge University Press, 1998)
60. K.A. Hoffmann, *Computational Fluid Dynamics for Engineers* (Engineering Education Systems, Austin, Texas, USA, 1989)
61. S.J. Billett, E.F. Toro, On the accuracy and stability of explicit schemes for multidimensional linear homogeneous advection equations. J. Comp. Phys. **131**, 247–250 (1997)
62. E.F. Toro, S.J. Billett, Centred TVD schemes for hyperbolic conservation laws. IMA J. Numer. Anal. **20**, 47–79 (2000)
63. E.F. Toro, A. Chakraborty, Development of an Approximate Riemann Solver for the Steady Supersonic Euler Equations. Aeronaut. J. **98**, 325–339 (1994)
64. E.F. Toro, A weighted average flux method for hyperbolic conservation laws. Proc. Roy. Soc. Lond. **A423**, 401–418 (1989)
65. E.F. Toro, The weighted average flux method applied to the time-dependent Euler equations. Phil. Trans. Roy. Soc. Lond. **A341**, 499–530 (1992)
66. E.F. Toro, Riemann problems and the WAF method for solving two-dimensional shallow water equations. Phil. Trans. Roy. Soc. Lond. **A338**, 43–68 (1992)
67. S.J. Billett, A class of upwind methods for conservation laws. Ph.D. thesis, College of Aeronautics, Cranfield University, UK (1994)
68. S.J. Billett, E.F. Toro, WAF-type schemes for multidimensional hyperbolic conservation laws. J. Comp. Phys. **130**, 1–24 (1997)
69. S.J. Billett, E.F. Toro, Unsplit WAF-type schemes for three dimensional hyperbolic conservation laws, in *Numerical Methods for Wave Propagation Toro* ed. by E.F., J. F. Clarke (Kluwer Academic Publishers, 1998), pp. 75–124
70. S.J. Billett, E.F. Toro, Implementing a three-dimensional finite volume WAF-type scheme for the Euler equations (1996)
71. P. Colella, Multidimensional upwind methods for hyperbolic conservation laws. J. Comput. Phys. **87**, 171–200 (1990)
72. M. Fey, R. Jeltsch, A new multidimensional euler scheme. Technical Report 92–09, SAM, Eidgenössische Technische Hochschule (ETH), Zürich, Switzerland (1992)
73. M. Fey, R. Jeltsch, S. Müller, The influence of a source term, an example: chemically reacting hypersonic flow. Technical Report 92–06, SAM, Eidgenössische Technische Hochschule (ETH), Zürich, Switzerland (1992)
74. R.J. LeVeque, High resolution finite volume methods on arbitrary grids via wave propagation. J. Comput. Phys. **78**, 36–63 (1988)
75. R.J. LeVeque, Simplified multidimensional flux limiter methods, in *Numerical Methods in Fluid Dynamics 4: Proceedings of the 1992 International Conference on Numerical Methods in Fluids* ed. by M.J. Baines, K.W. Morton (Reading, 1993), pp. 173–189
76. E.F. Toro, W. Hu, Centred unsplit finite volume schemes for multidimensional hyperbolic conservation laws, in *Godunov Methods: Theory and Applications*, ed. by E.F. Toro. (Kluwer Academic/Plenum Publishers, 2001), pp. 899–906
77. E.F. Toro, On Glimm–related schemes for conservation laws. Technical Report MMU–9602, Department of Mathematics and Physics, Manchester Metropolitan University, UK (1996)

# Chapter 14
# ADER High-Order Methods

**Abstract** This chapter is concerned with advanced, numerical methods for evolutionary partial differential equations, following the ADER framework. The methodology is an unlimited-order, non-linear fully discrete one-step extension of Godunov's method that operates in the finite volume and discontinuous Galerkin finite element frameworks. The approach is applicable to multidimensional systems in conservative and non-conservative forms, on structured and unstructured meshes. The building block of ADER is the generalised Riemann problem $GRP_m$, admitting stiff and non-stiff source terms, with initial data consisting of reconstruction polynomials of arbitrary degree $m$. The resulting ADER schemes are of order $m + 1$ in both space and time. Here we review some key aspects of the methodology and give appropriate references. The presentation is general, so as to include any system of hyperbolic balance laws and extensions, but we also identify specific applications to the shallow water equations of interest in this book. The chapter is concluded with two examples that highlight the key point of very high-order methods; that is, for small errors high-order ADER methods are orders-of-magnitude more efficient than low-order methods.

## 14.1 Introduction

There is evidence that numerical methods of very high order of accuracy, in both space and time, for solving evolutionary partial differential equations, are orders-of-magnitude cheaper than lower-order methods for attaining a prescribed, small error. Such methods are therefore mandatory for computationally-demanding, ambitious scientific and technological applications. In this chapter we present the ADER methodology, a class of one-step, fully discrete non-linear methods that includes all order of accuracy in space in time, in a single framework, either finite volumes or discontinuous Galerkin (DG) finite elements. ADER was originally introduced in [1–3]. See also [4, 5] for early applications of the ADER approach in the field of acoustics.

In 1959 Godunov communicated his method to solve the Euler equations of gas dynamics [6], thus establishing the Godunov's school of thought, regarding the construction of numerical methods for approximating solutions of hyperbolic equations. The building block of the original first-order Godunov upwind method is the conventional piece-wise constant data Riemann problem, for which many solvers are available in the literature [7], as well as extensions, analysis and applications [8]. The ADER methodology, succinctly presented in this Chapter, is a high-order extension of the Godunov method, for which the building block is the generalised Riemann problem $GRP_m$. This is a piece-wise smooth data Cauchy problem in which the equations admit source terms, stiff or non-stiff, and the initial data is represented by non-linear reconstructed polynomials of arbitrary degree $m$, or some other functions. The $GRP_m$ results in ADER schemes of $(m + 1)$-th order of accuracy in both space and time. As $m$ is arbitrary, there is no theoretical accuracy barrier in the ADER approach. In the early application of Dumbser and Munz to linear acoustics, they implemented ADER schemes up to order 24-th in space and time [5]. The ADER methodology, in the finite volume and discontinuous Galerkin finite elements frameworks, applies to both structured [9, 10] and unstructured meshes [11–15].

The key ingredient of ADER, in both the finite volume and DG frameworks, is the solution of the generalised Riemann problem $GRP_m$. Sometimes in the literature, ADER is described as a time dicretization scheme. This is erroneous, as the $GRP_m$ intimately couples space and time at the level of the intercell numerical flux. Cell-based reconstructed functions, such as polynomials, provide the piece-wise smooth initial conditions for the $GRP_m$. The high-order ADER intercell numerical flux is computed from the solution of $GRP_m$. The numerical source is computed from a cell-volume integral evaluated on a time-evolved reconstruction polynomial within the control volume, using a strategy analogous to that for the numerical flux. For finite volume ADER schemes, the polynomial data results from non-linear spatial reconstruction methods [13, 16–23]. For ADER schemes in the DG framework the polynomial data is readily available [5, 13, 14, 24].

Solvers for the generalised Riemann problem $GRP_m$ may be classified in various ways. A first distinction is whether the solver is semi-analytical or fully numerical. Semi-analytical methods [1, 2, 25] construct a high-order polynomial in time right at the cell interface. To this end, use is made of the Cauchy-Kovalevskaya (CK) procedure [26, 27], whereby high-order time derivatives of the unknown vector are expressed as functionals of high-order spatial derivatives. The CK procedure is, algebraically, a demanding task to perform, for which symbolic manipulators must be used. Recently, a simplification to the CK procedure was put forward by Montecinos and Balsara [28]; see also [29]. There are also fully numerical solvers [15, 30]. These avoid the use of the CK procedure, which renders the schemes more easily applicable.

Another distinction is whether $GRP_m$ solvers can deal adequately with stiff source terms in the governing PDEs. The numerical $GRP_m$ solver of Dumbser et al. [15] was the first scheme to successfully deal with stiff source terms, making it possible to reconcile two classically contradictory requirements: stiffness and very high order of accuracy. Also, the locally implicit, semi-analytical solver of Montecinos and Toro [25] overcomes this limitation, though still requiring the use of the CK procedure. See also [28, 29]. Much of the material of this Chapter is based on the recent review [31].

## 14.2 ADER in the Finite Volume Framework

In this Section we present the essential elements of the ADER numerical approach to solve hyperbolic balance laws to high order of accuracy in both space and time.

### 14.2.1 Preliminaries

We wish to solve systems of time-dependent partial differential equations (PDEs)

$$\partial_t \mathbf{Q}(\mathbf{x},t) + \mathcal{A}(\mathbf{Q}(\mathbf{x},t)) = \mathbf{S}(\mathbf{Q}(\mathbf{x},t)) + \mathcal{D}(\mathbf{Q}(\mathbf{x},t)), \mathbf{x} \in \mathcal{R}^3, \ t > 0, \quad (14.1)$$

along with initial and boundary conditions. $\mathbf{Q}$ is the vector of unknowns; $\mathcal{A}$ is a first-order advection differential operator in multiple space dimensions; $\mathcal{D}$ is a dissipative differential operator and $\mathbf{S}$ is the vector of source terms. Equations (14.1) include systems written in non-conservative form. In the rest of this chapter we shall consider the special case of an $N \times N$ system of hyperbolic balance laws in one space dimension

$$\partial_t \mathbf{Q}(x,t) + \partial_x \mathbf{F}(\mathbf{Q}(x,t)) = \mathbf{S}(\mathbf{Q}(x,t)), \ x \in \mathcal{R}, \ t > 0. \quad (14.2)$$

Here $\mathbf{Q}, \mathbf{F}(\mathbf{Q}), \mathbf{S}(\mathbf{Q})$ define conserved variables, fluxes and sources, respectively, with

$$\mathbf{Q} = [q_1, q_2, \ldots, q_N]^T, \ \mathbf{F}(\mathbf{Q}) = [f_1, f_2, \ldots, f_N]^T, \ \mathbf{S}(\mathbf{Q}) = [s_1, s_2, \ldots, s_N]^T. \quad (14.3)$$

Consider the space-time control volume $V_i^n = [x_{i-\frac{1}{2}}, x_{i+\frac{1}{2}}] \times [t_n, t_{n+1}]$, with $\Delta x = x_{i+\frac{1}{2}} - x_{i-\frac{1}{2}}$ and $\Delta t = t_{n+1} - t_n$. Exact integration of (14.2) in $V_i^n$ gives the formula

$$\mathbf{Q}_i^{n+1} = \mathbf{Q}_i^n - \frac{\Delta t}{\Delta x}(\mathbf{F}_{i+\frac{1}{2}} - \mathbf{F}_{i-\frac{1}{2}}) + \Delta t \mathbf{S}_i, \quad (14.4)$$

with definitions

$$\left. \begin{aligned} \mathbf{Q}_i^n &= \frac{1}{\Delta x} \int_{x_{i-\frac{1}{2}}}^{x_{i+\frac{1}{2}}} \mathbf{Q}(x, t_n) dx, \\ \mathbf{F}_{i+\frac{1}{2}} &= \frac{1}{\Delta t} \int_{t_n}^{t_{n+1}} \mathbf{F}(\mathbf{Q}(x_{i+\frac{1}{2}}, t))dt, \\ \mathbf{S}_i &= \frac{1}{\Delta t \Delta x} \int_{t_n}^{t_{n+1}} \int_{x_{i-\frac{1}{2}}}^{x_{i+\frac{1}{2}}} \mathbf{S}(\mathbf{Q}(x,t))dxdt. \end{aligned} \right\} \quad (14.5)$$

Formula (14.4) is exact with definitions (14.5). Finite volume numerical methods emerge from (14.4) if integrals in (14.5) are suitably approximated, leading to $\mathbf{F}_{i+\frac{1}{2}}$, the **numerical flux**, and to $\mathbf{S}_i$, the **numerical source**. These emerge from the second and third lines in (14.5) as suitable approximations to the respective integrals. The cell integral average $\mathbf{Q}_i^n$ is computed explicitly only at the initial time from the initial conditions; for later times this results automatically from formula (14.4), which becomes a one-step numerical method for evolving spatial cell averages in time.

**Example: the first-order Godunov method.** For the homogeneous case $\mathbf{S}(\mathbf{Q}) = \mathbf{0}$, the classical first-order Godunov method [6] results if the integrand for the flux integral for $\mathbf{F}_{i+\frac{1}{2}}$ is assumed to be the similarity solution $\mathbf{G}_{i+\frac{1}{2}}(\hat{x}/\tau)$ of the conventional Riemann problem

$$\left. \begin{array}{l} \text{PDEs: } \partial_t \mathbf{Q} + \partial_x \mathbf{F}(\mathbf{Q}) = \mathbf{0}, \quad x \in \mathcal{R}, \quad t > 0, \\ \text{ICs: } \quad \mathbf{Q}(x, 0) = \begin{cases} \mathbf{Q}_i^n & \text{if } x < x_{i+\frac{1}{2}}, \\ \mathbf{Q}_{i+1}^n & \text{if } x > x_{i+\frac{1}{2}}, \end{cases} \end{array} \right\} \tag{14.6}$$

in local coordinates

$$\hat{x} = x - x_{i+\frac{1}{2}}, \qquad \tau = t - t_n. \tag{14.7}$$

As $\mathbf{G}_{i+\frac{1}{2}}(\hat{x}/\tau)$ is constant on the interface $\hat{x}/\tau = 0$, the Godunov numerical flux becomes

$$\mathbf{F}_{i+\frac{1}{2}}^{God} = \frac{1}{\Delta t} \int_0^{\Delta t} \mathbf{F}(\mathbf{G}_{i+\frac{1}{2}}(0)) dt = \mathbf{F}(\mathbf{G}_{i+\frac{1}{2}}(0)). \tag{14.8}$$

In the Godunov method, the data at time $t_n$ is furnished by a set of $M$ cell integral averages $\{\mathbf{Q}_i^n\}_{i=1}^M$ in the computational domain, given in the first line of (14.5). This naturally results in discontinuous data for (14.6) at the cell interface $x_{i+\frac{1}{2}}$. See Chaps. 6, 7 and 11. See also [7, 32, 33] for background on the Godunov method as well as exact and approximate solvers for the conventional Riemann problem (14.6).

### 14.2.2 The ADER Approach to High Order

The ADER method is a two-fold generalization of the first-order Godunov method (14.4), (14.8), namely: (i) the equations admit source terms, and (ii) accuracy in space and time is arbitrary. Instead of the piece-wise constant data for the Godunov method at time $t_n$, ADER makes use of a set of polynomials $\{\mathbf{P}_i(x)\}_{i=1}^M$ of arbitrary degree $m$, as illustrated in Fig. 14.1 for three consecutive cells. The resulting ADER method, based on the same one-step formula (14.4) as for Godunov's method, will be of arbitrary accuracy $m + 1$ in both space and time for the full inhomogeneous PDEs.

There are two items to be determined in the numerical update formula (14.4), the numerical flux and the numerical source. For computing the numerical flux, instead of solving the classical Riemann problem (14.6), the ADER numerical flux is computed from the solution of a **Generalised Riemann problem** $GRP_m$ for hyperbolic balance laws at the interface $x = x_{i+\frac{1}{2}}$ defined as

## 14.2 ADER in the Finite Volume Framework

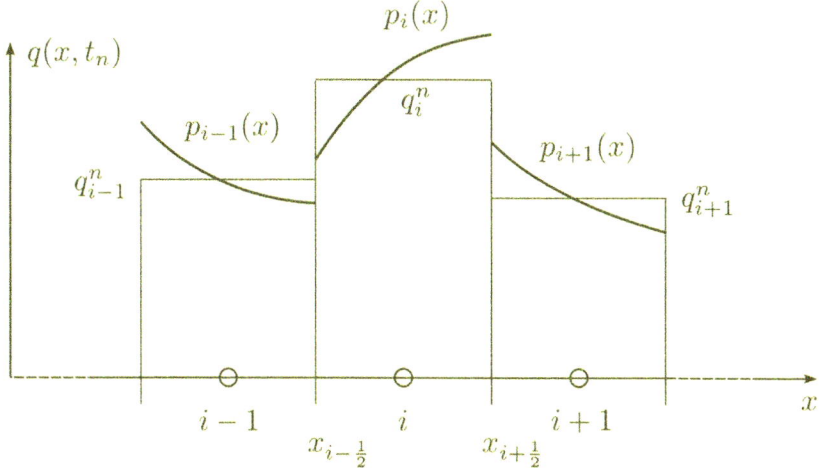

**Fig. 14.1** Single component $p(x)$ of reconstructed polynomials $\mathbf{P}_l(x)$ in cells $I_{i-1}$, $I_i$ and $I_{i+1}$. Such polynomials are the initial conditions for a generalised Riemann problem of the type (14.9) at the interface $x = x_{i+\frac{1}{2}}$ between two neighbouring cells (courtesy of Prof. LO Müller, University of Trento, Italy)

$$GRP_m : \begin{cases} \text{PDEs: } \partial_t \mathbf{Q} + \partial_x \mathbf{F}(\mathbf{Q}) = \mathbf{S}(\mathbf{Q}) \,, \ x \in \mathcal{R} \,, \ t > 0 \,, \\ \text{ICs: } \mathbf{Q}(x,0) = \begin{cases} \mathbf{P}_i(x) & \text{if } x < x_{i+\frac{1}{2}} \,, \\ \mathbf{P}_{i+1}(x) & \text{if } x > x_{i+\frac{1}{2}} \,. \end{cases} \end{cases} \quad (14.9)$$

The $GRP_m$ is a two-fold generalisation of the conventional Riemann problem (14.6), namely (i) the PDEs admit source terms $\mathbf{S}(\mathbf{Q})$ and (ii) the initial conditions are two polynomials $\mathbf{P}_i(x)$ and $\mathbf{P}_{i+1}(x)$ of arbitrary degree $m$ provided by non-linear spatial reconstruction algorithms, such as ENO or WENO [16, 19, 21]. The non-linear character of the reconstruction polynomials is necessary in order to circumvent Godunov's theorem [6]. Compare the Cauchy problems (14.6) and (14.9) for the Godunov and ADER methods, respectively. Figure 14.2 shows the wave structure of a typical generalised Riemann problem displayed in the $x$-$t$ plane. Once the solution $\mathbf{J}_{i+\frac{1}{2}}(\tau)$ to $GRP_m$ (14.9) along the interface is obtained, the ADER numerical flux results from approximating the flux integral in (14.5) via a suitable quadrature rule, namely

$$\mathbf{F}_{i+\frac{1}{2}}^{Ader} = \frac{1}{\Delta t} \int_0^{\Delta t} \mathbf{F}(\mathbf{J}_{i+\frac{1}{2}}(\tau)) d\tau \approx \sum_{o=1}^{\hat{\alpha}} \omega_o \mathbf{F}(\mathbf{J}_{i+\frac{1}{2}}(\gamma_o \Delta t)). \quad (14.10)$$

Recommended choices for the computational parameters in (14.10) are $\hat{\alpha} = 3$ and

$$\gamma_1 = \frac{1}{2} - \frac{\sqrt{15}}{10}, \quad \gamma_2 = \frac{1}{2}, \quad \gamma_3 = \frac{1}{2} + \frac{\sqrt{15}}{10}, \quad \omega_1 = \frac{5}{18}, \omega_2 = \frac{8}{18}, \quad \omega_3 = \frac{5}{18}. \quad (14.11)$$

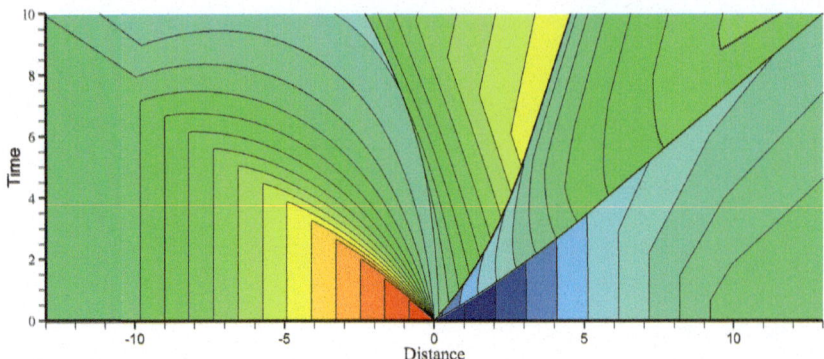

**Fig. 14.2** Solution of a generalised Riemann problem (14.9) for the one-dimensional Euler equations represented in the $x$-$t$ plane. Note the curved characteristics and wave paths. ADER makes use of the solution of $GRP_m$ at the interface $x = x_{i+\frac{1}{2}}$, locally at $x = 0$, (along the $t$ axis) to evaluate the numerical flux to be used in (14.4) (courtesy of Dr. VA Titarev, the Russian Academy of Sciences, Moscow, Russia)

The computation of the numerical source $S_i$ is described in Sect. 14.3.3.

There are by now several methods available to solve the generalised Riemann problem (14.9). In this Chapter we review in some detail three of the existing methods and give references for alternative solvers. All the $GRP_m$ solvers reviewed here, seek an approximation $\mathbf{J}_{i+\frac{1}{2}}(\tau)$ to the solution of (14.9) at the interface $x = x_{i+\frac{1}{2}}$ to be used in the computation of the numerical flux $\mathbf{F}_{i+\frac{1}{2}}^{Ader}$ (14.10) for the one-step formula (14.4).

**Remark 1** (Stability of ADER) The ADER scheme in the finite volume framework, in one space dimension has linear stability condition $0 \leq |c| \leq C_{lim} = 1$, **for all orders of accuracy.** Admittedly, this may be contrary to expectation.

## 14.3 The Toro-Titarev Solver for $GRP_m$

To compute the ADER numerical flux $\mathbf{F}_{i+\frac{1}{2}}^{Ader}$ (14.10) the Toro-Titarev (TT) solver [2] seeks an approximation $\mathbf{J}_{i+\frac{1}{2}}(\tau)$ to $GRP_m$ (14.9) at the interface, departing from the LeFloch-Raviart expansion [34]

$$\mathbf{J}_{i+\frac{1}{2}}(\tau) = \mathbf{Q}(x_{i+\frac{1}{2}}, 0_+) + \sum_{l=1}^{m} \left[ \partial_t^{(l)} \mathbf{Q}(x_{i+\frac{1}{2}}, 0_+) \right] \frac{\tau^l}{l!} \,. \qquad (14.12)$$

The leading term in (14.12) is defined as

$$\mathbf{Q}(x_{i+\frac{1}{2}}, 0_+) = \lim_{\tau \to 0_+} \mathbf{Q}(x_{i+\frac{1}{2}}, \tau) \,. \qquad (14.13)$$

## 14.3 The Toro-Titarev Solver for $GRP_m$

Determining (14.12) involves two distinct steps: computing the leading term (14.13) and computing the higher-order terms in (14.12), as explained in what follows.

### 14.3.1 Flux Leading Term

The leading term is computed by solving the **conventional** Riemann problem

$$\begin{aligned} \text{PDEs:} & \quad \partial_t \mathbf{Q} + \partial_x \mathbf{F}(\mathbf{Q}) = \mathbf{0}, \quad x \in \mathcal{R}, \quad t > 0, \\ \text{ICs:} & \quad \mathbf{Q}(x, 0) = \begin{cases} \mathbf{P}_i(x_{i+\frac{1}{2}}) & \text{if } x < x_{i+\frac{1}{2}}, \\ \mathbf{P}_{i+1}(x_{i+\frac{1}{2}}) & \text{if } x > x_{i+\frac{1}{2}}. \end{cases} \end{aligned} \quad (14.14)$$

The initial conditions are **piece-wise constant**, given by limiting values of $\mathbf{P}_i(x)$ and $\mathbf{P}_{i+1}(x)$ from left and right respectively. Denoting by $\mathbf{G}^{(0)}_{i+\frac{1}{2}}(\hat{x}/\tau)$ the similarity solution of (14.14) we set

$$\mathbf{Q}(x_{i+\frac{1}{2}}, 0_+) = \mathbf{G}^{(0)}_{i+\frac{1}{2}}(0) . \quad (14.15)$$

Exact or approximate solvers can be used to find $\mathbf{G}^{(0)}_{i+\frac{1}{2}}(\hat{x}/\tau)$, the sampling of which gives the **Godunov state** $\mathbf{G}^{(0)}_{i+\frac{1}{2}}(0)$. See Chap. 6 for an exact solver for the shallow water equations and Chap. 11 for approximate solvers.

### 14.3.2 Flux Higher-Order Terms

The higher-order terms in (14.12) involve the following steps:

1. The Cauchy-Kovalevskaya (CK) procedure to express time derivatives as functionals of spatial derivatives:

$$\partial_t^{(l)} \mathbf{Q}(x, t) = \mathbf{T}^{(l)} \left( \partial_x^{(0)} \mathbf{Q}, \partial_x^{(1)} \mathbf{Q}, \dots, \partial_x^{(l)} \mathbf{Q} \right) . \quad (14.16)$$

$\mathbf{T}^{(l)}$ is specific to the particular PDE system of interest. As a very simple example consider the PDE $\partial_t q + \lambda \partial_x q = 0$, with $\lambda$ constant. Clearly, from the PDE the first-order time derivative is $\partial_t q = -\lambda \partial_x q$. Also, the second-order time derivative is $\partial_t^2 q = (-\lambda)^2 \partial_x^2 q$. In general, for any integer $k > 2$, it can be proved by induction that $\partial_t^k q = (-\lambda)^k \partial_x^k q$.
In the more general case, the pending problem in (14.16) is to determine the arguments of the functional $\mathbf{T}^{(l)}$, which are spatial derivatives.

2. Evolution equations for spatial derivatives of order $l$, for $l = 1, \dots, m$:

$$\partial_t (\partial_x^{(l)} \mathbf{Q}(x, t)) + \mathbf{A}(\mathbf{Q}) \partial_x (\partial_x^{(l)} \mathbf{Q}(x, t)) = \mathbf{K}^{(l)} . \quad (14.17)$$

Here $\mathbf{A}(\mathbf{Q})$ is the Jacobian matrix of the PDEs in (14.9) and the source term $\mathbf{K}^{(l)}$ arises from the CK procedure.

3. Simplifications to (14.17) involve (i) a linearisation of the PDEs with a constant Jacobian matrix and (ii) neglecting the source terms, namely

$$\mathbf{A}(\mathbf{Q}) \approx \mathbf{A}(\mathbf{G}_{i+\frac{1}{2}}^{(0)}(0)) = \mathbf{A}_{i+\frac{1}{2}}^{(0)} \quad \text{and} \quad \mathbf{K}^{(l)} = \mathbf{0} . \tag{14.18}$$

The above simplifications can be justified rigorously for some special cases [35]. For the general case, empirical evidence suggests that, for smooth solutions, such simplifications do retain the theoretically aimed order of accuracy $m + 1$.

4. Solution of the conventional, linearised Riemann problems for spatial derivatives

$$\left.\begin{array}{l} \partial_t(\partial_x^{(l)}\mathbf{Q}(x,t)) + \mathbf{A}_{i+\frac{1}{2}}^{(0)} \partial_x(\partial_x^{(l)}\mathbf{Q}(x,t)) = \mathbf{0} , \\ \partial_x^{(l)}\mathbf{Q}(x,0) = \begin{cases} \mathbf{P}_i^{(l)}(x_{i+\frac{1}{2}}) & \text{if } x < x_{i+\frac{1}{2}} , \\ \mathbf{P}_{i+1}^{(l)}(x_{i+\frac{1}{2}}) & \text{if } x > x_{i+\frac{1}{2}} , \end{cases} \end{array}\right\} \tag{14.19}$$

where $\mathbf{P}_i^{(l)}(x) = \partial_x^{(l)}\mathbf{Q}(x,t)$; likewise $\mathbf{P}_{i+1}^{(l)}(x)$. The similarity solution of (14.19) in local coordinates is denoted as $\mathbf{G}_{i+\frac{1}{2}}^{(l)}(\hat{x}/\tau)$, with Godunov states $\mathbf{G}_{i+\frac{1}{2}}^{(l)}(0)$, for $l = 1, \ldots, m$.

5. The complete solution. Having determined the leading term (14.15) and all the arguments of $\mathbf{T}^{(l)}$ in (14.16) via (14.19), the final solution reads

$$\mathbf{J}_{i+\frac{1}{2}}(\tau) = \mathbf{G}_{i+\frac{1}{2}}^{(0)}(0) + \sum_{l=1}^{m} \left[\mathbf{T}^{(l)}\left(\mathbf{G}_{i+\frac{1}{2}}^{(0)}(0), \ldots, \mathbf{G}_{i+\frac{1}{2}}^{(l)}(0)\right)\right] \frac{\tau^l}{l!} . \tag{14.20}$$

The approximate solution $\mathbf{J}_{i+\frac{1}{2}}(\tau)$ to the $GRP_m$ (14.9) is inserted in (14.10) to compute the high-order ADER numerical flux for use in the one-step conservative formula (14.4). Further details are found in [2, 3] and in Chaps. 19 and 20 of [7]. To complete the scheme for the inhomogeneous system (14.2) we still need to determine the numerical source in (14.4).

### 14.3.3 The Numerical Source

The numerical source in (14.4) follows from the volume integral in (14.5), namely

$$\mathbf{S}_i = \frac{1}{\Delta t \Delta x} \int_0^{\Delta t} \int_{x_{i-\frac{1}{2}}}^{x_{i+\frac{1}{2}}} \mathbf{S}(\mathbf{Q}_i(x,t)) dx dt , \tag{14.21}$$

where $\mathbf{Q}_i(x,t)$ is an approximate solution in the control volume $V_i^n = [x_{i-\frac{1}{2}}, x_{i+\frac{1}{2}}] \times [t_n, t_{n+1}]$. For a second-order scheme, the mid-point integration rule in space and time

## 14.3 The Toro-Titarev Solver for $GRP_m$

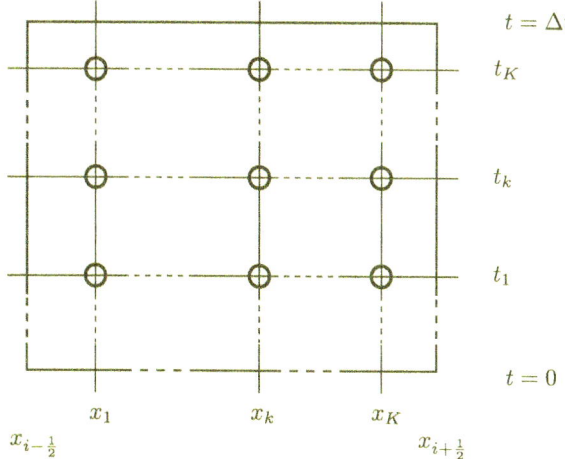

**Fig. 14.3** Numerical source. Integration points within the control volume $V_i^n$ to compute the source term to high-order of accuracy

applies, for which one requires the solution value at a single point $(x_i, \frac{1}{2}\Delta t)$ in the volume $V_i^n$. For orders of accuracy higher than 2, one needs an appropriate number of spatial positions inside the cell $I_i$ and time levels within the time interval $[0, \Delta t]$, as illustrated in Fig. 14.3.

For a generic, fixed, spatial integration point $x_\alpha$ within the cell $I_i$, as depicted in Fig. 14.3, we write a time power series expansion of the form

$$\mathbf{Q}(x_\alpha, \tau) = \mathbf{Q}(x_\alpha, 0) + \sum_{k=1}^{m} \partial_t^{(k)} \mathbf{Q}(x_\alpha, 0) \frac{\tau^k}{k!}. \tag{14.22}$$

This expansion is entirely analogous to the LeFloch-Raviart expansion (14.12) used for the solution of the $GRP_m$ at the interface between two cells. Here, however, the expansion is performed in the interior of the relevant cell $I_i$ and the leading term is simply the reconstruction polynomial at the position $x_\alpha$, at which point the function is continuous. Recall that at the local time $t = 0$, the solution coincides with the spatial reconstruction polynomial, that is $\mathbf{Q}_i(x, 0) = \mathbf{P}_i(x)$, a continuous function. Therefore, there is not need to solve a conventional Riemann problem to find the leading term, it is simply given as

$$\mathbf{Q}(x_\alpha, 0) = \mathbf{P}_i(x_\alpha). \tag{14.23}$$

The higher-order terms in (14.22) are computed via the Cauchy-Kovalesvkaya procedure, as for the $GRP_m$, whereby time derivatives in (14.22) are expressed as functionals of space derivatives and of the source terms $\mathbf{S}(\mathbf{Q})$, namely

$$\partial_t^{(l)} \mathbf{Q}(x, t) = \mathbf{T}^{(l)} \left( \partial_x^{(0)} \mathbf{Q}, \partial_x^{(1)} \mathbf{Q}, \ldots, \partial_x^{(l)} \mathbf{Q} \right). \tag{14.24}$$

The arguments of the functionals (14.24) are well defined in the entire cell $I_i$, as the solution at the local time $t = 0$ is represented by the smooth reconstructed polynomial $\mathbf{P}_i(x)$, with well-defined spatial derivatives

$$\mathbf{P}_i^{(l)}(x) \equiv \frac{d^l}{dx^l}\mathbf{P}_i(x) \,, \quad x \in [x_{i-\frac{1}{2}}, x_{i+\frac{1}{2}}] \,. \tag{14.25}$$

Therefore

$$\mathbf{Q}(x_\alpha, \tau) = \mathbf{P}_i(x_\alpha) + \sum_{k=1}^{m} \mathbf{T}^{(k)}\left(\mathbf{P}_i^{(0)}(x_\alpha), \mathbf{P}_i^{(1)}(x_\alpha), \ldots, \mathbf{P}_i^{(k)}(x_\alpha)\right) \frac{\tau^k}{k!}. \tag{14.26}$$

Expression (14.26) is capable of reproducing all necessary space-time grid points within the control volume $V_i^n$, illustrated in Fig. 14.3. There are several integration rules available for computing the multidimensional integral (14.21); see [36], for example. Using a Gaussian quadrature rule we obtain

$$\mathbf{S}_i = \sum_{\alpha=1}^{m_\alpha} \sum_{\beta=1}^{m_\beta} \omega_\alpha \omega_\beta \mathbf{S}(\mathbf{Q}(x_\alpha, \gamma_\beta \Delta t)) \,. \tag{14.27}$$

Here $x_\alpha = x_{i-\frac{1}{2}} + \gamma_\alpha \Delta x$, where $\gamma_s$ and $\omega_s$ are properly scaled nodes and weights of the rule.

An ADER-TT scheme, for systems of hyperbolic balance laws (14.2) in conservation law form, of arbitrary order of accuracy in space and time has been presented. The one-step scheme (14.4) uses the numerical flux (14.10), with integrand (14.20), and the numerical source (14.27).

In the next section we study arbitrary high-order ADER methods in conjunction with the Harten, Engquist, Osher and Chakravarthy (HEOC) solver for the $GRP_m$ [17].

## 14.4 The HEOC Solver for $GRP_m$

The HEOC (Harten, Engquist, Osher and Chakravarthy) solver for the $GRP_m$ defined in (14.9) emerges from a re-interpretation proposed by Castro and Toro [37] of the high-order method of Harten et al. [16, 17] as an ADER method. By doing so a corresponding solver for the $GRP_m$ emerges. The HEOC solver consists of a **time evolution** step and an evolved-data **interaction step**. Figure 14.4 illustrates the two steps, evolution in the bottom and interaction in the top part.

## 14.4 The HEOC Solver for $GRP_m$

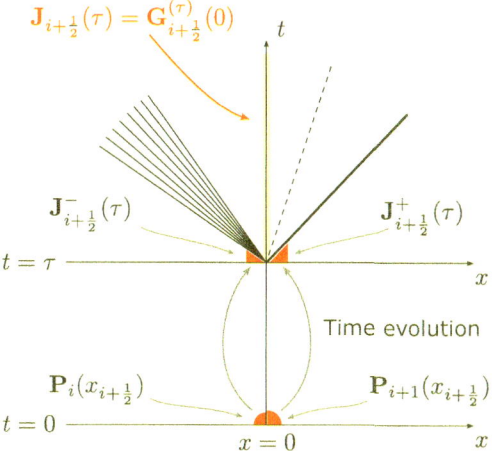

**Fig. 14.4** HEOC solver for $GRP_m$. Bottom part illustrates data evolution step. Limiting data values $\mathbf{P}_i(x_{i+\frac{1}{2}})$ and $\mathbf{P}_{i+1}(x_{i+\frac{1}{2}})$ from left and right reconstruction polynomials are evolved in time via expansions (14.28) to arbitrary time $\tau$. Top part illustrates the evolved-data interaction step. The similarity solution $\mathbf{G}^{(\tau)}_{i+\frac{1}{2}}(\hat{x}/\tau)$ of the conventional Riemann problem (14.32) with piece-wise constant data $\mathbf{J}^-_{i+\frac{1}{2}}(\tau)$ and $\mathbf{J}^+_{i+\frac{1}{2}}(\tau)$, right at the interface is found. The HEOC solution of $GRP_m$ at time $t = \tau$ is given by $\mathbf{G}^{(\tau)}_{i+\frac{1}{2}}(0)$

### 14.4.1 Time-Evolution Step for the Flux

The time-evolution step evolves the limiting values of the data left and right of the interface via two conventional Taylor expansions

$$\left.\begin{aligned}\mathbf{J}^-_{i+\frac{1}{2}}(\tau) &= \mathbf{Q}(x^-_{i+\frac{1}{2}}, 0) + \sum_{l=1}^{m} \left[\partial^{(l)}_t \mathbf{Q}(x^-_{i+\frac{1}{2}}, 0)\right] \frac{\tau^l}{l!}\,, \\ \mathbf{J}^+_{i+\frac{1}{2}}(\tau) &= \mathbf{Q}(x^+_{i+\frac{1}{2}}, 0) + \sum_{l=1}^{m} \left[\partial^{(l)}_t \mathbf{Q}(x^+_{i+\frac{1}{2}}, 0)\right] \frac{\tau^l}{l!}\,,\end{aligned}\right\} \quad (14.28)$$

with

$$\mathbf{Q}(x^-_{i+\frac{1}{2}}, 0) = \mathbf{P}_i(x_{i+\frac{1}{2}})\,, \quad \mathbf{Q}(x^+_{i+\frac{1}{2}}, 0) = \mathbf{P}_{i+1}(x_{i+\frac{1}{2}})\,. \quad (14.29)$$

See bottom part of Fig. 14.4. Through the application of the CK procedure (14.16) we express all time derivatives in (14.28) as functionals of spatial derivatives of left and right reconstruction polynomials, with

$$\mathbf{P}^{(l)}_i(x_{i+\frac{1}{2}}) = \frac{d^l}{dx^l}\mathbf{P}_i(x_{i+\frac{1}{2}}), \quad \mathbf{P}^{(l)}_{i+1}(x_{i+\frac{1}{2}}) = \frac{d^l}{dx^l}\mathbf{P}_{i+1}(x_{i+\frac{1}{2}})\,. \quad (14.30)$$

Therefore

$$\begin{aligned}\partial_t^{(l)}\mathbf{Q}(x_{i+\frac{1}{2}}^-,0) &= \mathbf{T}^{(l)}\left(\mathbf{P}_i^{(0)}(x_{i+\frac{1}{2}}),\mathbf{P}_i^{(1)}(x_{i+\frac{1}{2}}),\ldots,\mathbf{P}_i^{(l)}(x_{i+\frac{1}{2}})\right), \\ \partial_t^{(l)}\mathbf{Q}(x_{i+\frac{1}{2}}^+,0) &= \mathbf{T}^{(l)}\left(\mathbf{P}_{i+1}^{(0)}(x_{i+\frac{1}{2}}),\mathbf{P}_{i+1}^{(1)}(x_{i+\frac{1}{2}}),\ldots,\mathbf{P}_{i+1}^{(l)}(x_{i+\frac{1}{2}})\right),\end{aligned} \quad (14.31)$$

which are then substituted into (14.28) to completely determine the states $\mathbf{J}_{i+\frac{1}{2}}^-(\tau)$ and $\mathbf{J}_{i+\frac{1}{2}}^+(\tau)$ either side of the interface at time $\tau$. See Fig. 14.4.

### 14.4.2 Data Interaction Step for the Flux

In the data interaction step one solves the classical Riemann problem

$$\begin{aligned}\text{PDEs:}\quad &\partial_t\mathbf{Q}+\partial_x\mathbf{F}(\mathbf{Q})=\mathbf{0}, \\ \text{ICs:}\quad &\mathbf{Q}(x,0)=\begin{cases}\mathbf{J}_{i+\frac{1}{2}}^-(\tau) & \text{if } x<x_{i+\frac{1}{2}}, \\ \mathbf{J}_{i+\frac{1}{2}}^+(\tau) & \text{if } x>x_{i+\frac{1}{2}}.\end{cases}\end{aligned} \quad (14.32)$$

See top part of Fig. 14.4. Denoting the similarity solution of (14.32) in local coordinates as $\mathbf{G}_{i+\frac{1}{2}}^{(\tau)}(\hat{x}/\tau)$ we obtain an approximate solution of the $GRP_m$ (14.9) at the interface $x = x_{i+\frac{1}{2}}$, at an arbitrary time $t = \tau$, with $0 < \tau \le \Delta t$, as

$$\mathbf{J}_{i+\frac{1}{2}}(\tau) = \mathbf{G}_{i+\frac{1}{2}}^{(\tau)}(0). \quad (14.33)$$

Figure 14.4 illustrates the HEOC solver, with its data evolution step in time via two Taylor expansions, followed by the solution of a conventional Riemann problem, whose data is the time evolved data left and right. The corresponding ADER numerical flux follows as in (14.10), with an appropriate sequence of times $\tau_1, \tau_2, \ldots, \tau_k$ in (14.33) for the numerical integration to the right order.

### 14.4.3 The Numerical Source

The numerical source in (14.4) follows from the volume integral in (14.5), namely

$$\mathbf{S}_i = \frac{1}{\Delta t \Delta x}\int_0^{\Delta t}\int_{x_{i-\frac{1}{2}}}^{x_{i+\frac{1}{2}}}\mathbf{S}(\mathbf{Q}_i(x,t))dxdt. \quad (14.34)$$

The determination of $\mathbf{S}_i$ in the ADER-HEOC methodology is exactly the same as for the ADER-TT method. See Eq. (14.27) and details in Sect. 14.3.3.

### 14.4.4 Variations of the HEOC Solver

There are several possible variations of the ADER scheme in conjunction with the HEOC solver for the $GRP_m$. Examples are given below.

1. **Riemann problems for time derivatives**. Castro and Toro [37] also proposed a variation of the HEOC solver for the $GRP_m$ just presented, whereby the solution is obtained from a sequence of conventional Riemann problems for **time derivatives**, directly. Such solver was called the CT solver for the $GRP_m$.
2. **Time reconstruction**. The CT solver for the $GRP_m$, directly based on time derivatives, inspired the **time reconstruction** (TR) solver for the $GRP_m$, which eliminates the need for the Cauchy-Kovalevskaya procedure altogether. Preliminary results obtained by Dematté and collaborators [30] are very encouraging.
3. **Direct use of a classical upwind flux**. Instead of solving the sequence of classical Riemann problems (14.32) to find a sequence of Godunov states and compute a sequence of fluxes (14.33) one may use directly, for each element of the sequence, a classical numerical flux, such as HLL, HLLC, Osher, or Roe, based on the states left and right states $\mathbf{J}^-_{i+\frac{1}{2}}(\tau)$ and $\mathbf{J}^+_{i+\frac{1}{2}}(\tau)$. See Chaps. 10 and 11.
4. **ADER-HEOC based on FORCE**. Instead of using an upwind flux above, one may use the centred FORCE flux based on the left and right states $\mathbf{J}^-_{i+\frac{1}{2}}(\tau)$ and $\mathbf{J}^+_{i+\frac{1}{2}}(\tau)$. Montecinos [38] successfully used the low-dissipation FORCE-$\alpha$ scheme [39]. See Chaps. 10 and 11.

## 14.5 The Montecinos-Toro Solver for $GRP_m$

This solver for the $GRP_m$, identified as MT, is locally implicit [40] and operates in both the Toro-Titarev and HEOC frameworks. A distinguishing feature of this solver is that is applicable to systems of balance laws with stiff source terms. The previous two solvers do not possess this property. In the TT framework the TM solver departs from an **implicit Taylor series expansion** right at the interface

$$\mathbf{Q}(x_{i+\frac{1}{2}}, \tau) = \mathbf{Q}(x_{i+\frac{1}{2}}, 0_+) - \sum_{l=1}^{m} \left[ \partial_t^{(l)} \mathbf{Q}(x_{i+\frac{1}{2}}, \tau) \right] \frac{(-\tau)^l}{l!} . \quad (14.35)$$

This is the implicit counterpart of the LeFloch-Raviart [34] expansion (14.12). The leading term $\mathbf{Q}(x_{i+\frac{1}{2}}, 0_+) = \mathbf{G}^{(0)}_{i+\frac{1}{2}}(0)$ is as in Sects. 14.3 and 14.4, and

$$\partial_t^{(l)} \mathbf{Q}(x_{i+\frac{1}{2}}, \tau) = \mathbf{T}^{(l)}(\mathbf{Q}(x_{i+\frac{1}{2}}, \tau), \partial_x \mathbf{Q}(x_{i+\frac{1}{2}}, \tau), ..., \partial_x^{(l)} \mathbf{Q}(x_{i+\frac{1}{2}}, \tau)) . \quad (14.36)$$

The CK functional $\mathbf{T}^{(l)}$, which corresponds to the CK functional (14.16), involves spatial derivatives of $\mathbf{Q}(x_{i+\frac{1}{2}}, \tau)$ at time $\tau$. In [40], two approaches were proposed to estimate these derivatives, also using implicit Taylor series expansions.

In the **RITA** method (Reduced Implicit Taylor Approach) for evolving $\partial_x^{(l)} \mathbf{Q}(x_{i+\frac{1}{2}}, \tau)$, one only takes $k$ terms up to reaching $l + k = m$, namely

$$\partial_x^{(l)} \mathbf{Q}(x_{i+\frac{1}{2}}, \tau) = \partial_x^{(l)} \mathbf{Q}(x_{i+\frac{1}{2}}, 0_+) - \sum_{r=1}^{m-l} \frac{(-\tau)^r}{r!} \partial_t^{(r)} (\partial_x^{(l)} \mathbf{Q}(x_{i+\frac{1}{2}}, \tau)), \quad (14.37)$$

for $l = 1, \ldots, m$. The leading term is approximated as

$$\partial_x^{(l)} \mathbf{Q}(x_{i+\frac{1}{2}}, 0_+) = \mathbf{G}_{i+\frac{1}{2}}^{(l)}(0). \quad (14.38)$$

Then

$$\left. \begin{array}{l} \partial_x^{(l)} \mathbf{Q}(x_{i+\frac{1}{2}}, \tau) = \\ \mathbf{G}_{i+\frac{1}{2}}^{(l)}(0) - \sum_{r=1}^{m-l} \frac{(-\tau)^r}{r!} \mathbf{D}^{(l+r)}(\mathbf{Q}(x_{i+\frac{1}{2}}, \tau), \ldots, \partial_x^{(r+l)} \mathbf{Q}(x_{i+\frac{1}{2}}, \tau)), \end{array} \right\} \quad (14.39)$$

where

$$\left. \begin{array}{l} \mathbf{D}^{(l+r)}(\mathbf{Q}(x_{i+\frac{1}{2}}, \tau), \ldots, \partial_x^{(r+l)} \mathbf{Q}(x_{i+\frac{1}{2}}, \tau)) = \\ \partial_x^{(l)} \mathbf{T}^{(r)}(\mathbf{Q}(x_{i+\frac{1}{2}}, \tau), \ldots, \partial_x^{(r)} \mathbf{Q}(x_{i+\frac{1}{2}}, \tau)). \end{array} \right\} \quad (14.40)$$

Here (14.35) and (14.39) form a local non-linear algebraic system for $\mathbf{Q}(x_{i+\frac{1}{2}}, \tau)$ and all its spatial derivatives. Thus the solution $\mathbf{Q}(x_{i+\frac{1}{2}}, \tau)$ is available for computing the numerical flux as in (14.10).

In the **CITA** method (Complete Implicit Taylor Approach) one uses implicit Taylor series with $m$ terms for all spatial derivatives, which are evolved according to

$$\left. \begin{array}{l} \partial_x^{(l)} \mathbf{Q}(x_{i+\frac{1}{2}}, \tau) = \\ \mathbf{G}_{i+\frac{1}{2}}^{(l)}(0) - \sum_{r=1}^{m} \frac{(-\tau)^r}{r!} \mathbf{D}^{(l+r)}(\mathbf{Q}(x_{i+\frac{1}{2}}, \tau), \ldots, \partial_x^{(r+l)} \mathbf{Q}(x_{i+\frac{1}{2}}, \tau)), \end{array} \right\} \quad (14.41)$$

where $\partial_x^{(l)} \mathbf{Q}(x_{i+\frac{1}{2}}, \tau) = \mathbf{0}$ for all $l > m$. Here (14.35) and (14.41) form a non-linear algebraic system for $\mathbf{Q}(x_{i+\frac{1}{2}}, \tau)$ and all its spatial derivatives. Thus the solution $\mathbf{Q}(x_{i+\frac{1}{2}}, \tau)$ is available for computing the numerical flux as in (14.10).

**Remark 2** A simplification of the CK procedure in the context of implicit Taylor series was proposed by Montecinos and Balsara [28], thus significantly improving the efficiency of the method. See also [29, 38] for further improvements.

The full description of the Montecinos-Toro solver in the HEOC framework is found in [40] as is the construction of the numerical source.

## 14.6 Supplementary Topics

Here we briefly discuss some supplementary issues and give appropriate references for further study.

### 14.6.1 Other Solvers for $GRP_m$

There are by now several methods available to solve the generalised Riemann problem $GRP_m$, wit $m$ denoting the degree of the data reconstruction polynomials. In Sects. 14.3, 14.4 and 14.5 we described in some detail three of these, whose main feature was its semi-analytical character relying on a time-series expansion of order $m$. Here we discuss alternative solvers.

The Dumbser-Enaux-Toro Solver (DET) solver [15] is a fully numerical scheme that circumvents the Cauchy-Kovalevskaya procedure and is able to deal with stiff source terms. The DET solver is akin to the HEOC solver, with a data-evolution step and an evolved-data interaction step. See Sect. 14.4. DET performs the data-evolution step numerically for the entire polynomials on each side of the interface. The evolved-data interaction step is as for the HEOC solver, employing a conventional Riemann solver at the interface at each time-integration point. With the evolved polynomial in time available in each cell, one can evaluate the volume integral to also define the numerical source. This is a very general and effective solver that has been applied to many challenging problems, including geophysical problems. Full details are found in the original reference [15]; see also [31].

Alternative methods to solve the $GRP_m$ have been communicated by Goetz and Iske [41], Goetz and Dumbser [41] and Goetz et al. [42]. More recently, a simplification to the CK procedure was put forward by Montecinos and Balsara [28, 29]. Dematté and collaborators have communicated a fully numerical solver, called DTMT, that also avoids the CK procedure [30]. This solver is inspired in the Castro-Toro solver [37], which interacts time derivatives directly at the interface via a linearised Riemann-problem solver for time derivatives. In the DTMT solver the time derivatives are recovered from a time reconstruction procedure based on the time history of the numerical solution. This solver is the subject of current investigations.

### 14.6.2 A Note on Spatial Reconstruction

Spatial reconstruction is first of all motivated by the need to achieve high order of accuracy. But then, due to Godunov's theorem, see Chap. 9, the spatial reconstruction must be non-linear to circumvent Godunov's theorem, for solving hyperbolic PDEs. To this end, two major methodologies have been put forward in the last few decades, namely Total Variation Diminishing (TVD) methods and Essentially Non-Oscillatory methods. TVD methods cannot achieve accuracy greater than one, for smooth solutions with extrema [43, 44]. The latter methods are more general and depart from non-linear spatial polynomial reconstructions based on cell averages arising in finite volume methods. TVD methods are studied exhaustively in [7] and have been applied in this book in previous chapters, see Chap. 12. There are essentially two approaches to spatial reconstruction, briefly reviewed below

The first, called ENO (Essentially Non-Oscillatory), was introduced by Harten and collaborators [16–18]. The second approach is called WENO (Weighted Essentially Non-Oscillatory) and was first introduced by Liu, Osher and Chan [18] and further developed by Jiang and Shu [19]. A more recent version of WENO is due Dumbser and Käser [13, 15, 21]. Non-linear reconstruction methods are applicable to a wide variety of fields in science and engineering [20]. For further details on spatial reconstruction we recommend the recent book [31] and the up-to-date list of references on the subject.

The construction of ENO polynomials departs from a given set of knots $P_i = (x_i, \bar{q}_i)$ in a corresponding stencil, where $\bar{q}_i$ denotes the integral cell average in cell $I_i$. See Figs. 14.5 and 14.6. The discussion also applies to point values $q_i = q(x_i)$. ENO polynomials have two basic properties:

1. **Locality.** Each knot $P_i = (x_i, \bar{q}_i)$ will have its own associated polynomial. No global polynomial is used; as is well known, global, high-degree polynomials are in general highly oscillatory.
2. **Adaptivity (non-linearity).** The polynomial coefficients depend on the choice of the **stencil**, which in turn depends on the behaviour of the cell-average data $\{\bar{q}_i\}$ in various potential stencils. This procedure makes polynomial interpolation a non-linear procedure.

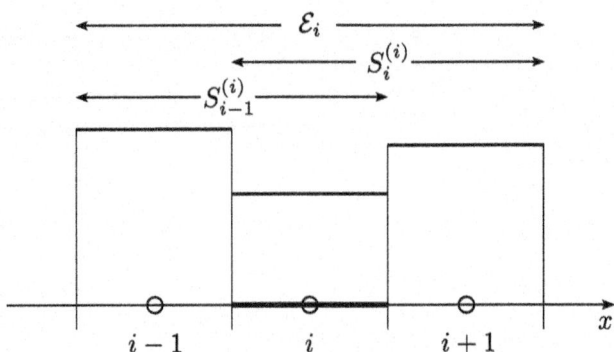

**Fig. 14.5** Two candidate stencils $S_{i-1}^{(i)}$ and $S_i^{(i)}$ for contructing a 1st-degree ENO polynomial $p_i(x)$ for cell $I_i$. The extended stencil $\mathcal{E}_i$ of $2m + 1 = 3$ cells is also shown

## 14.6 Supplementary Topics

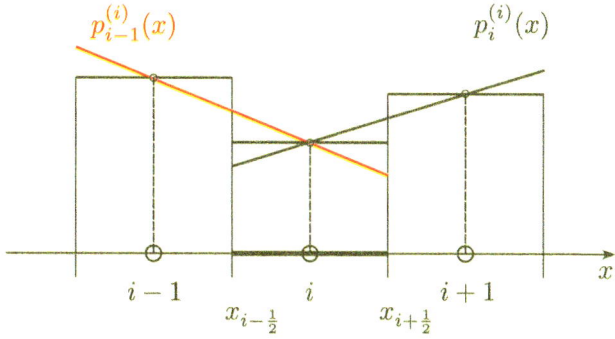

**Fig. 14.6** Two candidate ENO polynomials: $p_{i-1}^{(i)}(x)$ in stencil $S_{i-1}^{(i)}$ (left) and $p_i^{(i)}(x)$ in stencil $S_i^{(i)}$ (right). Note that the 1st-degree polynomials pass exactly through the knots to uphold conservation; this property does not hold in general

### Piece-Wise 1st-Degree ENO Polynomial

The simplest non-trivial piece-wise polynomial representation in the interval $[a, b]$ is based on polynomials of degree $m = 1$, that is, straight lines. As illustrated in Fig. 14.5 for the 1st-degree polynomial $p_i(x)$ in the generic cell $I_i$, in addition to cell $I_i$ one needs one more cell, either left or right. In terms of knots, there are two candidates: $P_{i-1} = (x_{i-1}, \bar{q}_{i-1})$ on the left and $P_{i+1} = (x_{i+1}, \bar{q}_{i+1})$ on the right. Therefore, there are two possible **stencils** that would allow the construction of a candidate ENO polynomial $p_i(x)$, namely

$$\left.\begin{aligned} \text{Left stencil:} \quad & S_{i-1}^{(i)} = \{I_{i-1}, I_i\}, \\ \text{Right stencil:} \quad & S_i^{(i)} = \{I_i, I_{i+1}\}. \end{aligned}\right\} \tag{14.42}$$

The task at hand is to find the stencil with the smoothest data.

In general, the search for the ENO stencil and thus the ENO polynomial of degree $m$ will determine an **extended stencil** $\mathcal{E}_i$ of $2m + 1$ cells. For the 1st-degree case depicted in Fig. 14.5 the extended stencil $\mathcal{E}_i$ has $2m + 1 = 3$ cells and contains the two candidate stencils $S_{i-1}^{(i)} = \{I_{i-1}, I_i\}$ and $S_i^{(i)} = \{I_i, I_{i+1}\}$. Each stencil in (14.42) generates a 1st degree polynomial, as illustrated in Fig. 14.6. By expressing each polynomial in Newton form we may write

$$\left.\begin{aligned} \text{Left polynomial:} \quad & p_{i-1}^{(i)}(x) = \bar{q}_i + q[x_{i-1}, x_i](x - x_i), \\ \text{Right polynomial:} \quad & p_i^{(i)}(x) = \bar{q}_i + q[x_i, x_{i+1}](x - x_i). \end{aligned}\right\} \tag{14.43}$$

That is

$$p_i(x) = \bar{q}_i + \begin{Bmatrix} q[x_{i-1}, x_i] \\ q[x_i, x_{i+1}] \end{Bmatrix} (x - x_i), \tag{14.44}$$

with two possibilities for choosing the coefficient of $(x - x_i)$ given in terms of divided differences, either $q[x_{i-1}, x_i]$ or $q[x_i, x_{i+1}]$.

**Divided differences.** Here we recall the definition of divided differences. For a set of knots $(x_i, q_i)$, for $i = 0, 1, \ldots, m$, divided differences are defined as follows

$$q[x_0] = q_0 \tag{14.45}$$

and the $k$-th order divided difference, for $k = 1, 2, \ldots$, is defined recursively as

$$q[x_0, x_1, \ldots, x_k] = \frac{q[x_1, x_2, \ldots, x_k] - q[x_0, x_1, \ldots, x_{k-1}]}{x_k - x_0} . \tag{14.46}$$

For details see any book of elementary numerical analysis, e.g. [36].

Returning to the polynomial (14.44), Fig. 14.6 shows the two candidate 1st degree polynomials: $p_{i-1}^{(i)}(x)$ and $p_i^{(i)}(x)$. Out of the two available candidates, ENO chooses the polynomial (14.44) with the smallest coefficient of $(x - x_i)$ in absolute; that is the divided difference with the **smallest absolute value**. Thus, in algorithmic form we write

$$p_i(x) = \bar{q}_i + q[x_j, x_{j+1}](x - x_i) , \tag{14.47}$$

with

$$j = \begin{cases} i - 1 & \text{if } |q[x_{i-1}, x_i]| < |q[x_i, x_{i+1}]| , \\ i & \text{if } |q[x_{i-1}, x_i]| \geq |q[x_i, x_{i+1}]| . \end{cases} \tag{14.48}$$

Then, a composite, piece-wise first degree polynomial $p(x)$ may be defined from (14.47)–(14.48) as

$$p(x) = \begin{cases} p_i(x) & \text{if } x \in I_i = [x_{i-\frac{1}{2}}, x_{i+\frac{1}{2}}] , \\ 0 & \text{if } x \notin I_i = [x_{i-\frac{1}{2}}, x_{i+\frac{1}{2}}] . \end{cases} \tag{14.49}$$

In the composite polynomial, for each cell $I_i$ one takes the **restriction** of $p_i(x)$, even if $p_i(x)$ is also defined outside cell $I_i$.

We note that the adaptivity of the ENO stencil search cannot be applied directly to the first cell $I_1$ and the last cell $I_M$ of the computational domain. This difficulty is resolved in practice by imposing the desired boundary conditions, which usually results in an enlarged domain from adding a suitable number of fictitious cells to those in the actual computational domain.

**Slopes and Total Variation Diminishing (TVD) Methods**

We can also write polynomials (14.47)–(14.48) as

$$p_i(x) = \bar{q}_i + \Delta_i(x - x_i) , \tag{14.50}$$

14.6 Supplementary Topics

where $\Delta_i$ is the **slope**, a divided difference, for which there are two choices. Large divided differences are associated with large derivatives and possible oscillations. Thus, effectively, the ENO method just described selects the polynomial with the smallest slope in absolute value, namely

$$\Delta_i^{ENO} = \begin{cases} \Delta_{i-\frac{1}{2}} & \text{if } |\Delta_{i-\frac{1}{2}}| < |\Delta_{i+\frac{1}{2}}|, \\ \Delta_{i+\frac{1}{2}} & \text{if } |\Delta_{i-\frac{1}{2}}| \geq |\Delta_{i+\frac{1}{2}}|. \end{cases} \quad (14.51)$$

For a scalar mesh function $\{q_k\}$ the slopes in (14.51) are

$$\Delta_{i-\frac{1}{2}} = \frac{q_i^n - q_{i-1}^n}{x_i - x_{i-1}} \quad \text{and} \quad \Delta_{i+\frac{1}{2}} = \frac{q_{i+1}^n - q_i^n}{x_{i+1} - x_i}. \quad (14.52)$$

TVD methods find suitable slopes $\Delta_i$ so as to guarantee an oscillation-free method. See Chap. 12; further details on TVD methods are found for example in [7, 43], and references therein.

Some final remarks are in order. First, higher-order ENO methods, to ensure conservation, require special procedures, omitted here. Second, WENO methods utilise linear combinations of potential ENO polynomials. Third, reconstruction is usually performed in terms of characteristic variables, rather than conserved or physical variables. For further details see [31].

### 14.6.3 ADER in the DG Framework

The discontinuous Galerkin (DG) finite element framework [45, 46] was originally proposed as a semi-discrete scheme, in which the time discretisation is performed through non-linear solvers for ordinary differential equations (ODEs). The overall accuracy is determined by the accuracy of the ODE solver, usually third order. Dumbser [12] formulated DG schemes in the fully discrete ADER framework, see also [14, 47–50]. In this manner the the resulting ADER-DG schemes have no accuracy barrier, as in the finite volume ADER methodology. In order to deal with shock waves in DG schemes a novel a posteriori subcell limiter has been introduced in [49, 51]. Extensions of the ADER finite volume and ADER-DG method to *moving* unstructured triangular, tetrahedral and polygonal meshes, including topology changes can be found, for example, in [51–56].

In the next section we specialise the ADER methodology in the finite volume framework to second-order of accuracy.

## 14.7 Examples: Second-Order ADER Schemes

In this section we illustrate the ADER approach in the finite volume framework as applied to the inhomogeneous system (14.2) but in the simpler setting of second-order schemes, including source terms. Recall that the update formula, as for the higher-order case, is

$$\mathbf{Q}_i^{n+1} = \mathbf{Q}_i^n - \frac{\Delta t}{\Delta x}(\mathbf{F}_{i+\frac{1}{2}} - \mathbf{F}_{i-\frac{1}{2}}) + \Delta t \mathbf{S}_i \ . \tag{14.53}$$

We need to compute the numerical flux $\mathbf{F}_{i+\frac{1}{2}}$ and the numerical source $\mathbf{S}_i$. Both items require the basic step of **spatial reconstruction**. For second-order methods one requires a piece-wise linear (first-degree) polynomial ($m = 1$) $\mathbf{P}_i(x)$ in each cell $I_i$, of the form

$$\mathbf{P}_i(x) = \mathbf{Q}_i^n + (x - x_i)\Delta_i \ , \quad i = 1, 2, \ldots, M \ , \tag{14.54}$$

where $\Delta_i$ is the slope of $\mathbf{P}_i(x)$ (a vector) in cell $I_i = [x_{i-\frac{1}{2}}, x_{i+\frac{1}{2}}]$, chosen from an ENO criterion, for example. Here, for simplicity, we assume a component-by-component reconstruction procedure. A better choice is reconstruction in terms of characteristic variables, not studied here. We shall construct two second-order ADER methods, which are distinguished by the way the corresponding generalised Riemann problem $GRP_1$ is solved. In the first method we use the Toro-Titarev solver, see Sect. 14.3, and in the second method we use the HEOC solver, see Sect. 14.4.

### 14.7.1 ADER2 with TT Solver for $GRP_1$

The second-order ADER method in conjunction with the TT solver for the $GRP_1$, denoted as ADER2-TT, assumes a piece-wise linear reconstruction, say ENO, and requires the numerical flux and the numerical source in (14.53).

#### The Numerical Flux

We now seek a time-dependent approximate solution $\mathbf{J}_{i+\frac{1}{2}}(\tau)$ of the generalised Riemann problem $GRP_1$ (14.9) for the case $m = 1$, right at the interface, along the $t$-axis, of the form

$$\mathbf{J}_{i+\frac{1}{2}}(\tau) = \mathbf{Q}(x_{i+\frac{1}{2}}, 0_+) + \tau \partial_t \mathbf{Q}(x_{i+\frac{1}{2}}, 0_+) \ . \tag{14.55}$$

The leading term is defined as the time limit from above

$$\mathbf{Q}(x_{i+\frac{1}{2}}, 0_+) = \lim_{\tau \to 0_+} \mathbf{Q}(x_{i+\frac{1}{2}}, \tau) \ . \tag{14.56}$$

## 14.7 Examples: Second-Order ADER Schemes

The task at hand is to compute the leading term $\mathbf{Q}(x_{i+\frac{1}{2}}, 0_+)$ and the higher-order term $\tau \partial_t \mathbf{Q}(x_{i+\frac{1}{2}}, 0_+)$ in (14.55). The leading term of the expansion requires the solution of the conventional Riemann problem (piece-wise constant data and no source term)

$$GRP_0: \begin{cases} \text{PDE: } \partial_t \mathbf{Q} + \partial_x \mathbf{F}(\mathbf{Q}) = \mathbf{0}, \quad x \in \mathcal{R}, t > 0, \\ \text{IC: } \mathbf{Q}(x, 0) = \begin{cases} \mathbf{P}_i(x_{i+\frac{1}{2}}) & \text{if } x < x_{i+\frac{1}{2}}, \\ \mathbf{P}_{i+1}(x_{i+\frac{1}{2}}) & \text{if } x > x_{i+\frac{1}{2}}. \end{cases} \end{cases} \quad (14.57)$$

Denoting the similarity solution of (14.57), in local coordinates, by $\mathbf{G}^{(0)}_{i+\frac{1}{2}}(\hat{x}/\tau)$ the leading term (14.56) is given by the Godunov state $\mathbf{G}^{(0)}_{i+\frac{1}{2}}(0)$ right at the interface, that is

$$\mathbf{Q}(x_{i+\frac{1}{2}}, 0_+) = \mathbf{G}^{(0)}_{i+\frac{1}{2}}(0). \quad (14.58)$$

In this step one can use any state solver for the conventional Riemann problem (14.57), exact or approximate. In effect, this step is entirely analogous to the first-order Godunov scheme. See Chap. 10.

We now consider the higher-order term of the expansion $\tau \partial_t \mathbf{Q}(x_{i+\frac{1}{2}}, 0_+)$ in (14.55). This involves the following steps:

1. The **Cauchy-Kovalevskaya (CK) procedure** is applied to the governing equations (14.2), to express the time derivative in terms of the remaining terms, namely

$$\partial_t \mathbf{Q}(x, t) = -\partial_x \mathbf{F}(\mathbf{Q}) + \mathbf{S}(\mathbf{Q}) = -\mathbf{A}(\mathbf{Q})\partial_x \mathbf{Q} + \mathbf{S}(\mathbf{Q}), \quad (14.59)$$

where the chain rule has been used to obtain the Jacobian matrix $\mathbf{A}(\mathbf{Q})$.

2. **Evolution equation for the space derivative**. Now the unknown in (14.59) is the spatial gradient $\partial_x \mathbf{Q}$, right at the interface. To determine it, we first write an evolution equation for $\partial_x \mathbf{Q}$, namely

$$\partial_t(\partial_x \mathbf{Q}) + \mathbf{A}(\mathbf{Q})\partial_x(\partial_x \mathbf{Q}) = \mathbf{K}. \quad (14.60)$$

3. **Simplifications to the evolution equation**. We next make two simplifications to system (14.60), namely the system is linearised and the source term $\mathbf{K}$ is neglected. That is

$$\mathbf{A}(\mathbf{Q}) \approx \mathbf{A}^{(0)}_{i+\frac{1}{2}} = \mathbf{A}(\mathbf{G}^{(0)}_{i+\frac{1}{2}}(0)), \quad \mathbf{K} = \mathbf{0}. \quad (14.61)$$

4. **Linearised Riemann problem for spatial derivatives**. The resulting linearised conventional Riemann problem from (14.60)–(14.61) for the spatial derivative reads

PDEs: $\partial_t(\partial_x \mathbf{Q}) + \mathbf{A}_{i+\frac{1}{2}}^{(0)} \partial_x(\partial_x \mathbf{Q}) = \mathbf{0}$ ,

IC: $\partial_x \mathbf{Q}(x, 0) = \begin{cases} \Delta_i \equiv \dfrac{d}{dx} \mathbf{P}_i(x) & \text{if } x < x_{i+\frac{1}{2}} , \\ \Delta_{i+1} \equiv \dfrac{d}{dx} \mathbf{P}_{i+1}(x) & \text{if } x > x_{i+\frac{1}{2}} . \end{cases}$ (14.62)

We solve (14.62) in local coordinates to obtain the solution $\mathbf{G}_{i+\frac{1}{2}}^{(1)}(\hat{x}/\tau)$. To find the Godunov state right at the interface, this solution is sampled, giving $\mathbf{G}_{i+\frac{1}{2}}^{(1)}(0)$ and therefore

$$\partial_x \mathbf{Q}(x_{i+\frac{1}{2}}, 0_+) = \mathbf{G}_{i+\frac{1}{2}}^{(1)}(0) .$$ (14.63)

5. **The complete solution and the flux**. By collecting results from (14.55), (14.58) and (14.63), the complete sought solution becomes

$$\mathbf{J}_{i+\frac{1}{2}}(\tau) = \mathbf{G}_{i+\frac{1}{2}}^{(0)}(0) + \tau[-\mathbf{A}_{i+\frac{1}{2}}^{(0)} \mathbf{G}_{i+\frac{1}{2}}^{(1)}(0) + \mathbf{S}(\mathbf{G}_{i+\frac{1}{2}}^{(0)}(0))] .$$ (14.64)

From the flux integral in (14.5) the numerical flux can be determined by inserting the approximate solution $\mathbf{J}_{i+\frac{1}{2}}(\tau)$ under the integral, namely

$$\mathbf{F}_{i+\frac{1}{2}} = \frac{1}{\Delta t} \int_0^{\Delta t} \mathbf{F}(\mathbf{J}_{i+\frac{1}{2}}(\tau)) d\tau .$$ (14.65)

Integration in (14.65) to second order (mid-point rule) gives the numerical flux as

$$\mathbf{F}_{i+\frac{1}{2}} = \mathbf{F}\left( \mathbf{G}_{i+\frac{1}{2}}^{(0)}(0) + \frac{1}{2}\Delta t[-\mathbf{A}_{i+\frac{1}{2}}^{(0)} \mathbf{G}_{i+\frac{1}{2}}^{(1)}(0) + \mathbf{S}(\mathbf{G}_{i+\frac{1}{2}}^{(0)}(0))] \right) .$$ (14.66)

Note that for higher-order ADER methods ($m \geq 2$), one generally performs numerical integration in (14.65) to the required accuracy, in order to determine the numerical flux. The numerical flux for the ADER2-TT scheme just described, in addition to the non-linear spatial reconstruction, requires the solution of one non-linear conventional Riemann problem for the leading term of the $GRP_1$ solution at the interface, and the solution of one linear Riemann problem for the gradients to find the higher order term.

**The flux knows of the source**. Note that the numerical flux $\mathbf{F}_{i+\frac{1}{2}}$ in (14.66) has a contribution from the source term $\mathbf{S}(\mathbf{Q})$ via the Cauchy-Kovalevskaya procedure (14.64). In other words, there is coupling between the flux and the source, as it should be.

## 14.7 Examples: Second-Order ADER Schemes

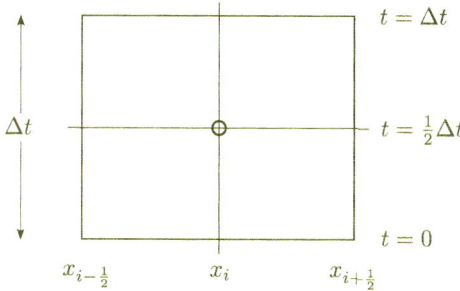

**Fig. 14.7** The numerical source is obtained by approximating the volume integral in (14.5) in the control volume $V_i^n$ through the mean-point rule in space and time to second-order accuracy; it is enough to find an approximation to the integrand at the mid point $(x_i, \frac{1}{2}\Delta t)$

### The Numerical Source

To compute the numerical source $S_i$ in (14.53) from the volume integral in (14.5) we need to know the integrand, that is the function $\mathbf{Q}(x, t)$ in the space-time control volume $V_i^n$ in (14.5). All we know in $V_i^n$ is the initial condition at the current time $t = t_n$, locally at $t = 0$. As a matter of fact the initial condition is represented by the reconstruction polynomial $\mathbf{P}_i(x)$ at time $t = t_n$. Given the second-order accuracy of the scheme, a mid-point rule evaluation of the volume integral in (14.5) gives the sought second-order approximation.

To achieve this we need an approximate solution at the mid point $(x_i, \frac{1}{2}\Delta t)$, as depicted in Fig. 14.7. For any smooth function $\mathbf{Q}(x, t)$, at local time zero, with $x = x_k \in I_i$, we may write the truncated time Taylor series expansion

$$\mathbf{Q}(x_k, \tau) = \mathbf{Q}(x_k, 0) + \tau \partial_t \mathbf{Q}(x_k, 0) . \tag{14.67}$$

In particular, at $x = x_i$, after applying the Cauchy-Kovalevskaya procedure we obtain

$$\mathbf{Q}\left(x_i, \frac{1}{2}\Delta t\right) = \mathbf{Q}_i^n + \frac{1}{2}\Delta t \partial_t \mathbf{Q}\left(x_i, \frac{1}{2}\Delta t\right) = \mathbf{Q}_i^n + \frac{1}{2}\Delta t[-\mathbf{A}(\mathbf{Q}_i^n)\Delta_i + \mathbf{S}(\mathbf{Q}_i^n)] , \tag{14.68}$$

where the slope $\Delta_i$ can be chosen from an ENO criterion (14.51) or a TVD criterion, as in Chap. 12. From the space-time integral (14.5), approximated by the mid-point rule in space and time, we obtain the numerical source as

$$\mathbf{S}_i = \mathbf{S}\left(\mathbf{Q}_i^n + \frac{1}{2}\Delta t[-\mathbf{A}(\mathbf{Q}_i^n)\Delta_i + \mathbf{S}(\mathbf{Q}_i^n)]\right) . \tag{14.69}$$

**The source knows of the flux**. Note that the numerical source $\mathbf{S}_i$ in (14.69) has a contribution from the flux $\mathbf{F}(\mathbf{Q})$ via the Cauchy-Kovalevskaya procedure. In other words, there is coupling between the flux and the source, as it should be.

The complete second-order ADER2-TT scheme is given by the update formula (14.53) with numerical flux (14.66) and numerical source (14.69).

### 14.7.2 ADER2 with HEOC Solver for $GRP_1$

Here we present a second-order accurate ADER scheme ($m = 1$) based on the HEOC solver for $GRP_1$; see Sect. 14.4 for the general case. The resulting scheme is denoted as ADER2-HEOC. The scheme is based on the update formula (14.53), which requires the numerical flux and the numerical source.

**The Numerical Flux**

To compute the numerical flux $\mathbf{F}_{i+\frac{1}{2}}$ in (14.53) from the flux integral in (14.5) we solve the generalised Riemann problem (14.9) to find the approximate solution $\mathbf{J}_{i+\frac{1}{2}}(\tau)$, where $\mathbf{P}_i(x)$ and $\mathbf{P}_{i+1}(x)$ are reconstructed polynomial vectors of degree $m = 1$ in cells $I_i$ and $I_{i+1}$, respectively. The flux computation is accomplished through two distinct steps: (i) data-evolution and (ii) evolved-data interaction via a conventional Riemann problem; see Fig. 14.8.

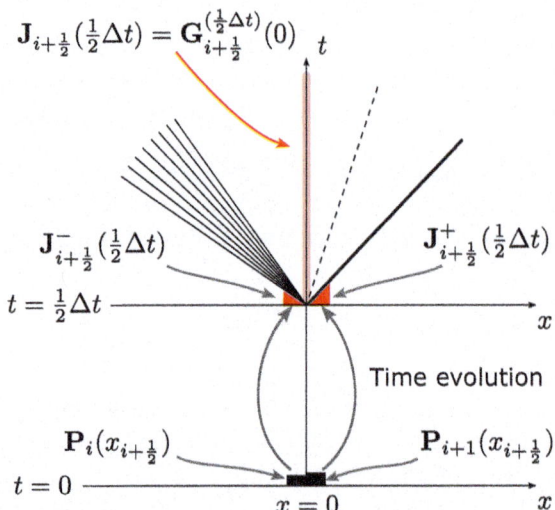

**Fig. 14.8** Solution of $GRP_1$ (14.9) (for $m = 1$) by the HEOC method. The left and right limiting polynomial values at the interface are evolved in time as shown by the vertical curved arrows either side of the interface (bottom part). The time-evolved values are the initial conditions for the interaction step (top part) via a conventional Riemann problem, whose solution structure is represented by straight characteristics. The Godunov state is then the sought solution of $GRP_1$ at the interface. Compare with Fig. 14.4

## Data Evolution in the HEOC Solver

The $GRP_m$ defined in (14.9) has initial conditions consisting of two first-degree polynomials ($m = 1$) polynomials $\mathbf{P}_i$ and $\mathbf{P}_{i+1}$ in cells $I_i$ and $I_{i+1}$ respectively. At the interface $x_{i+\frac{1}{2}}$, the left and right limiting values of the reconstruction polynomials, often called **boundary extrapolated values**, are defined as

$$\mathbf{Q}(x^-_{i+\frac{1}{2}}, 0) = \mathbf{P}_i(x_{i+\frac{1}{2}}), \quad \mathbf{Q}(x^+_{i+\frac{1}{2}}, 0) = \mathbf{P}_{i+1}(x_{i+\frac{1}{2}}). \tag{14.70}$$

See Fig. 14.8. In the HEOC solver, these boundary values are evolved in time using time Taylor series expansions as follows

$$\left. \begin{array}{l} \mathbf{J}^-_{i+\frac{1}{2}}(\tau) = \mathbf{Q}(x^-_{i+\frac{1}{2}}, 0) + \tau \partial_t \mathbf{Q}(x^-_{i+\frac{1}{2}}, 0), \\ \mathbf{J}^+_{i+\frac{1}{2}}(\tau) = \mathbf{Q}(x^+_{i+\frac{1}{2}}, 0) + \tau \partial_t \mathbf{Q}(x^+_{i+\frac{1}{2}}, 0). \end{array} \right\} \tag{14.71}$$

Fig. 14.8 illustrates, via the vertical curved arrows, the time evolution up to time $t = \frac{1}{2}\Delta t$, of the limiting values $\mathbf{Q}(x^-_{i+\frac{1}{2}}, 0)$ from the left cell $I_i$ and $\mathbf{Q}(x^+_{i+\frac{1}{2}}, 0)$ from the right cell $I_{i+1}$. After using the Cauchy-Kovalevskaya procedure, the time derivatives in (14.71) can be expressed as

$$\left. \begin{array}{l} \partial_t \mathbf{Q}(x^-_{i+\frac{1}{2}}, 0) = -\partial_x \mathbf{F}(\mathbf{Q}(x^-_{i+\frac{1}{2}}, 0)) + \tau \mathbf{S}(\mathbf{Q}(x^-_{i+\frac{1}{2}}, 0)), \\ \partial_t \mathbf{Q}(x^+_{i+\frac{1}{2}}, 0) = -\partial_x \mathbf{F}(\mathbf{Q}(x^+_{i+\frac{1}{2}}, 0)) + \tau \mathbf{S}(\mathbf{Q}(x^+_{i+\frac{1}{2}}, 0)). \end{array} \right\} \tag{14.72}$$

The flux gradients in (14.72) to second-order may be approximated as follows

$$\left. \begin{array}{l} \partial_x \mathbf{F}(\mathbf{Q}(x^-_{i+\frac{1}{2}}, 0)) = \dfrac{\mathbf{F}(\mathbf{Q}(x^-_{i+\frac{1}{2}}, 0)) - \mathbf{F}(\mathbf{Q}(x^+_{i-\frac{1}{2}}, 0))}{\Delta x}, \\ \partial_x \mathbf{F}(\mathbf{Q}(x^+_{i+\frac{1}{2}}, 0)) = \dfrac{\mathbf{F}(\mathbf{Q}(x^-_{i+\frac{3}{2}}, 0)) - \mathbf{F}(\mathbf{Q}(x^+_{i+\frac{1}{2}}, 0))}{\Delta x}. \end{array} \right\} \tag{14.73}$$

Then, the evolved boundary values left and right of the interface, at time $\tau = \frac{1}{2}\Delta t$, after using (14.71), (14.72) and (14.73), become

$$\left. \begin{array}{l} \mathbf{J}^-_{i+\frac{1}{2}}(\frac{1}{2}\Delta t) = \mathbf{Q}(x^-_{i+\frac{1}{2}}, 0) - \dfrac{1}{2}\dfrac{\Delta t}{\Delta x}\left[\mathbf{F}(\mathbf{Q}(x^-_{i+\frac{1}{2}}, 0)) - \mathbf{F}(\mathbf{Q}(x^+_{i-\frac{1}{2}}, 0))\right] \\ \quad + \frac{1}{2}\Delta t \mathbf{S}(\mathbf{Q}(x^-_{i+\frac{1}{2}}, 0)) \end{array} \right\} \tag{14.74}$$

and

$$\left.\begin{aligned}\mathbf{J}^+_{i+\frac{1}{2}}(\tfrac{1}{2}\Delta t) &= \mathbf{Q}(x^+_{i+\frac{1}{2}},0) - \frac{1}{2}\frac{\Delta t}{\Delta x}\left[\mathbf{F}(\mathbf{Q}(x^-_{i+\frac{3}{2}},0)) - \mathbf{F}(\mathbf{Q}(x^+_{i+\frac{1}{2}},0))\right] \\ &\quad + \tfrac{1}{2}\Delta t \mathbf{S}(\mathbf{Q}(x^+_{i+\frac{1}{2}},0)).\end{aligned}\right\} \quad (14.75)$$

The next step involves the interaction of the two evolved data states (14.74)–(14.75) right at the interface at time $\tau = \tfrac{1}{2}\Delta t$, see Fig. 14.8 (top part).

**Evolved-Data Interaction**

In order to find the numerical flux right at the interface $x_{i+\frac{1}{2}}$ at time $\tau = \tfrac{1}{2}\Delta t$, the next stage is to resolve the interaction of the evolved left data $\mathbf{J}^-_{i+\frac{1}{2}}(\tfrac{1}{2}\Delta t)$ and right data $\mathbf{J}^+_{i+\frac{1}{2}}(\tfrac{1}{2}\Delta t)$ in (14.74) and (14.75) respectively, via the solution of the conventional (piece-wise-constant data, homogeneous) Riemann problem

$$GRP_0: \begin{cases} \text{PDEs: } \partial_t \mathbf{Q} + \partial_x \mathbf{F}(\mathbf{Q}) = \mathbf{0}, \\ \text{ICs: } \mathbf{Q}(x,0) = \begin{cases} \mathbf{J}^-_{i+\frac{1}{2}}(\tfrac{1}{2}\Delta t) & \text{if } x < x_{i+\frac{1}{2}}, \\ \mathbf{J}^+_{i+\frac{1}{2}}(\tfrac{1}{2}\Delta t) & \text{if } x > x_{i+\frac{1}{2}}. \end{cases} \end{cases} \quad (14.76)$$

The structure of the solution of this conventional Riemann problem at the interface, at the half time $\tau = \tfrac{1}{2}\Delta t$ in Fig. 14.8 is represented by the wave configuration made up of straight characteristics. Let us denote the similarity solution of (14.76) in local coordinates as $\mathbf{G}^{\frac{1}{2}\Delta t}_{i+\frac{1}{2}}(\hat{x}/\tau)$. The corresponding sought Godunov state right at the interface is $\mathbf{G}^{\frac{1}{2}\Delta t}_{i+\frac{1}{2}}(0)$, which is obtained after sampling the solution. Finally, the numerical flux follows directly from the solution of (14.76) and becomes

$$\mathbf{F}_{i+\frac{1}{2}} = \frac{1}{\Delta t}\int_0^{\Delta t} \mathbf{F}(\mathbf{G}^{\frac{1}{2}\Delta t}_{i+\frac{1}{2}}(0))dt = \mathbf{F}(\mathbf{G}^{\frac{1}{2}\Delta t}_{i+\frac{1}{2}}(0)). \quad (14.77)$$

Note that stages (14.76) and (14.77) are just like in the conventional first-order Godunov method.

**The Numerical Source**

The numerical source for this version of the ADER scheme is identical to that for the ADER2-TT scheme of the previous section, namely

$$\mathbf{S}_i = \mathbf{S}\left(\mathbf{Q}^n_i + \frac{1}{2}\Delta t[-\mathbf{A}(\mathbf{Q}^n_i)\Delta_i + \mathbf{S}(\mathbf{Q}^n_i)]\right). \quad (14.78)$$

Therefore the complete numerical scheme ADER2-HEOC via the update formula (14.53) is defined by the numerical flux (14.77) and the numerical source (14.78).

As seen in Sect. 14.4, in the data-interaction step, one may use any available upwind flux directly. In this case ADER2-HEOC reproduces the MUSCL-Hancock scheme, with one difference, ADER2-HEOC also handles the source term to second-order accuracy. A FORCE-type centred flux can also be used, in which case ADER2-HEOC reproduces the second-order centred FLIC scheme [57].

We remark that the ADER scheme with the HEOC solver for the $GRP_1$ preserves the coupling between the numerical flux and the numerical source, as seen with the Toro-Tiratev solver for the $GRP_1$.

## 14.8 ADER Applied to Shallow Water Flows

There are numerous applications of the high-order ADER approach to the shallow water equations. Here I mention a few. One of the earliest applications of ADER, in both the finite volume and the DG frameworks, to the shallow water equations was carried out by Castro [58]. Vignoli et al. [59] applied high-order ADER schemes to the shallow water equations with variable bed elevation. Castro et al. [60] formulated ADER schemes in the finite volume and discontinuous Galerkin finite element frameworks for wave propagation problems with a space-time adaptation technique on unstructured meshes. Applications included tsunami and seismic wave propagation problems.

Dumbser et al. [47] developed a new family of ADER type well-balanced, path-conservative quadrature-free one-step finite volume and discontinuous Galerkin finite element schemes on unstructured meshes for solving hyperbolic equations with non-conservative products and stiff source terms. Applications to various types of geophysical problems were reported. Major novelties of this work include the well-balanced property, non-conservative products and stiff source terms. Navas-Montillo and Murillo [61] developed a well-balanced ADER scheme for the shallow water equations, with variable bed elevation.

More recently, Toro et al. [62] reported an application of ADER to the two-dimensional shallow water equations in conjunction with the flux vector splitting scheme of Toro and Vazquez [63]. Schemes of up to fifth-order in space an time on unstructured meshes were implemented. Simulations included tsunami wave propagation in the Pacific Ocean. Guerrero et al. [64] have recently presented ADER-DG schemes with a-posteriori subcell finite volume limiters. They applied the scheme to the multi-layer shallow water equations with a density dependent pressure function, allowing for an arbitrary number of layers. The scheme is arbitrarily high order accurate in both space and time; this allows the detailed capturing of solution features, even on very coarse meshes. Moreover, the schemes are also well-balanced. Numerical solutions are reported to compare very well with with experimental data. This paper represents the state-of-the art on sophisticated numerical methods applied to sophisticated extensions of the conventional (one layer) shallow water equations. Ferrari and Vacondio [65] implemented and second-order ADER-type scheme. Novelties include a TVD non-linear constraint and implementation in CUDA language to exploit the computational power of modern Graphic Processing Units.

## 14.9 High Accuracy is Efficiency

When computing solutions to partial differential equations, one almost invariably aims for accurate numerical solutions, that is, small errors as measured against exact solutions. Aiming for small errors is of fundamental importance when relying on PDEs to understand the physics they embody. Attempting to interpret the physical meaning of numerical output plagued with truncation errors would be futile. The question is how to proceed to compute accurate solutions. There are essentially two distinct tools at our disposal: fine meshes and high-order accurate numerical methods. As the design and implementation of highly accurate numerical methods may be discouraging, it might be tempting to consider low-order methods used on very fine meshes to compute solutions with small errors. Clearly, for a fixed mesh and a fixed output time, a low-order method, say second-order, will have a lower cost in CPU time than a high-order method, sixth order, say. On the other hand, the error $\epsilon_{ho}$ from the high-order method will be lower than the error $\epsilon_{lo}$ from a lower-order method. The question is how much smaller is $\epsilon_{ho}$ compared with $\epsilon_{lo}$. In other words, for a fixed error $\epsilon$, will a high-order method attain $\epsilon$ on a coarse mesh at a lower cost than the low-order method on a fine mesh? The answer is yes, as we illustrate through two examples.

### 14.9.1 Efficiency for the Linear Advection Equation

In this example we solve the linear advection equation $\partial_t q + \lambda \partial_x q = 0$, with constant $\lambda$, on a fixed-length domain, with periodic boundary conditions and a smooth profile as initial condition. Figure 14.9 shows computed results on the *cost-error* plane from three methods, namely the first order Godunov upwind method and the ADER method in third and fifth-order modes. Consider the line corresponding to the Godunov method. Each symbol (triangle) corresponds to a complete computation, giving an error value on the *error*-axis and a CPU cost on the *cost*-axis. Starting from the top left corner, as the mesh is refined the error reduces and the CPU cost increases. After 11 computations the cost is already high and therefore we assume that linear extrapolation is justified; see prolongation of line. For the prescribed error $\epsilon = 10^{-8}$ (thick horizontal line) the extrapolated CPU cost is given by the intersection of the extrapolated line with the fixed-error horizontal line. The CPU cost turns out to be a very high number, in seconds, as is the number of required computing cells. See numbers in the Figure. In contrast, the third-order ADER method attains the same

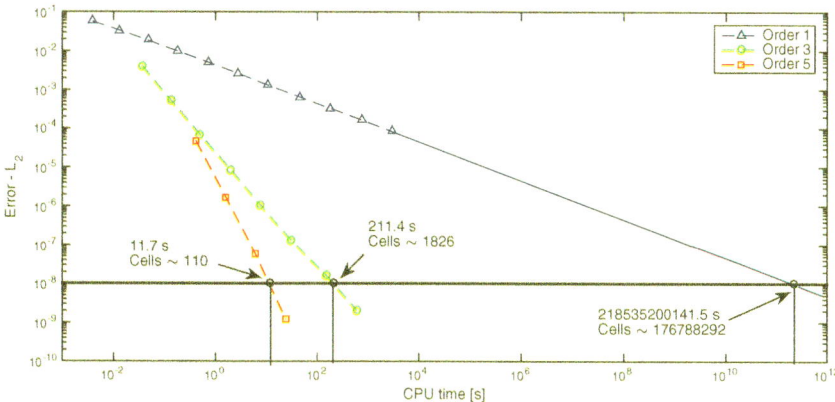

**Fig. 14.9** Efficiency illustrated via the linear advection equation (Courtesy of Dr. C Contarino, Computational Life, Italy)

error in $CPU = 211.4\ s$, while the fifth-order ADER method does so in $CPU = 11.4\ s$. From this example we see that high-order methods are much more efficient than low-order methods, by orders of magnitude in fact.

### 14.9.2 Efficiency for the Euler Equations

Here we assess efficiency of high-order numerical methods through a test problem for the one-dimensional, compressible Euler equations for $x \in [-1, 1]$; with $\gamma = 1.4$; initial conditions: pressure $p = 2$, velocity $u = 1$, density $\rho = 2 + (sin(\pi x))^4$; periodic boundary conditions and output time $T_{out} = 2$. The exact solution is $p = u = 1$ and $\rho = 2 + (sin(\pi(x - t)))^4$ [66]. For background on the Euler equations see [7]. For three fixed error values in the $L_2$ norm, Fig. 14.10 shows computational costs (bars) against order of accuracy, from ADER schemes of $2nd$ to $6th$ order of accuracy. The results were obtained with the Toro-Titarev solver for the $GRP_m$ of Sect. 14.3, in the finite volume framework, with the exact Riemann solver for the leading term [7]. The WENO scheme [19] was used for the non-linear spatial reconstruction. For a symbolic algorithm for the computation of the linear weights for a WENO polynomial of arbitrary degree $m$ see [31].

As seen in Fig. 14.10, the computational cost decreases sharply as the order of accuracy is increased, putting in evidence the efficiency of high-order methods. For the prescribed error $\epsilon_1 = 10^{-7}$ the second-order ADER method costs 15333 times more than the $6th$-order ADER method. For $\epsilon_2 = 10^{-8}$ the second-order ADER method costs 54346 times more than the $6th$-order ADER method and for $\epsilon_3 = 10^{-9}$ the second-order ADER method costs 192620 times more than the $6th$-order ADER

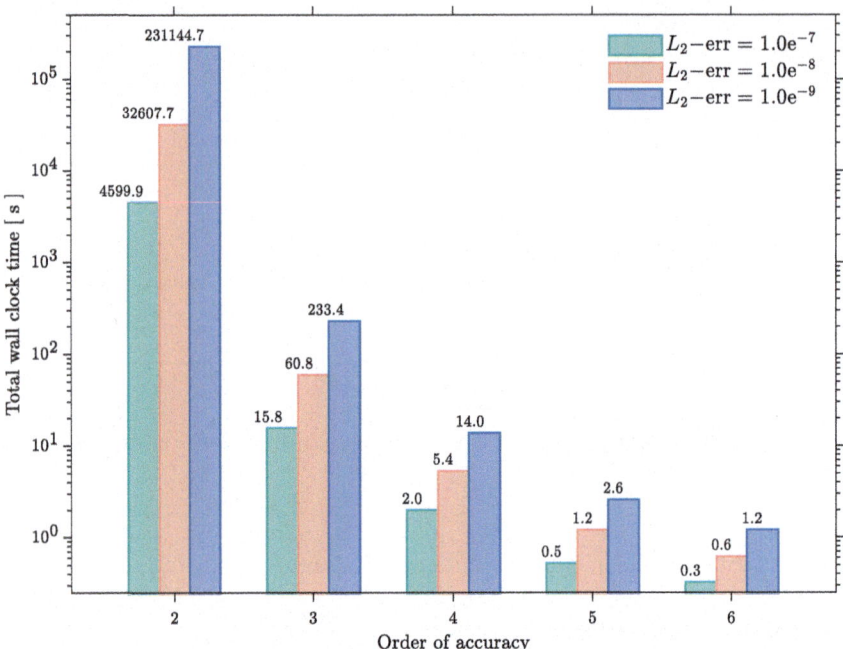

**Fig. 14.10** Efficiency gains from higher-order schemes for the one-dimensional, compressible Euler equations, with CFL coefficient $c_{cfl} = 0.95$ for all schemes. Computational cost (bars) against order of accuracy from 2nd to 6th order in space and time for three prescribed errors: $\epsilon_1 = 10^{-7}$, $\epsilon_2 = 10^{-8}$ and $\epsilon_3 = 10^{-9}$. CPU times for the first-order method (Godunov) are excluded from the graph; they are *estimated* to be more than 37 thousand years for $\epsilon_1$, more than 5 million years for $\epsilon_2$ and more than 800 million years for $\epsilon_3$. Note the exponential decay of the computational cost with increasing order of accuracy

method, that is more than five orders of magnitude. The smaller the prescribed error $\epsilon$, the larger the efficiency gains from very-high order methods.

The costs for the first-order scheme (the Godunov method) are not included in the plot. These are prohibitively large to calculate directly. Instead, we computed a few cases for relatively coarse meshes and then fitted least-squares lines to these data, followed by extrapolation to the prescribed errors to get an *estimate* of the costs in CPU time. The results are the following: For $\epsilon_1 = 10^{-7}$ cost in CPU time is $c \approx 37025$ years; for $\epsilon_2 = 10^{-8}$ cost in CPU time is $c \approx 5495100$ years and for $\epsilon_3 = 10^{-9}$ cost in CPU time is $c \approx 815587135$ years. Therefore, if accurate computed solutions are sought, low-order methods with fine meshes are not an option.

From the overall exercise, we observe that for a fixed error $\epsilon$, the cost $c$ in CPU time decays exponentially with the order of accuracy $m + 1$, namely

$$c = \alpha e^{-\beta(m+1)}, \quad \alpha = \alpha(\epsilon; K), \quad \beta = \beta(\epsilon; K), \quad (14.79)$$

with $K$ representing a set of computational parameters, such as the output time, the particular solver used for $GRP_m$ and the reconstruction method.

## 14.10 Concluding Remarks and Further Reading

This chapter has dealt with an advanced topic on numerical methods for solving evolutionary partial differential equations, to simulate, in particular, fluid flow phenomena, such as shallow water flows.

We focused on the family of very-high order ADER schemes. These are fully discrete algorithms of arbitrary (unlimited) accuracy in both space and time that operate in both the finite volume and the discontinuous Galerkin finite element frameworks. Recall that the first-order Godunov method is based on the classical Riemann problem $GRP_0$, with piece-wise constant data and no source terms. ADER is based on the Generalised Riemann problem $GRP_m$; here $m$ is the degree of the reconstruction polynomials that furnish the data to $GRP_m$, and source terms are included. Three methods to solve $GRP_m$ were presented in some detail, including a locally implicit scheme able to handle stiff source terms, thus yielding ADER schemes able to reconcile stiffness with very high order of accuracy.

The ADER numerical methodology is very general. ADER methods have been applied to multidimensional problems on structured and unstructured meshes, to solve hyperbolic equations in conservative and non-conservative form; to solve parabolic equations; even just diffusion-reaction PDEs. A large range of applications have been tackled, from shallow water flows to compressible single and multiphase problems; aero acoustics, tsunami and seismic wave propagation; dispersive systems; the Einstein equations of general relativity; the entire human circulation, and many more.

Efficiency is the unique feature of very high-order ADER methods. In a nutshell, if one aims at computing solutions that are close to the true solution of the PDEs so that these reliably reveal the physics they embody, then (i) low-order methods, being **prohibitively expensive**, are impractical to use, and (ii) high-order methods are the only option.

For further reading we recommend the following references.

1. Toro EF, Millington RC and Nejad LAM. **Towards Very High-Order Godunov Schemes.** In Godunov Methods: Theory and Applications. Edited Review, EF Toro (Editor) 2001, Kluwer Academic/Plenum Publishers, pp 905–937. [1]. This paper contains the first comprehensive account of the ADER schemes up to 1999.
2. Toro EF and Titarev VA. **Solution of the Generalised Riemann Problem for Advection-Reaction Equations.** Proc. Roy. Soc. London A. 2002, Vol. 458, pp 271–281. [2]
3. Titarev VA and Toro EF. **ADER: Arbitrary High Order Godunov Approach.** J. Scientific Computing 2002, Vol. 17, pp 609–618. [3]

4. EF Toro. **Riemann Solvers and Numerical Methods for Fluid Dynamics. A Practical Introduction**. Springer, third edition, 2009 [7]. Chapters 19 and 20 contain an elementary introduction to most ingredients of the ADER approach and associated references up to 2009.
5. Eleuterio F. Toro, Vladimir Titarev, Michael Dumbser, Armin Iske, Claus R. Goetz, Cristóbal E. Castro, Gino I. Montecinos and Riccardo Dematté. **The ADER approach for approximating hyperbolic equations to very high accuracy**. XVIII International Conference on Hyperbolic Problems: Theory, Numerics, Applications. Editors: C. Parés, T. Morales de Luna, M.L. Muñoz, M. J. Castro. Springer, 2024. [31].

# References

1. E.F. Toro, *Shock-Capturing Methods for Free-Surface Shallow Flows* (Wiley and Sons Ltd, 2001)
2. E.F. Toro, V.A. Titarev, Solution of the generalised Riemann problem for advection-reaction equations. Proc. Roy. Soc. London A **458**, 271–281 (2002)
3. V.A. Titarev, E.F. Toro, ADER: arbitrary high order Godunov approach. J. Sci. Comput. **17**, 609–618 (2002)
4. T. Schwartzkopff, C.D. Munz, E.F. Toro, ADER: high-order approach for linear hyperbolic systems in 2D. J. Sci. Comput. **17**, 231–240
5. M. Dumbser, C.D. Munz, ADER discontinuous Galerkin schemes for aeroacoustics. Comptes Rendus Mécanique **33**, 683–687 (2005)
6. S.K. Godunov, Finite difference methods for the computation of discontinuous solutions of the equations of fluid dynamics. Mat. Sb. **47**, 271–306 (1959)
7. E.F. Toro, *Riemann Solvers and Numerical Methods for Fluid Dynamics*, A Practical Introduction, 3rd edn. (Springer, 2009)
8. E.F. Toro, R.C. Millington, L.A.M. Nejad, Towards very high–order Godunov schemes, in *Godunov Methods: Theory and Applications*. Edited Review, ed. by E.F. Toro (Kluwer Academic/Plenum Publishers, 2001), pp. 905–937
9. V.A. Titarev, E.F. Toro, ADER schemes for three-dimensional hyperbolic systems. J. Comput. Phys. **204**, 715–736 (2005)
10. E.F. Toro, V.A. Titarev, ADER schemes for scalar hyperbolic conservation laws with source terms in three space dimensions. J. Comput. Phys. **202**(1), 196–215 (2005)
11. M. Käser, A. Iske, Adaptive ADER schemes for the solution of scalar non-linear hyperbolic problems. J. Comput. Phys. **205**, 486–508 (2005)
12. M. Dumbser, *Arbitrary High Order Schemes for the Solution of Hyperbolic Conservation Laws in Complex Domains*. Ph.D. thesis, Institut für Aero- un Gasdynamik, Universität Stuttgart, Germany (2005)
13. M. Dumbser, M. Käser, V.A. Titarev, E.F. Toro, Quadrature-Free non-oscillatory finite volume schemes on unstructured meshes for nonlinear hyperbolic systems. J. Comput. Phys. **226**(8), 204–243 (2007)
14. M. Dumbser, D. Balsara, E.F. Toro, C.D. Munz, A unified framework for the construction of one-step finite-volume and discontinuous Galerkin schemes. J. Comput. Phys. **227**, 8209–8253 (2008)
15. M. Dumbser, C. Enaux, E.F. Toro, Finite volume schemes of very high order of accuracy for stiff hyperbolic balance laws. J. Comput. Phys. **227**(8), 3971–4001 (2008)
16. A. Harten, S. Osher, Uniformly high-order accurate nonoscillatory Schemes I. SIAM J. Numer. Anal. **24**(2), 279–309 (1987)

17. A. Harten, B. Engquist, S. Osher, S.R. Chakravarthy, Uniformly high order accuracy essentially non-oscillatory Schemes III. J. Comput. Phys. **71**, 231–303 (1987)
18. X.D. Liu, S. Osher, T. Chan, Weighted essentially non-oscillatory schemes. J. Comput. Phys. **115**, 200–212 (1994)
19. G.S. Jiang, C.W. Shu, Efficient implementation of weigthed ENO schemes. Technical Report ICASE 95-73, NASA Langley Research Center, Hampton, USA (1995)
20. C.W. Shu, High order weighted nonoscillatory schemes for convection dominated problems. SIAM Rev **51**, 82–126 (2009)
21. M. Dumbser, M. Käser, Arbitrary high order non-oscillatory finite volume schemes on unstructured meshes for linear hyperbolic systems. J. Comput. Phys. **221**(2), 693–723 (2007)
22. D. Levy, G. Puppo, G. Russo, Central WENO schemes for hyperbolic systems of conservation laws. Math. Model. Numer. Anal. **33**, 547–571 (1999)
23. M. Dumbser, W. Boscheri, M. Semplice, G. Russo, Central weighted ENO schemes for hyperbolic conservation laws on fixed and moving unstructured meshes. SIAM J. Sci. Comput. 348:A2564–A2591 (2017)
24. M. Dumbser, M. Käser, E.F. Toro, An arbitrary high order discontinuous Galerkin method for elastic waves on unstructured meshes V: local time stepping and $p$-adaptivity. Geophys J Int **171**, 695–717 (2007)
25. E.F. Toro, C.E. Castro, B.J. Lee, A novel numerical flux for the 3D Euler equations with general equation of state. J. Comput. Phys. **303**, 80–94 (2015)
26. I.G. Petrovskii, *Partial Differential Equations* (London Iliffe Books Ltd., 1967)
27. L.C. Evans. *Partial Differential Equations* (American Mathematical Society, 2002)
28. G.I. Montecinos, D.S. Balsara, A simplified Cauchy-Kovalevskaya procedure for the local implicit solution of generalized Riemann problems of hyperbolic balance laws. Comput. Fluids **202**, 104490 (2020)
29. G.I. Montecinos, A. Santacá, M. Celant, L.O. Müller, E.F. Toro, ADER scheme with a simplified solver for the generalized Riemann problem and an average ENO reconstruction procedure. Application to blood flow. Comput. Fluids **248**, 105685 (2022)
30. R. Dematté, V.A. Titarev, G.I. Motecinos, E.F. Toro, ADER methods for hyperbolic equations with a time-reconstruction solver for the generalized Riemann problem. The scalar case. Commun. Appl. Math. Comput. **2**, 369–402 (2020)
31. E.F. Toro, V. Titarev, M. Dumbser, A. Iske, C.R. Goetz, C.E. Castro, G.I. Montecinos, R. Dematté, The ADER approach for approximating hyperbolic equations to very high accuracy, in *Hyperbolic Problems: Theory, Numerics, Applications*, vol. I, ed. by C. Parés, M.J. Castro, T. Morales de Luna, M.L. Muñoz-Ruiz, pp. 83–108 (Springer, 2024)
32. E.F. Toro, *Riemann Solvers and Numerical Methods for Fluid Dynamics. A Practical Introduction* (Springer, 1997)
33. E.F. Toro, *Riemann Solvers and Numerical Methods for Fluid Dynamics. A Practical Introduction, Second Edition* (Springer, 1999)
34. P. Le Floch, P.A. Raviart, An asymptotic expansion for the solution of the generalized riemann problem. Part 1: general theory. Ann. Inst. Henri Poincaré. Analyse non Lineáre **5**(2), 179–207 (1988)
35. C.R. Goetz, A. Iske, Approximate solutions of generalized Riemann problems for nonlinear systems of hyperbolic conservation laws. Math. Comput. **85**, 35–62 (2016)
36. M.J. Maron, R.J. Lopez, *Numerical Analysis* (Wadsworth, 1991)
37. C.E. Castro, E.F. Toro, Solvers for the high-order Riemann problem for hyperbolic balance laws. J. Comput. Phys. **227**, 2481–2513 (2008)
38. G.I. Montecinos, A universal centered high-order method based on implicit Taylor series expansion with fast second order evolution of spatial derivatives. J. Comput. Phys. **443**, 110535 (2021)
39. E.F. Toro, B. Saggiorato, S. Tokareva, A. Hidalgo, Low-dissipation centered schemes for hyperbolic equations in conservative and non-conservative form. J. Comput. Phys. **416**(109545) (2020)

40. E.F. Toro, G.I. Montecinos, Implicit, semi-analytical solution of the generalised Riemann problem for stiff hyperbolic balance laws. J. Comput. Phys. **303**, 146–172 (2015)
41. C.R. Goetz, M. Dumbser, A novel solver for the generalized Riemann problem based on a simplified LeFloch-Raviart expansion and a local space-time discontinuous Galerkin formulation. J. Sci. Comput. **69**(2), 805–840 (2016)
42. C.R. Goetz, D.S. Balsara, M. Dumbser, A family of HLL-type solvers for the generalized Riemann problem. Comput. Fluids **169**(2), 201–212 (2018)
43. P.K. Sweby, High resolution schemes using flux limiters for hyperbolic conservation laws. SIAM J. Numer. Anal. **21**, 995–1011 (1984)
44. S.P. Spekreijse, Multigrid solution of monotone second-order discretizations of hyperbolic conservation laws. Math. Comput. **49**(179), 135–155 (1987)
45. B. Cockburn, C.W. Shu, TVB Runge–Kutta local projection discontinuous Galerkin method for conservation Laws II: general framework. Math. Comput. **52**(–):411 (1989)
46. B. Cockburn, C.W. Shu, The Runge–Kutta discontinuous Galerkin method for conservation laws. J. Comput. Phys. 141(–):199 (1998)
47. M. Dumbser, M.J. Castro, C. Parés, E.F. Toro, ADER schemes on unstructured meshes for nonconservative hyperbolic systems: applications to geophysical flows. Comput. Fluids **38**(9), 731–1748 (2009)
48. M. Dumbser, M.J. Castro, C. Parés, E.F. Toro, A. Hidalgo, FORCE schemes on unstructured meshes II: non-conservative hyperbolic systems. Comput. Methods Appl. Mech. Eng. **199**(9–12), 625–647 (2010)
49. M. Dumbser, O. Zanotti, R. Loubère, S. Diot, A posteriori subcell limiting of the discontinuous Galerkin finite element method for hyperbolic conservation laws. J. Comput. Phys. **278**, 47–75 (2014)
50. S. Busto, M. Dumbser, C. Escalante, S. Gavrilyuk, N. Favrie, On high order ADER discontinuous Galerkin schemes for first order hyperbolic reformulations of nonlinear dispersive systems. J. Sci. Comput. **87**, 48 (2021)
51. W. Boscheri, M. Semplice, M. Dumbser, Central WENO Subcell finite volume limiters for ADER discontinuous Galerkin schemes on fixed and moving unstructured meshes. Commun. Comput. Phys. **25**, 311–346 (2019)
52. W. Boscheri, M. Dumbser, A direct Arbitrary-Lagrangian-Eulerian ADER-WENO finite volume scheme on unstructured tetrahedral meshes for conservative and non-conservative hyperbolic systems in 3d. J. Comput. Phys. **275**, 484–523 (2014)
53. W. Boscheri, M. Dumbser, Arbitrary-Lagrangian-Eulerian discontinuous Galerkin schemes with a posteriori subcell finite volume limiting on moving unstructured meshes. J. Comput. Phys. **346**, 449–479 (2017)
54. M. Dumbser, W. Boscheri, M. Semplice, G. Russo, Central weighted ENO schemes for hyperbolic conservation laws on fixed and moving unstructured meshes. SIAM J. Sci. Comput. **39**(6), A2564–A2591 (2017)
55. E. Gaburro, W. Boscheri, S. Chiocchetti, C. Klingenberg, V. Springel, M. Dumbser, High order direct Arbitrary-Lagrangian-Eulerian schemes on moving Voronoi meshes with topology changes. J. Comput. Phys. **407**, 109167 (2020)
56. E. Gaburro, A unified framework for the solution of hyperbolic PDE systems using high order direct arbitrary-lagrangian-eulerian schemes on moving unstructured meshes with topology change. Arch. Computat. Methods Eng. **28**, 1249–1321 (2021)
57. E.F. Toro, S.J. Billett, Centred TVD schemes for hyperbolic conservation laws. IMA J. Numer. Anal. **20**, 47–79 (2000)
58. C.E. Castro, *High-Order ADER FV/DG Numerical Methods for Hyperbolic Equations*. Ph.D. thesis, Department of Civil and Environmental Engineering, University of Trento, Italy (2007)
59. G. Vignoli, V.A. Titarev, E.F. Toro, ADER schemes for the shallow water equations in channel with irregular bottom elevation. J. Comput. Phys. **227**, 3212–3243 (2008)
60. C.E. Castro, E.F. Toro, M. Käser, ADER scheme on unstructured meshes for shallow water: simulation of tsunami waves. Geophys. J. Int. **189**, 1505–1520 (2012)

61. A. Navas-Montilla, J. Murillo, 2D well-balanced augmented ADER schemes for the shallow water equations with bed elevation and extension to the rotating frame. J. Comput. Phys. **372**, 316–348 (2018)
62. E.F. Toro, C.E. Castro, D. Vanzo, A. Siviglia, A flux-vector splitting scheme for the shallow water equations extended to high-order on unstructured meshes. Int. J. Numer. Methods Fluids (2022). https://doi.org/10.1002/fld.5099
63. E.F. Toro, M.E. Vázquez-Cendón, Flux splitting schemes for the Euler equations. Computers and Fluids **70**, 1–12 (2012)
64. E. Guerrero-Fernández, M.J. Castro-Díaz, M. Dumbser, T. Morales, de Luna, An Arbitrary High Order Well-Balanced ADER-DG Numerical Scheme for the Multilayer Shallow-Water Model with Variable Density. Journal of Scientific Computing **9**, 52 (2022). https://doi.org/10.1007/s10915-021-01734-2
65. A. Ferrari, R. Vacondio, An augmented HLLEM ADER numerical model parallel on GPU for the porous Shallow Water Equations. Comput. Fluids. (2022) https://doi.org/10.1016/j.compfluid.2022.105360
66. E.F. Toro, A multi-stage numerical flux. Appl. Numer. Math. **56**, 1464–1479 (2006)

# Chapter 15
# DAM-BREAK Modelling

**Abstract** Methods presented in this book are applied to simulate dam-break problems. Dams are man-made water reservoirs storing large volumes for the supply of drinking water, irrigation, hydroelectric power generation, navigation and flood control, for example. Historically, dams have been the centre of catastrophic events over the years, usually associated with dam-wall failure, or massive slides of material into the water reservoir. Understanding of the physical phenomena involved in such failures is of fundamental importance in the prevention and management of such events. Numerical simulation of the ensuing water flow downstream of the dam is nowadays a standard practice in environmental and hydraulic engineering. In this Chapter we present two case studies that are relevant to the subject of dam-break mathematical modelling, combining numerical, theoretical and experimental methods. The first case concerns a circular dam (2D) for which a radial reference 1D solution is available for rigorous assessment of the performance of the numerical methods. High-resolution numerical computations using methods from this book capture the physical aspects of wave propagation combining the primary outward travelling circular shock with the inward travelling circular rarefaction wave, that on reflection from the centre generates a complex wave system. The second case is based on the activities of the CADAM European Workshop on Dam-Break Modelling (1996–1999). Here we study preactical, laboratory physical models, which were specifically designed for validating numerical software intended for dam-break simulation of real events. A detailed experimental and computational study is presented for a chosen laboratory model.

## 15.1 Introduction

Dams are man-made reservoirs storing large volumes of water contained by natural terrain elevations, such as a narrow canyon or valley, and man-made walls. Dams are, in the main, part of overall water-resource projects for the purpose of drinking-water supply, irrigation, hydroelectric power generation, navigation and flood control. In addition to their benefits to society, dams have been the centre of catastrophic events over the years. These may be the direct consequence of sudden failure of the dam walls. Such failures in turn induce water flow that causes devastation along its path in valleys, with considerable damage to property and loss of life. This kind of water flow may also be caused by displacement of large amounts of water by massive slides of material adjacent to the dam into the reservoir; this causes enormous waves that can completely overtop the dam.

This was the case of the Vajont dam in northern Italy in 1963, in which an estimated number of 2000 people died as a result of the water flow along the valley, downstream of the dam [1]. Interestingly, in this particular case the man-made walls survived intact the overtopping event. Another well-known case in Europe is the Malpasset dam break that occurred in France in 1959 and caused the death of 421 people. This case is regarded as unique in that there was total failure of the man-made dam wall. Details of the Malpasset accident and modern dam-break simulation techniques are discussed by Hervouet and Petitjean [2].

Fortunately, these events are rare. However, when they occur the consequences are disastrous. In order to prevent such phenomena from taking place, or to understand how the process will evolve when it actually happens, an understanding of the physical phenomena involved is of fundamental importance. The processes associated with dam failure are varied and exceedingly complex. Examples of such processes include sliding, bridging, sediment/debris transport and wave propagation in very complex domains. As part of risk analysis for civil protection, inundation maps are produced. These are currently based on numerical simulation of the wave propagation process. An important item of information is the *wave arrival time*. Accuracy and reliability are of paramount importance in this application of mathematical modelling and numerical methods, for the lives of many people may be at risk.

For general background on dams see Chap. 6 of the book by Roberson et al. [3]. See also the review article by García-Navarro [4], who discusses numerical issues related to dam-break simulation. Other recommended works are Glaister [5, 6], Toro [7], Molinaro [8], Alcrudo and García-Navarro [9], García-Navarro [4], García-Navarro and Alcrudo [10], Bellos et al. [11], Soulis [12], Betamio and Franco [13], Alcrudo and García-Navarro [14], Ambrosi [15], Fraccarollo and Toro [16, 17], Nujic [18], Sleigh et al. [19], Lauber and Hager [20, 21], Mingham and Causon [22], Aureli et al. [23], Delis et al. [24], Ferrari et al. [25] and Manenti et al. [1]. I particularly recommend the article by Hervouet and Petitjean [2].

In recognition of the importance of dam-break modelling, a workshop on the subject was initiated by Electricité de France in Paris in 1996. The activity was later formalised under the name of CADAM; it was funded by the European Community and involved about 30 scientists and engineers from about 10 European countries. Background information on some of the CADAM activities is found in the CADAM Report edited by Morris et al. [26]. The author is indebted to all colleagues who took part in CADAM, and particularly to those who contributed with experimental data and numerical simulations.

The time-dependent, one- and two-dimensional shallow water equations studied in this book are currently accepted as a possible basic mathematical model to study the phenomena of wave propagation induced by dam-breaking events. The task is to solve these equations numerically, suitably augmented with the appropriate terms

## 15.2 Circular Dam: Computation of Wave Phenomena

that account for various additional processes. In this Chapter we study some problems that are related to dam-break modelling. They are simplified versions of real events but they do contain some of the key elements involved in the numerical simulation of real-life events. In Sect. 15.2 we study the wave propagation phenomena associated with the sudden collapse of an idealised two-dimensional circular dam. We carry out a systematic numerical study of the time evolution of the associated wave system. We study this problem by solving a one-dimensional inhomogeneous model arising from the cylindrical symmetry of the problem and by solving the full two-dimensional equations. Systematic comparisons are made between models and numerical methods used. This example should be used as a test case for all numerical methods intended for application to realistic dam-break problems and associated numerical software. Section 15.3 is based on the activities of the CADAM European Workshop on Dam-Break Modelling (1996–1999). Here we study more realistic two-dimensional physical models, which were specifically designed for validating numerical software intended for dam-break simulation. For these problems, numerical results using some of the numerical methods presented in this book are compared with experimental data.

### 15.2 Circular Dam: Computation of Wave Phenomena

Here we consider an idealised circular dam with horizontal bottom and study the wave propagation phenomena associated with the sudden collapse of the entire circular dam wall.

#### 15.2.1 Geometry, Equations and Methods

We assume the dam to be enclosed by an infinitely thin circular wall of radius $R = 2.5$ m in a square computational domain of 40 m × 40 m with centre at the point $(x_c, y_c)$, with $x_c = 20$ m, $y_c = 20$ m. As initial conditions we set $u(x, y, 0) = v(x, y, 0) = 0$ throughout the domain, and for the water depth we take

$$h(x, y, 0) = \begin{cases} h_{ins} = 2.5 \text{ m} & \text{if } (x - x_c)^2 + (y - y_c)^2 \leq R^2 \ , \\ h_{out} = 0.5 \text{ m} & \text{if } (x - x_c)^2 + (y - y_c)^2 > R^2 \ . \end{cases} \quad (15.1)$$

The circular dam wall is assumed to collapse instantaneously. This is like a two-dimensional Riemann problem for the system of PDEs consisting of the two-dimensional shallow water equations on a horizontal bottom, namely

$$\partial_t \mathbf{Q} + \partial_x \mathbf{F}(\mathbf{Q}) + \partial_y \mathbf{G}(\mathbf{Q}) = \mathbf{0} \ . \quad (15.2)$$

Here **Q**, **F(Q)** and **G(Q)** are the vectors of conserved variables and fluxes in the $x$- and $y$-directions respectively, namely

$$\mathbf{Q} = \begin{bmatrix} h \\ hu \\ hv \end{bmatrix}, \; \mathbf{F(Q)} = \begin{bmatrix} hu \\ hu^2 + \frac{1}{2}gh^2 \\ hu\,v \end{bmatrix}, \; \mathbf{G(Q)} = \begin{bmatrix} hv \\ hvu \\ hv^2 + \frac{1}{2}gh^2 \end{bmatrix}, \quad (15.3)$$

where $u(x, y, t)$ and $v(x, y, t)$ are respectively the $x$- and $y$-components of velocity, $h(x, y, t)$ is the *depth*, related to the *free surface* $s(x, y, t)$ and the *bed* elevation $b(x, y)$ via

$$s(x, y, t) = b(x, y) + h(x, y, t). \quad (15.4)$$

Here we take a horizontal bed throughout the domain, $b(x, y) = 0$. To solve the two-dimensional equations we use the WAF method as described in Chap. 12, along with the HLLC approximate Riemann solver discussed in Chap. 11 and a second-order dimensional splitting scheme discussed in Chap. 13. For most computations reported here we use a mesh of $200 \times 200$ cells. For the calculations reported we use a safety coefficient $S_c = 0.9$ and since the splitting scheme used has stability limit $C_{lim} = 1$ in two-dimensions, the CFL coefficient used is therefore $C_{cfl} = 0.9$.

As the problem has cylindrical symmetry along the radial direction $r$, one can derive an inhomogeneous one-dimensional system, namely

$$\partial_t \mathbf{Q} + \partial_r \mathbf{F(Q)} = \mathbf{S(Q)}, \quad (15.5)$$

where

$$\mathbf{Q} = \begin{bmatrix} h \\ hu \end{bmatrix}, \; \mathbf{F(Q)} = \begin{bmatrix} hu \\ hu^2 + \frac{1}{2}gh^2 \end{bmatrix}, \; \mathbf{S(Q)} = -\frac{1}{r}\begin{bmatrix} hu \\ hu^2 \end{bmatrix}. \quad (15.6)$$

Here $u = u(r, t)$ is used also to represent the radial velocity. In order to cross-check our numerical results, we also carry out numerical computations on the inhomogeneous one-dimensional system (15.5)–(15.6) and compare results with those obtained from the full two-dimensional system (15.2)–(15.3). In order to produce reference radial solutions, we solve (15.5)–(15.6) with a very fine mesh of $M = 1000$ cells. Results may be regarded as *exact solutions*. Various methods discussed in this book were used, all producing similar results. Here we report computations on (15.5)–(15.6) using the WAF and the Random Choice Methods. The WAF method is used in conjunction with the HLLC Riemann solver described in Chap. 10; the CFL coefficient was taken as $C_{cfl} = 0.9$, and the limiter used was SUPERBEE. The geometric source terms in (15.6) are treated with the standard second-order splitting scheme given in Chap. 13.

## 15.2.2 Computational Results

Figure 15.1 shows the initial free-surface elevation above the *ambient state* of water of uniform depth $h(x, y, 0) = 0.5$ m, at time $t = 0$, given in (15.1). In setting up the initial conditions, cells cut by the circle defining the infinitely thin circular dam wall are assigned values proportional to the areas inside and outside the circle.

Figure 15.2 shows the computed free-surface position at time $T_{out} = 0.4$ s. Clearly visible are an outward-propagating circular shock wave and an inward-propagating circular rarefaction wave. At the time the solution is displayed, the rarefaction is about to reach the centre of the dam. Note the very pronounced depth gradient behind the

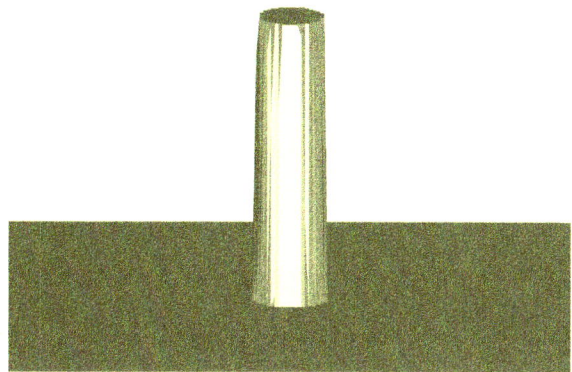

**Fig. 15.1** Circular dam-break problem. Free-surface position at the initial time $t = 0$ given by (15.1)

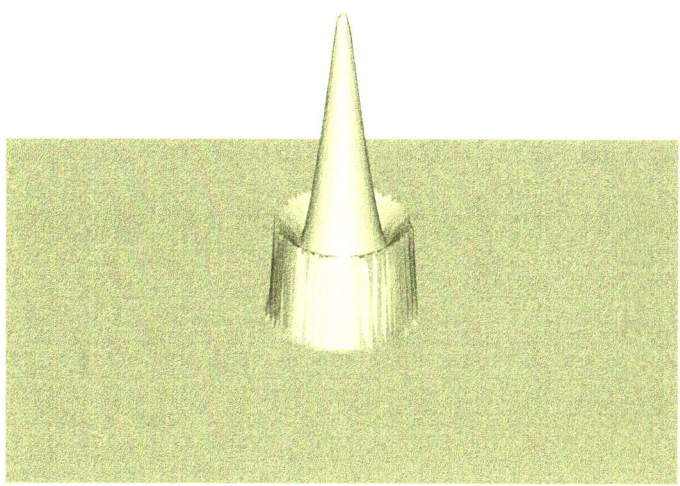

**Fig. 15.2** Circular dam-break problem. Top view of computed free-surface position at time $T_{out} = 0.4$ s

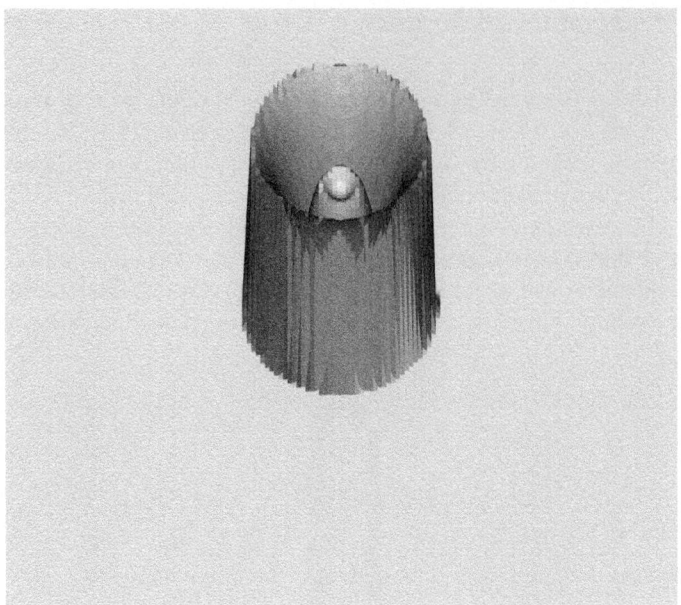

**Fig. 15.3** Circular dam-break problem. Top view of computed free-surface position at time $T_{out} = 0.7$ s

shock. See also Fig. 15.13, which shows free surface and particle velocity on a slice of the profile of Fig. 15.2 through the centre.

Figure 15.3 shows the computed free-surface distribution at time $T_{out} = 0.7$ s, as seen from the top. The circular rarefaction has by now imploded into the centre and has reflected, producing a very pronounced fall of the free-surface elevation in the vicinity of the centre. Note the *dip* in free-surface elevation right in the centre; this feature is difficult to resolve numerically.

Figure 15.4 displays the computed free-surface distribution at time $T_{out} = 1.4$ s. The circular shock has propagated further in the outward direction. The reflected inner circular rarefaction has *overexpanded the flow* to the point that the free-surface position has fallen well below the *ambient depth* initially outside the circular dam, depicted by the horizontal plane; Fig. 15.5 shows a bottom view of the same result. The free surface is very close to zero and a *secondary circular shock* has been formed.

Figure 15.6 shows a top view of the computed free-surface distribution at time $T_{out} = 3.5$ s and Fig. 15.7 shows a corresponding bottom view of the same result. The primary circular shock has propagated further away from the centre and the secondary circular shock has propagated towards the centre.

Figure 15.8 shows a top view of the computed free-surface distribution at time $T_{out} = 4.7$ s, shortly after the secondary shock has imploded into the centre; this shock has reflected from the centre and is now propagating in the outward direction; Fig. 15.9 shows a bottom view of the computed free-surface distribution at the same time; see also the slice through the centre shown in Fig. 15.14.

## 15.2 Circular Dam: Computation of Wave Phenomena

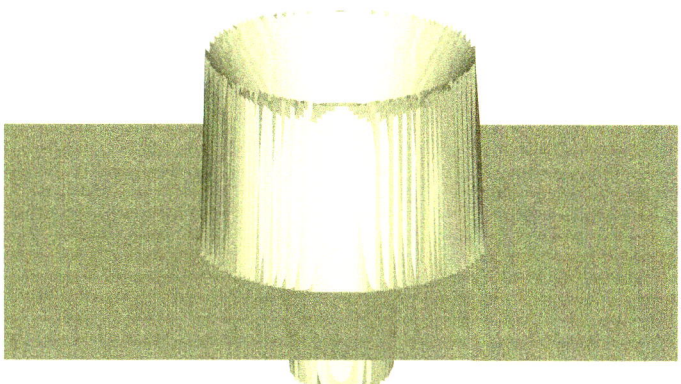

**Fig. 15.4** Circular dam-break problem. Side view of computed free-surface position at time $T_{out} = 1.4$ s. Note free surface below ambient state

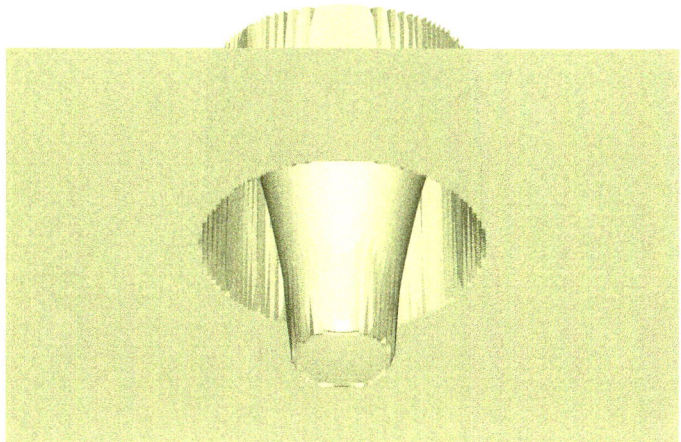

**Fig. 15.5** Circular dam-break problem. Bottom view of computed free-surface position at time $T_{out} = 1.4$ s. Note secondary circular shock below ambient state. Compare with Fig. 15.4

A representation of the complete wave process may be constructed by considering the distribution $h = h(r, t)$ of free-surface elevation across a slice through the centre of the dam. Figure 15.10 shows a wave diagram on the $r$-$t$ plane and Fig. 15.11 shows a *three-dimensional* version of this wave diagram. Note the orientation in Fig. 15.11, where $x$ measures radial distance $r$, $y$ measures time and $z$ measures free-surface elevation $h$.

The complete wave process may be summarised as follows: a primary circular shock wave emerges and propagates outwards. The strength of this shock, and consequently its speed, decreases with time; this can be seen from the *curved* shock wave paths in Figs. 15.10 and 15.11. In addition to the primary circular shock, a circular rarefaction emerges from the dam-break process and propagates towards the centre. This rarefaction wave reflects from the centre as a rarefaction wave and causes a rapid

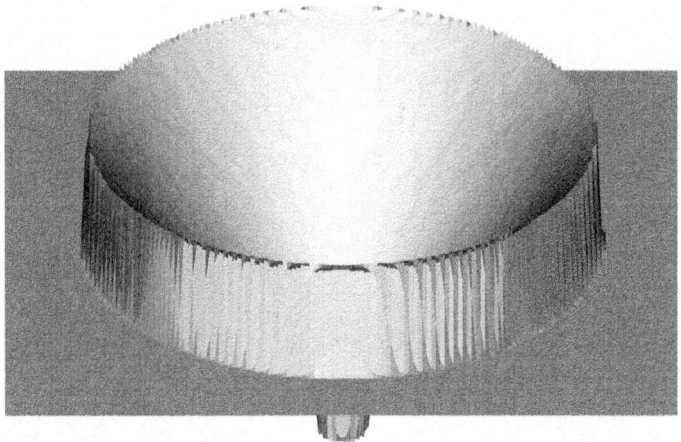

**Fig. 15.6** Circular dam-break problem. Top/side view of computed free-surface position at time $T_{out} = 3.5$ s

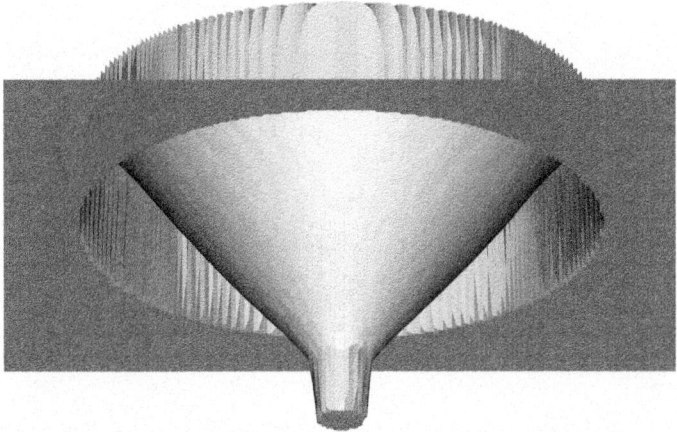

**Fig. 15.7** Circular dam-break problem. Bottom view of computed free-surface position at time $T_{out} = 3.5$ s. Secondary shock has propagated towards the centre; compare with Fig. 15.6

fall of the free-surface elevation near the centre. The values of the depth here are very close to zero. This *overexpansion of the flow* causes the formation of a *secondary circular shock wave*; see the inner part of Fig. 15.10. This shock wave is initially swept (slightly) in the outward direction, then it slows down, comes to a halt and begins to propagate towards the centre at an accelerating pace; note the curved shock paths in the middle of Figs. 15.10 and 15.11. Eventually, this secondary shock implodes into the centre, then reflects and begins to propagate in the outward direction.

The computed results show that after the implosion of the secondary shock there is no more wave activity around the centre of the dam. This situation is different from the gas dynamics analogue, where a *pulsating* shock wave system consisting

15.2 Circular Dam: Computation of Wave Phenomena

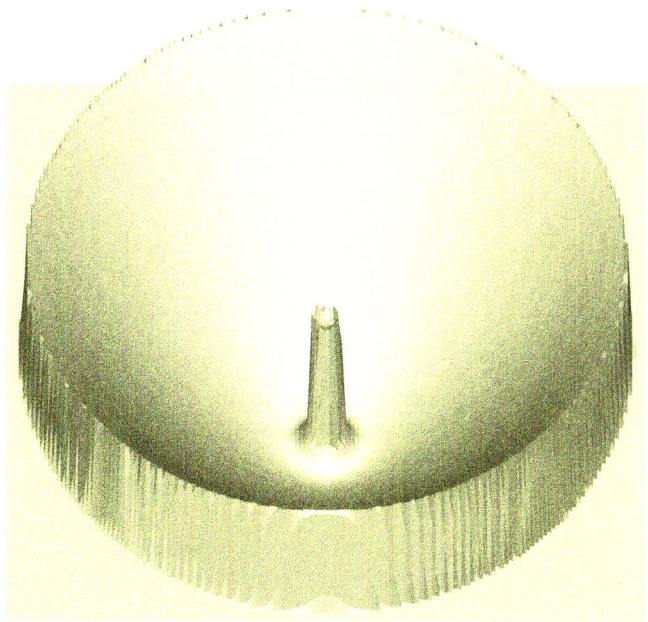

**Fig. 15.8** Circular dam-break problem. Top view of computed free-surface position at time $T_{out} = 4.7$ s. Secondary shock has imploded and reflected from the centre

**Fig. 15.9** Circular dam-break problem. Bottom view of computed free-surface position at time $T_{out} = 4.7$ s; compare with Fig. 15.8

of several implosions and explosions may be generated in the so-called *near field*, which is the region in the vicinity of the centre of the dam. Figure 15.12 shows the time history of free-surface elevation at the centre (left plot) and at a point 4 m away from the centre (right plot). The main features seen at the centre are the strong

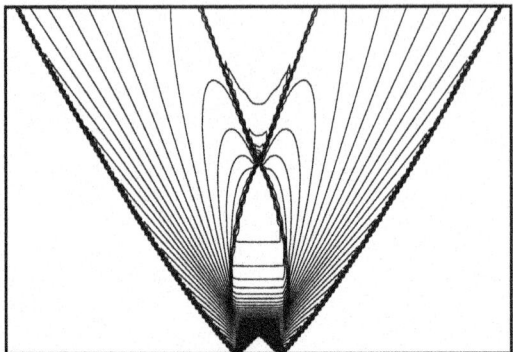

**Fig. 15.10** Circular dam-break problem. Wave diagram on $r$-$t$ plane of free-surface elevation on a slice through the centre of the dam. Compare with Fig. 15.11

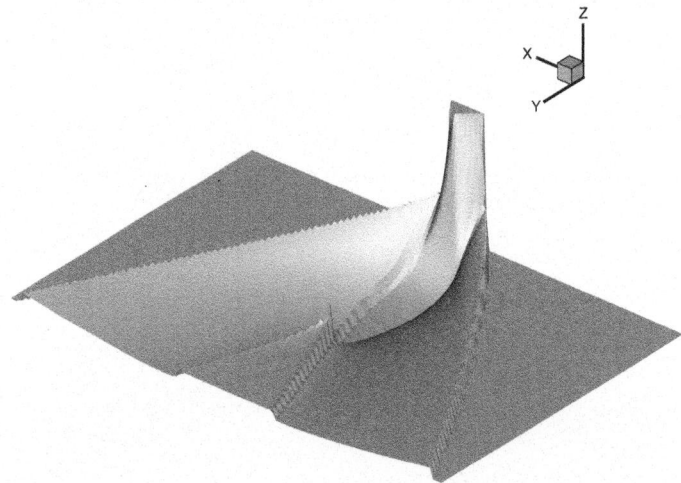

**Fig. 15.11** Circular dam-break problem. Three-dimensional $r$-$t$-$h$ representation of wave diagram of Fig. 15.10. Here $x$ is radial distance $r$, $y$ is time and $z$ is free-surface elevation $h$

rarefaction reaching the centre and the imploded secondary shock wave at a time of about $t = 4.25$ s. After the secondary shock wave, the ambient conditions are restored at the centre. At the position 4 m away from the centre (right plot of Fig. 15.12) one sees the passage of two shock waves: the primary shock and the secondary shock after the implosion.

In order to cross-check the computations, I have systematically compared numerical solutions to the full two-dimensional model (15.2)–(15.3) against solutions from the radial one-dimensional inhomogeneous model (15.5)–(15.6) computed with various numerical methods. I am satisfied that the numerical methods used compute the correct solution. Figures 15.13 and 15.14 show a comparison of two-dimensional

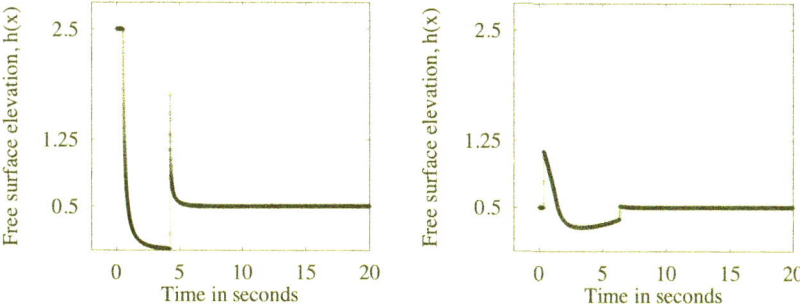

**Fig. 15.12** Circular dam-break problem. Radial solution (symbols and line) computed by the WAF method. Left: time variation of depth at the centre (20, 20). Right: time variation of depth at (20, 24)

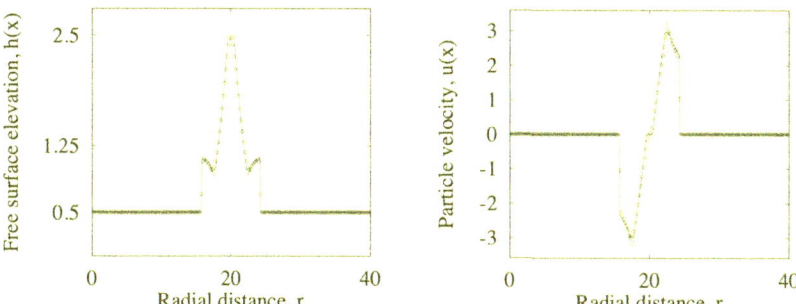

**Fig. 15.13** Circular dam-break problem. Two-dimensional solution (symbol) and radial solution 1D (line) at time $T_{out} = 0.4$ s. The split 2D WAF method uses a mesh of $200 \times 200$ cells and the 1D WAF in the radial direction uses 200 cells

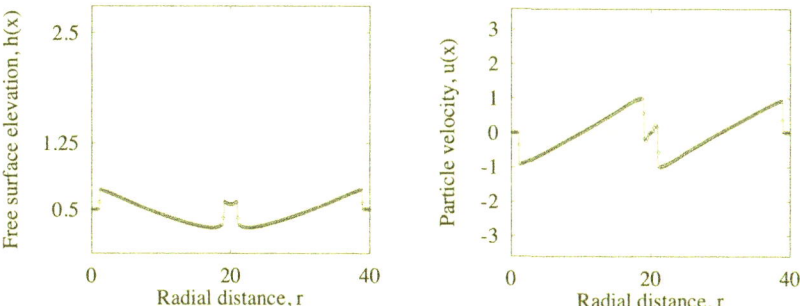

**Fig. 15.14** Circular dam-break problem. Two-dimensional solution (symbol) and radial solution 1D (line) at time $T_{out} = 4.7$ s. The split 2D WAF method uses a mesh of $200 \times 200$ cells and the 1D WAF in the radial direction uses 200 cells

computations for (15.2)–(15.3) against radial 1D computations for (15.5)–(15.6). The agreement is very satisfactory.

In this section we have applied some of the numerical methods studied in this book to an idealised problem that is related to dam-break modelling. The detailed computational study of the circular dam break problem has revealed a fascinating wave phenomenon involving the propagation of a circular shock in the outer direction and a rich wave interaction process near the centre of the dam. Figures 15.10 and 15.11 summarise the time evolution of the complete wave interaction process.

In the next section we carry out numerical computations on a more realistic dam-break problem, for which there are detailed experimental measurements. We compare numerical and experimental results.

## 15.3 Physical Models: Experiments and Numerics

Here we study dam-break like problems arising from the activities of the European Workshop on Dam-Break Modelling (CADAM, Concerted Action on Dam-Break Modelling, 1996–1999). The one and two-dimensional shallow water equations, and the computational methods studied in this book are put to use to simulate some well designed dam-break test problems. Comparison of computational results against experimental measurements is carried out.

### 15.3.1 Introduction

This section uses some of the numerical methods presented in this book and applies them to solve the shallow water equations in order to simulate solutions to some selected test problems that are related to dam-break modelling and for which experimental measurements are available. Within the framework of the European Workshop on Dam-Break Modelling (CADAM, Concerted Action on Dam-Break Modelling, 1996–1999), various test problems were designed with the specific aim of assessing mathematical models and numerical methods intended for applications to dam-break modelling in realistic scenarios. All tests chosen are generally gross simplifications of real-life dam-break situations but they do retain crucial elements likely to cause challenges to models, to numerical methods or to both, in real-life situations. Physical models were then constructed and laboratory measurements were produced. Interestingly, in the very design of the laboratory experiments, numerical simulations were used intensively.

In this section we report on just one of the CADAM physical models to illustrate the applicability of the two-dimensional shallow water equations and the applicability of the numerical methods presented in this book.

## 15.3.2 The Problem: A Dam with Channel Bend at 45°

Here we select the CADAM test 1, which we now describe. The construction of the model and the experiments were carried out by Professor Y. Zech and collaborators at the Catholic University of Louvain, Belgium; see Soares et al. [27] and Morris et al. [26] for more details.

The domain of interest consists of a rectangular reservoir connected to a channel containing a 45° bend, as depicted in Fig. 15.15. The rectangular reservoir has dimensions 2.39 m × 2.44 m; this is connected perpendicularly to a channel of constant rectangular cross-section of width 0.495 m and of length about 8 m, with a bend of 45°. Both the reservoir and the channel are horizontal and are connected by a vertical step of height 0.33 m, as depicted in Fig. 15.16. With reference to the origin $(0, 0)$ in the $x$-$y$ plane, the geometry of the physical model is given by the ten points of Table 15.1. Gauges to measure free-surface elevation as a function of time are placed at 9 positions and denoted as $GK$, for $K = 1, \ldots, 9$; the coordinates of these positions are given in Table 15.2.

## 15.3.3 Numerical Methods and Computational Geometry

Here we solve the two-dimensional shallow water equations (15.2)–(15.3) by the *unsplit* first-order finite volume Godunov scheme, in conjunction with the HLLC

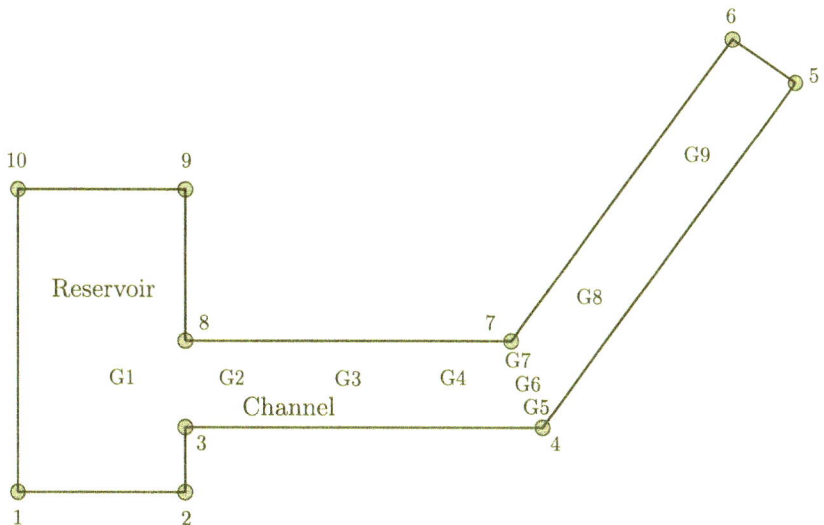

**Fig. 15.15** Top view of physical model consisting of a rectangular reservoir connected to a channel with a 45° bend. The vertical step in section 3–8 connects the bottom of the reservoir to the bottom of the channel; see Fig. 15.16

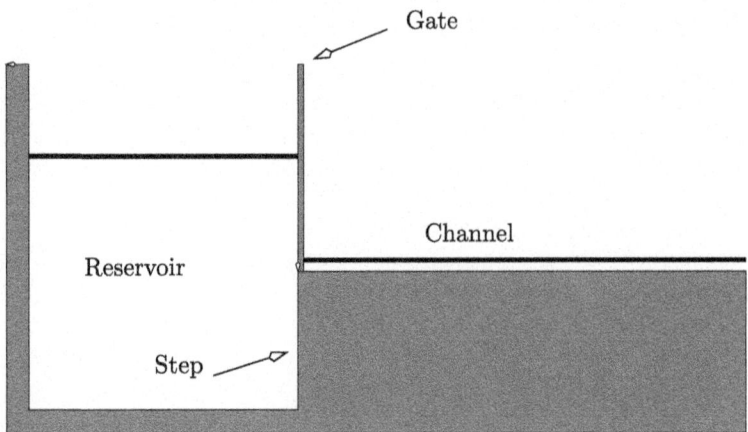

**Fig. 15.16** Side view of physical model near the rectangular reservoir. The vertical step in section 3-8 of Fig. 15.15 connects the bottom of the reservoir to the bottom of the channel

**Table 15.1** Coordinates for ten points defining the physical model in Fig. 15.15

| Point number $k$ | $x_k$, m | $y_k$, m |
| --- | --- | --- |
| 1 | 0.0 | 0.0 |
| 2 | 2.39 | 0.0 |
| 3 | 2.39 | 0.445 |
| 4 | 6.64 | 0.445 |
| 5 | 9.575 | 3.38 |
| 6 | 9.225 | 3.73 |
| 7 | 6.435 | 0.94 |
| 8 | 2.39 | 0.94 |
| 9 | 2.39 | 2.44 |
| 10 | 0.0 | 2.44 |

**Table 15.2** Coordinates for the nine gauges $GK$ in Fig. 15.15

| Gauge number $K$ | $x_k$, m | $y_k$, m |
| --- | --- | --- |
| 1 | 1.59 | 0.69 |
| 2 | 2.74 | 0.69 |
| 3 | 4.24 | 0.69 |
| 4 | 5.74 | 0.69 |
| 5 | 6.74 | 0.72 |
| 6 | 6.65 | 0.80 |
| 7 | 6.56 | 0.89 |
| 8 | 7.07 | 1.22 |
| 9 | 8.13 | 2.28 |

15.3 Physical Models: Experiments and Numerics

approximate Riemann solver. See Chaps. 11 and 13 for details on the numerical methods used. As to the mesh generation approach, we adopt the Cartesian-Cut Cell method [28, 29]. Details of this meshing approach as applied to the two-dimensional shallow water equations are found in the unpublished report [30]. In essence, one generates a Cartesian mesh on the rectangle determined by the minima and maxima of the coordinates defining the geometry. Cells are then divided into two classes: Cartesian cells not cut by the boundary and Cartesian cells cut by the boundary. The Cartesian cells are treated by the methods described in Chap. 13. The cut cells must be treated properly by accounting for the shape of the boundary, which is not aligned with the Cartesian directions; see Chap. 10.

The power of the approach is that the grid generation process becomes trivial, independently of the complexity of the geometry. However, a considerable effort is required to correctly deal with the cut cells, their connectivity to the rest of the mesh and application of boundary conditions. One difficulty is that there will invariably be very small cut cells that set a stringent requirement on the stability of the (explicit) numerical method. We resolve this difficulty through a *cell-merging* approach.

The step of Fig. 15.16 is not included here in the reported computed results. We simply assume that the bed elevation is given by the channel elevation. This assumption has an effect on the results; the rarefaction propagating into the reservoir after the gate is withdrawn has different speeds and amplitudes to that on the real, deeper reservoir. In addition to the above assumption, we do not include friction terms in the present results. The collective experience, see for example Morris et al. [26], is that bed and wall friction terms have an important effect on the results.

The initial conditions for the problem are zero velocity everywhere and the water free-surface elevation takes on two values, namely $h(x, y, 0) = 0.25$ m within the reservoir and $h(x, y, 0) = 0.01$ m in the channel. The solution was computed using two meshes, a coarse mesh of $100 \times 70$ cells and a fine mesh of $300 \times 200$ cells.

### *15.3.4 Comparison of Numerical Results with Experiments*

Figure 15.17 shows the computed free-surface elevation on the complete physical model at time $T_{out} = 0.5$ s. Note the orientation of the coordinate axes; at this time most of the action takes place in the vicinity of the gate position, Fig. 15.18 shows a contour plot of free-surface elevation at time $T_{out} = 0.5$ s in the region close to the gate; a bore propagates into the channel and a rarefaction travels into the reservoir.

Figure 15.19 shows the computed free-surface elevation at time $T_{out} = 4.0$ s, and the counterplot of Fig. 15.20 shows details of the flow in the vicinity of the channel bend. By this time the leading bore has reached the bend, it has reflected and diffracted, and continued to travel along the second section of the channel. Note the complicated flow structure near the corner where a shock interacts with a strong rar-

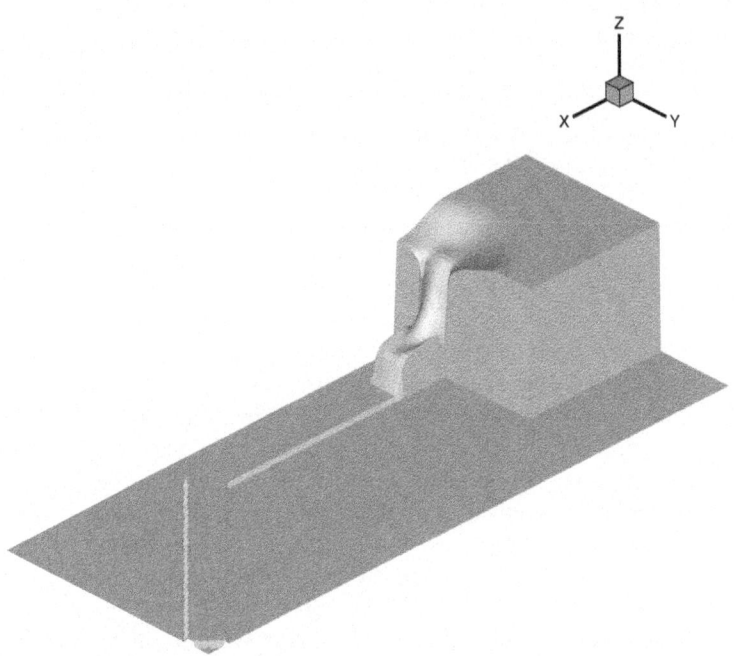

**Fig. 15.17** Computed free-surface elevation at time $T_{out} = 0.5$ s. Note axis orientation

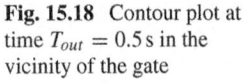

**Fig. 15.18** Contour plot at time $T_{out} = 0.5$ s in the vicinity of the gate

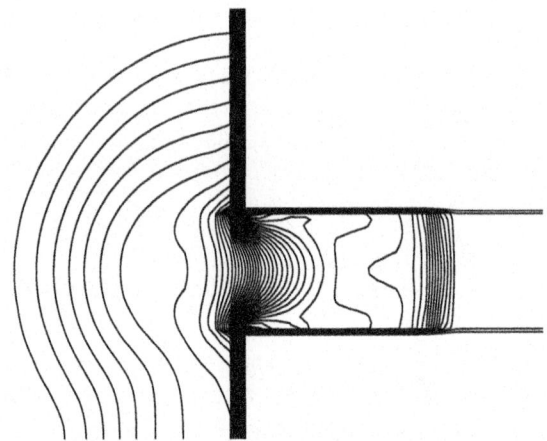

efaction. Figure 15.21 shows the computed free-surface elevation at the output time $T_{out} = 40.0$ s. A this time the maximum free-surface elevation is about one-fifth of the initial value within the reservoir. Note the complex, virtually steady pattern of shock waves in the channel section containing gauges 8 and 9. Figure 15.22 shows a contour plot of the free-surface elevation at time $T_{out} = 40.0$ s in the second section

## 15.3 Physical Models: Experiments and Numerics

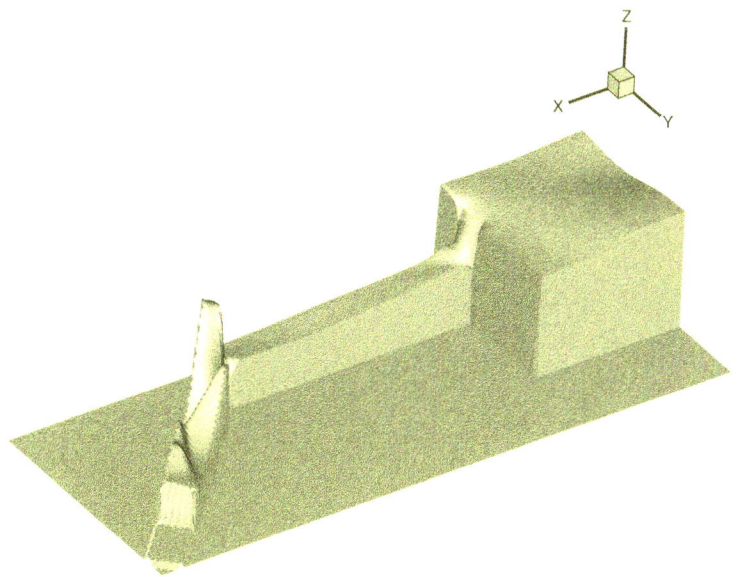

**Fig. 15.19** Computed free-surface elevation at time $T_{out} = 4.0$ s. Note axis orientation

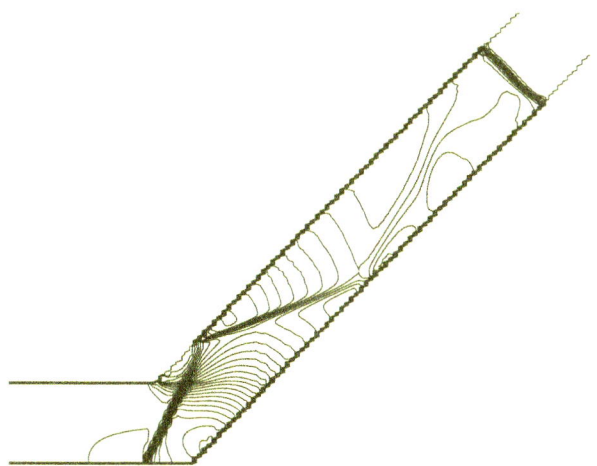

**Fig. 15.20** Contour plot at time $T_{out} = 4.0$ s in the vicinity of the 45° bend

of the channel; compare with Fig. 15.21. It is seen that very small variations in the location of the gauges result in significant variations in measured or computed free-surface elevations.

Figures 15.23, 15.24, 15.25, 15.26, 15.27, 15.28, 15.29, 15.30 and 15.31 show time histories of free-surface elevation at the nine positions $GK$ shown in Fig. 15.15. Comparison is made between experimentally measured results (symbols) and com-

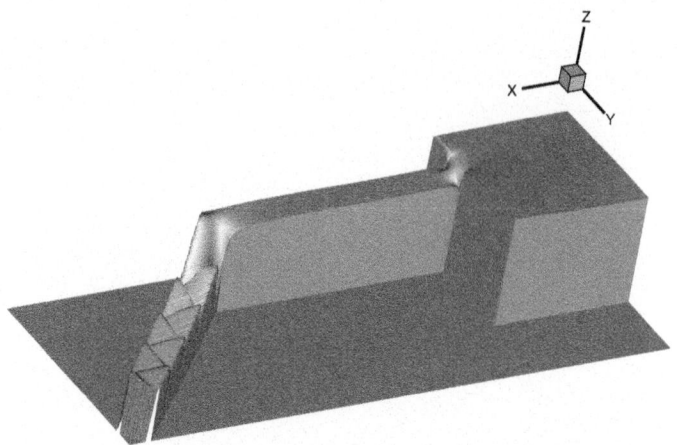

**Fig. 15.21** Computed free-surface elevation at time $T_{out} = 40.0$ s. Mesh $M = 300 \times 200$

**Fig. 15.22** Contour plot of computed free-surface elevation at time $T_{out} = 40.0$ s near 45° bend. Mesh $M = 300 \times 200$

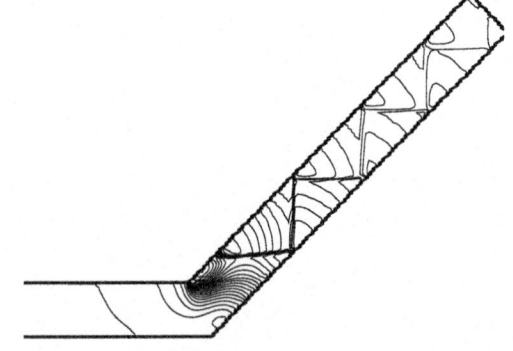

**Fig. 15.23** Physical dam-break model. Computed free-surface elevation (line) and experimental measurements (symbols) at gauge 1. Mesh used $M = 300 \times 200$

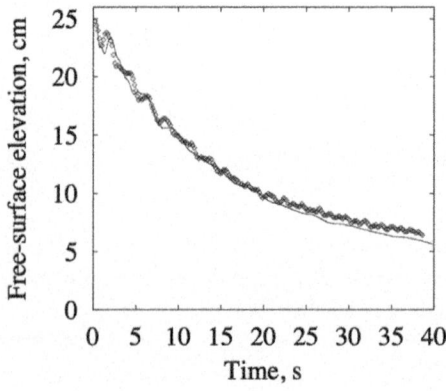

15.3 Physical Models: Experiments and Numerics

**Fig. 15.24** Physical dam-break model. Computed free-surface elevation (line) and experimental measurements (symbols) at gauge 2. Mesh used $M = 300 \times 200$

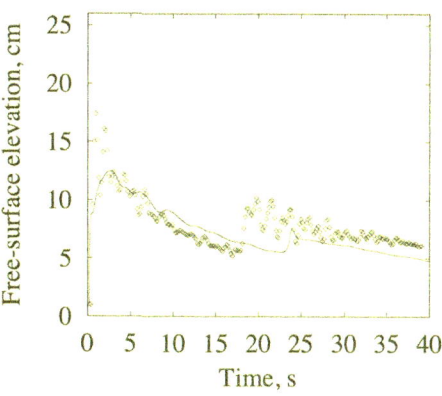

**Fig. 15.25** Physical dam-break model. Computed free-surface elevation (line) and experimental measurements (symbols) at gauge 3. Mesh used $M = 300 \times 200$

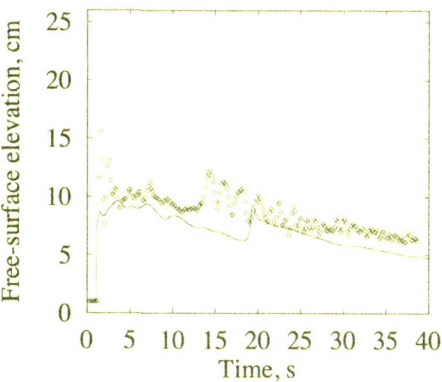

**Fig. 15.26** Physical dam-break model. Computed free-surface elevation (line) and experimental measurements (symbols) at gauge 4. Mesh used $M = 300 \times 200$

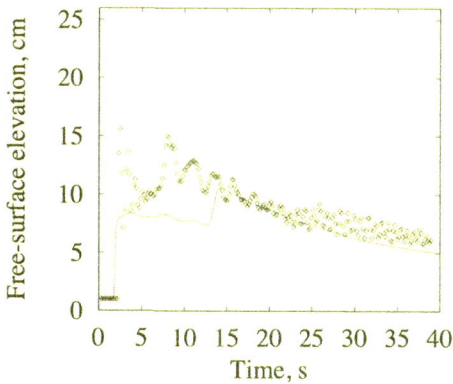

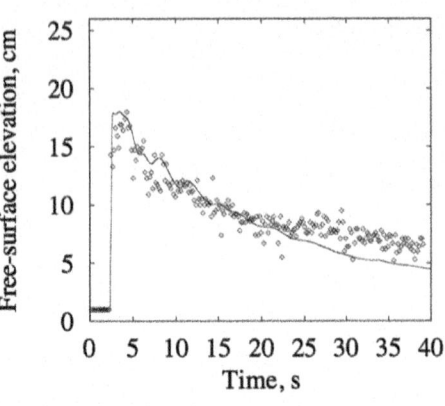

**Fig. 15.27** Physical dam-break model. Computed free-surface elevation (line) and experimental measurements (symbols) at gauge 5. Mesh used $M = 300 \times 200$

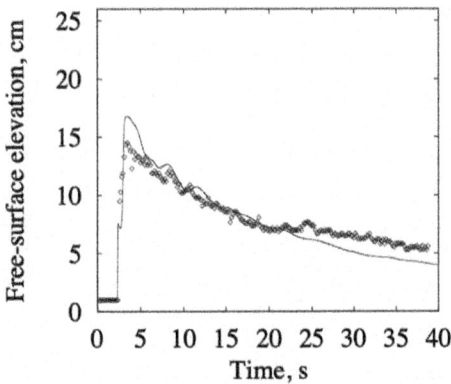

**Fig. 15.28** Physical dam-break model. Computed free-surface elevation (line) and experimental measurements (symbols) at gauge 6. Mesh used $M = 300 \times 200$

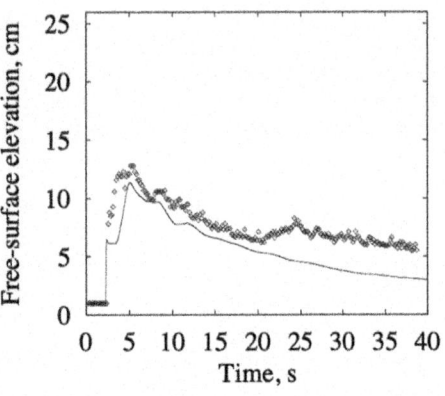

**Fig. 15.29** Physical dam-break model. Computed free-surface elevation (line) and experimental measurements (symbols) at gauge 7. Mesh used $M = 300 \times 200$

**Fig. 15.30** Physical dam-break model. Computed free-surface elevation (line) and experimental measurements (symbols) at gauge 8. Mesh used $M = 300 \times 200$

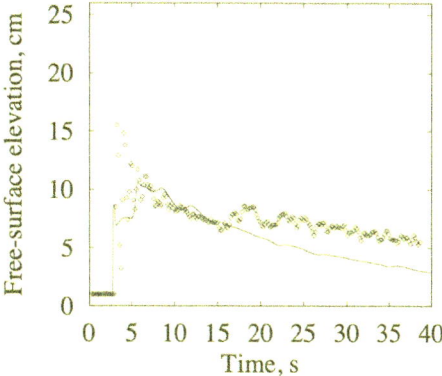

**Fig. 15.31** Physical dam-break model. Computed free-surface elevation (line) and experimental measurements (symbols) at gauge 9. Mesh used $M = 300 \times 200$

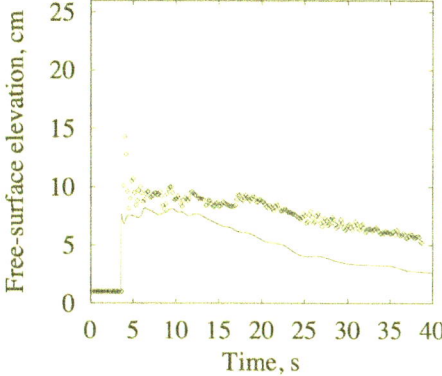

puted results (line). The comparison of results for gauge 1, shown in Fig. 15.23, is very satisfactory, although the amplitude and speed of the rarefaction waves in the reservoir do not agree very well. Figure 15.24 shows results for gauge 2 of Fig. 15.15; it shows the arrival of the first shock travelling downstream of the reservoir and the arrival of the second shock reflected from the channel bend. The agreement for the second shock is not very good; the numerical shock propagates upstream towards the reservoir more slowly than the experimentally observed shock. It is here where bed and wall friction play a role. In our computations the shock travels against a faster flow than in the real situation, which results in its slower speed. Figure 15.25 shows results for gauge 3 of Fig. 15.15. As for gauge 2, the agreement between numerics and experiment is good for the arrival of the first shock but unsatisfactory for the second shock. The long-time trend in the computation for all three gauges so far is satisfactory, in that it agrees well with experiment. All the above comments also apply to the results of gauge 4 shown in Fig. 15.26.

The results for gauges 5 to 7, shown in Figs. 15.27, 15.28 and 15.29, are very satisfactory, considering that the gauges are positioned right at the channel bend, where the two-dimensional effects are very strong. In general the coarse mesh results, omitted here, do not appreciably differ from these fine mesh results. However, for gauge 7 the coarse mesh results are visibly inaccurate just behind the shock. The comparison for gauges 8 and 9, shown in Figs. 15.30 and 15.31, are generally quite satisfactory, except for the long-time behaviour; it appears as if the numerical model *empties* the reservoir and channel more *easily*. We remark, however, that comparisons at these two positions are difficult given the character of the flow there.

There are several aspects of the exercise that would require a detailed discussion. Neglecting friction terms is obviously a significant simplification of the physics of the problem. These can be easily implemented as source terms, if desired. The elimination of the step between the reservoir and the channel will also have an effect on the results. With the step in place, waves in the reservoir will have different speeds to those in the present computations, and reflected waves from the reservoir and the emptying precess will also be affected. Then, the initial depth $h(x, y, 0) = 0.01$ m, rather than zero, will also affect the speed of propagation of waves. See Chap. 11 on the challenges of wet/dry fronts. Finally the numerical method with Cartesian Cut Cell approach is first-order accurate. This can be easily remedied by using the better methodologies discussed in this book.

## 15.4 Conclusions

We have studied two physical problems that are relevant to the understanding of the wave propagation phenomena associated with the collapse of dams. The first problem concerns the wave phenomena emerging from the rupture of a cylindrical dam, which is simulated by solving the two-dimensional shallow water equations using a second-order Godunov type numerical method, whose main components have been presented in detail in this book. Given the radial symmetry of the problem, a one-dimensional system with geometric source terms along the radial direction, is derived; this model furnishes an additional solution that is used to compare with the full two-dimensional computations in order to assess the quality of the solutions. The high-resolution computations reveal a wave system comprising the expected shock wave travelling in the outer radial direction along with an intriguing wave phenomena in the *near field*, close to the centre of the circular dam.

The second problem is a physical dam-break model, chosen from the work of the European Workshop on Dam-Break Modelling (CADAM, Concerted Action on Dam-Break Modelling, 1996–1999). The numerical results have been compared with the experimental measurements, and agreement between numerics and experiment is, on the whole, satisfactory. Improved numerical simulations can, in principle, be obtained by adding bed and wall friction terms, as source terms, to the basic two-dimensional shallow water equations. A better treatment of the step joining the reservoir and the channel may also result in more accurate simulations. Both classes of test problems may be of use in an engineering post-graduate teaching setting, concerned with dam-break modelling. More generally, they may also be useful in a research-laboratory setting concerned with geophysical fluid dynamics.

# References

1. S. Manenti, E. Pierobon, M. Gallati, S. Sibilla, L. D'alpaos, E. Macchi, S. Todeschini, Vajont disaster: smoothed particle hydrodynamics modeling of the postevent 2D experiments. J. Hydraul. Eng. **05015007**, 142 (2015)
2. J.M. Hervouet, A. Petitjean, Malpaset dam-break revisited with two-dimensional computations. J. Hydraulic Res. **37**(6), 777–788 (1999)
3. J.A. Roberson, J.J. Cassidy, M.H. Chaudhry, *Hydraulic Engineering* (Houghton Mifflin Company, Boston, 1988)
4. P. García-Navarro, Dam-Break Flow Simulation, in *Numerical Modelling of Hydrodynamical Systems*, ed. by P. Garcia-Navarro, E. Playan (University of Zaragoza, Spain, 1999), pp.27–56
5. P. Glaister, Difference Schemes for the Shallow Water Equations. Technical Report 9/87, Department of Mathematics, University of Reading, England (1987)
6. P. Glaister, Approximate Riemann solutions of the shallow water equations. J. Hydraulic Res. **26**(3), 293–306 (1988)
7. E.F. Toro, Riemann problems and the WAF method for solving two-dimensional shallow water equations. Phil. Trans. R. Soc. London **A338**, 43–68 (1992)
8. P. Molinaro, Dam–Break analysis: a state of the art, in *Computational Water Resources*, ed. by B. Bensaro, C.M.P. Ovazar (1991)
9. F. Alcrudo, P. García-Navarro, Flux difference splitting for 1D open channel flow equations. Int. J. Num. Meth. Fluids **14**, 1009–1018 (1992)
10. P. García-Navarro, F. Alcrudo, 1D open channel flow simulation using TVD McCormack scheme. J. Hydraulic Eng. ASCE **118**, 1359–1373 (1992)
11. C.W. Bellos, J.V. Soulis, J.G. Sakkas, Experimental results of two-dimensional dam-break flows. J. Hydraulic Res. **30**(2), 225–252 (1992)
12. J. Soulis, Computation of two dimensional dam-break flood flows. Int. J. Numer. Meth. Fluids **14**, 631–664 (1992)
13. A. Betamio de Almeida, B. Franco, Modelling of dam break flows, in *Computer Modelling of Free Surface and Pressurised Flows*, ed. by Chaudhry, Mays NATO ASI Serie E: Applied Sciences, vol. 274 (1993)
14. F. Alcrudo, P. García-Navarro, A high resolution Godunov-type scheme in finite volumes for the 2D shallow water equations. Int. J. Num. Meth. Fluids **16**, 489–505 (1993)
15. D. Ambrosi, Approximation of shallow water equations by Riemann solvers. Int. J. Num. Meth. Fluids **20**, 157–168 (1995)
16. L. Fraccarollo, E.F. Toro, A shock–capturing method for two dimensional dam–break problems. *Proceedings of the Fifth International Symposium in Computational Fluid Dynamics*, Sendai, Japan (1993)
17. L. Fraccarollo, E.F. Toro, Experimental and numerical assessment of the shallow water model for two-dimensional dam-break type problems. J. Hydraulic Res. **33**, 843–864 (1995)
18. M. Nujic, Efficient implementation of non-oscillatory schemes for the computation of free-surface flows. J. Hydraulic Res. **33**, 101–111 (1995)
19. P.A. Sleigh, P.H. Gaskell, M. Berzins, N. Wright, An unstructured finite volume algorithm for predicting flow in rivers and estuaries. Comput. Fluids **27**, 479–508 (1998)
20. G. Lauber, W.H. Hager, Experiments to dambreak wave: horizontal channel. J. Hydraulic Res. **36**(3), 291–307 (1998)

21. G. Lauber, W.H. Hager, Experiments to dambreak wave: sloping channel. J. Hydraulic Res. **36**(5), 761–773 (1998)
22. C.G. Mingham, D.M. Causon, Calculation of unsteady bore diffraction using a high-resolution finite volume method. J. Hydraulic Res. **38**, 49–56 (2000)
23. F. Aureli, P. Mignos, M. Tomirotti, Numerical simulation and experimental verification of dam-break flows with shocks. J. Hydraulic Res. **38**, 197–206 (2000)
24. A.I. Delis, C.P. Skeels, S.C. Ryrie, Evaluation of some approximate Riemann solvers for transient open channel flows. J. Hydraulic Res. **38**, 217–231 (2000)
25. A. Ferrari, L. Fraccarollo, M. Dumbser, E.F. Toro, A. Armanini, A new 3D parallel SPH scheme for free surface flows. J. Fluid Mech. **663**, 456–477 (2010)
26. M. Morris, J.C. Galland, P. Balabanis. *Concerted Action on Dam–Break Modelling. Proceedings of the CADAM Meeting at Wallingford, UK, 2 to 3 March 1999.* European Commission (1999). ISBN 92–828–7108–8
27. S. Soares Frazão, X. Sillen, Y. Zech, Dan–break flow through sharp bends. physical models and 2d boltzmann model validation. European Commission (1999). ISBN 92–828–7108–8
28. M. Aftosmis, M.J. Berger, J. Melton, Robust and efficient cartesian mesh generation for component-based geometry. AIAA J. **36**(6), 952–960 (1998)
29. J.F. Clarke, S. Karni, J.J. Quirk, L.G. Simmons, P.L. Roe, E.F. Toro, Numerical computation of two-dimensional, unsteady detonation waves in high energy solids. J. Comput. Phys. **106**, 215–233 (1993)
30. E.F. Toro, A cartesian-cut cell approach for solving the two–dimensional shallow water equations in arbitrary geometries. *Unpublished* (2000)

# Chapter 16
# Mach Reflection in Tsunamis

**Abstract** This chapter is partly motivated by the tsunami event occurred in Japan on 12th July 1993. The event devastated parts of the Okushiri island, where observed maximum run-up water elevations were of the order of 30 m. The number of fatalities reported was 239. Over several decades oceanographers have observed the occurrence of Mach reflection of tsunami bores, which may help explain the devastating power of tsunamis. Based on the two-dimensional shallow water equations and the numerical methods studied in this book, this chapter presents an analytical and numerical study of bore reflection phenomena, confirming that both regular and Mach reflection patterns are possible in shallow water flows.

## 16.1 Introduction

This chapter presents a numerical and analytical study of the bore reflection patterns that occur when a bore reflects from a solid vertical wall placed at an angle to the bore propagation direction. For more than six decades it has been known, though not widely known, to oceanographers and other scientists, that free-surface gravity waves reflected from vertical walls, placed at an *angle to the wave propagation direction*, exhibit characteristics similar to those of shock waves in compressible media. See for instance the pioneering works of Perroud [1] and Wiegel [2], who observed the occurrence of *Mach reflection* in laboratory studies. Based on these observations it has also been conjectured that it is the occurrence of Mach reflection that might explain the unusual increases in wave amplitude and the unexpected devastating power of tsunamis when they reach coastal areas.

The laboratory studies of Perroud [1] established that for solitary waves three possible wave reflection patterns may occur. The occurrence of such patterns depends on the *angle of incidence*, which is defined as *the angle formed by the direction of propagation of the wave and the solid vertical wall placed in its path*. For angles of incidence $\theta_I$ greater than 35°–45°, *regular reflection* occurs; for angles $\theta_I$ greater than 20° and less than 40°–45°, *Mach reflection* occurs. The third possibility, essentially a special case of Mach reflection, is realised for angles of incidence $\theta_I$ of less than about 20°. This type of reflection is characterised by a curved Mach stem and, in the

© The Author(s), under exclusive license to Springer Nature Switzerland AG 2024
E. F. Toro, *Computational Algorithms for Shallow Water Equations*,
https://doi.org/10.1007/978-3-031-61395-1_16

author's words, the *absence of a reflected wave*. Wiegel [2] carried out laboratory experiments designed to study the occurrence of Mach reflection for the case of periodic waves. He found that for the *shallow water* regime Mach reflection did occur, with similar characteristics to those of solitary waves. He also tried to establish a relationship between the occurrence of Mach reflection and the unusual behaviour of tsunamis at Hilo (Hawaii). He reproduced photographic evidence from the actual event of features remarkably similar to a *Mach stem*. He concluded that Mach reflection may provide a plausible explanation for the destructive power of the Chilean tsunami of 22nd May 1960, which caused devastation not only around the Chilean coast but also in very distant places such as Hawaii and Japan, about 24 hours later. The experimental observations of Wiegel and collaborators prompted a good deal of theoretical work. See for example the works of Miles [3, 4], Melville [5], Yue and Mei [6], Pedersen [7] and Tanaka [8]. These works have concentrated exclusively on solitary waves and periodic waves in the shallow water regime. It is interesting to note that perhaps the most important case, namely that of *bore reflection patterns*, has not been widely studied. This is somehow surprising because the phenomenon of shock-wave Mach reflection has been studied very extensively in the case of gas dynamics. See for instance the textbook of Ben-Dor [9] on the very subject and the many references therein. It appears as if the first reported attempts to study the phenomenon of bore Mach reflection are due to Krehl and van der Geest [10], Takayama et al. [11] and Toro et al. [12]. Krehl and van der Geest [10] observed Mach reflection experimentally, and the first theoretical attempt using the two-dimensional non-linear shallow water equations is due to Takayama et al. [11], Toro et al. [12], and Mingham and Causon [13]. See also the more recent paper of Dumbser et al. [14], in which the sophisticated numerical methods introduced in Chap. 14 of this book are used.

The study of the reflection of free-surface gravity waves from vertical walls constitutes an important area of basic fluid mechanics; it is also important from the viewpoint of applications to various environmental disciplines, such as oceanography, coastal and hydraulic engineering, when designing coastal defences against tsunamis, for instance. Tsunamis are free-surface water waves produced by submarine and other disturbances. As they approach coastal areas, they undergo the effects of refraction and diffraction; their amplitude increases unexpectedly, with the associated destructive power as they impinge on breakwaters and other man-made structures. In actual fact, the motivation for the work of Takayama et al. [11] and Toro et al. [12] was the occurrence of a tsunami in Japan on 12th July 1993, which devastated parts of the Okushiri island. Observed maximum run-up elevations were of the order of 30 m and 239 fatalities were reported. The material of this chapter is based on the preliminary results reported in [11, 12]. I am grateful to Dr. M. Olim (Seagate Technology, USA) and Professor K. Takayama and his collaborators (Shock Wave Research Centre, Tohoku University, Sendai, Japan) for their contributions to the subject.

Here, we first establish theoretically that Mach reflection of bores in shallow water is possible. Also, criteria for transition between regular reflection and Mach reflection are identified. Numerical computations are carried out to verify some of the theoretical findings. In Sect. 16.2 we define the problem. In Sect. 16.3 we report on an analytical study of bore reflection from vertical walls placed at an angle to the propagation direction of an isolated bore. Section 16.4 deals with numerical computation of bore Mach reflection. In Sect. 16.5 we draw conclusions and discuss further work on the subject.

## 16.2 The Problem

Shock reflection phenomena in compressible media are by now well documented, see for instance [9]. To introduce bore reflection phenomena in free-surface gravity waves, let us first consider a horizontal open channel of rectangular cross-section. Figure 16.1 shows a top view of the situation of interest. Water flows from left to right between two vertical walls; one of them bends inwards at $x = a$ by an angle $\theta_W$. Assume now that a vertical gate positioned at $x = s$ separates two levels of stationary water of depth $h_1$ (deep water) and $h_0$ (shallow water). Sudden removal of the gate, vertically say, begins the process called *dam-break problem*, which is analogous to the *shock-tube problem* in gas dynamics. Removal of the gate generates a depression (or rarefaction) wave moving into the deep water region and a bore (or shock) propagating in the positive $x$-direction. For background information see Chaps. 1, 5 and 6. We denote the bore speed by $S$ and the celerity ahead of the bore by $a_0$, with $a_0 = \sqrt{gh_0}$, where $g$ is the acceleration due to gravity; the bore *Froude number* is defined as $F_{rs} = S/a_0$. At a later time, the right-travelling bore will impinge on the section of the right wall (at the bottom of Fig. 16.1) that is bent at an angle $\theta_W$, called the *wall angle*, and will produce a bore reflection pattern. In the language of oceanographers, one defines the *angle of incidence* $\theta_I$ as the *angle formed*

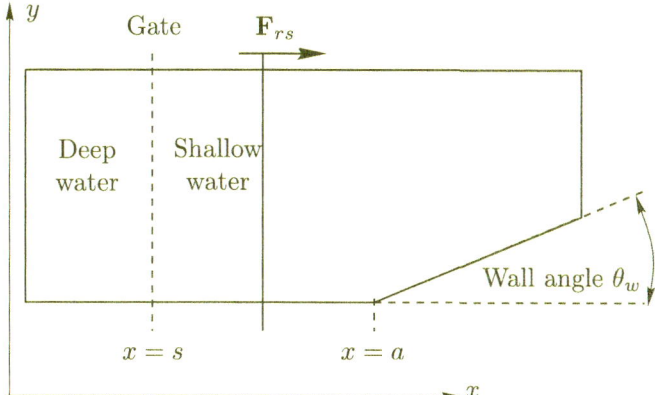

**Fig. 16.1** Top view of an open horizontal channel of rectangular cross-section enclosed by two vertical walls. A vertical gate at $x = s$ separates two constant levels of stationary water. The right-hand vertical wall (bottom of diagram) is assumed to bend inwards at $x = a$ by an angle $\theta_W$. Sudden removal of the gate produces a left-running rarefaction and a right-running shock of Froude number $F_{rs}$

by *the direction of the wave advance and the vertical reflecting wall*; zero incidence corresponds to a wall parallel to the direction of wave propagation, from which no reflection takes place of course. So defined, the angle of incidence is identical to the *wall angle* $\theta_W$ depicted in Fig. 16.1. In the gas dynamics literature, the angle of incidence is defined as the angle formed by the wave path, or front, and the wall. Here we shall call this angle the *wave angle* and it will be denoted by $\theta_0$; see Eq. (16.3).

The particular wave-reflection pattern that emerges will depend on the Froude number $F_{rs}$ and the wall angle $\theta_W$. For the rest of the discussion here we simply assume an isolated right-travelling bore as depicted in Fig. 16.1; no reference to the rarefaction is necessary. Two possible outcomes are depicted in Figs. 16.2 and 16.3.

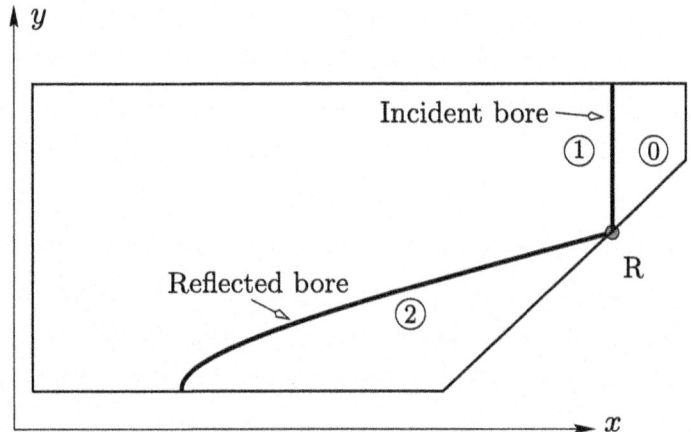

**Fig. 16.2 Regular reflection:** incident and reflected bore meet at $R$ on the wall

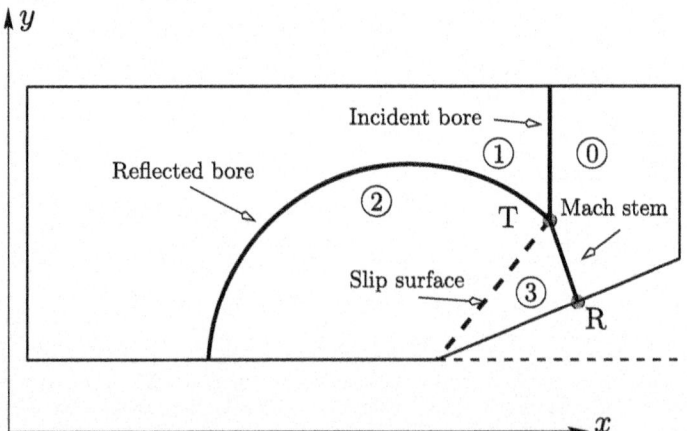

**Fig. 16.3 Mach reflection:** incident bore, reflected bore, Mach stem and slip surface (or slipstream) meet at triple point $T$ away from the wall

For wall angles greater than a critical angle $\theta_{crit}$ the reflection is **regular reflection** and the corresponding typical pattern is shown in Fig. 16.2. For wall angles $\theta_W$ less than $\theta_{crit}$, the reflection is **Mach reflection**, of which Fig. 16.3 shows a typical wave pattern. Regular reflection is characterised by the presence of a pair of bores, namely the incident bore, present at the start of the process, and a new bore, the reflected curved bore. The two bores meet at a point $R$, which lies on the reflecting vertical wall. See Fig. 16.2.

The Mach reflection pattern of Fig. 16.3 shows three bores: the incident bore, the reflected curved bore, and a new bore called the *Mach stem*. All three bores meet at a point $T$, called the *triple point*, which lies at a finite distance away from the reflecting vertical wall. From the triple point there emerges another wave, called the *slip surface* or *slipstream*; this carries a jump in velocity and separates particles into two classes. One class comprises particles processed by the Mach stem and the other class comprises particles first processed by the incident bore and then by the reflected bore; the water free-surface elevation remains continuous across the slip surface.

## 16.3 Analytical Study

The experimental evidence available for solitary waves and periodic waves suggests that the phenomenon of interest here does take place in shallow flows. We therefore assume that the problem can be approximated by the time-dependent, non-linear two-dimensional shallow water equations, namely

$$\partial_t \mathbf{Q} + \partial_x \mathbf{F}(\mathbf{Q}) + \partial_y \mathbf{G}(\mathbf{Q}) = \mathbf{0} , \qquad (16.1)$$

where

$$\mathbf{Q} = \begin{bmatrix} h \\ hu \\ hv \end{bmatrix}, \quad \mathbf{F}(\mathbf{Q}) = \begin{bmatrix} hu \\ hu^2 + \frac{1}{2}gh^2 \\ huv \end{bmatrix}, \quad \mathbf{G}(\mathbf{Q}) = \begin{bmatrix} hv \\ huv \\ hv^2 + \frac{1}{2}gh^2 \end{bmatrix}. \qquad (16.2)$$

Here $h$ is the water depth (free-surface position), $u$ is the $x$-component of velocity, $v$ is the $y$-component of velocity and $g$ is the acceleration due to gravity, assumed constant, $g = 9.81$ m/s$^2$. The bottom distribution is assumed to be horizontal and friction terms are neglected; therefore no source terms $\mathbf{S}(\mathbf{Q})$ are present in Eq. (16.1). Background on the equations and numerical methods is found in preceding chapters, e.g., Chaps. 1, 4 and 15.

The analysis applied herein follows closely the method of analysis accepted in the field of shock waves, see for example [9]. Some of the terminology used here is also borrowed from gas dynamics. An oblique bore propagating in water of constant depth $h_0$ at a constant velocity $u_0$ is assumed. A Galilean transformation may therefore be performed to place the origin of the coordinate system at the moving bore. In the new coordinate system, water of depth $h_0$ flows obliquely into the stationary bore at velocity $u_0$. The angle between the bore and the streamline entering it from region 0 is denoted as $\theta_0$ and is called here the *wave angle*; in the gas dynamics literature it is called the angle of incidence. Note the following relationship between wave angle and wall angle (identical to angle of incidence here):

$$\theta_I \equiv \theta_W = \tfrac{1}{2}\pi - \theta_0 \,. \tag{16.3}$$

The water emerges on the other side of the bore, i.e., in region 1, having changed its depth to $h_1$ and its velocity magnitude to $u_1$. The streamlines which pass through the bore are deflected, and the angle of deviation from the original flow direction is denoted as $\delta_0$ and is called the *deflection angle*.

### 16.3.1 Oblique Bore Relations

By analysing the steady-state equations of conservation of mass and momentum across the oblique bore, one may derive expressions for the surface height and velocity vector behind the bore in terms of those ahead of it. For background on the Rankine–Hugoniot jump conditions see Chap. 2. The corresponding equations are

$$h_0 u_0 \sin \theta_0 = h_1 u_1 \sin(\theta_0 - \delta_0) \,, \tag{16.4}$$

$$u_0 \cos \theta_0 = u_1 \cos(\theta_0 - \delta_0) \,, \tag{16.5}$$

$$h_0 u_0^2 \sin^2 \theta_0 + \tfrac{1}{2} h_0^2 = h_1 u_1^2 \sin^2(\theta_0 - \delta_0) + \tfrac{1}{2} h_1^2 \,. \tag{16.6}$$

Equations (16.4), (16.5) and (16.6) may be rewritten to obtain the following expression for the *jump* of surface height across the bore:

$$\frac{h_1}{h_0} = \gamma_0 = \frac{1}{2}\left(-1 + \sqrt{1 + 8 Fr_{0,n}^2}\right) \,, \tag{16.7}$$

where $Fr_{0,n}^2 = Fr_0^2 \sin^2 \theta_0$ and the (incident) bore Froude number is $Fr_0 = u_0/\sqrt{gh_0}$. It is important to notice from Eq. (16.7) that a jump in the surface height across the oblique bore exists, i.e., $\gamma_0 > 1$, only when $Fr_{0,n} > 1$. The conservation equations may also be used to derive expressions for the deflection angle

$$\delta_0 = \theta_0 - \tan^{-1}\left(\frac{1}{\gamma_0} \tan \theta_0\right) \tag{16.8}$$

and for the velocity jump

$$\frac{u_1}{u_0} = \frac{1}{\gamma_0} \frac{\sin \theta_0}{\sin(\theta_0 - \delta_0)} \,. \tag{16.9}$$

From Eqs. (16.7), (16.8) and (16.9) it is obvious that the only parameters affecting the oblique bore relations are the incident Froude number, $Fr_0$, and the wave angle, $\theta_0$. For given $\theta_0$ and $Fr_0$, the deflection angle and the flow properties behind the oblique bore are uniquely determined.

Before proceeding to the analysis, we identify the various regions of interest and establish the notation. An oblique bore interacting with a solid vertical wall may undergo either a *regular reflection* or a *Mach reflection*. For an incident bore of given Froude number and wave angle smaller than some critical value $\theta_{crit}$, that is $\theta_0 < \theta_{crit}$, regular reflection occurs. These two bores in regular reflection intersect at the reflection point $R$, which is located on the solid vertical wall, see Fig. 16.2. The two discontinuities divide the flow field in the vicinity of the reflection point into three regions, which we label 0 in the undisturbed water; 1 behind the incident bore, and 2 behind the reflected bore. For wave angles larger than the critical value, $\theta_0 > \theta_{crit}$, regular reflection is no longer physically possible and Mach reflection occurs. The wave configuration typical of a Mach reflection consists of four discontinuities: the incident bore, the reflected bore, the Mach stem and the *slipstream*; see Fig. 16.3. The slipstream separates the water which passed through the incident and the reflected bores from the water that passed though the Mach stem. While all properties change abruptly across the first three discontinuities, only the velocity magnitude changes across the slipstream. The four discontinuities intersect at the triple point $T$. The discontinuities divide the flow field near the triple point into four regions: 0 the undisturbed water; 1 behind the incident bore; 2 behind the reflected bore, and 3 between the Mach stem, the slipstream and the solid vertical wall; see Fig. 16.3.

### 16.3.2 Regular Reflection

To analyse the flow field in the vicinity of the reflection point, the origin of the coordinate system is transformed to the reflection point. Assuming that an incident oblique bore of known $Fr_{0,n}$ and $\theta_0$ undergoes a regular reflection from a solid wall, the conditions behind the reflected bore, i.e., in region 2, may be calculated by changing the indices in Eqs. (16.7), (16.8) and (16.9) to relate the properties in region 1 to those in region 2, namely

$$\frac{h_2}{h_1} = \gamma_1 = \frac{1}{2}\left(-1 + \sqrt{1 + 8Fr_{1,n}^2}\right), \qquad (16.10)$$

$$\delta_1 = \theta_1 - \tan^{-1}\left(\frac{1}{\gamma_1}\tan\theta_1\right), \qquad (16.11)$$

and

$$\frac{u_2}{u_1} = \frac{1}{\gamma_1}\frac{\sin\theta_1}{\sin(\theta_1 - \delta_1)}, \qquad (16.12)$$

where

$$Fr_{1,n}^2 = Fr_1^2 \sin^2 \theta_1 = \frac{Fr_{0,n}^2 \sin^2 \theta_1}{\gamma_0^3 \sin^2(\theta_0 - \delta_0)}. \tag{16.13}$$

The reflected bore must deflect the flow so it is parallel to the solid wall, that is

$$\delta_0 = \delta_1. \tag{16.14}$$

Equations (16.7)–(16.14) constitute a system of eight algebraic equations with ten unknowns: $Fr_0$, $Fr_1$, $\gamma_0$, $\gamma_1$, $\theta_0$, $\theta_1$, $\delta_0$, $\delta_1$, $u_1/u_0$ and $u_2/u_1$. For given $Fr_0$ and $\theta_0$ these equations may be solved, although for some combinations of $\theta_0$ and $Fr_0$ no surface height jump across the incident or the reflected bores is possible, due to either $Fr_{0,n} < 1$ or $Fr_{1,n} < 1$, while for other combinations the system of equations may possess no real solutions. Complex solutions of this system are obviously non physical.

### 16.3.3  Transition from Regular to Mach Reflection

If for a given $Fr_{0,n}$, one gradually increases the value of $\theta_0$ starting with a value for which Eqs. (16.7)–(16.14) possess two sets of real solutions, then the two sets of solutions approach each other and eventually merge into a single solution at some limiting value of $\theta_0 = \theta_0^d$, beyond which no real solutions exist. As long as real solutions to this set of equations can be found, one can calculate the Froude number behind the reflected bore, $Fr_2$. At $\theta_0 = \theta_0^d$ one finds that $Fr_2 < 1$. By gradually decreasing $\theta_0$ from $\theta_0^d$ one can find a value $\theta_0 = \theta_0^c$ at which $Fr_2 = 1$, and $Fr_2 > 1$ for $\theta_0 < \theta_0^c$. The physical significance of the value of $Fr_2$ is that for $Fr_2 > 1$ the signals which originated downstream from the reflection point cannot catch up with the reflection point. The opposite holds for $Fr_2 < 1$. These two values, $\theta_0^d$ and $\theta_0^c$, constitute the basis of the most often used criteria for *regular to Mach reflection transition* in gas dynamics. In the context of shallow water waves, we call these the *detachment* and the *celerity* criteria, hence the superscripts $d$ and $c$. We find it convenient to express the results in terms of the *angle of incidence*, or equivalently in terms of the *wall angle* $\theta_W$, which is related to the *wave angle* $\theta_0$ via Eq. (16.3). The values of $\theta_W^d$ and $\theta_W^c$ and the corresponding total surface height jumps are shown in Table 16.1 for various values of incident normal bore Froude number (squared).

Table 16.1 shows that the transition angles predicted by these two criteria are not identical but are fairly close. This is also observed in gas dynamics. Figure 16.4 shows plots of transition wall angles against $F_{rs}^2$ according to the detachment and the celerity criteria. Each of the transition curves divides the plane formed by the Froude number and the wall angle into two regions. For wall angles greater than $\theta_W^c$ and $\theta_W^d$, *regular reflection* is predicted; for wall angles less than $\theta_W^c$ and $\theta_W^d$, *Mach reflection* is predicted. In the next Section we carry out some numerical computations that partially validate these theoretical predictions.

## 16.4 Numerical Study

**Table 16.1** Transition angles $\theta_W^d$ and $\theta_W^c$ as functions of $Fr_{0,n}^2$. Also shown are corresponding total surface height jumps

| $Fr_{0,n}^2$ | $\theta_W^d$ | $h_{20,d}$ | $\theta_W^c$ | $h_{20,c}$ |
|---|---|---|---|---|
| 2 | 51.2 | 2.50 | 51.9 | 2.38 |
| 3 | 54.2 | 3.72 | 54.9 | 3.52 |
| 4 | 55.1 | 4.81 | 55.8 | 4.55 |
| 5 | 55.5 | 5.81 | 56.2 | 5.50 |
| 6 | 55.6 | 6.75 | 56.4 | 6.39 |
| 7 | 55.6 | 7.64 | 56.4 | 7.24 |
| 8 | 55.6 | 8.49 | 56.4 | 8.05 |
| 9 | 55.6 | 9.31 | 56.3 | 8.83 |
| 10 | 55.5 | 10.10 | 56.3 | 9.58 |

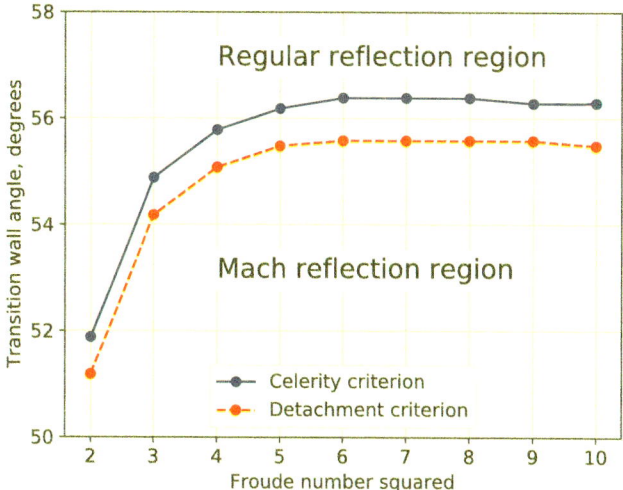

**Fig. 16.4** Mach reflection transition criteria. Transition wall angle against Froude number squared. Celerity transition criterion is shown blue symbols and solid line, while the detachment transition criterion is shown by red symbols and dotted line. The celerity criterion gives larger transition angles than the detachment criterion. Both transition criteria divide the complete region into two subregions: regular reflection above and Mach reflection below

## 16.4 Numerical Study

Here we present a numerical study of regular and Mach reflection wave patterns. We solve the two-dimensional shallow water Eqs. (16.1)–(16.2) by the weighted average flux (WAF) method discussed in Chap. 12 in conjunction with a second-order splitting scheme, as discussed in Chap. 13. We verify, numerically, the existence of bore Mach reflection. We consider a horizontal channel, see Fig. 16.1, of length $L = 10.0$ m with the apex of the reflecting vertical wall situated at $x = 3.0$ m; the initial bore is situated at $x = 1.5$ m. Ahead of the bore the water depth is chosen to be $h_0 = 0.1$ m and the velocity vector is zero. Computations are carried out on a mesh of $300 \times 200$ cells in the $x$- and $y$-directions, respectively. The CFL coefficient is 0.9. We consider an initial bore of bore Froude number $F_{rs} = 2.0$ and two wall angles, namely $\theta_W = 60.0°$ and $\theta_W = 25.0°$. Solutions are displayed at time $T_{out} = 3.5$ s, using half the mesh of the computations.

Figure 16.5 shows the computed free-surface elevation for the wall angle $\theta_w = 60.0°$ at time $T_{out} = 3.5$ s. According to the theory, this case corresponds to *regular reflection*; see Table 16.1 and Fig. 16.4. This is confirmed by the numerical result; the triple point lies on the reflecting wall; note the orientation of the coordinate axes. Figures 16.7, 16.8, 16.9 and 16.10 show contour plots of free-surface elevation $h$, $x$-velocity component $u$, $y$-velocity component $v$ and magnitude of the velocity vector; 80 levels were used.

Figure 16.6 shows the computed free-surface elevation for the wall angle $\theta_w = 25.0°$ at time $T_{out} = 3.5$ s. According to the theory, this case corresponds to *Mach reflection*; see Table 16.1 and Fig. 16.4. This is confirmed by the result of Fig. 16.6; note the orientation of coordinate axes. The Mach stem is clearly seen; the triple point lies a finite distance away from the wall. Figures 16.11, 16.12, 16.13 and 16.14 show

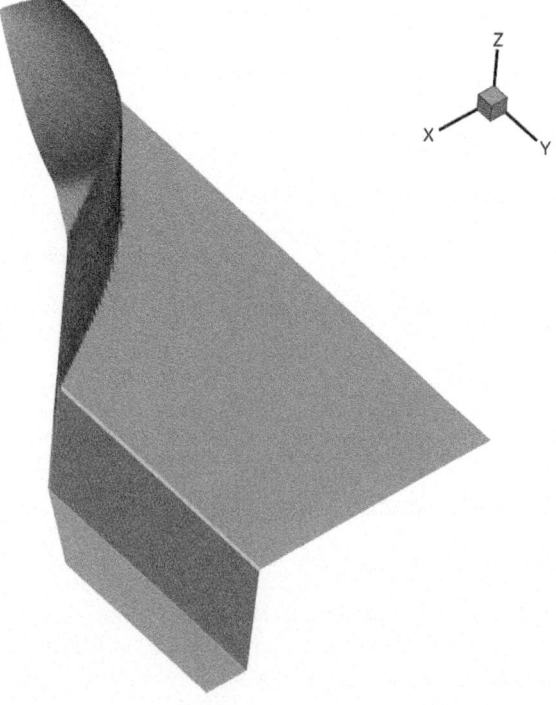

**Fig. 16.5** Computed bore regular reflection for bore Froude number $F_{rs} = 2.0$ and wall angle $\theta_W = 60.0°$. Free surface elevation at time $T_{out} = 3.5$ s

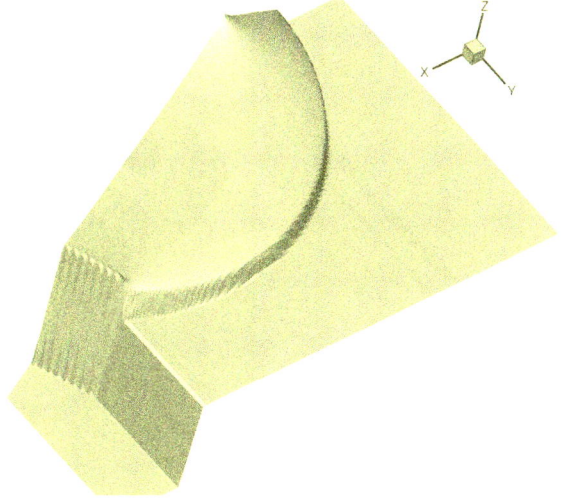

**Fig. 16.6** Computed bore Mach reflection for bore Froude number $F_{rs} = 2.0$ and wall angle $\theta_W = 25.0°$. Free-surface elevation at time $T_{out} = 3.5$ s

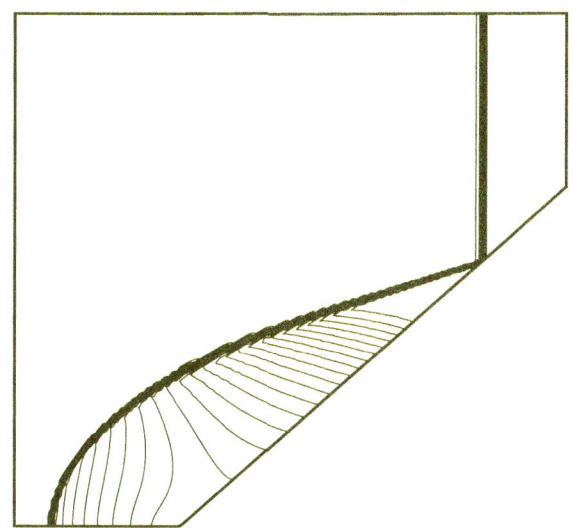

**Fig. 16.7** Computed bore regular reflection for bore Froude number $F_{rs} = 2.0$ and wall angle $\theta_W = 60.0°$. Contour plot of free-surface elevation at $T_{out} = 3.5$ s

contour plots of free-surface elevation $h$, $x$-velocity component $u$, $y$-velocity component $v$ and magnitude of the velocity vector; 80 levels were used. The *slipstream* that separates two portions of flow of different velocities is clearly seen in Figs. 16.12, 16.13 and 16.14.

**Fig. 16.8** Computed bore regular reflection for bore Froude number $F_{rs} = 2.0$ and wall angle $\theta_W = 60.0°$. Contour plot of $x$ component of velocity $u$ at $T_{out} = 3.5$ s

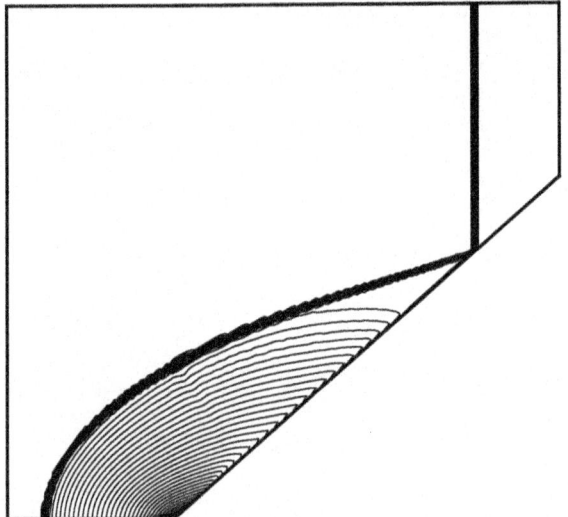

**Fig. 16.9** Computed bore regular reflection for bore Froude number $F_{rs} = 2.0$ and wall angle $\theta_W = 60.0°$. Contour plot of $y$ component of velocity $v$ at time $T_{out} = 3.5$ s

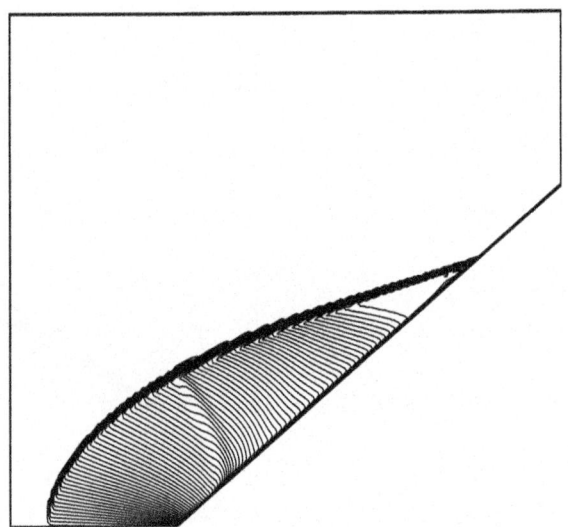

According to the theory of the previous section, transition from Mach to regular reflection takes place at $\theta_W = 55.1°$ when using the detachment criterion and at $\theta_W = 55.8°$ when using the celerity criterion. A desirable task is to find numerically the transition angles $\theta_W^n$ and to compare them with the values given by the detachment and celerity criteria, see Table 16.1 and Fig. 16.4.

## 16.4 Numerical Study

**Fig. 16.10** Computed bore regular reflection for bore Froude number $F_{rs} = 2.0$ and wall angle $\theta_W = 60.0°$. Contour plot of velocity magnitude at $T_{out} = 3.5$ s

**Fig. 16.11** Computed bore Mach reflection for bore Froude number $F_{rs} = 2.0$ and wall angle $\theta_W = 25.0°$. Contour plot of free-surface elevation at time $T_{out} = 3.5$ s

**Fig. 16.12** Computed bore regular reflection for bore Froude number $F_{rs} = 2.0$ and wall angle $\theta_W = 25.0°$. Contour plot of $x$ component of velocity $u$ at $T_{out} = 3.5$ s

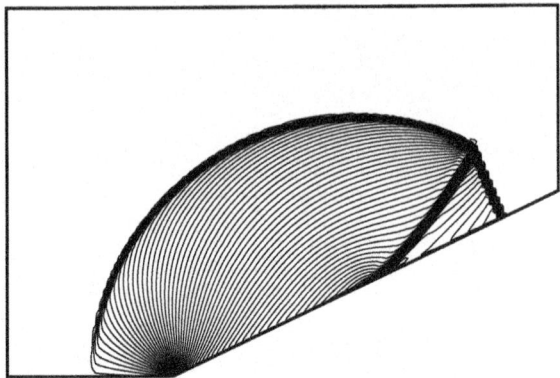

**Fig. 16.13** Computed bore Mach reflection for bore Froude number $F_{rs} = 2.0$ and wall angle $\theta_W = 25.0°$. Contour plot of $y$ component of velocity $v$ at time $T_{out} = 3.5$ s

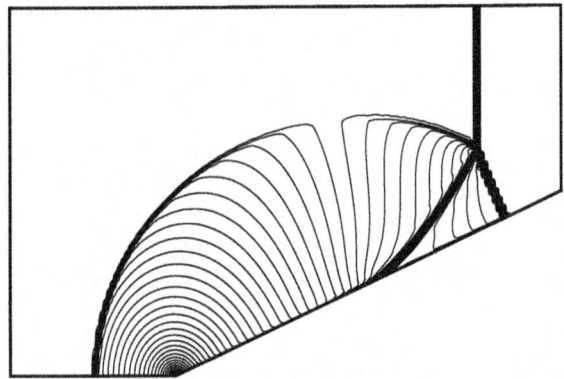

**Fig. 16.14** Computed bore regular reflection for bore Froude number $F_{rs} = 2.0$ and wall angle $\theta_W = 25.0°$. Contour plot of velocity magnitude at $T_{out} = 3.5$ s

## 16.5 Closing Remarks

The occurrence of bore Mach reflection in shallow water flows has been established both theoretically and numerically. Also, transition criteria between regular and Mach reflection have been put forward. A potentially attractive research programme could include a systematic experimental, theoretical and computational study of the relationship between the two proposed transition criteria. In addition to the scientific value, the results of such study would also have practical significance in hydraulics, oceanography and coastal engineering. Beyond shallow water, it is almost certain that the phenomenon of Mach reflection will also take place in other *shallow-flow* situations, such as in debris flow and heavy gas dispersion.

# References

1. P.H. Perroud, The solitary wave reflection along a straight vertical wall at oblique incidence. Technical Report IER Technical Report, 99–3, University of California, Berkeley, USA (1957)
2. R.L. Wiegel, Water wave equivalent of mach reflection, in *Proceedings of the 9th Conference Coastal Engineering, ASCE, Chap. 6* (ASCE, 1964), pp. 82–102
3. J. Miles, Obliquely interacting solitary waves. J. Fluid Mech. **79**, 157–169 (1977)
4. J. Miles, Resonantly interacting solitary waves. J. Fluid Mech. **79**, 171–179 (1977)
5. W.K. Melville, On the mach reflection of a solitary wave. J. Fluid Mech. **98**, 285–297 (1980)
6. D.K.P. Yue, C.C. Mei, Forward diffraction of stokes waves by a thin wedge. J. Fluid Mech. **99**, 33–52 (1980)
7. G. Pedersen, Three-dimensional wave patterns generated by moving disturbances at transcritical speeds. J. Fluid Mech. **196**, 39–63 (1988)
8. M. Tanaka, Mach reflection of a large-amplitude solitary wave. J. Fluid Mech. **248**, 637–661 (1993)
9. G. Ben-Dor, *Shock Wave Reflection Phenomena* (Springer, 1992)
10. P. Krehl, M. van der Geest, The discovery of the Mach reflection effect and its demonstration in the auditorium. Shock Waves **1**, 3–15 (1991)
11. K. Takayama, Y. Miura, M. Olim, T. Saito, E.F. Toro, Mach reflection of water waves in conjunction with Tsunamis at the Okushiri Island, in *Symposium on Shock Waves, Tokyo, Japan, January 1994*
12. E.F. Toro, M. Olim, K. Takayama, Unusual increase in tsunami wave amplitude at the Okushiri Island: mach reflection of shallow water waves, in *Proceedings of the 22nd International Symposium on Shock Waves*, ed. by Ball, Hillier, Roberts, vol. II (University of Southampton, UK, 1999), pp. 1207–1212
13. C.G. Mingham, D.M. Causon, Calculation of unsteady bore diffraction using a high-resolution finite volume method. J. Hydraul. Res. **38**, 49–56 (2000)
14. M. Dumbser, M.J. Castro, C. Parés, E.F. Toro, ADER schemes on unstructured meshes for nonconservative hyperbolic systems: applications to geophysical flows. Compute. Fluids **38**(9), 731–1748 (2009)

# Chapter 17
# Concluding Remarks

**Summary of This Book**

This book is a contribution to the broad field of computational geophysical fluid dynamics, focused specifically on numerical algorithms for solving the shallow water equations. The overall subject, covered in 16 chapters, is highly interdisciplinary, involving fluid mechanics, the physics of wave propagation, mathematics of partial differential equations, numerical analysis of hyperbolic PDEs, the engineering aspects of specific applications, and computing. We have endeavoured to maintain a *healthy balance* between the various scientific disciplines involved to keep the material accessible to all scientists working in these various fields. Fluid mechanics, the physics of wave propagation and the mathematical character of the governing shallow water equations provide the building block for the entire book; see Chaps. 1 to 5. Analytical techniques are deployed in Chaps. 6 to 8 to obtain closed-form solutions to the full non-linear equations; these constitute solid bases for subsequent chapters on the foundations of numerical analysis for the shallow water equations, in Chaps. 9 to 11. The construction of 10 approximate Riemann solvers in Chap. 11 rests heavily on the contents of previous chapters. The Godunov theorem, stated and proved in Chap. 9, establishes the need for constructing non-linear methods, if higher-accuracy is aimed for. Second-order non-linear methods based on TVD and ENO criteria were developed in Chap. 12. Then, two major issues were addressed in Chap. 13, namely source terms and multiple spatial dimensions. In this book we have provided concepts and methods that are fundamental ingredients for advanced higher-order schemes in the frameworks of finite volume and discontinuous Galerkin finite element methods. The finite volume methodology is developed in Chap. 14 for constructing non-linear, fully discrete higher-order upwind and centred methods for solving the non-linear time-dependent two-dimensional shallow water equations. The last two chapters illustrate the applicability of both the shallow water equations and the computational methods to solve them, studied in this book. In Chap. 15 we have presented two case studies that are relevant to the subject of mathematical modelling of dam-break problems, combining numerical, theoretical and experimental methods. The first case study concerns a two-dimensional circular, idealised dam for which a radial reference 1D solution is available for rigorous assessment of the performance of numerical methods. The second case study is based on the activities of the CADAM European Workshop on Dam-Break Modelling (1996–1999). We considered realistic two-dimensional laboratory physical models, specifically designed for validating numerical software intended for simulating real dam-break events. Chapter 16 deals with a fundamental physical problem, namely the occurrence of **Mach reflection** of tsunami bores, which may help explain the devastating power of tsunamis. By using the two-dimensional shallow water equations and the numerical methods studied in this book, we performed an analytical and numerical study of bore reflection phenomena, confirming that both **regular reflection** and **Mach reflection** patterns are possible in shallow water flows. Through the computations we also confirmed the existence of two regions separated by a *transition line*, namely a region of regular reflection and another of Mach reflection. Our computational results are in line with the two theoretical transition criteria established in this Chapter, the **celerity transition criterion** and the **detachment transition criterion**.

The contents of this book provide solid bases for various direct applications in hydraulics, coastal and environmental engineering, for example. Ambitious applications involving complex, time-dependent three-dimensional free-surface flows, such as in oceanic and atmospheric flows, will require further developments, both in terms of mathematical models and computational algorithms. Some concepts studied in this book may be extended to other computational approaches not dealt with here. In what follows we address some of these issues in more detail.

**Potential Applications and Further Developments**

**The Riemann Problem.** One of the central themes of this book is the Riemann problem for the shallow water equations, for which exact and approximate solvers were presented. The use of the Riemann problem goes beyond the topics studied in this book. For example, Riemann problem solutions can be utilised in theoretical studies of simple shallow water models. The Riemann problem can be used to implement boundary conditions, even if the numerical methodology used for the interior of the computational domain is not based on the Riemann problem. The Riemann problem solution can also be used in other, very different, numerical approaches, such as **front tracking**; see for example Swartz and Wendroff [1], Risebro and Tveito [2], Grove [3], Langseth [4], LeVeque and Shyue [5], Garaizar and Trangenstein [6] and Glimm et al. [7], [8]. The Riemann problem can also be used in meshless approaches, such as the **Smooth Particle Hydrodynamics (SPH)** methodology; see for example Ben Moussa [9], Ben Moussa and Vila [10], Vila [11], Ferrari et al. [12–14]. General background on SPH and applications are found in Gingold and Monagham [15], Lucy [16], Monagham [17] and Randles and Libertsky [18]. See also the more recent works of Colagrossi and Landrinni [19], Antuono et al. [20] and Fang et al. [21]. The Riemann problem solution has also been introduced in **discontinuous Galerkin finite element methods** (DG). See the early, seminal papers of Cockburn and collaborators [22, 23]; see also van der Vegt et al. [24]. Traditionally,

DG schemes have been used in the framework of semi-discrete methods. For the fully-discrete ADER schemes in the DG framework see the works of Dumbser and collaborators [25–32]. See also the recent review paper of Toro et al. [33].

**Tsunami Waves and Shallow Water Fluid Dynamics.** The mathematical models and the numerical methods to solve them presented in this book are applicable to the simulation of the dynamics of water flows in ocean tides, breaking of waves on shallow beaches, roll waves in open channels, flood waves in rivers, surges, dam-break wave modelling, and various specific applications in environmental fluid dynamics. Naturally, each specific application will require suitable ingredients to complete an adequate model. Much of the extra work needed will concern the source terms, to include for instance, friction terms and chemical-reaction terms. Transport of pollutant will require suitable extensions of the basic shallow water system to include transport equations for chemical species. A prominent area of study concerns the propagation of tsunami waves in oceans, for which there are numerous applications of shallow water type models. See Castro, Toro and Käser [34, 35], LeVeque et al. [36], Williamson et al. [37], Oportus et al. [38], Macías et al. [39], Ongaro et al. [40], Toro et al. [41], and Berger and LeVeque [42], to name but a few.

**Debris Flow and Sediment Transport.** Debris flow and sediment transport are specific areas in environmental engineering in which the methods of this book, suitably extended, can be applied. Following Iverson [43], *'debris flows occur when masses of poorly sorted sediment, agitated and saturated with water, surge down slopes in response to gravitational attraction'*. Debris flow modelling is an area of increasing interest in environmental disciplines. There are currently various mathematical models of the *shallow water type* for this type of problems, thus the numerical methods studied in this book could be extended to this area of applications. Fundamentals on debris flow modelling are discussed by Savage [44], Savage and Smith [45], Savage and Hutter [46, 47], Hutter et al. [48], Takahashi [49–51], Iverson [43, 52–54], Iverson et al. [55], Iverson and Denlinger [56], Denlinger and Iverson [57], Dumbser et al. [29], Armanini and Michiue [58], Vanzo et al. [59], Fernández-Nieto and Vigneaux [60] and and Siviglia et al. [61].

**Heavy-Gas Dispersion.** The modelling of heavy-gas dispersion in complex terrain is, potentially, another area of application of the methods of this book. Certain industrial processes involve storing hazardous liquefied gases, which may accidentally be released into populated areas. An important topic of research and development in industrial and government agencies is concerned with risk assessment of accidental release of these heavier-than-air gases. This area is commonly known as heavy-gas dispersion. Much experimental and modelling work is involved and useful mathematical models are of the shallow-water type; the methods presented in this book can therefore be extended to simulate such flows. Some references on the subject include Britter [62], Chan et al. [63], Eidsvik [64], Kukkonen and Nikmo [65], Webber et al. [66] and Chiapolino et al. [67]. A useful source of information is the PhD thesis of Hankin [68] and the many references therein. An interesting, recent work concerned with experimental and computational issues in heavy-gas dispersion is due to Ma et al. [69].

**Water Flow in Complex Channel Networks.** Computing water flows in complex networks of shallow-water channels is a straightforward area of application of the methods of this book. There are open-channel flow situations in which the computational domain consists of networks of essentially one-dimensional channels connected by multi-dimensional junctions. In this case one may deploy the one-dimensional shallow water equations for the one-dimensional channels and the full two-dimensional equations at the junctions, locally. An efficient hybrid approach has been developed by Bellamoli, Müller and Toro [70] and shown to result in huge computational savings. For further developments see also Liu et al. [71] and Kivva et al. [72].

**Implicit Methods.** All the methods studied in this book are **explicit** for the advection terms. For the source terms there is a choice of schemes for ODEs, explicit and implicit. The explicit schemes for the advection terms are simple to implement and they perform very well for time-dependent problems involving significant wave interaction, and for which time accuracy is needed. For large-scale problems the restrictive *stability condition* of these methods renders them inefficient. Implicit methods are then a possible choice. *Implicit upwind methods* for the shallow water equations have been put forward. Early communications include Vázquez-Cendón [73], Bermúdez et al. [74] and Anastasiou and Chan [75]. Implicit methods for the shallow water equations using other approaches have been put forward by Casulli [76] and by Casulli and Cheng [77]. Casulli has also extended his methodology to more general free-surface flows [78]. See also Stansby and Lloyd [79]. More recent works include sophisticated semi-implicit methods of high-order of accuracy for the shallow water equations. See for example, Busto and Dumbser [80], and Boscheri, Tavelli and Castro [81]. These more recent schemes rely on the flux vector splitting method of Toro and Vázquez [82] to separate all pressure terms responsible for fast waves, from the slowly-moving advection waves. See also Gómez-Bueno et al. [83].

**Well-balanced Schemes.** Not studied in this book are refinements to the schemes that make them **well-balanced**. This issue arises in the presence of source terms, particularly of the **geometric type**, such as in water flows in spatial domains with variable bathymetry. In steady state or nearly steady state regimes, the equations express **balance** between the flux gradient and the source term. Numerical methods experience difficulties in respecting this balance. Pioneering work in this area goes back to Bermúdez and Vázquez [84], amongst others. This topic has become an important area of research in many application areas. More recent works relevant to the shallow water equations include LeVeque [85], Vázquez [86], Fernández et al. [87], Castro et al. [88], Dumbser et al. [29], Canestrelli et al. [89], Castro et al. [90], Delestre and Lagrée [91], Müller et al. [92], Müller and Toro [93], Navas-Montilla and Murillo [94], Castro and Parés [95], Ghitti et al. [96], Arpaia et al. [97], Guerrero et al. [98], Barsukow and Berberich [99], Pimentel et al. [100], Martaud and Berthon [101], González-Tabernero et al. [102], and many more.

**Multilayer Shallow Water Models.** The basic shallow water equations considered in this book can be extended to multilayer shallow water systems. See the early work of Baines [103] on two-layer models in the context of atmospheric flows. There are some mathematical issues that require careful consideration; the full system of equations is reported to be of mixed elliptic-hyperbolic type; hence the initial value problem is ill-posed; see Castro et al. [104, 105]. This mathematical difficulty is also found in models for compressible multiphase flows and is currently the subject of much research. See Scoz, Bertazzi and Toro [106] for a mixed elliptic-hyperbolic model arising in the modelling of physiological flows. Multilayer shallow water equations represent a significant step forward in improving modelling capabilities of shallow water type systems, making then suitable for simulating large scale geophysical flows. The subject is very dynamic; examples of works in the literature include Gavrilyuk and Kazakova [107], Fernández-Nieto et al. [108, 109], Chesnokov et al. [110] and Guerrero et al. [98], to name but a few.

**Geophysical Fluid Dynamics, Closing Remarks**

Conventionally, *geophysical fluid dynamics* is concerned with the large scale dynamics of gases in the atmosphere and of water in oceans. In these scenarios, additional issues enter the discussion, such as the quasi-spherical shape of the computational domain and the effects of earth rotation, with its associated Coriolis forces. The basic two-dimensional, one-layer shallow water equations are an approximation to the three-dimensional atmosphere or the ocean. The approximation retains basic features of planetary waves and still plays an important role in providing benchmark solutions to assess methodologies intended for more complex mathematical models. However, the physical complexity of three-dimensional oceanic and atmospheric flows requires more comprehensive mathematical models than those embodied by the one-layer shallow water equations of this book. Clearly, within the shallow-water framework, multilayer shallow water type models discussed above can capture more physics. Furthermore, multilayer shallow water models are currently undergoing significant development in the fields of oceanic and atmospheric dynamics, where the underlying concepts and methodologies applicable to the basic one-layer shallow water equations of this book remain relevant.

**The Library NUMERICA**

Most of the numerical methods presented in this book have been applied to solve selected test problems. Use has been made of codes in the HYPER-WAT library of source codes for teaching, research and applications. This is a sub-library of the NUMERICA library [111]. For details, the interested reader may contact the author, Professor E. F. Toro, Email: eleuterio.toro@unitn.it. Full details about the NUMERICA library are found in:

http://www.eleuteriotoro.com.

# References

1. B.K. Swartz, B. Wendroff, AZTEC: a front tracking code based on Godunov's method. Appl. Numer. Math. **2**, 385–397 (1986)
2. N.H. Risebro, A. Tveito, Front tracking applied to nonstrictly hyperbolic systems of conservation laws. SIAM J. Sci. Stat. Comput. **12**, 1401–1419 (1991)
3. J.W. Grove, Application of front tracking to the simulation of shock refraction and unstable mixing. J. Appl. Numer. Math. **14**, 213–237 (1994)
4. J.O. Langseth, On an Implementation of a Front Tracking Method for Hyperbolic Conservation Laws. Technical Report 1994–1996, Department of Informatics, University of Oslo, Norway (1994)
5. R.J. LeVeque, K.M. Shyue, Two-Dimensional front tracking based on high resolution wave propagation methods. J. Comput. Phys. **123**, 354–368 (1996)
6. F.X. Garaizar, J. Trangenstein, Front tracking for shear bands in an antiplane shear model. J. Comput. Phys. 54–69 (1997)
7. J. Glimm, M.J. Graham, J.W. Grove, X.-L. Li, T.M. Smith, D. Tan, F. Tangerman, Q. Zhang, Front tracking in two and three dimensions. Comput. Math. Applic. **35**, 1–11 (1998)
8. J. Glimm, J.W. Grove, X.-L. Li, K.M. Shyue, Q. Zhang, Y. Zeng, Three dimensional front tracking. SIAM J. Sci. Comput. **19**, 703–727 (1998)
9. B. Ben Moussa, Meshless particle methods: recent developments for non–linear conservation laws in bounded domain, in *Godunov Methods: Theory and Applications (Edited Review)*, ed. by E.F. Toro (Kluwer Academic/Plenum Publishers, 2001)
10. B. Ben Moussa, J.P. Vila, Convergence of SPH method for scalar non–linear conservation laws. SIAM J. Numer. Anal. **37**(3), 863–887 (2000)
11. J.P. Vila, On particle weighted methods and smooth particle hydrodynamics. Math. Model. Methods Appl. Sci. **9**, 161–209 (1999)
12. A. Ferrari, M. Dumbser, E.F. Toro, A. Armanini, A new stable version of the SPH method in lagrangian coordinates. Commun. Comput. Phys. **4**(2), 378–404 (2008)
13. A. Ferrari, M. Dumbser, E.F. Toro, A. Armanini, A new 3D parallel SPH scheme for free surface flows. Comput. Fluids **38**, 1203–1217 (2009)
14. A. Ferrari, L. Fraccarollo, M. Dumbser, E.F. Toro, A. Armanini, A new 3D parallel SPH scheme for free surface flows. J. Fluid Mech. **663**, 456–477 (2010)
15. R.A. Gingold, J.J. Monagham, Smooth particle hydrodynamics: theory and application to non-spherical stars. Mon. Not. R. Astron. Soc. **181**, 375–389 (1977)
16. L. Lucy, A numerical approach to the testing of the fission hypothesis. Astron. J. **82**, 1013–1024 (1977)
17. J.J. Monagham, Why particle methods work. SIAM J. Sci. Stat. Comput. **3**(4), 422–433 (1982)
18. R.W. Randles, L.D. Libertsky, Smooth particle hydrodynamics, some recent improvements and applications. Comput. Methods Applic. Mech. Eng. **139**, 375–408 (1996)
19. A. Colagrossi, M. Landrinni, Numerical simulation of interfacial flows by smoothed particle hydrodynamics. J. Comput. Phys. **191**(2), 448–475 (2003)
20. M. Antuono, A. Colagrossi, S. Marrone, Numerical diffusive terms in weakly-compressible SPH schemes. Comput. Phys. Commun. **183**(12), 2570–2580 (2003)
21. X.-L. Fang, A. Colagrossi, P.-P. Wang, A.-M. Zhang, An accurate and robust axisymmetric SPH method based on Riemann solver with applications in ocean engineering. *Ocean Engineering*, p. 110369 (2022)
22. B. Cockburn, C.W. Shu, TVB Runge–Kutta local projection discontinuous Galerkin method for conservation laws II: general framework. Math. Comput. **52**(–), 411– (1989)
23. B. Cockburn, C.W. Shu, The Runge–Kutta discontinuous Galerkin method for conservation laws. J. Comput. Phys. **141**(–), 199– (1998)
24. J.J.W. van der Vegt, van der Ven H., O.J. Boelens, Discontinuous Galerkin methods for partial differential equations, in *Godunov Methods: Theory and Applications (Edited Review)*, ed. by E.F. Toro (Kluwer Academic/Plenum Publishers, 2001)

25. M. Dumbser, C.D. Munz, ADER discontinuous Galerkin schemes for aeroacoustics. Comptes Rendus Mécanique **33**, 683–687 (2005)
26. M. Dumbser, M. Käser, V.A. Titarev, E.F. Toro, Quadrature-Free non-oscillatory finite volume schemes on unstructured meshes for nonlinear hyperbolic systems. J. Comput. Phys. **226**(8), 204–243 (2007)
27. M. Dumbser, M. Käser, E.F. Toro, An arbitrary high order discontinuous Galerkin method for elastic waves on unstructured meshes V: local time stepping and $p$-adaptivity. Comput. Phys. Commun. **171**, 695–717 (2007)
28. M. Dumbser, D. Balsara, E.F. Toro, C.D. Munz, A unified framework for the construction of one-step finite-volume and discontinuous Galerkin schemes. J. Comput. Phys. **227**, 8209–8253 (2008)
29. M. Dumbser, M.J. Castro, C. Parés, E.F. Toro, ADER schemes on unstructured meshes for nonconservative hyperbolic systems: applications to geophysical flows. Comput. Fluids **38**(9), 731–1748 (2009)
30. M. Dumbser, M.J. Castro, C. Parés, E.F. Toro, A. Hidalgo, FORCE schemes on unstructured meshes II: non-conservative hyperbolic systems. Comput. Methods Appl. Mech. Eng. **199**(9–12), 625–647 (2010)
31. M. Dumbser, O. Zanotti, R. Loubère, S. Diot, A posteriori subcell limiting of the discontinuous Galerkin finite element method for hyperbolic conservation laws. J. Comput. Phys. **278**, 47–75 (2014)
32. S. Busto, M. Dumbser, C. Escalante, S. Gavrilyuk, N. Favrie, On high order ADER discontinuous Galerkin schemes for first order hyperbolic reformulations of nonlinear dispersive systems. J. Sci. Comput. **87**, 48 (2021)
33. E.F. Toro, V. Titarev, M.Dumbser, A. Iske, C.R. Goetz, and C.E. Castro, G.I. Montecinos, R. Dematté, The ADER approach for approximating hyperbolic equations to very high accuracy, in *Hyperbolic Problems: Theory, Numerics, Applications*, vol. I, ed. by C. Parés, M.J. Castro, T. Morales de Luna, M.L. Muñoz-Ruiz, pp. 83–108 (Springer, 2024)
34. C.E. Castro, M. Käser, E.F. Toro, Spacetime adaptive numerical methods for geophysical applications. Philos. Trans. R. Soc. London A **367**, 4613–4631 (2009)
35. C.E. Castro, E.F. Toro, M. Käser, ADER scheme on unstructured meshes for shallow water: simulation of tsunami waves. Geophys. J. Int. **189**, 1505–1520 (2012)
36. R.J. LeVeque, D.L. George, Tsunami modelling with adaptively refined finite volume methods. Acta Numerica (2011). https://doi.org/10.1017/S0962492911000043:211-289
37. A.L. Williamson, D. Rim, L.M. Adams, R.J. LeVeque, D. Melgar, F.I. González, A source clustering approach for efficient inundation modeling and regional scale probabilistic tsunami hazard assessment. Front. Earth Sci. **8**, 591663 (2020)
38. M. Oportus, R. Cienfuegos, A. Urrutia, R. Aránguiz, P.A. Catalán, M.A. Hube, Ex post analysis of engineered tsunami mitigation measures in the town of Dichato. Chile. Nat. Hazards **103**, 367–406 (2020)
39. J. Macías, M.J. Castro, C. Escalante, Performance assessment of the Tsunami-HySEA model for NTHMP tsunami currents benchmarking. Laboratory data. Coastal Eng. **158**, 103667 (2020)
40. T.E. Ongaro, M. de'Michieli Vitturi, M. Cerminara, A. Fornaciai, L. Nannipieri, M. Favalli, B. Calusi, J. Macías, M.J. Castro, S. Ortega, J.M. González-Vida, C. Escalante, Modeling tsunamis generated by submarine landslides at Stromboli Volcano (Aeolian Islands, Italy): a numerical benchmark study. Front. Earth Sci. **9**, 628652 (2021)
41. E.F. Toro, C.E. Castro, D. Vanzo, A. Siviglia, A flux-vector splitting scheme for the shallow water equations extended to high-order on unstructured meshes. Int. J. Numer. Methods Fluids (2022). https://doi.org/10.1002/fld.5099
42. M.J. Berger, R.J. LeVeque, Towards adaptive simulations of dispersive tsunami propagation from an asteroid impact, in *Proceedings of the International Congress of Mathematicians*, pp. 5056–5071 (EMS Press, 2022)
43. R.M. Iverson, The physics of debris flows. Rev. Geophys. **35**(3), 245–296 (1997)
44. S.B. Savage, The mechanics of rapid granular flows. Adv. Appl. Mech. **24**, 289–366 (1984)

45. S.B. Savage, W.K. Smith, A model for the plastic flow of landslides. US Geol. Surv. Prof. Pap. 1385 (1986)
46. S.B. Savage, K. Hutter, The motion of a finite mass of granular materials down a rough incline. J. Fluid Mech. **199**, 177–215 (1989)
47. S.B. Savage, K. Hutter, The dynamics of avalanches of granular materials from initiation to runout, I analysis. Acta Mech. **86**, 201–223 (1991)
48. K. Hutter, B. Svendsen, D. Rickenmann, Debris-Flow modelling: a review. Continuum Mech. Themodyn. **8**, 1–35 (1996)
49. T. Takahashi, The mechanical characteristic of debris flow. J. Hydraul. Division Am. Soc. Civ. Eng. **104**, 1153–1169 (1978)
50. T. Takahashi, Debris flow on prismatic open channel. J. Hydraul. Division Am. Soc. Civ. Eng. **106**, 381–396 (1980)
51. T. Takahashi, Debris flow. Annu. Rev. Fluid Mech. **13**, 57–77 (1981)
52. R.M. Iverson, A constitutive equation for mass movement behaviour. J. Geol. **93**, 143–160 (1985)
53. R.M. Iverson, Unsteady, nonuniform landslide motion, 1, theoretical dynamics and the steady Datum State. J. Geol. **94**, 1–15 (1986)
54. R.M. Iverson, Unsteady, nonuniform landslide motion, 2, linearised theory and the kinematics of transient response. J. Geol. **94**, 349–364 (1986)
55. R.M. Iverson, M.E. Reid, R.G. LaHusen, Debris-Flow mobilisation from landslides. Annu. Rev. Earth Planet. Sci. **25**, 85–138 (1997)
56. R.M. Iverson, R.P. Denlinger, Flow of variably fluidized granular masses across 3–D terrain: 1. Coulomb mixture theory. J. Geophys. Res. (2000) (in press)
57. R.P. Denlinger, R.M. Iverson, Flow of variably fluidized granular masses across 3-D Terrain: 2. Numerical predictions and experimental tests. J. Geophys. Res. **106**(B1), 553–566 (2001)
58. A. Armanini, M. Michiue, *Recent Developments on Debris Flows*. Lecture Notes in Earth Sciences, vol. 64 (Springer, 1997)
59. D. Vanzo, A. Siviglia, E.F. Toro, Pollutant transport by shallow water equations on unstructured meshes: hyperbolization of the model and numerical solution via a novel flux splitting scheme. J. Comput. Phys. **321**, 1–20 (2016)
60. E.D. Fernández-Nieto, P. Vigneaux, Some remarks on avalanches modelling: an introduction to shallow flows models. Technical Report HAL Id: hal-01066445, HAL. https://hal.science/hal-01066445
61. A. Siviglia, D. Vanzo, E.F. Toro, A splitting scheme for the coupled Saint Venant-Exner model. Adv. Water Res. **159**, 104062 (2022)
62. R.E. Britter, Atmospheric dispersion of dense gases. Annu. Rev. Fluid Mech. **21**, 317–344 (1989)
63. S.T. Chan, H.C. Rodean, D.L. Ermak, Numerical simulations of atmospheric releases of heavy gases over variable terrain. Technical Report UCRL–87256, Lawrence Livermore Laboratory, University of California, Livermore, California, USA (1982)
64. K.J. Eidsvik, A model for heavy gas dispersion in the atmosphere. Atmos. Environ. **14**, 769–777 (1980)
65. J. Kukkonen, J. Nikmo, Modelling heavy gas cloud transport in sloping terrain. J. Hazardous Mater. **31**, 155–176 (1992)
66. D.M. Webber, S.J. Jones, D. Martin, A model of the motion of a heavy gas cloud release on a uniform slope. J. Hazardous Mater. **33**, 101–122 (1993)
67. A. Chiapolino, S. Courtiaud, E. Lapébie, R. Saurel, Modeling heavy-gas dispersion in air with two-layer shallow water equations. Fluids **6**(1) (2021). https://doi.org/10.3390/fluids6010002
68. R.K.S. Hankin, *Heavy Gas Dispersion Over Complex Terrain*, Ph.D. thesis, University of Cambridge, UK (1997)
69. Y. Ma, A. Li, J. Che, T. Wang, C. Yang, L. Che, J. Liu, Investigation of heavy gas dispersion characteristics in a static environment: spatial distribution and volume flux prediction. Build. Environ. **242**, 110501 (2023)

70. F. Bellamoli, L.O. Müller, E.F. Toro, A numerical method for junctions in networks of shallow-water channels. Appl. Math. Comput. **337**, 190–213 (2018)
71. X. Liu, A. Chertock, A. Kurganov, K. Wolfkill, One-Dimensional/Two-Dimensional coupling approach with quadrilateral confluence region for modeling river systems. J. Sci. Comput. **81**, 297–1328 (2019)
72. S. Kivva, M. Zheleznyak, O. Pylypenko, V. Yoschenko, Open water flow in a wet/dry multiply-connected channel network: a robust numerical modeling algorithm. Pure Appl. Geophys. **177**, 342103458 (2020)
73. M.E. Vázquez-Cendón, Estudio de Esquemas Descentrados para su Aplicación a las Leyes de Conservación Hiperbólicas con Términos de Fuente. Ph.D. thesis, Departamento de Matemáticas Aplicadas, Universidad de Santiago de Compostela, España (1994)
74. L. Bermúdez, A. Dervieux, J.A. Desideri, M.E. Vázquez, Upwind schemes for the two-dimensional shallow water equations with variable depth using unstructured meshes. Comput. Methods Appl. Mech. **155**, 49–72 (1998)
75. K. Anastasiou, C.T. Chan, Solution of the 2D shallow water equations using the finite volume method on unstructured triangular meshes. Int. J. Numer. Meth. Fluids **24**, 1225–1245 (1997)
76. V. Casulli, Semi-Implicit finite difference methods for the two-dimensional shallow water equations. J. Comput. Phys. **86**, 56–74 (1990)
77. V. Casulli, R.T. Cheng, Semi-Implicit finite difference methods for three-dimensional shallow water flow. Int. J. Numer. Methods Fluids **15**, 629–648 (1992)
78. V. Casulli, A semi-implicit finite difference method for non-hydrostatic free-surface flow. Int. J. Numer. Methods Fluids **30**, 425–440 (1999)
79. P.K. Stansby, P.M. Lloyds, A semi-implicit lagrangian scheme for 2D shallow water flow with a two-layer turbulence model. Intern. J. Numer. Methods Fluids **20**, 115–133 (1995)
80. S. Busto, M. Dumbser, A staggered semi-implicit hybrid finite volume/finite element scheme for the shallow water equations at all Froude numbers. Appl. Numer. Math. **175**, 108–132 (2022)
81. W. Boscheri, M. Tavelli, C.E. Castro, An all Froude high order IMEX scheme for the shallow water equations on unstructured Voronoi meshes. Appl. Numer. Math. **185**, 311–335 (2023)
82. E.F. Toro, A. Siviglia, Simplified blood flow model with discontinuous vessel properties: analysis and exact solutions. *Modelling Physiological Flows Series: Modelling, Simulation and Applications*, ed. by D. Ambrosi, A. Quarteroni, G. Rozza (Springer, 2012), pp. 19–39. ISBN 978-88-470-1934-8
83. I. Gómez-Bueno, S. Boscarino, M.J. Castro, C. Parés, G. Russo, Implicit and semi-implicit well-balanced finite-volume methods for systems of balance laws. Appl. Numer. Math. **184**, 18–48 (2023)
84. L. Bermúdez, M.E. Vázquez, Upwind methods for hyperbolic conservation laws with source terms. Comput. Fluids **23**, 1049–1071 (1994)
85. R.J. LeVeque, Balancing source terms and flux gradients in high-resolution godunov methods. J. Comput. Phys. **146**, 346–365 (1998)
86. M.E. Vázquez-Cendón, Improved treatment of source terms in upwind schemes for the shallow water equations in channels with irregular geometry. J. Comput. Phys. **148**, 497–526 (1999)
87. E.D. Fernández-Nieto, D. Bresch, J. Monnier, A consistent intermediate wave speed for a well-balanced HLLC solver. C. R. Acad. Sci. Paris **346**, 795–800 (2008)
88. M.J. Castro, J.M. Gallardo, J.A. López, C. Parés, Well-Balanced high order extensions of Godunov's method for semi-linear balance laws. SIAM J. Numer. Anal. **46**, 1012–1039 (2008)
89. A. Canestrelli, A. Siviglia, M. Dumbser, E.F. Toro, Well-balanced high-order centred schemes for non-conservative hyperbolic systems. Applications to shallow water equations with fixed and mobile bed. Adv. Water Res. **32**, 834–844 (2009)
90. M.J. Castro, A. Pardo, C. Parés, E.F. Toro, On some fast well-balanced first order solvers for nonconservative systems. Math. Comput. **79**(271), 1427–1472 (2010)
91. O. Delestre, P.-Y. Lagrée, A well-balanced finite volume scheme for blood flow simulation. Int. J. Numer. Methods Fluids **72**, 177–205 (2013)

92. L.O. Müller, E.F. Toro, Well-balanced high-order solver for blood flow in networks of vessels with variable properties. Int. J. Numer. Methods Fluids **29**(12), 1388–1411 (2013)
93. L.O. Müller, C. Parés, E.F. Toro, Well-balanced high-order numerical schemes for one-dimensional blood flow in vessels with varying mechanical properties. J. Comput. Phys. **242**(7), 53–85 (2013)
94. A. Navas-Montilla, J. Murillo, 2D well-balanced augmented ADER schemes for the shallow water equations with bed elevation and extension to the rotating frame. J. Comput. Phys. **372**, 316–348 (2018)
95. M.J. Castro, C. Parés, Well-balanced high-order finite volume methods for systems of balance laws. J. Sci. Comput. **82**, 48 (2020). https://doi.org/10.1007/s10915-020-01149-5
96. B. Ghitti, C. Berthon, M.H. Le, E.F. Toro, A fully well-balanced scheme for the 1D blood flow equations with friction source term. J. Comput. Phys. **421**, 109750 (2020)
97. L. Arpaia, M. Ricchiuto, A.G. Filippini, R. Pedreros, An efficient covariant frame for the spherical shallow water equations: well balanced DG approximation and application to tsunami and storm surge. Ocean Model. **169**, 101915 (2022)
98. E. Guerrero-Fernández, M.J. Castro-Díaz, M. Dumbser, T.M. de Luna, An arbitrary high order well-balanced ADER-DG numerical scheme for the multilayer shallow-water model with variable density. J. Sci. Comput. **9**, 52 (2022). https://doi.org/10.1007/s10915-021-01734-2
99. W. Barsukow, J.P. Berberich, A well-balanced active flux method for the shallow water equations with wetting and drying. Commun. Appl. Math. Comput. (2022). https://doi.org/10.1007/s42967-022-00241-x
100. E. Pimentel-García, L.O. Müller, E.F. Toro, C. Parés, High-order fully well-balanced numerical methods for one-dimensional blood flow with discontinuous properties. J. Comput. Phys. **475**, 111869 (2023)
101. L. Martaud, C. Berthon, Fully well-balanced entropy stable Godunov numerical schemes for the shallow water equations with the topography source term. Technical Report HAL Id: hal-04394378, HAL open science (2024)
102. V. González-Tabernero, M.J. Castro, J.A. García-Rodríguez, High-order well-balanced numerical schemes for one-dimensional shallow-water systems with Coriolis terms. Appl. Math. Comput. **469**(128528) (2024)
103. P.G. Baines, *Topographic Effects in Stratified Flows* (Cambridge University Press, 1995)
104. M. Castro, J.T. Frings, S. Noelle, C. Parés, G. Puppo, On the hyperbolicity of two- and three-layer shallow water equations, in *Proceedings of the International Conference on Hyperbolic Problems* (American Mathematical Society, MI, 2010), pp. 657–664
105. M.J. Castro, E.D. Fernández-Nieto, J.M. González-Vida, C. Parés-Madroñal, Numerical treatment of the loss of hyperbolicity of the two-layer shallow-water system. J. Sci. Comput. **48**(1), 16–40 (2011)
106. A. Scoz, L. Bertazzi, E.F. Toro, On well-posedness of a mathematical model for cerebrospinal fluid in the optic nerve sheath and the spinal subarachnoid space. Appl. Math. Comput. **413**, 126625 (2022)
107. S. Gavrilyuk, M. Kazakova, Hydraulic jumps in two-layer flows with a free surface. Technical Report HAL Id: hal-00956266, HAL open science (2014)
108. E.D. Fernández-Nieto, E. Koné, T. Chacón-Rebollo, A multilayer method for the hydrostatic Navier-Stokes equations: a particular weak solution. J. Sci. Comput. **60**, 408–437 (2014)
109. E.D. Fernández-Nieto, M. Parisot, Y. Penel, J. Sainte-Marie, A hierarchy of dispersive layer-averaged approximations of Euler equations for free-surface flows. Commun. Math. Sci. **16**(5), 1169–1202 (2018)
110. A.A. Chesnokov, S.L. Gavrilyuk, V.Y. Liapidevskii, Mixing and nonlinear internal waves in a shallow flow of a three-layer stratified fluid. Phys. Fluids **34**, 075104 (2022)
111. E.F. Toro, *NUMERICA, A Library of Source Codes for Teaching, Research and Applications* (Numeritek Ltd., 1999). https://www.numeritek.com

# Index

**A**

Acceleration due to gravity, 5
Accuracy theorem, 172
ADER-DG, 317, 318, 335, 395
ADER finite volume, 317, 318
ADER2-HEOC, 340
ADER scheme, 326, 338, 340, 342
ADER2-TT, 336
Admissible shock, 34
Advection equation, 12
Advection operator, 288
Advection-reaction splitting, 283, 285, 288, 289
Advection system, 286
Advection term, 291, 292
Amplification factor, 297, 304
Angle of incidence, 377, 379, 384
Approximate Riemann solver, 202, 225, 236

Approximate-State Riemann solver, 228
Asymptotically stable system, 294
Atmospheric pressure, 6

**B**

Backward Euler method, 295
Balance law, 189
Bathymetry, 1, 2, 4, 66
Bed friction, 373, 374
Bed slope, 145
Bed variation, 111
Bed wetting, 251
Bi-orthonormality, 70
Body force, 3, 5
Bore, 9, 86, 130

Bore reflection, 377, 379
Bore reflection patterns, 377
Bottom, 7
Boundary condition, 6, 127, 200
Boundary extrapolation, 269
Breadth, 11
Breadth variation, 287
Burgers' equation, 33, 41, 91, 107

**C**

Canonical form, 44
Canonical path, 247
Canonical system, 47
Cartesian-Cut Cell method, 365
Cartesian mesh, 210, 214
Cauchy-Kovalevskaya procedure, 169, 325, 327, 337, 339, 341
Cauchy problem, 16, 29
Celerity, 42, 52, 68, 142
Celerity criterion, 384, 388
Celerity transition criterion, 394
Cell average, 196, 215
Cell merging, 367
Centred method, 208, 292
Centred numerical source, 197
Centred TVD method, 275
CFL coefficient, 356
CFL number, 167
Channel networks, 396
Characteristic curve, 16, 45, 46, 49
Characteristic field, 74
Characteristic form, 44
Characteristic limiting, 273
Characteristic polynomial, 21, 42, 68, 293

© The Editor(s) (if applicable) and The Author(s), under exclusive license to Springer Nature Switzerland AG 2024
E. F. Toro, *Computational Algorithms for Shallow Water Equations*,
https://doi.org/10.1007/978-3-031-61395-1

Characteristic speed, 28, 41, 44, 45, 53, 74
Characteristic variable, 22, 23, 42, 43, 45, 47, 54, 292
Chilean tsunami 1960, 378
Circular dam, 355
Circular rarefaction wave, 357, 358
Circular shock wave, 357, 358
CIR scheme, 166
Coefficient of numerical viscosity, 174, 175
Complete Riemann solver, 243
Compressible material, 3
Computational cell, 3, 301
Concave flux, 28
Concerted Action on Dam-Break Modelling (CADAM), 354, 364
Conservation law, 1–3, 8, 9, 76, 107, 190
Conservation of mass, 8
Conservation principle, 2, 3
Conservative form of a scheme, 176
Conservative formula, 215, 262, 302
Conservative formulation, 67, 109
Conservative method, 11
Conserved variable, 9, 67, 71
Consistency of numerical flux, 215
Contact discontinuity, 85, 88, 100, 103, 124, 131, 132, 136, 228, 232, 233
Continuity equation, 7
Control volume, 3, 31
Conventional Riemann problem, 270
Convex flux, 28
Courant number, 167, 172, 239, 264, 291
Critical flow, 141, 217, 218, 228–231, 238, 266
Critical rarefaction, 141, 229, 238, 266, 267

Cross-sectional width, 11
Cylindrical symmetry, 356


**D**
Dam, 353, 354
Dam-break modelling, 353, 362, 364
Dam-break problem, 52, 86, 87, 113, 355, 364
Debris flow, 395
Debris transport, 354
Decoupled equations, 43, 44
Deflection angle, 382
Density, 3, 4
Dependent variable, 3
Depression, 86
Depth function, 118, 119
Depth positivity condition, 120, 121, 131, 134, 136, 143, 146, 230

Detachment criterion, 384, 388
Detachment transition criterion, 394
Determinant, 42, 68
Diagonalisable system, 43
Diagonalisation, 22, 43
Diagonal matrix, 43
Difference equation, 165
Differential form, 1–3, 190
Differential operator, 165, 166
Diffusion coefficient, 15
Dimensional splitting, 284, 298
Discharge, 11
Discontinuous Galerkin finite element method, 394
Discontinuous Galerkin finite elements, 225

Discontinuous solution, 9, 10
Divergence operator, 2
Divided differences, 334
DOT conservative scheme, 243–245
DOT non-conservative, 247, 248
DOT numerical flux, 244
Dot product, 2
Dry bed, 114, 119, 129–131, 134–136, 140, 142
Dry-bed Riemann problem, 142, 143
Dumbser-Enaux-Toro (DET) solver, 331
Dynamical condition, 6


**E**
Efficiency, 330, 344, 345
Eigenstructure, 21, 42, 56, 66, 71, 74, 78
Eigenvalue, 21, 42, 56–58, 68, 71, 73, 74, 78, 82, 293, 294
Eigenvector, 21, 42, 43, 57, 58, 72, 74, 79, 82
Elliptic equations, 74, 79
ENO method, 331
Entropy fix, 238
Entropy fix for the Roe solver, 238
Entropy fix of Harten and Hyman, 238
Entropy glitch, 217, 218
Entropy satisfying method, 141
Entropy-violating shock, 217
Equation of state, 81
Euler equations, 81
Euler method, 291, 295
Exact Riemann solver, 113, 115, 141
Exact splitting (for sources), 289
Explicit method, 167, 215
Explicit scheme, 295

# Index

## F
Fictitious state, 200
Finite difference, 164, 165
Finite volume, 3, 76, 301
Finite volume formula, 32
Finite volume method, 191
First-order splitting scheme, 300
Fluctuation, 246, 248
Flux, 9, 77
Flux limiter, 266, 273
Flux Vector Splitting, 81, 206
FORCE-$\alpha$, 248–250
FORCE flux, 209, 275
FORCE method, 177, 209, 216
FORCE non-conservative, 248
Free surface, 4, 5, 7, 11, 52
Friction slope, 11
Fromm method, 178
Front tracking, 394
Froude number, 78, 104, 105, 383, 384
FTCS method, 167, 168
Fullydiscrete method, 211

## G
Galilean transformation, 381
General form of a scheme, 170
Generalised Riemann problem, 268, 336, 337, 340
Generalised solution, 10, 33
Genuinely non-linear field, 75, 111
Geometric source term, 356
Geophyical fluid dynamics, 397
Glimm's method, 146, 204
Godunov flux, 194, 225
Godunov's centred method, 177
Godunov state, 194
Godunov's theorem, 172, 173, 331
Godunov's upwind method, 166, 167, 171, 193, 216
Gradient operator, 2

## H
Head of rarefaction wave, 92, 123, 124
Heavy-gas dispersion, 395
HEOC solver for $GRP_m$, 326, 329, 340, 341
HLLC Riemann solver, 233, 356, 365
HLL Riemann solver, 232
Homogeneity property, 80, 81
Homogeneous equation, 9, 44, 47, 54, 140, 145, 286, 290
Homogeneous system, 76

Hyperbolic equation, 22, 42, 74
Hyperbolicity, 73, 76
Hyperbolic system, 43, 73, 74, 79, 80

## I
Implicit method, 215, 295, 396
Incident bore, 381
Incomplete Riemann solver, 243
Independent variable, 3
Inhomogeneous equation, 54, 140, 145, 285–288
Initial condition, 48, 52
Initial-value problem, 16, 45, 48, 49
Integral average, 263
Integral form, 1, 2, 10, 31, 76, 189, 190
Integration path, 241
Integration volume, 263
Intersection point of integration paths, 241
Inundation map, 354

## J
Jacobian matrix, 67, 82, 236, 293, 294

## K
Kinematic condition, 6

## L
Lax entropy condition, 34
Lax–Friedrichs flux, 208, 235, 275
Lax-Friedrichs method, 168, 169, 177
Lax-Wendroff flux, 209
Lax-Wendroff method, 169, 170, 177
LeFloch-Raviart expansion, 325
Left eigenvector, 69, 70, 72, 73, 79
Leibniz's formula, 8
Limiter function, 265
Linear advection equation, 16, 44, 45, 47, 49, 163, 265
Linearised shallow water equations, 39, 50, 56
Linearly degenerate field, 75, 228, 262
Linear model, 41
Linear scheme, 171
Linear stability analysis, 291
Linear system, 42
Linear theory, 5

## M
Mach reflection, 377, 378, 381, 383, 385, 386, 394

Mach stem, 378, 381, 383
Malpasset dam, 354
Mass conservation, 2, 3
Mesh spacing, 164
Mesh speed, 235
MINBEE limiter, 266, 274
Modified equation, 185
Momentum conservation, 2, 3
Momentum equation, 8
Monotone scheme, 172, 210
Monotonicity condition, 175
Montecinos-Toro solver for $GRP_m$, 329, 331
Moving reflective boundary condition, 201
Multilayer shallow water, 396
MUSCL-Hancock method, 268

### N
Near field, 360
Newton-Raphson method, 114, 121
Non-concave flux, 28
Non-conservative formulation, 39, 71, 109, 144
Non-conservative method, 11, 12, 246
Non-conservative system, 245
Non-convex flux, 28
Non-critical rarefaction, 266
Non-linear shallow water theory, 5
Non-staggered grid, 204, 205
Non-uniqueness, 34
Normal flux component, 9, 77
Numerical flux, 176, 192, 215, 265, 319, 342

NUMERICA Library, 397
Numerical operator, 165
Numerical software, 353, 355, 394
Numerical source, 192, 196, 320, 324, 328, 339, 342

### O
Oblique bore, 381–383
Oblique bore relations, 382
Okushiri island tsunami (1993), 377, 378
Ordinary differential equation, 16
Osher-Solomon approximate Riemann solver, 239–242
Outward unit normal vector, 76
Overtopping, 354

### P
Parabolic equation, 15

Particle velocity, 86, 88, 104, 105, 131
Passive scalar, 122, 135, 262, 266
Path-conservative method, 246
Periodic waves, 378
Phase angle, 304
Phase plane, 75
Phase space, 75
Physical flux, 91, 215
Physical model, 364
Physical variable, 4, 71
Point value, 196
Pollutant transport, 12, 228, 262
Pressure, 1, 3
Pressure positivity condition, 137
Primary circular shock wave, 358, 359
Primitive variable, 4, 11, 71
Primitive variable formulation, 78, 107, 229

Primitive variable method, 246
Primitive variable Riemann solver, 229
Principal part, 66

### Q
Quasi-linear form, 12, 28, 40, 41, 66, 71, 81, 111

### R
Radial solution, 356
Radial velocity, 356
Random Choice Method, 146, 204, 206, 216, 356
Random number, 204, 206
Random sampling, 204
Rankine-Hugoniot conditions, 11, 33, 91, 108, 131, 382
Rarefaction shock, 34, 141, 217
Rarefaction wave, 85, 86, 88–90, 92–94, 103

Reaction system, 286
Reaction term, 291, 292
Reference direction, 212
Reflected bore, 381, 384
Reflective boundary condition, 201
Regular reflection, 377, 378, 381, 383, 386, 394
Riemann invariant, 90, 111, 132
Riemann problem, 19, 36, 49–51, 60, 86–88, 110, 111, 115, 136, 394
Right eigenvector, 69, 70, 72, 73
Risk analysis, 354
River flow, 2, 10
Roe approximate Riemann solver, 236

Index 407

Roe average, 236–238
Roe matrix, 236
Rotated frame, 212
Rotated Riemann problem, 213
Rotational invariance, 76, 77, 211
Rotation matrix, 77
Runge–Kutta method, 295, 296
Rusanov flux, 235

**S**

Safety coefficient, 202
Saint Venant equations, 10, 81, 111
Sampling the solution, 121
Scaling factor, 43, 69, 70
Secondary circular shock wave, 358, 360
Second-order splitting scheme, 300
Sediment transport, 354, 395
Semidiscrete method, 210, 211, 296
Shallow water, 55
Shallow water equations, 343
Shear wave, 85, 88, 90, 101, 124, 136
Shock-fitting, 246
Shock formation, 30
Shock Froude number, 104, 105
Shock speed, 104, 105, 108, 109
Shock strength, 109
Shock wave, 86, 91, 104, 105, 107–109, 130

Similarity solution, 88, 122
Simple waves, 90
Simultaneous update scheme, 284
Singularity, 50
SLIC scheme, 275
Slipstream, 381, 383, 387
Slip surface, 381, 383
Smooth Particle Hydrodynamics (SPH), 394

Solitary wave, 377, 378
Sonic flow, 217, 228
Sonic point, 242
Sonic rarefaction, 238
Source operator, 288
Source term, 4, 9, 11, 12, 44, 53–55, 76, 145, 285, 286, 289, 374
Space-marching, 80
Spatial reconstruction, 331
Species equation, 12
Splitting scheme (for sources), 287
Spurious oscillations, 141, 265
Stability, 284, 302
Stability condition, 174, 206, 303, 304
Stability of a scheme, 168

Stability of ODEs, 294
Staggered-grid, 210
Star region, 52, 61, 63, 88, 121, 130, 140, 141
Steady flow, 74, 78–80, 82
Stiffness, 284
Stiffness ratio, 294
stiff ODEs, 294
Streamline, 382
Strict hyperbolicity, 42, 74
Structured mesh, 210, 301
SUPERBEE limiter, 266, 273, 356
Supercritical flow, 74, 78–80

**T**

Tail of rarefaction wave, 92, 123, 124, 132, 134, 135
Tangential velocity, 88, 89, 93, 94, 97, 100, 101, 104, 105
Telescopic property, 215, 216
Tensor, 3
Time-marching, 80
Time step, 164, 202
Toro-Titarev (TT) solver for $GRP_m$, 322, 329, 336
Toro-Vázquez splitting, 207, 396
Total Variation Diminishing (TVD) method, 261, 331
Transcritical rarefaction, 141
Transmissive boundary condition, 201
Trapezoidal method, 295, 297
Trial function, 303
Triple point, 381
Truncation error, 166, 172, 175
Tsunami waves, 377, 378, 395
TVD Runge–Kutta method, 296
Two-rarefaction approximation, 120, 121, 140
Two-Rarefaction Riemann solver, 231
Two-shock approximation, 140
Two-Shock Riemann solver, 231

**U**

Unphysical oscillations, 265
Unsplit finite volume scheme, 301
Unstructured mesh, 210, 212, 220, 301
Upwinded numerical source, 197
Upwinding, 166, 195
Upwinding source term, 290
Upwind method, 206, 208

**V**

Vacuum, 110, 129, 131
Vajont dam, 354
Van Albada limiter, 266, 273
Van Leer limiter, 266, 273
Variable bathymetry, 144
Variable bed, 39, 41, 53, 54
Velocity, 3
Vertical acceleration, 1, 6
Viscous form of a scheme, 174
Von Neumann stability, 302

**W**

WAF limiter, 266
WAF method, 263, 309, 356
WAF state, 264
WAF TVD flux, 265
WAF unsplit scheme, 308
Wall angle, 379, 380, 384
Wall friction, 373, 374
Warming-Beam method, 178
Wave angle, 382–384

Wave length, 5
Wave speed estimate, 234
Wave strength, 237
Weakly hyperbolic equations, 74
Weakly hyperbolic system, 22
Weak shock wave, 109
Weak solution, 9, 33
Well-balanced scheme, 284, 290, 396
WENO method, 332
Wet bed, 42, 74, 114, 129–131, 134, 140
Wet/dry front, 113, 114, 126, 129, 134–137, 142–144, 235, 251, 252, 254, 255, 267

Wetted cross-sectional area, 10

**X**

X-sweep, 299

**Y**

Y-sweep, 299

The manufacturer's authorised representative in the EU is Springer Nature Customer Service Centre GmbH, Europaplatz 3, 69115 Heidelberg, Germany. If you have any concerns regarding our products, please contact ProductSafety@springernature.com

Printed and bound by CPI Group (UK) Ltd, Croydon, CR0 4YY

26/03/2026

02078941-0008